Name	Symbol	Atomic Number	Atomic Weight	
Neodymium	Nd	60	144.2	
Neon	Ne	10	20.1	
Neptunium	Np	93	237.048	a,b
Nickel	Ni	28	58.6934(2)	
Niobium	Nb	41	92.90638(2)	
Nitrogen	N	7	14.00674(7)	c,d
Nobelium	No	102	(259)	a
Osmium	Os	76	190.2(1)	c
Oxygen	O	8	15.9994(3)	c,d
Palladium	Pd	46	106.42(1)	c
Phosphorus	P	15	30.973762(4)	
Platinum	Pt	78	195.08(3)	
Plutonium	Pu	94	(244)	a
Polonium	Po	84	(209)	a
Potassium	K	19	39.0983(1)	
Praseodymium	Pr	59	140.90765(3)	
Promethium	Pm	61	(145)	a
Protactinium	Pa	91	231.03588(2)	a
Radium	Ra	88	226.0254	a,b
Radon	Rn	86	(222)	a
Rhenium	Re	75	186.207(1)	
Rhodium	Rh	45	102.90550(3)	
Rubidium	Rb	37	85.4678(3)	c
Ruthenium	Ru	44	101.07(2)	c
Samarium	Sm	62	150.36(3)	c
Scandium	Sc	21	44.955910(9)	
Selenium	Se	34	78.96(3)	
Silicon	Si	14	28.0855(3)	d
Silver	Ag	47	107.8682(2)	c
Sodium	Na	11	22.989768(6)	
Strontium	Sr	38	87.62(1)	c,d
Sulfur	S	16	32.066(6)	c,d
Tantalum	Ta	73	180.9479(1)	
Technetium	Tc	43	(98)	a
Tellurium	Te	52	127.60(3)	c
Terbium	Tb	65	158.92534(3)	
Thallium	Tl	81	204.3833(2)	
Thorium	Th	90	232.0381(1)	a,c
Thulium	Tm	69	168.93421(3)	
Tin	Sn	50	118.710(7)	c
Titanium	Ti	22	47.88(3)	
Tungsten	W	74	183.85(3)	
Unnilhexium	Unh	106	(263)	a
Unnilpentium	Unp	105	(262)	a
Unnilquadium	Unq	104	(261)	a
Unnilseptium	Uns	107	(262)	a
Uranium	U	92	238.0289(1)	a,c,e
Vanadium	V	23	50.9415(1)	
Xenon	Xe	54	131.29(2)	c,e
Ytterbium	Yb	70	173.04(3)	c
Yttrium	Y	39	88.90585(2)	
Zinc	Zn	30	65.39(2)	
Zirconium	Zr	40	91.224(2)	c

[a] Element has no stable nuclides. However, three such elements (Th, Pa, and U) do have a characteristic terrestrial isotopic composition, and for these an atomic weight is tabulated.

[b] Element for which the value A_r is that of the radioisotope of longest half-life.

[c] Geological specimens are known in which the element has an isotopic composition outside the limits for normal material. The difference between the atomic weight of the element in such specimens and that given in the table may exceed the implied uncertainty.

[d] Range in isotopic composition of normal terrestrial material prevents a more precise $A_r(E)$ being given; the tabulated $A_r(E)$ value should be applicable to any normal material.

[e] Modified isotopic compositions may be found in commercially available material because it has been subjected to an undisclosed or inadvertent isotopic separation. Substantial deviations in atomic weight of the element from that given in the table can occur.

REAGENT
CHEMICALS

REAGENT CHEMICALS

CHEMICALS

EIGHTH EDITION

AMERICAN CHEMICAL SOCIETY SPECIFICATIONS

Official from April 1, 1993

American Chemical Society
Washington, DC
1993

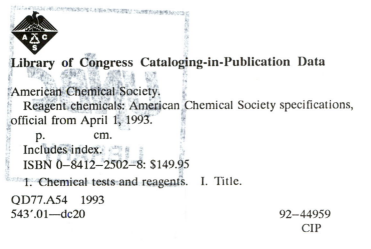

Library of Congress Cataloging-in-Publication Data

American Chemical Society.
   Reagent chemicals: American Chemical Society specifications, official from April 1, 1993.
        p.   cm.
   Includes index.
   ISBN 0–8412–2502–8: $149.95

   1. Chemical tests and reagents.   I. Title.

QD77.A54   1993
543′.01—dc20   92–44959
                                                                    CIP

The paper used in this publication meets the minimum requirements of American National Standard for Information Sciences—Permanence of Paper for Printed Library Materials, ANSI Z39.48–1984.   ∞

CONTENTS

Prepared by the
COMMITTEE ON ANALYTICAL REAGENTS
(1987–1992)

Clarence Lowery, Chairman 1985–1991
Paul A. Bouis, Chairman 1992–

Alfred J. Barnard (deceased)
Kishor D. Desai
Jack P. Fletcher
Jerry R. Hale
Benjamin D. Halpern
Norman C. Jamieson
Richard S. Juvet
Richard A. Link
Loren C. McBride
Rajendra V. Mehta
John R. Moody

Anthony D. Pietrzykowski
Charles J. Pouchert
Joseph M. Rao
Nancy S. Simon
Vernon A. Stenger
Cyrus M. Strauss
Samuel M. Tuthill
Frank G. Walthall
Ruth H. Wantz
Donald H. Wilkins
Charles M. Wilson

William E. Schmidt, Secretary

PREFACE

The ACS Committee on Analytical Reagents evolved from a Committee on the Purity of Chemical Reagents that was established in 1903. Then analysts were disturbed by the quality of reagents available and the discrepancies between labels and the actual purity of the materials. The Committee's role in resolving these issues expanded rapidly after its 1921 publication of specifications for ammonium hydroxide and for hydrochloric, nitric, and sulfuric acids.

Specifications appeared initially in *Industrial and Engineering Chemistry* and later in its *Analytical Edition*. In 1941 the existing specifications were reprinted in a single pamphlet. Revisions and new specifications were later gathered into a book, the 1950 edition of *Reagent Chemicals*. Two editions per decade followed except during the 1970s, when only one edition was published. The word "specification", as used in this book, means the individual entry for each reagent chemical. The specification has two components, the requirements (usually expressed in numerical terms) and the tests (instructions for the analytical methods to be used in determining conformity to the requirements).

The commonplace introduction of instrumentation into analytical laboratories, beginning in the late 1950s, resulted in dramatic improvements in the sensitivity and accuracy of analytical measurements. In line with these improved instruments, the requirements for reagent chemicals and the tests used to measure their purity needed to be improved. Under the capable leadership of Stanley Clabaugh and later Vernon A. Stenger, Samuel M. Tuthill, and Wallace Rohrbough, these improvements were made during the 1960s and 1970s. The first instrumental technique, flame emission spectroscopy, was introduced into the compendium in 1961. It was followed by other appropriate techniques in an effort to develop test methods that would be as accurate and cost-effective as possible.

The changes in this edition are aimed at updating the general test methods and making the book easier to read. Format changes include a new requirements layout and a continuation of the trend toward detailed general methods in the front of the book, with only reagent-specific test conditions under each individual chemical.

Updated methods include gas chromatography (in which capillary columns are now used), water determination (in which coulometric methods have been added), and the replacement of flame emission techniques by atomic absorption for metal determinations. In addition, the general methods for chromatography and atomic absorption have been extensively revised to reflect their practical use in today's typical analytical laboratory. New items in this edition are the elimination of boiling points and density requirements, the use of liquid chromatography in assay determinations, and 24 new reagents. Also added are sections on determining detection limits and on the preparation and standardization of volumetric solutions.

Assay requirements and methods have been added for most of the reagents, along with a description of what the methods actually measure. A systematic elimination or simplification of tedious classical procedures has been initiated. As an example, the test for substances not precipitated by ammonium sulfide has been replaced by atomic absorption. Classical procedures, however, will continue to be the cornerstone of inorganic reagent assays because of their inherent higher precision and accuracy for the determination of major components in reagent chemicals.

The choice of nomenclature, abbreviations, units, and editing policies has been generally guided by current ACS practices. Structural formulas are now shown for most organic chemicals. Atomic weights have been recalculated using the 1989 IUPAC Table of Standard Atomic Weights. The membership of the Committee since the publication of the 7th edition has been

Alfred J. Barnard	1981–1986	Anthony D. Pietrzykowski	1986–1991
(Deceased)		Charles J. Pouchert	1978–1991
Paul A. Bouis	1987–	Joseph M. Rao	1991–
Chairman	1992–	William E. Schmidt	1967–
Kishor D. Desai	1981–	Secretary (nonvoting)	
Jack P. Fletcher	1987–1990	Nancy S. Simon	1991–
Jerry R. Hale	1990–1992	Vernon A. Stenger	1962–
Benjamin D. Halpern	1988–	Chairman	1967–1973
Norman C. Jamieson	1982–	Consultant	1992–
Richard S. Juvet	1986–	Cyrus M. Strauss	1982–1987
Richard A. Link	1991–	Samuel M. Tuthill	1958–
Clarence Lowery	1974–	Chairman	1974–1980
Chairman	1985–1991	Consultant	1982–
Loren C. McBride	1983–	Frank G. Walthall	1982–1991
Rajendra V. Mehta	1992–	Ruth H. Wantz	1991–
John R. Moody	1986–	Donald H. Wilkins	1983–1988
		Charles M. Wilson	1988–

Clarence Lowery served as chairman of the Committee during most of the revision. He was succeeded by Paul A. Bouis in the spring of 1992.

Special acknowledgment goes to Barbara Barr, Larry Becker, Nick Csikai, Gene Desotelle, Joe Gammell, Tom Hummel, Mike Kertis, George W. Limpert, James D. McLean, Peter Okolovitch, Tom Schware, Virgil T. Turkelson, Richard M. Van Effen, Steven R. Villaseñor, Bruce Wille, Phillip K. Williams, and Preston D. Wright for their valuable contributions in developing some of the analytical procedures that have greatly improved the quality of the test methods.

The Committee normally meets in the spring and fall of each year. After each meeting, changes of an immediate nature are reported in *Chemical and Engineering News*. As such changes accumulate and as changes of a nonurgent nature (including new reagents, specifications, and test methods) are approved, it is expected that they will appear in *Analytical Chemistry*. Such publications will be announced in *Chemical and Engineering News*.

The Committee plans at least one formal supplement, sized to fit this book, before the next full edition is published. Interim changes published in *Chemical and Engineering News* or *Analytical Chemistry* will be included. The availability of such supplements will be announced in *Chemical and Engineering News*. Cards for requesting them, supplied in the back of this book, should be filled in and returned. This request is best handled when the book is first received, but it can be done upon announcement of the supplements.

The Committee has formal written operating procedures that describe and govern its operations. Interested parties may obtain a copy of these procedures by addressing their request to:

Secretary, ACS Committee on Analytical Reagents
c/o Books Department
American Chemical Society
1155 16th Street, NW
Washington, DC 20036

The Committee urges that any errors observed be reported, invites constructive criticism, and welcomes suggestions, particularly for new reagents and improved test methods. Communications on these subjects should be sent to the secretary at this address. Anyone interested in serving on the committee should also contact the secretary.

DEFINITIONS, PROCEDURES, STANDARDS, AND SPECIFICATIONS

REQUIREMENTS AND TESTS

The specifications prepared by the Committee on Analytical Reagents of the American Chemical Society are intended to serve for reagents to be used in precise analytical work of a general nature. It is recognized that there may be special uses for which reagents conforming to other, or more rigorous, specifications may be needed. Therefore, where known and where feasible, some of the specifications herein include requirements and tests for certain specialized uses. However, it is impossible to include specifications for all such uses, and thus there may be occasions when it will be necessary for the analyst to further purify reagents known to have special purity requirements for certain uses.

The requirements and the details of tests are based on published work, on the experience of members of the Committee in the examination of reagent chemicals on the market, and on studies of the tests made by members of the Committee. The limits and procedures are designed for application to reagents in freshly opened containers. Reagents in containers of extended age, in containers subject to constant changes in humidity or headspace gas content (as by repetitive opening and closing of the container), or subjected to potential inadvertent contamination by repeated opening of the container may not conform to the designated requirements. Where the possibility of change due to age, humidity, light, or headspace contamination is recognized, the specification usually contains a warning; nonetheless, the analyst is cautioned to take appropriate steps to ensure the

1

continued purity of the reagents, especially after opening the container.

In determining quality levels to be defined by new or revised specifications, the Committee is guided by the following general principles. When a specification is first prepared, it will usually be based on the highest level of purity (of the reagent to which it applies) that is competitively available in the United States. Generally, the term "competitively available" is understood to mean that the material is available from two or more producers. If a significantly higher level of purity subsequently becomes available on the same competitive basis, the specification will generally be revised accordingly.

Because the requirements of a specification relating to the content of designated impurities must necessarily be expressed in terms of maximum permissible limits, products conforming to the specification will normally contain less than the maximum permissible proportion of some or all of these impurities. A given preparation of a reagent chemical that has less than the maximum content of one or more impurities permitted by the specification is, therefore, not considered as of higher quality than that defined by the specification.

A lower permissible limit for a given impurity will be adopted only if it is significantly different from the one it is intended to supersede. In general, a new requirement for an impurity whose content is not greater than 0.01% will not be considered significantly different unless it decreases the maximum permissible content of the impurity by at least 50%. This principle will also be approximated in the revision of those requirements defined by the term, "Passes test".

Tests as written are considered to be applicable only to the accompanying requirements. Modification of a requirement, especially if the change is toward a higher level of purity, will necessitate reconsideration, and often revision, of the test to ensure its validity.

The assays and tests described herein constitute the methods upon which the ACS specifications for reagent chemicals are based. The analyst is not prevented, however, from applying alternative methods of analysis that produce results of at least equal reliability. In the event of doubt or disagreement concerning a substance purported to comply with the ACS specifications, only the methods described herein are applicable.

SOLVENTS FOR SPECIAL PURPOSES

For some solvents, the 6th and 7th editions of *Reagent Chemicals* had separate specifications defining them as either "suitable for use in ultraviolet spectrophotometry", "suitable for use in determining pesticide residues", or "suitable for use in high-performance liquid chromatography". In this edition, these special-use reagent chemicals have been treated in a single integrated presentation. The seller shall designate in product labeling the suitability for one or more of these special uses on the basis of the relevant requirements and tests.

DISCLAIMER

The reagent chemicals included herein may be hazardous substances, and the use of such reagent chemicals and the application of the various test methods may involve hazardous substances, operations, and equipment. The American Chemical Society (ACS) and the ACS Committee on Analytical Reagents do not purport in this book or in any other publication to specify minimum legal standards or to address all of the risks and safety problems associated with reagent chemicals, their use, or the methods prescribed for testing them.

No warranty, guarantee, or representation is made by ACS or the ACS Committee on Analytical Reagents as to the accuracy or sufficiency of the information contained herein, and ACS and the ACS Committee on Analytical Reagents assume no liability or responsibility in connection therewith. It is the responsibility of whoever uses the reagent chemicals and/or the testing methods set forth in this book to establish appropriate safety and health practices and to determine the applicability of any regulatory standards and/or limitations. Users of this book should consult and comply with pertinent local, state, and federal laws and should consult legal counsel if there are any questions or concerns about the applicable laws, safety issues, and reagent chemicals or the testing methods set forth herein.

CONTAINERS

The container is the device that holds the reagent and that is, or may be, in direct contact with the reagent. The closure is part of the container.

The container in which a reagent is sold and/or stored must be suitable for its intended purpose and should not interact physically or chemically with the contained reagent so as to alter its quality (within a reasonable period of time) beyond the requirements of the specification.

Containers for solids normally have wide mouths to facilitate both the filling of the container and the removal of the contents. Containers for liquids normally have narrow mouths so that the contents may be easily poured into other, frequently smaller, containers.

Prior to its being filled, the container should be free from extraneous particulate matter, otherwise clean, and dry if necessary.

INTERPRETATION OF REQUIREMENTS

The requirements of reagent chemicals can be divided into two main classes: an assay or quantitative determination of the principal or active constituent and the determination of the impurities or minor constituents. In some cases physical properties are specified.

Method Detection Limits

The purity of reagent chemicals continues to improve, driven by customer demand and the evolution of manufacturing processes and analytical technology. A clear and practical definition of analytical detection limits is vital to the accurate and precise determination of purity. An excellent review and definition of detection limits has recently been published.*

The section covering the determination of detection limits is reproduced in the following paragraphs. The original work provides other details, including the use of control charts to measure and monitor the accuracy and precision of any analytical procedure.

INTRODUCTION

Detection limits are controversial, principally because of inadequate definition and confusion of terms. Frequently, the instrumental detection limit is used for the method detection limit and vice versa. Whatever term is used, most analysts agree that the smallest amount that can be detected above the noise in a procedure and within a stated confidence limit is the detection limit. The confidence limits are set so that probabilities of both Type I and Type II errors (see below) are acceptably small.

Current practice identifies several detection limits, each of which has a defined purpose. These are the instrument detection limit (IDL), the lower limit of detection (LLD), the method detection limit (MDL), and the limit of quantitation (LOQ). Occasionally the instru-

---

Standard Methods for the Examination of Water and Wastewater, 17th ed., 1989, pp 1–18. (Reproduced with permission. Copyright 1989 American Public Health Association, American Water Works Association, and Water Pollution Control Federation.)

ment detection limit is used as a guide for determining the MDL. The relationship among these limits is approximately IDL:LLD:MDL:LOQ = 1:2:4:10.

DETERMINING DETECTION LIMITS

An operating analytical instrument usually produces a signal (noise) even when no sample is present or when a blank is being analyzed. Because any quality assurance (QA) program requires frequent analysis of blanks, the mean and standard deviation become well known; the blank signal becomes very precise, i.e., the Gaussian curve of the blank distribution becomes very narrow. The IDL is the constituent concentration that produces a signal noise greater than three standard deviations of the mean noise level or that can be determined by injecting a standard to produce a signal that is five times the signal-to-noise ratio. The IDL is useful for estimating the constituent concentration or amount in an extract needed to produce a signal to permit calculating an estimated method detection limit.

The LLD is the amount of constituent that produces a signal sufficiently large that 99% of the trials with that amount will produce a detectable signal. Determine the LLD by multiple injections of a standard at near-zero concentration (concentration no greater than five times the IDL). Determine the standard deviation by the usual method. To reduce the probability of a Type I error (false detection) to 5%, multiply σ by 1.645 from a cumulative normal probability table. Also, to reduce the probability of a Type II error (false nondetection) to 5%, double this amount to 3.290. As an example, if 20 determinations of a low-level standard yielded a standard deviation of 6 μg/L, the LLD is $3.29 \times 6 = 20$ μg/L.

The MDL differs from the LLD in that samples containing the constituent of interest are processed through the complete analytical method. The method detection limit is greater than the LLD because of extraction efficiency and extract concentration factors. The MDL can be achieved by experienced analysts operating well-calibrated instruments on a nonroutine basis. For example, to determine the MDL, add a constituent to reagent water, or to the matrix of interest, to make the concentration near the estimated MDL. Analyze seven portions of this solution and calculate the standard deviation(s). From a table of the one-sided t distribution select the value of t for $7 - 1 = 6$ degrees of freedom and at the 99% level; this value is 3.14. The product $3.14 \times \sigma$ is the desired MDL.

Although the LOQ is useful within a laboratory, the practical quantitation limit (PQL) has been proposed as the lowest level achievable among laboratories within specified limits during routine laboratory operations. The PQL is significant because different laboratories will produce different MDLs even though using the same analytical procedures, instruments, and sample matrices. The PQL is about five times the MDL and represents a practical and routinely achievable detection limit with a relatively good certainty that any reported value is reliable.

Rounding Procedures

For comparison of analytical results with requirements for assays and impurities, the observed or calculated values are rounded to the number of decimal places carried in the requirement. The method and the rounding procedure are in accord with ASTM (American Society for Testing and Materials) Practice E 29, for "Indicating Which Places of Figures Are to Be Considered Significant in Specified Limiting Values".

The procedure is as follows: If the digit following the last place to be retained is not equal to 5, round to the nearest number; if the digit to be dropped is 5 or 5 followed by zeros, round to an even number. This rounding procedure is illustrated in the accompanying table.

Requirement	Observed Value	Rounded Value	Pass / Fail
Not less than 98%	97.6	98	pass
	97.5	98	pass
	97.4	97	fail
Not less than 98.0%	97.95	98.0	pass
	97.94	97.9	fail
Not more than 0.01%	0.014	0.01	pass
	0.015	0.02	fail
	0.016	0.02	fail
Not more than 0.02%	0.015	0.02	pass
	0.025	0.02	pass
	0.026	0.03	fail

(Observe that the foregoing procedure does not conform to the common electronic calculator and computer procedure of rounding up when the digit to be dropped is 5 or 5 followed by zeros.) The rounded value should be obtained in a single step by direct rounding of the most precise value available and not in two or more steps of successive rounding. For example, 97.5487 rounds to 97.5 against a requirement of 97.6 and not in two possible steps of 97.55 and then 97.6.

The formula weights and factors for computing results are based on the 1989 International Atomic Weights shown on the inside of the front cover of this book. The formula weights are rounded to two decimal places.

Assay Requirements

Assay requirements are included for most of the reagent chemicals in this book. An assay value, in the sense used herein, is the content or concentration of a stated major component in the reagent. Unless otherwise specified, assay requirements are on an as-is basis (i.e., without drying, ignition, or other pretreatment of the sample).

Unless described in great detail and carried out with exceptional skill, available assay methods seldom are accurate enough to permit using a weighed quantity of a reagent so assayed in an exacting stoichiometric operation. This use of reagent chemicals should be limited to those designated as standards (for example, acidimetric or reductometric standards) because especially exacting assay methods are provided for such reagents.

Except in the case of standards, assays, through their minimum and maximum limits, mainly serve to assure acceptable consistency of the strength of reagents offered in the marketplace. They are particularly useful, for example, in the requirements for acid–water systems to control strength; for alkalies to limit the content of water and carbonate; for oxidizing and reducing substances that may change strength during storage; and for hydrates to control, within reasonable limits, deviations in the amount of water from that indicated in the formula. If, however, it should be necessary to use such reagents in stoichiometric operations, the user should ascertain the exact values to employ.

Impurity Requirements

Requirements for impurities are expressed in the following ways: (1) as numerical limits; (2) in terms of the expression "Passes test" with an accompanying approximate numerical limit; or (3) in terms of the expression "Passes test" without an approximate numerical limit. The distinction among these forms of expression is based on the Committee's opinion as to the relative quantitative significance of the prescribed test methods. The methods given for determining conformity to requirements of the first type are considered to yield, in competent hands, what are usually thought of as "quantitative" results, whereas those of the second type can be expected to yield only approximate values. Those in class 3 give definitions that cannot be expressed in numbers. It is obvious, however, that these distinctions as to quantitative significance cannot be sharp and that even the numerically expressed requirements are not all defined with equal accuracy. The final and essential definition of any requirement must, therefore, reside in the prescribed test method rather than in its numerical expression.

If a test method yields results that are adequately reproducible on repeated trials in different laboratories, it offers a satisfactory definition of the content of an impurity whether or not the result can be expressed by a number. Although the Committee has endeavored to base requirements, so far as possible, on test methods that meet this criterion, a considerable number are based on essentially undefined statements such as "no turbidity", "no color", or "the color shall not be completely discharged in ... minutes". Although some of the requirements of this kind could be replaced by others based on quantitative comparisons or measurements, to do so would require more costly or time-consuming procedures than appear at this time to be justified. The approach to an ultimate goal of replacing in every instance the word "none", or its equivalent, by the expression "maximum allowable ... ." therefore is limited both by deficiencies of knowledge and by practical considerations of expediency.

Unlisted Impurities

The primary objective of the Committee in preparing reagent specifications is to assure the user of the strength, quality, and purity of the reagents. It is, however, manifestly impossible to include in each specification a test for every impurity and contaminant that may be present. The Committee recognizes that for certain uses more stringent or additional requirements may be appropriate, and for such uses additional testing beyond that described in the specifications should be employed by the user. The Committee's intent in establishing the specifications is to recognize the common uses for which the reagent is employed and to establish requirements that are consistent both with these uses and with the manufacturing processes and quality of the available reagents. The presence of moisture, either as water of crystallization or as an adventitious impurity, falls within the purview of contamination unless permitted by the requirements in the applicable specification.

While tests for foreign particulate matter are not usually included in the specifications for solid reagents, such matter constitutes contamination. Similarly, although tests for clarity are not usually included in the specifications for liquid reagents or for solutions of solid reagents, the presence of haze, turbidity, or foreign particulate matter also constitutes contamination.

In some instances residual amounts of substances that have been added as aids in the process of purification may be present. An example is the use of complexing agents to keep certain metal ions in solution during recrystallization. These substances not only are impurities but may interfere with the tests.

Certain reagents, such as desiccants and indicators, have requirements that assure suitability for their intended use. Such reagents may contain impurities that do not interfere with their intended use, but that may make these reagents unsuitable for other uses.

When the Committee becomes aware of an unlisted impurity that affects adversely the known or specified uses of a reagent, a new requirement is added to the specifications, provided a suitable method of test is available. Users of reagents can protect themselves against the effects of unlisted impurities on a specific analytical procedure by applying, ad hoc, an appropriate suitability test.

Added Substances

Unless otherwise specified for an individual reagent chemical, the reagents described in this volume may contain suitable preservatives or stabilizers, intentionally added to retard or inhibit natural processes of deterioration. Such preservatives or stabilizers may be regarded as suitable only if the following conditions are met:

1. They do not exceed the minimum quantity required to achieve the desired effect.
2. They do not interfere with the tests and assays prescribed for the individual reagents.
3. The presence of any added substance must be declared on the label of the individual package. Unless of a proprietary nature, the name and concentration of any added substance should be stated on the label.

Identity Requirements

Identity requirements and tests are not included in the specifications. If there is any question as to the identity of a chemical, identity can be ascertained by appropriate analytical methods.

Particle Size for Granular Materials

When a mesh or a mesh range is stated on the label for a reagent chemical, the label shall include reference to a coarse sieve and to a finer sieve. The sieve number (mesh) is related to the sieve opening as indicated in the accompanying table.

Sieve No. (mesh)	Sieve Opening (mm)
2	9.52
4	4.76
8	2.38
10	2.00
20	0.84
30	0.59
40	0.42
50	0.297
60	0.250
80	0.177

When tested according to the following procedures, at least 95% of the material shall pass through the coarse sieve and at least 70% shall be retained on the finer sieve. This requirement applies only to materials that are 60-mesh or coarser.

Procedure.   The sieves used in this procedure shall be those known as the U.S. Standard Sieve Series. Details of the standardization of such sieves can be found in ASTM E 11, Specification for Wire-Cloth Sieves for Testing Purposes.

Place 25 to 100 g of the material to be tested upon the appropriate coarser standard sieve, which is mounted above the finer sieve, to which a close-fitting pan is attached. Place a cover on the coarser sieve and shake the stack in a rotary horizontal direction and vertically by tapping on a hard surface for not less than 20 min or until sifting is practically complete. Weigh accurately the amount of material remaining on each sieve.

An alternate procedure may be used in which the screening through the standard sieves is carried out in a mechanical sieve shaker. This shaker reproduces the circular and tapping motion given to the testing sieves in hand sifting, but with a mechanical action. Follow the directions provided by the manufacturer of the shaker.

PRECAUTIONS FOR TESTS

The descriptions of the individual tests are intended to give all essential details without repetition of considerations that should be obvious to an experienced analyst. A few suggestions are given for precautions and procedures that are particularly applicable to the routine testing of reagent chemicals.

Samples for Testing

To eliminate accidental contamination or possible change in composition, samples for testing must be taken from freshly opened containers.

ACCURACY OF MEASUREMENTS

In specifying the weights or volumes of sample to be used in the individual test procedures, it is intended, unless otherwise specified in the individual procedure, that the accuracy of measurement be such that the amount of sample used is within 2.0% of the stated amount. Thus, where a 10-g (mL) sample is specified, the amount actually taken for analysis must be between 9.8 and 10.2 g (mL). Similarly, where a test procedure directs that a solution be diluted to a specific volume or that a specified volume of solution be used, it is intended that the volume actually be within 2.0% of the stated amount.

Where the term "pipet" is used in a verbal sense, it is intended that the specified volume be taken in a volumetric pipet conforming to the tolerances accepted by NIST (National Institute of Standards and Technology) (*see* "The Calibration of Small Volumetric Laboratory Glassware", NBSIR 74–461, December 1974).

The addition of small volumes of liquid reagents is generally stated to the nearest 0.05 mL (0.05 mL, 0.10 mL, 0.15 mL, etc.). Use of the term "drop" to represent 0.05 mL is avoided.

Reagents

Reagents used in the testing should conform to ACS specifications. Reagents not covered by ACS specifications should be of the best grade obtainable and should be examined carefully for interfering impurities.

Blank Tests

Many of the tests are for minute quantities of the impurities sought. Hence complete blank tests must be made covering the water and other reagents used in each step of the tests—including, for example, filtration and ignition. Frequently, however, the directions do stipulate a control, a blank, or other device that corrects for possible impurities in the water and other reagents (*see* page 69).

GENERAL DIRECTIONS AND PROCEDURES

The general directions and/or procedures for the most common ACS requirements and tests are described in this section. To conserve space in the specifications for particular reagent chemicals, these procedures are not repeated in the individual test descriptions. Page references to this section are cited for most of the following tests: gravimetric methods (insoluble matter and residue after ignition); measurements of physical properties (boiling range, color, density, freezing point, and melting point); colorimetric and turbidimetric determinations of selected impurities (ammonium, arsenic, chloride, heavy metals, iron, nitrate, nitrogen compounds, phosphate, silicate, and sulfate); and certain instrumental methods (atomic absorption spectrophotometry, potentiometry, polarography, electrometric end-point detection, and chromatography).

Gravimetric Method for Determining Small Amounts of Impurities

In many of the reagent specifications it is directed that a precipitate or residue be collected and either dried or ignited so as to provide certain information concerning the purity of the reagent. Except where it is directed otherwise in the specific reagent tests, these directions and precautions will be used in collecting, drying, and igniting precipitates and residues.

GENERAL CONSIDERATIONS

It is imperative that the analyst use the best techniques in performing any of the operations included in this book. Exceptional cleanliness and protection against accidental contamination from dirt and fumes are rigid requirements for acceptable results. The choice of equipment is generally left to the discretion of the analyst, but it must provide satisfactory precision and accuracy. All weighings must be determined with an uncertainty of not more than ± 0.0002 g.

When the use of a tared container is specified, the container will be carried through a series of operations identical to those used in the procedure, including drying, igniting, cooling in a suitable desiccator, and weighing. The length of the drying or ignition and the temperature employed must be the same as specified in the procedure in which the tared equipment is to be used. Where it is directed to dry or ignite to constant weight, two successive weighings may differ by not more than ± 0.0002 g, the second weighing following a second drying or ignition period.

In those operations wherein a filtering crucible is specified, a fritted-glass crucible, a porous porcelain crucible, or a crucible having a sponge platinum mat may be used. Certain solutions may attack the filtering vessel; for example, both fritted glass and porcelain are affected by strongly alkaline solutions. Even platinum sponge mats are attacked by hydrochloric acid unless the crucible is first washed with boiling water to remove oxygen.

COLLECTION OF PRECIPITATES AND RESIDUES

In many of the tests the amount of residue or precipitate may be so small as to escape easy detection. Therefore, the absence of a weighable residue or precipitate must never be assumed. However small, it must be properly collected, washed, and dried or ignited. The size and type of the filter paper to be used is selected according to the amount of precipitate to be collected, not the volume of solution to be filtered.

Often when ammonia is added in slight excess to precipitate the R_2O_3 group of elements, the amount of precipitate produced is so small that it is hard to see and collect. If a small amount of a suspension of ashless filter paper pulp is added toward the end of the period of digestion on the steam bath or hot plate, the flocculation and collection of the hydroxides are facilitated.

Some precipitates have a strong tendency to "creep"; others, like magnesium ammonium phosphate, stick fast to the walls of the container. The use of the rubber-tip "policeman" is recommended, sometimes supplemented by a small segment of ashless filter paper. The analyst should guard against the possibility of any of the precipitate escaping beyond the upper rim of the filter paper.

Precipitates, especially if recently produced, may be partially dissolved unless proper precautions are taken with both their formation and subsequent washing. For example, a large excess of ammonia should be avoided when the R_2O_3 hydroxides are precipitated because aluminum hydroxide is measurably soluble in excess ammonia.

In general, precipitates should not be washed with water alone but with water containing a small amount of common ion to decrease the solubility of the precipitate. It is much better to wash with several small portions of washing solution than with fewer and larger portions. Calcium oxalate precipitates should be washed with dilute (about 0.1%) ammonium oxalate solution; therefore, mixed precipitates that may contain calcium oxalate and magnesium ammonium phosphate should be washed with a 1% ammonia solution containing also 0.1% ammonium oxalate.

IGNITION OF PRECIPITATES

The proper conditions for obtaining the precipitate are given in the individual test directions. The precipitate is collected on an ashless filter paper of suitable size and porosity (for example, a paper of 2.5-μm retention for a precipitate made up of fine particles), the residue and paper are thoroughly washed with the proper solution, and the moist filter paper is folded about the residue. The filter paper is then placed in a suitable tared preconditioned crucible, and the paper is dried. The drying may be accomplished by using a 105°C oven, by heating on a hot plate, with an infrared lamp, or by the careful use of an Argand burner. When dry, the paper is charred at the lowest possible temperature, the charred paper is destroyed by gentle ignition, and the crucible is finally ignited at 800 ± 25°C for 15 min. The ignited crucible is cooled in a suitable desiccator and weighed.

INSOLUBLE MATTER

The intent of these tests is to determine the amount of insoluble foreign matter (e.g., filter fibers and dust particles) present under test conditions designed to dissolve completely the substance being tested.

Procedure. Prepare a solution of the sample as specified in the individual test directions. Unless otherwise specified, heat to boiling in a covered beaker and digest on a steam bath (or a low-temperature hot plate) for 1 h. Filter the hot solution through a suitable tared, medium-porosity (10–15-μm) filtering crucible. Unless otherwise specified, wash the beaker and filter thoroughly with hot water, dry at 105°C, cool in a desiccator, and weigh.

RESIDUE AFTER EVAPORATION

This test is designed to determine the amount of any higher boiling impurities or nonvolatile dissolved material that may be present in a reagent chemical. It is used chiefly in testing organic solvents and some acids.

The preferred container for the evaporation of almost all reagents is a platinum dish. However, in many cases other containers (for example, dishes of porcelain, silica, or aluminum) may be found suitable. In a few cases, such as when strong oxidants are evaporated, platinum is contraindicated.

A steam bath usually is specified in the tests in this book for evaporations, but the analyst should be alert to the possibility that steam may contain volatile water-treatment chemicals that could condense on the evaporating dish while it is on the steam bath and affect its weight.

The residue is not dried to constant weight in this test because continued heating may slowly volatilize some of the high-boiling impurities. The drying conditions specified will, however, yield reproducible and reliable results so that drying to constant weight is not necessary.

Procedure.   Place the quantity of reagent specified in the individual test description in a suitable tared container (*see* General Considerations, page 12). Evaporate the liquid gently so that boiling does not occur, and in a hood, protected from any possibility of contamination. Unless otherwise specified, dry the residue in an oven at 105°C for 30 min. Cool the container in a suitable desiccator and weigh. Calculate the percent residue from the weight of the residue and the weight of the sample.

RESIDUE AFTER IGNITION

These tests are designed and intended to determine the amount of nonvolatile inorganic material that may be present in a reagent. They are applied to those inorganic reagents that can be sublimed or volatilized without decomposition and to various organic reagents. Among those reagent chemicals so tested are ammonium salts, mercury salts, and some inorganic acids.

The interpretation of the instruction to "ignite" has varied widely. Directions to "ignite gently", "ignite at cherry redness", and "ignite at a low red heat" are not specific enough to ensure the same procedure by different analysts.

After considerable study of the various residues involved, a temperature of 800 ± 25°C for 15 min has been adopted. This should assure constant conditions for conversion of the residues to the desired composition without causing appreciable loss of the impurities themselves.

The procedure cited in the following paragraph is not inflexible. In particular, certain organic reagents are difficult to volatilize, and slight variations in the test have been suggested where appropriate. Reagents with a high water content should be dried before ignition to prevent loss of sample. It is left to the discretion of the analyst as to the method of heating to be used, whether a hot plate, Argand burner, or infrared lamp. The final ignition at higher temperature should be done under oxidizing conditions.

Procedure. Ignite the quantity of reagent specified in the individual test description in a tared preconditioned crucible or dish in a well-ventilated hood, protected from air currents. Heating should be gentle and slow at first, and continued at a rate such that 1 to 2 h is required to volatilize inorganic samples completely or to char organic samples thoroughly. If the sample is a liquid, evaporate completely by heating gently without boiling. Cool the crucible and moisten the residue with 0.5 mL of sulfuric acid, unless specified otherwise. Ignite the crucible until white fumes of sulfur trioxide cease to evolve—then, finally, ignite at 800 ± 25°C for 15 min. Cool in a suitable desiccator, weigh, and calculate the percentage residue.

"SUBSTANCES NOT PRECIPITATED BY __"

These procedures are designed and intended to determine those soluble impurities that remain after the principal constituent has been removed by precipitation, as prescribed in the reagent specification.

In these tests a solution of the reagent in question is treated with a particular precipitating agent, the precipitate is removed by filtration, and a specified volume of the filtrate is used for the determination. The aliquot portion is transferred to a tared dish on a steam bath (or a low-temperature hot plate), and the liquid is evaporated without boiling or spattering. A moist residue, often an ammonium salt, may remain. When volatilizing this salt, great care must be taken to avoid spattering, which may cause a loss of some of the residue. After the salt has been volatilized, the residue is ignited as directed in the particular test description, cooled in a suitable desiccator, and weighed.

WEIGHTS OF PRECIPITATES AND RESIDUES

With few exceptions, the weight of the sample used in the test will ensure that at least 0.001 g of the impurity to be determined will be present if the sample fails the test.

Measurement of Physical Properties

BOILING RANGE

The boiling-range procedure is essentially an empirical method. Hence adherence to the specified procedural details is required to obtain consistent and reproducible results. The recommended method is that developed by ASTM, and the analyst is referred to ASTM D 1078–86, Test Method for Distillation Range of Volatile Organic Liquids, for information concerning the description of the apparatus and the proper performance of the test. If the liquid to be tested is extremely volatile or produces noxious fumes during distillation, suitable cooling and venting facilities must be provided.

COLOR (APHA)*

The color intensity of liquids may be estimated rapidly by using platinum–cobalt standards, as described in ASTM standard method D 1209-84, Test Method for Color of Clear Liquids (Platinum–Cobalt Scale). This method is particularly applicable to those materials in which the color-producing bodies have light-absorption characteristics nearly identical with those in the standards. Colors having hues other than light yellow or reddish yellow cannot be determined with these standards.

The platinum–cobalt color standards contain carefully controlled amounts of potassium chloroplatinate and cobaltous chloride. Each platinum–cobalt color unit is equivalent to 1 mg of platinum per liter of solution (1 ppm), and the standards are named accordingly. For example, the No. 20 platinum–cobalt standard contains 20 ppm of platinum. These platinum–cobalt standards are also called APHA and Hazen standards.

Apparatus.   The apparatus for this measurement consists of a series of Nessler tubes, a color comparator, and a light source. These tubes must match each other with respect to the color of the glass and height of the graduation mark. They must be fitted with suitable closures to prevent loss of liquid by evaporation and contamination of the standards by dust or dirt. The most commonly used type of Nessler tube is the 100-mL tall-form tube. However, for some sam-

---

*Standard Methods for the Examination of Water and Wastewater, 17th ed., American Public Health Association, American Water Works Association, and Water Pollution Control Federation, Washington, DC, 1989, p 2-2.

ples, particularly those having darker colors, a better color match may be obtained by using 100-mL short-form Nessler tubes.

For the most accurate estimation of color, it is desirable to use a comparator constructed so that white light is reflected from a white glass plate with equal intensity through the longitudinal axes of the tubes being compared. The tubes are shielded so that no light enters the tubes from the side. In most cases, satisfactory estimates of the color can be obtained by holding the sample and standards close to each other over a white plate.

The best source of light is generally considered to be diffuse daylight. However, for general routine analyses, the use of a titrating lamp equipped with a "daylight"-type fluorescent tube is satisfactory.

Preparation of APHA No. 500 Platinum–Cobalt Standard.   Add approximately 500 mL of reagent water to a 1000-mL volumetric flask, add 100 mL of hydrochloric acid, and mix well. Weigh 1.245 g of potassium chloroplatinate (K_2PtCl_6) and 1.000 g of cobalt chloride hexahydrate ($CoCl_2 \cdot 6H_2O$) to the nearest milligram, and transfer to the flask. Swirl until complete dissolution is effected, dilute to the mark with reagent water, and mix thoroughly. The spectral absorbance of this APHA No. 500 standard must fall within the limits given below when measured in a suitable spectrophotometer, with a 1-cm light path and reagent water as the reference liquid in a matched cell.

Wavelength (nm)	Absorbance
430	0.110 to 0.120
455	0.130 to 0.145
480	0.105 to 0.120
510	0.055 to 0.065

Preparation of APHA Platinum–Cobalt Standards.   Prepare the required color standards by diluting the volume of APHA No. 500 platinum–cobalt standard listed in the following table with reagent water to a total volume of 100 mL. The use of a buret is recommended in measuring the No. 500 standard. This series of standards is usually sufficient to permit an experienced analyst to make color comparisons with the necessary precision and accuracy. If a more exact estimate of color is desired, additional standards may be prepared to supplement those given by using proportional amounts of the No. 500 platinum–cobalt standard.

APHA Pt–Co Color Standard Preparation

APHA Pt–Co Color Standard Number	APHA No. 500 Pt–Co Standard (mL to dilute to 100 mL)
0	0.00
1	0.20
3	0.60
5	1.00
10	2.00
15	3.00
18	3.60
20	4.00
25	5.00
30	6.00
35	7.00
40	8.00
50	10.00
60	12.00
70	14.00
80	16.00
90	18.00
100	20.00
120	24.00
140	28.00
160	32.00
180	36.00
200	40.00
250	50.00
300	60.00
350	70.00
400	80.00
450	90.00
500	100.00

Procedure.   Transfer 100 mL of the sample to a matched 100-mL tall-form Nessler tube. (If the sample is turbid, filter or centrifuge before filling the tube to remove visible turbidity.) Compare the color of the sample with the colors of the series of platinum–cobalt standards in matching Nessler tubes. View vertically down through the tubes against a white background. Report as the color the number of the APHA standard that most nearly matches the sample. In the event that the color lies midway between two standards, report the darker of the two.

DENSITY AT 25°C

Density (g/mL) may be determined by any reliable method in use in the testing laboratory, such as by a pycnometer or a density meter. A satisfactory method uses a 5-mL U-shaped bicapillary pycnometer described in *Ind. Eng. Chem., Anal. Ed.* **1944,** *16*, 55 and adopted by ASTM (*Anal. Chem.* **1950,** *22*, 1452).

Procedure.   Tare a pycnometer, the volume of which has been accurately determined at 25°C. Fill the pycnometer with the liquid to be tested and immerse in a constant-temperature (25 ± 0.1°C) water bath for at least 30 min. Observe the volume of liquid at this temperature. Remove the pycnometer from the bath, wipe it dry, and weigh.

$$D \text{ at } 25°C = \frac{W}{V} + 0.0010$$

where

D = density in g/mL

W = weight, in grams, of liquid in pycnometer at 25°C

V = volume, in mL, of liquid in pycnometer at 25°C

0.0010 = correction factor to compensate for buoyancy

The correction of 0.0010 for air buoyancy is accurate to ±0.0001 g/mL if the density varies from 0.7079 g/mL (anhydrous ethyl ether) to 1.585 g/mL (carbon tetrachloride) and the pycnometer is weighed at a temperature of from 20° to 30°C at an atmospheric pressure of from 720 to 775 mm of mercury.

FREEZING POINT

Place 15 mL of sample in a test tube (20 × 150 mm) in which is centered a thermometer traceable to NIST. The sample tube is centered by corks in an outer tube about 38 × 200 mm. Cool the whole apparatus, without stirring, in a bath of shaved or crushed ice with water enough to wet the outer tube. When the temperature is about 3°C below the normal freezing point of the reagent, stir to induce freezing and read the thermometer every half minute. The temperature that remains constant for 1 to 2 min is the freezing point.

MELTING POINT

Where required, the melting point may be determined by the capillary tube method. Calibration of the method by the use of reference standards is required. A satisfactory capillary tube method is as follows:

Apparatus. A suitable melting-point apparatus consists of a glass container for a bath of transparent fluid, an appropriate stirring device, a thermometer traceable to NIST, and a controlled source of heat. The requisite temperature is considered in selecting the bath fluid, but light paraffin is generally suitable, and certain liquid silicones are well adapted to the higher temperature ranges. The fluid is deep enough to permit immersion of the thermometer to its specified immersion depth such that the bulb is still about 2 cm above the bottom of the bath. The capillary tube is about 10 cm long and 1.0 ± 0.2 mm internal diameter, with walls 0.2–0.3 mm thick.

Procedure. Reduce the substance under test to a very fine powder, and render it anhydrous if it contains water of hydration by drying it as directed. If the substance contains no water of hydration, dry it over a suitable desiccant for not less than 16 h.

Charge a capillary glass tube, one end of which is heat-sealed, with sufficient quantity of the dry powder to form a column 2.5–3.5 mm high in the bottom of the tube when packed down as closely as possible by moderate tapping on a solid surface.

Heat the bath until the temperature is about 10°C below the expected melting point and is rising at a rate of 1 ± 0.5°C/min. Remove the thermometer and quickly attach the capillary tube to the thermometer by wetting both with a drop of the bath liquid, or otherwise, and adjust its height so that the material in the capillary tube is level with the thermometer bulb.

Replace the thermometer and attached capillary tube when the temperature is 5°C below the expected melting point, and continue heating until the solid coalesces and is completely melted. Record the temperature.

Colorimetry and Turbidimetry

Conditions and quantities for colorimetric or turbidimetric tests have been chosen to produce colors or turbidities that can be observed easily. These conditions and quantities approach but do not reach the minimum that can be observed. Five minutes, unless some other time is specified, must be allowed for the development of the colors or turbidities before the comparisons are made. If solutions of samples contain any turbidity or insoluble matter that might interfere with the later observation of colors or turbidities, they must be filtered before the addition of the reagent used to produce the color or turbidity. Conditions of the tests will vary from one reagent to another, but most of the tests fall close to the limits indicated in the following table.

Usual Limits for Colorimetric or Turbidimetric Standards

Sample	Quantity Used for Comparison (mg)	Approximate Volume (mL)
Ammonium (NH_4)	0.01	50
Arsenic (As)	0.002–0.003	3
Barium (Ba) chromate test	0.05–0.10	50
Chloride (Cl)	0.01	20
Copper(Cu)		
Dithizone extraction	0.006	10–20
Hydrogen sulfide	0.02	50
Pyridine thiocyanate	0.02	10–20
Heavy metals (as Pb)	0.02	50
Iron (Fe)		
1,10-Phenanthroline	0.01	25
Thiocyanate	0.01	50
Lead (Pb)		
Chromate	0.05	25
Dithizone	0.002	10–20
Sodium sulfide	0.01–0.02	50
Manganese (Mn)	0.01	50
Nitrate (NO_3)		
Brucine sulfate	0.005–0.02	50
Diphenylamine	0.005	25
Phosphate (PO_4)		
Direct molybdenum blue	0.02	25
Molybdenum blue ether		
extraction	0.01	25
Sulfate (SO_4)	0.04–0.06	12
Zinc (Zn)		
Dithizone extraction	0.006–0.01	10–20

AMMONIUM (TEST FOR AMMONIA AND AMINES)

Procedure.   Dissolve the specified quantity of sample in 60 mL of ammonia-free water in a Kjeldahl flask connected through a spray trap to a condenser, the end of which dips below the surface of 10 mL of 0.1 N hydrochloric acid. Add 20 mL of freshly boiled sodium hydroxide solution (10%), distill 35 mL, and dilute the distillate with water as directed. Add 2 mL of freshly boiled sodium hydroxide solution (10%), mix, add 2 mL of Nessler reagent, and again mix. Any color should not be darker than that produced in a standard containing the amount of ammonium ion (NH_4) specified in the individual test description in an equal volume of solution containing 2 mL of the sodium hydroxide solution and 2 mL of Nessler reagent.

ARSENIC

This colorimetric comparative procedure is based upon the reaction between silver diethyldithiocarbamate and arsine.

Interferences.   Metals or salts of metals such as chromium, cobalt, copper, mercury, molybdenum, nickel, palladium, and silver may interfere with the evolution of arsine.

Antimony, which forms stibine, may produce a positive interference in color development with silver diethyldithiocarbamate solution. Although potassium iodide and stannous chloride in the generator tend to repress the evolution of stibine at low levels of antimony, at higher levels repression is incomplete, a situation that can lead to erroneously high results for arsenic.

Interference from antimony, however, can be essentially eliminated by adding ferric ion to the generator. An arsenic test employing this addition is included in the standard for phosphoric acid, page 519.

Antimony can be determined, if desired, by atomic absorption or by differential pulse polarography. An atomic absorption procedure for antimony is included in the aforementioned standard for phosphoric acid.

Apparatus.   The apparatus (*see* Figure 1) consists of an arsine generator (a) fitted with a scrubber unit (c) and an absorber tube (e), with standard-taper or ground-glass ball-and-socket joints (b and d) between the units. Alternatively, any apparatus embodying the principle of the assembly described and illustrated may be used.

Sample Solution.   Dissolve the specified weight of sample in water and dilute with water to 35 mL, or use the solution prepared as directed in the individual reagent standard.

Procedure.   Unless otherwise specified in the individual test description, proceed as follows: To the sample add 20 mL of dilute sulfuric acid (1 + 4), 2 mL of potassium iodide reagent solution (16.5 g of KI per 100 mL), and 0.5 mL of stannous chloride reagent solution (4 g of $SnCl_2 \cdot 2H_2O$ in 10 mL of hydrochloric acid) and mix. Allow the

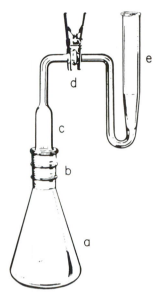

Figure 1. Arsenic test apparatus.

mixture to stand for 30 min at room temperature. Pack the scrubber tube (c) with two pledgets of cotton previously wetted with a satu- rated solution of lead acetate, freed from excess lead acetate solution by squeezing, and dried in a vacuum. Allow a small space between the two pledgets. Place 3.0 mL of diethyldithiocarbamate solution (1 g in 200 mL of freshly distilled pyridine) in the absorber tube (e) and 3.0 g of granulated zinc (No. 20 mesh) in the generator flask, and immediately connect the scrubber–absorber assembly to the flask. Place the generator flask in a water bath maintained at $25 \pm 3°C$, and swirl the flask gently at 10-min intervals. (The addition of a small amount of isopropyl alcohol to the generator flask may promote uniformity of the rate of gas evolution.) After 45 min, disconnect the tubing from the generator flask and transfer the silver diethyldithio- carbamate solution to a suitable color-comparison tube. Any red color in the silver diethyldithiocarbamate solution of the sample should not exceed that in a standard obtained from an amount of standard arsenic solution equivalent to the limit of arsenic specified under the individual reagent chemical, when treated with the same quantities of the same reagents and in the same manner.*

---

*The color comparison may be made in a photometer at the wavelength of maximum absorbance occurring between 535 and 540 nm. If a photometer is used, the silver diethyldithiocarbamate solution should be free from any trace of impurity (before the test is started). The use of capped cells is recommended.

CHLORIDE

The general test procedure cited here applies to the majority of reagent chemicals having a chloride specification.

Perform the test for the sample and the standard in glass tubes or cylinders of the same diameter, matched as closely as practicable. Use identical quantities of the same reagents for the sample and the standard solutions. If after dissolution the solution is not clear, filter it through a chloride-free filter paper. Prepare the filter paper by washing with water until the filtrate gives a negative test for chloride. The filter may also be prewashed with dilute nitric acid.

Experience has shown that visual turbidimetric comparisons are best made between solutions containing 0.01 mg of chloride ion in a volume of 20 mL.

Procedure.   Unless otherwise stated, dissolve the specified quantity of sample in water and dilute with water to 20 mL. If the substance is already in solution, dilute with water to 20 mL. Filter if necessary. For the standard, use 1 mL of the standard chloride solution (0.01 mg of Cl in 1 mL), and dilute with water to 20 mL. To each tube add 1 mL of nitric acid and 1 mL of silver nitrate reagent solution. Mix, allow to stand for 5 min protected from sunlight, and compare. The solutions can best be viewed visually against a black background. Any turbidity in the solution of the sample should not exceed that in the standard.

For samples yielding colored solutions dissolve the sample, prepare the standard, and treat the two solutions as directed in the specification. The comparison is best made by viewing both turbidities through the same depth and color of solution. Thus, superimpose a tube containing the specified quantity of sample in a volume of water equal to the total volume of the test over the tube containing the standard turbidity, and place a tube containing the same volume of water below the tube containing the sample and added reagents. The comparison tubes may be machine-made vials, long style, of about 20-mL capacity.

DITHIZONE TESTS

Because dithizone is extremely sensitive to many metal ions, all glassware used must be specially cleaned. After the usual cleaning, it should be thoroughly rinsed with warm dilute nitric acid (1 + 1) and finally rinsed with reagent water. All glassware used in the preparation and storage of reagents and in performing the tests must be made of lead- and zinc-free glass. Alternatively, for solution storage, polyethylene containers may be used.

HEAVY METALS (as Pb)

The heavy metals test is designed to limit the common metallic impurities (Ag, As, Bi, Cd, Cu, Hg, Mo, Pb, Sb, and Sn) that produce colors when sulfide ion is added to slightly acid solutions that contain them. In this test, such heavy metals are expressed "as lead" by comparing the color developed in a test solution with that developed in a lead standard/control solution under test conditions that have been established through collaborative study. As a result of such study, it has been determined that the optimum conditions for proper performance of the test are: pH of the solution between 3 and 4, total volume of 50 mL, use of freshly prepared hydrogen sulfide water, and use of a standard/control containing 0.02 mg of lead ion (Pb). These conditions are rarely modified in the individual procedures, and it is only after collaborative confirmation that the modifications are suitable for the individual reagent. The color comparisons should be made, using matched 50-mL color-comparison tubes, by viewing vertically over a white background. In some instances, an appropriate amount of the sample is added to the standard, thereafter called the control (*see* page 79), because the color developed in the test may be affected by the sample being tested. In other instances, no such effort occurs and the standard may be prepared without any of the sample. In general, the individual standard procedures provide only for the preparation of a sample solution. The remainder of the test directions, being the same for all reagents, are provided in this chapter.

Two general test procedures are provided. Method 1 is to be used unless otherwise specified in the individual standard. This method is used, generally, for those substances that yield clear, colorless solutions under the specified test conditions. Method 2 is used, generally, only for certain organic compounds.

Unless otherwise directed in the individual specification, the test shall be carried out as follows.

Procedure.

Method 1.

Test Solution.   Using the solution prepared as directed in the individual standard, adjust the pH to between 3 and 4 (by using a pH meter) with 1 N acetic acid or ammonium hydroxide (10% NH_3). Dilute with water to 40 mL, if necessary, and mix.

Standard / Control Solution.   To 20 mL of water, or to the portion of that solution specified in the individual standard, add

0.02 mg of lead ion (Pb).* Dilute with water to 25 mL and mix. Adjust the pH (within 0.1 unit) to the value established for the test solution (using a pH meter) with 1 N acetic acid or ammonium hydroxide (10% NH_3), dilute with water to 40 mL, and mix.

To each of the tubes containing the test solution and the standard/control solution, respectively, add 10 mL of freshly prepared hydrogen sulfide water and mix. Any brown color produced in the test solution within 5 min should not be darker than that produced in the standard/control solution.

Method 2.

Test Solution. Transfer the quantity of substance specified in the individual specification to a suitable crucible, add sufficient sulfuric acid to wet the sample, and carefully heat at a low temperature until thoroughly charred. (The crucible may be loosely covered with a suitable lid during the charring.) Add to the carbonized mass 2 mL of nitric acid and 0.25 mL of sulfuric acid, and heat cautiously until white fumes of sulfur trioxide are no longer evolved. Ignite, preferably in a muffle furnace, at 500 to 600°C until the carbon is burned off completely. Cool, add 4 mL of 6 N hydrochloric acid, and cover. Digest on the steam bath for 15 min, uncover, and slowly evaporate on the steam bath to dryness. Moisten the residue with 0.05 mL of hydrochloric acid, add 10 mL of hot water, and digest for 2 min. Add ammonium hydroxide (10%), dropwise, until the solution is just alkaline, and dilute with water to 25 mL. Adjust the pH of the solution to between 3 and 4 (by using a pH meter) with 1 N acetic acid, dilute with water to 40 mL, and mix.

Standard / Control Solution. To 20 mL of water, add 0.02 mg of lead ion (Pb), and dilute with water to 25 mL. Adjust the pH (within 0.1 unit) to the value established for the test solution (by using a pH meter) with 1 N acetic acid or ammonium hydroxide (10% NH_3), dilute with water to 40 mL, and mix.

To each of the tubes containing the test solution and the standard/control solution, respectively, add 10 mL of freshly prepared hydrogen sulfide water, and mix. Any brown color produced in the test solution within 5 min should not be darker than that produced in the standard/control solution.

---

*When need for a control solution is specified in the individual specification and a direction is given therein to add a weight of lead ion (Pb), no further lead shall be added. Proceed to the dilution or pH adjustment, as necessary.

IRON

Two procedures are provided for the determination of iron (Fe). Method 1, which employs thiocyanate, is usually performed in 50 mL of solution containing 2 mL of hydrochloric acid, to which 30 to 50 mg of ammonium peroxydisulfate crystals and 3 mL of ammonium thiocyanate reagent solution (about 30%) are added. These quantities of acid and thiocyanate were selected so that the production of the red color would not be affected significantly by anything except the iron. The use of peroxydisulfate to oxidize the iron prevents fading of the color and eliminates the need for immediate comparison with the standard. For certain alkaline earth salts, permanganate is used to oxidize the iron.

Method 2, which employs 1,10-phenanthroline, is used generally for phosphate salts. It is usually performed by adjusting the pH (when necessary) to between 4 and 6, adding 6 mL of hydroxylamine hydrochloride reagent solution (10%) and 4 mL of 1,10-phenanthroline reagent solution (0.1%), and diluting with water to 25 mL. Because development of the color may be slow, the comparison of the color produced is not made until 60 min after addition of the 1,10-phenanthroline solution. The hydroxylamine hydrochloride solution is the unacidified reagent solution and not the acidified solution used in the dithizone test for lead and other metals.

The two general procedures, which are referred to in the individual test requirements, are described here. Where a difference appears between the general test procedure and the directions set forth in the individual specification, the directions in the specification are to be followed.

Procedure.

Method 1 (Thiocyanate).

Dissolve the specified amount of sample in 40 mL of water or use the sample solution prepared as directed in the individual test description. Unless otherwise directed in the individual specification, add 2 mL of hydrochloric acid and, if necessary, dilute with water to 50 mL. Add 30 to 50 mg of ammonium peroxydisulfate crystals and 3 mL of ammonium thiocyanate reagent solution and mix. Any red color produced should not exceed that produced by 0.01 mg of iron (Fe) in an equal volume of solution containing the quantities of reagents used in the test.

Method 2 (1,10-Phenanthroline).

Dissolve the specified amount of sample in 10 mL of water, or use the sample solution prepared as directed in the individual test description. Add 6 mL of hydroxylamine hydrochloride reagent solution and 4 mL of 1,10-phenanthroline reagent solution and dilute with water to 25 mL. Any red color produced within 1 h should not exceed that produced by 0.01 mg of iron (Fe) in an equal volume of solution containing the quantities of reagents used in the test.

NITRATE

Four different tests are applied to the determination of nitrate: brucine sulfate, diphenylamine, indigo carmine, and phenoldisulfonic acid. General procedures are provided below for brucine sulfate and diphenylamine.

Brucine Sulfate.   It is recognized that the sensitivity of brucine sulfate for determining small amounts of nitrate varies with the conditions of each particular test. Therefore the tests are designed so that the sensitivity of the brucine sulfate for determining small amounts of nitrate is determined under the particular conditions of the test. The prescribed test must be followed carefully to ensure the development of a reproducible color.

Procedure.

Sample Solution A.   Dissolve the specified quantity or volume of the sample in 3 mL of water by heating in a boiling water bath. Dilute with brucine sulfate reagent solution to 50 mL.

Control Solution B.   To the specified volume of standard nitrate solution (0.01 mg of NO_3 in 1 mL), add the same quantity of sample as in sample solution A and sufficient water to make a volume of 3 mL. Dissolve the mixture by heating in a boiling water bath. Dilute with brucine sulfate reagent solution to 50 mL.

Blank Solution C.   Use 50 mL of brucine sulfate reagent solution.

Heat the three solutions in a preheated (boiling) water bath for 10 min. Cool rapidly in an ice bath to room temperature. Set a spectrophotometer at 410 nm and, using 1-cm cells, adjust the instrument to read zero absorbance with blank solution C in the light path, then determine the absorbance of sample solution A. Adjust the instrument to read zero absorbance with sample solution A in the light path and determine the absorbance of control solution B. The absorbance of sample solution A should not exceed that of control solution B.

Diphenylamine.   The prescribed tests must be followed carefully to ensure the development of a reproducible and stable color. Under slightly different conditions the oxidation reaction may produce a somewhat different color, which is not stable.

Procedure.   Place the specified quantity of the sample in a dry beaker. Cool the beaker thoroughly in an ice bath and add 22 mL of sulfuric acid that has been cooled to ice-bath temperature. Allow the mixture to warm to room temperature and swirl the beaker at intervals to effect gentle dissolution with slow evolution of the acid vapor. Prepare a standard by evaporating to dryness a solution containing the specified quantity of standard nitrate solution (0.01 mg of NO_3 in 1 mL) and 0.01 g of anhydrous sodium carbonate. Treat the residue exactly as the sample. Add 3 mL of diphenylamine reagent to each solution and digest on the steam bath for 90 min. Compare the color of the solutions visually. Any blue color produced in the solution of the sample should not exceed that in the standard.

NITROGEN COMPOUNDS: TEST FOR AMMONIA, AMINES, AND NITROGEN COMPOUNDS REDUCED BY ALUMINUM

Procedure.   Dissolve the specified quantity of sample in 60 mL of ammonia-free water in a Kjeldahl flask connected through a spray trap to a condenser, the end of which dips below the surface of 10 mL of 0.1 N hydrochloric acid. Add 10 mL of freshly boiled sodium hydroxide solution (10%) and 0.5 g of aluminum wire, in small pieces, to the Kjeldahl flask. Allow to stand for 1 h protected from loss of, and exposure to, ammonia. Distill 35 mL and dilute the distillate with water to 50 mL. Add 2 mL of freshly boiled sodium hydroxide solution (10%), mix, add 2 mL of Nessler reagent, and again mix. Any color should not be darker than that produced in a standard containing the amount of nitrogen (N) specified in the individual test description and treated exactly as is the sample.

PHOSPHATE

Three general procedures are provided to test for phosphate: direct molybdenum blue (method 1), extracted molybdenum blue (method 2), and precipitation (method 3). In method 2, conditions are established that allow only the phosphomolybdate to be extracted into ether, thus eliminating any interference from arsenate and silicate.

Procedure.

Method 1. Prepare the sample solution as directed in the individual test description. Add 1 mL of ammonium molybdate reagent solution and 1 mL of 4-(methylamino)phenol sulfate reagent solution, and allow to stand at room temperature for 2 h. Any blue color should not exceed that produced by 0.02 mg of phosphate ion (PO_4) in an equal volume of solution containing the quantities of reagents used in the test.

Method 2. Prepare the sample solution as directed in the individual test description. Transfer to a separatory funnel, add 35 mL of ether, shake vigorously, and allow the layers to separate. Draw off and discard the aqueous layer unless it is indicated in the individual specification that the aqueous layer should be saved for the silica determination. Wash the ether layer twice with 10 mL of dilute hydrochloric acid (1 + 9), drawing off and discarding the aqueous layer each time. Add 0.2 mL of a freshly prepared 2% solution of stannous chloride in hydrochloric acid. If the solution is cloudy, shake with a small amount of dilute hydrochloric acid (1 + 9) to clear. Any blue color should not exceed that produced by a standard containing the concentration of phosphate ion (PO_4) cited in the individual test description and treated exactly as is the sample.

Method 3. Dissolve 20 g of sample in 100 mL of dilute ammonium hydroxide (1 + 4). Prepare a standard containing 0.1 mg of phosphate ion (PO_4) in 100 mL of dilute ammonium hydroxide (1 + 4). To each solution add dropwise, with rapid stirring, 3.5 mL of ferric nitrate reagent solution (10%) and allow to stand about 15 min. If necessary, warm gently to coagulate the precipitate, but do not boil. Filter and wash several times with dilute ammonium hydroxide (1 + 9). Dissolve the precipitate by pouring 60 mL of warm dilute nitric acid (1 + 3) through the filter and catching the solution in a 250-mL glass-stoppered conical flask. Add 13 mL of ammonium hydroxide, warm the solution to 40°C, and add 50 mL of ammonium molybdate–nitric acid reagent solution. Shake vigorously for 5 min and allow to stand for 2 h at 40°C. Any yellow precipitate obtained from the sample should not exceed that obtained from the standard.

SILICATE

Small amounts of silica or silicate are determined by a molybdenum blue method. Conditions have been established that allow the extraction of the silicomolybdate and thus eliminate any interference from arsenate or phosphate.

SULFATE

Procedure A.

Method 1. Unless otherwise stated in the specification, dissolve (0.005/% limit) g of sample* in a minimum volume of water. Add 1 mL of dilute hydrochloric acid (1 + 19). If necessary, filter through a small filter, and wash with two 2-mL portions of water. Dilute to 10 mL and add 1 mL of barium chloride reagent solution. Any turbidity should not exceed that produced by 0.05 mg of sulfate ion (SO_4) in an equal volume of solution containing the quantities of reagents used in the test. Unless otherwise specified, compare 10 min after adding the barium chloride to the sample and standard solutions.

Method 2. Unless otherwise stated in the specification, dissolve (0.005/% limit) g of sample in water, and evaporate to dryness on a hot plate at low setting. Proceed as in method 1.

Method 3. Unless otherwise stated in the specification, add 10 mg of sodium carbonate to (0.005/% limit) g of sample, and evaporate to dryness. Proceed as in method 1.

Method 4. Unless otherwise stated in the specification, dissolve (0.005/% limit) g of sample in water, evaporate to 5 mL, add 1 mL of bromine water, and evaporate to dryness. Proceed as in method 1.

Standard Sulfate Solution (0.02 mg of SO$_4$ in 1 mL). Dissolve 0.148 g of anhydrous sodium sulfate, Na_2SO_4, in hydrochloric acid (3 + 1), and dilute with hydrochloric acid (3 + 1) to 100 mL. Dilute 20 mL of this solution with hydrochloric acid (3 + 1) to 1 L.

Procedure B. The sample is digested at 130 ± 10°C with a reducing mixture of red phosphorus, hydriodic acid, and iodine. The resulting hydrogen sulfide is swept by a stream of nitrogen into a reaction mixture of *N,N*-dimethyl-*p*-phenylenediamine and ferric sulfate, and the methylene blue produced is determined spectrophotometrically.

---

*Gives a sample size that provides 0.05 mg of SO_4 at limit. The standard also contains 0.05 mg of SO_4. To determine the sample weight in grams, divide 0.005 by the limit (in actual percent). For example, if the limit is 0.01%, 0.005 is divided by 0.01 to arrive at 0.5 g of sample.

Apparatus.   Nitrogen flow from a gas cylinder is regulated with a needle valve and a flowmeter. The gas is passed through a gas-scrubbing tower containing 5 g of mercuric chloride in 100 mL of 2% potassium permanganate to remove any traces of sulfur gases. Water should be added to the tower as needed to maintain the volume at approximately 10 mL. Tubing should be of a nonrubber material. The gas should be preheated to the temperature of the bath before it enters the reaction vessel by passing the gas through a metal tubing with an outside diameter of approximately 7 mm that is submerged in the bath.

An oil bath is required, in a suitable container, heated with stirring to the prescribed temperature ($130 \pm 10°C$), with a means of regulating the temperature. Silicone fluid in a 3-L stainless steel beaker fitted with a Teflon cover has been found satisfactory. Heat is applied with an electric heating mantle and controlled with a variable transformer. The oil is stirred with an electric or an air stirrer. The cover provides openings for a dial-type metal thermometer, gas tubing inlet and outlet, stirrer shaft, and the reaction vessel.

The reaction vessel is a borosilicate glass test tube fitted with a 24/40 standard taper outer joint with a total capacity of about 30 mL. A head about 13 cm high with a 24/40 standard taper inner joint fits the reaction vessel. An outlet tube emerges at the top of the head and turns at a right angle. An inlet tube, having an outside diameter of 8 mm, enters the side of the head and extends to the bottom of the reaction vessel. The preheated gas bubbles through the sample–reagent mixture that is being heated inside the bath. All fittings and tubing leading from the reaction vessel should be glass (glass ball joints are best).

The effluent gas from the reaction vessel should pass through a pair of scrubbing tubes, each the same size and construction as the reaction vessel. The gas is to enter each scrubber through a tube down the center, flow into the solution specified, and pass out the top into the next vessel. The first scrubber should contain sufficient phosphoric acid (1 + 1) just to cover the tip of the bubbler tube. (Drain and recharge when more than 1 inch of liquid accumulates, but do not interrupt during a day's run.) The second scrubber should contain 10 mL of water. (Recharge after running five samples.) The gas is finally bubbled through a delivery tube, having an outside diameter of 4 or 5 mm, into the absorbing solution in a 50-mL volumetric receiving flask.

Procedure.   Transfer 5 mL of absorbing solution to a 50-mL volumetric receiving flask, add 30 mL of water, and place the delivery tube in the flask. Add the sample solution and 5 mL of reducing mixture (*see* Reagents), previously shaken to produce a slurry, to the reaction vessel. Connect the reaction vessel to the scrubber assembly and place the vessel in the oil bath. Connect the delivery tube to the scrubber assembly, raise the receiving flask so that the delivery tube dips well below the surface of the acetate absorbing solution, and connect the gas lines. Adjust the nitrogen flow through the apparatus at a rate of 55 ± 10 mL/min and run for 25 min. Discard the first result of a day's run. Use the second result obtained by repeating the entire procedure to this point.

Disconnect the delivery tube but do not remove from the receiving flask. (*Caution: Take the reaction assembly apart in the hood, since noxious white fumes are produced by the hot reducing mixture.*) Add first 5 mL of *N,N*-dimethyl-*p*-phenylenediamine solution (*see* Reagents) through the delivery tube, then 1 to 2 mL of water, and swirl to mix. Avoid vigorous shaking. Add 1 mL of ferric sulfate solution (*see* Reagents) through the delivery tube, followed by 1 to 2 mL of water. Remove the delivery tube without further washing, stopper the flask, and shake vigorously. When the bubbles dissipate, dilute to volume with water and allow to stand in the dark for at least 10 min. Set a spectrophotometer at 665 nm and, using 1-cm cells, read the absorbance vs. water on the day that the standard is run. For the standards add 1.0, 2.0, and 3.0 mL (0.02, 0.04, and 0.06 mg), respectively, of the foregoing sulfate standard for procedure B to a reaction vessel and carry out the entire procedure used for the sample solution. Plot a curve of absorbance vs. sulfate concentration.

Reagents

1. **Reducing Mixture.**   Transfer 50 g of red phosphorus to a 2-L round-bottomed flask fitted with a 24/40 joint, add 500 mL of 47% hydriodic acid through a funnel, and place in an ice bath. Add 170 g of iodine gradually, with agitation, through the funnel. Wash the iodine into the flask with 100 mL of water and allow to stand overnight. Connect a water-cooled condenser to the flask in a heating mantle in the hood, insert a glass tube through the condenser to dip into the reagent in the flask, and bubble nitrogen through the reagent at a rate of 100 to 150 bubbles per minute. Heat to boiling, reflux for approximately 2 h, and cool to room temperature. Shake to produce a slurry before taking an aliquot for each analysis. If the reagent has stood for longer than a day without shaking, stir for 30 min before use.

2. **Absorbing Solution for Hydrogen Sulfide.**   Dissolve 50 g of zinc acetate dihydrate, $(CH_3COO)_2Zn \cdot 2H_2O$, and 12.5 g of

sodium acetate trihydrate, $CH_3COONa \cdot 3H_2O$, in water. Dilute with water to 1 L and filter. Any turbidity that forms on standing is caused by absorption of carbon dioxide from the air and does not interfere with the test.

3. **N,N-*Dimethyl*-p-*phenylenediamine* Solution.** Add 200 mL of sulfuric acid to approximately 700 mL of water. Cool to room temperature, add 1 g of N,N-dimethyl-p-phenylenediamine(p-aminodimethylaniline) monohydrochloride, and dilute with water to 1 L.

4. ***Ferric Sulfate Solution.*** Add 25 mL of sulfuric acid to 125 g of ferric sulfate n-hydrate, $Fe_2(SO_4)_3 \cdot nH_2O$. Add about 700 mL of water, digest on a steam bath to dissolve, cool to room temperature, and dilute with water to 1 L.

Atomic Absorption Spectroscopy

Trace metal analysis is routinely done by emission or absorption spectroscopy techniques. Inductively coupled plasma atomic emission spectroscopy (ICP–AES) and Atomic Absorption Spectroscopy (AAS) are the predominant instrumental analytical techniques used for trace metal analysis. ICP–AES is the technique of choice when rapid, quantitative, multielement analysis of solutions is required. AAS is used when accurate determinations of only a few elements in solution are required. AAS is the most cost-effective analytical technique for the trace metal specifications contained in this monograph. AAS is a very sensitive technique capable of parts per million (ppm) analysis with flame atomization, and parts per billion (ppb) analysis when furnace atomization is employed. AAS has few interferences, especially when compared to emission techniques. The detailed principles of AAS are given elsewhere. The purpose here is to describe how the technique should be used to assure that the specifications for the stated elements are met. A specific procedure is not given, because operating details will vary with instrument design. Trace metal analysis cited in this monograph can of course be done with emission techniques or graphite furnace (GFAA).

GENERAL BACKGROUND

In flame AAS the sample is aspirated into a high-temperature flame, usually air– or nitrous oxide–acetylene. After the solvent evaporates, the flame produces metal atoms that absorb light at wavelengths characteristic of the specific atom. The wavelength at which each element absorbs light is usually unique to that element. The amount of light absorbed is proportional to the concentration of the element present, a relationship referred to as Beer's law. For this law to be valid, the bandwidth of the radiation to be absorbed by the atom must be narrower than the absorption line for the absorbing species. The hollow cathode lamp used in AAS is an emission source whose line widths are typically one-tenth that of the absorption lines used in AAS. The optical system in AAS isolates the desired emission line from the lamp, focuses it through the sample to maximize the signal, separates it from background radiation, and finally passes it into the detector.

Modern AAS spectrometers are equipped with a turret to accomodate multiple hollow cathode lamp sources. The burner is slotted and will vary in design and path length according to the gases to be used. The monochromator will have a resolution of at least 0.1 nm. The signal from the photomultiplier tube detector is processed by a microprocessor and displayed on a digital readout or a terminal screen. A deuterium background corrector or equivalent is provided.

Specific analytical instrument conditions, such as the flame gases to be used, the wavelength, and the slit width, are given in the procedure for the individual reagent. The lamp and burner position and the flame characteristics should be optimized for a given determination.

SENSITIVITY AND DETECTION LIMITS

Sensitivity in AAS is defined as the concentration of a test element in an aqueous solution that will produce an absorption of 1% or 0.0044 absorbance units. It is normally expressed in $\mu g/mL$ or $\mu g/g$ per 1% absorption. Note that this expression is related to the slope of the calibration curve.

The sensitivity of the instrument is normally optimized during the method setup. A tabulation of expected absorbance vs. standard concentration is shown in Table I on page 37. Excellent detailed definitions and procedures for determining detection limits exist; in particular, *see* the reference cited on page 4 of this book. Several instrumental variables can be manipulated to improve the detection limit.

INSTRUMENTATION

The manufacturer's instruction manual should be followed for detailed operating procedures and for matters of care, cleaning, and calibration. It is assumed that the analyst is familiar with the texts available on atomic absorption spectrophotometry and is aware of the various interferences and sources of error. The individual procedures

Table I. Expected Absorbance of Standards

Element	Wave-Length (nm)	Sens. Check[a] (mg / L)	Linear Range[b] (mg / L)	Min.[c] (mg / L)	Min. (mg / 25 mL)	Max.[d] (mg / 25 mL)	Recommended Standard (mg / 25 mL)	Expected Absorbance Units
Antimony	217.6	25.0	30.0	2.50	0.063	0.750	0.10–0.20	0.03–0.06
Barium	553.6	20.0	20.0	2.00	0.050	0.500	0.10–0.20	0.04–0.08
Bismuth	223.1	20.0	20.0	2.00	0.050	0.500	0.20–0.40	0.08–0.16
Cadmium	228.8	1.5	2.0	0.15	0.004	0.050	0.02–0.04	0.11–0.21
Calcium	422.7	4.0	5.0	0.40	0.010	0.125	0.02–0.04	0.04–0.08
Cobalt	240.7	7.0	3.5	0.70	0.018	0.088	0.02–0.04	0.02–0.05
Copper	324.8	4.0	5.0	0.40	0.010	0.125	0.02–0.04	0.04–0.08
Iron	248.3	5.0	5.0	0.50	0.013	0.125	0.05–0.10	0.08–0.16
Lead	217.0	9.0	20.0	0.90	0.023	0.500	0.10–0.20	0.09–0.18
Lithium	670.8	2.0	3.0	0.20	0.005	0.075	0.01–0.02	0.04–0.08
Magnesium	285.2	0.3	0.5	0.03	0.001	0.013	0.005–0.01	0.13–0.27
Manganese	279.5	2.5	2.0	0.25	0.006	0.050	0.02–0.04	0.06–0.13
Molybdenum	313.3	30.0	40.0	3.00	0.075	1.000	0.50–1.00	0.13–0.27
Nickel	232.0	7.0	2.0	0.70	0.018	0.050	0.05–0.10	0.05–0.11
Potassium	766.5	2.0	2.0	0.20	0.005	0.050	0.02–0.04	0.08–0.16
Silver	328.1	2.5	4.0	0.25	0.006	0.100	0.02–0.04	0.08–0.19
Sodium	589.0	0.5	1.0	0.05	0.001	0.025	0.005–0.01	0.08–0.16
Strontium	460.7	5.0	5.0	0.5	0.013	0.125	0.05–0.10	0.08–0.16
Zinc	213.9	1.0	1.0	0.1	0.003	0.025	0.01–0.02	0.08–0.16

NOTE: This table is based on instrument-specific conditions. In some instances the recommended standard addition is lower than cited in this table. Dilution may be modified to meet maximum allowable limits.

[a] Sens. Check is the concentration giving approximately 0.2 AU.
[b] Linear Range is the upper concentration of linear range.
[c] Min. mg/L is the concentration giving 0.02 AU (sens. Check divided by 10).
[d] Max. is the upper limit in mg per 25 mL (linear range divided by 40).

take the following interferences into account insofar as possible. However, because of the differences among instruments, the analyst should be aware of these effects and check the particular instrument.

INTERFERENCES

Surface tension, viscosity, acid content, and solute content may affect the nebulization rate and influence the sensitivity by altering the concentration of absorbing species in the flame volume under observation. Oxide formation within the flame may make it desirable to add complexing or releasing agents. Reduction of ionization in the presence of alkali elements can cause serious errors when these elements are present in some samples and not in others. Flame gas absorption becomes appreciable at 250–300 nm and becomes increasingly severe at lower wavelengths. Scattering of radiant energy from particles in the flame is usually negligible above about 300 nm or at sample concentrations on the order of 10 g/L or less. Spurious absorbance by flame gas or scattering will bias results obtained by standard additions.

Where spurious absorption or other background correction is required, details are given in the procedure for an individual chemical. For an element with the stated resonance line in the ultraviolet region (< 350 nm), the correction can be performed with a deuterium discharge lamp. For an element with the stated line in the visible region (> 350 nm), the adjacent-line technique can be employed. In this method, the background absorption is determined at a wavelength adjacent, ±5 nm, to the line being used for the element of interest; that is, background absorption is measured at a wavelength where significant atomic absorption by that element is absent.

REAGENTS AND STANDARDS

Precautions should be taken to avoid contamination. Glassware should be carefully cleaned and stored. The highest purity water should be used, obtained preferably from a mixed-bed strong-acid, strong-base ion-exchange cartridge capable of producing water with an electrical resistivity of 12–15 megohm-cm. Teflon or plasticware can be substituted when leaching of metals from glassware is suspected.

Accurate standard solutions of elements to be determined are needed for the preparation of working standards. Suitable standard solutions are commercially available. These solutions should be traceable to NIST (National Institute of Standards and Technology) standards. Concentrations of 1 to 10 g/L are recommended for storage.

More dilute solutions and working standards may be adsorbed on the walls of glassware and should not be used for more than 24 h unless experience has shown that they are constant in concentration for a longer period.

ANALYSIS

Quantitative analysis by AAS depends on the applicability of Beer's law. In this monograph the method of standard additions is used for trace metal determination of reagents prepared as solutions. The method of standard additions is applicable at low concentrations where a linear relationship exists between signal and concentration. Most instruments can indicate the signal as transmission ($\%T$) or absorbance (A). Absorbance is directly proportional to concentration but $\%T$ is a logarithmic relationship.

The use of standard additions places the standards in a matrix similar to that of the sample. It is essential that the sample solution, along with the additions, have absorbances in the range for which linearity is expected, routinely 0.01 to 0.5 A. The use of standard addition does not eliminate interferences, but assures that the element of interest behaves similarly in the sample and the standard additions. A calibration line obtained for aqueous standards can be used if its slope and the slope of the standard addition line are identical. In that event, chemical (matrix) effects are negligible.

As previously indicated, reagent-specific conditions for AAS analysis can be found listed under the individual reagents. This information includes but is not limited to sample preparation, element wavelength, use of background correction, fuel–oxidant mixture, use of additives, and the recommended standard additions.

PROCEDURE

Prepare the sample for analysis by dissolving in water or by following a reagent-specific procedure. Prepare four solutions for standard additions: the reagent blank (which may be only water), the sample solution, and two standard additions. The recommended additions usually correspond to half and equal to the specification limit. The signal for the reagent blank preparation can be subtracted from the signals of the three other solutions or the autozero feature of the instrument employed. The three results can be treated graphically or mathematically to determine the analyte concentration in the sample. When several elements are to be determined, they can be added simultaneously to the standard addition solutions. A separate set of standard additions may be required if different sample weights must be used to stay in the linear absorbance (A) range. The following section provides an example of a typical procedure.

REQUIREMENTS

MAXIMUM ALLOWABLE

Calcium (Ca) 0.01%
Manganese (Mn) 0.001%
Potassium (K) 0.005%
Sodium (Na) 0.005%
Strontium (Sr) 0.005%

TESTS

CALCIUM (Ca), MANGANESE (Mn), POTASSIUM (K), SODIUM (Na), AND STRONTIUM (Sr)

Sample Stock Solution.   Dissolve 10.0 g of sample with water in a 100-mL volumetric flask and dilute to the mark with water (0.10 g/mL).

Standard Additions.   To a set of three 25-mL volumetric flasks add 2.0 mL (0.2 g) of sample stock solution. To two of the flasks add the specified amounts of calcium ion (Ca) and sodium ion (Na) from the following table. Add 2 mL of 5% potassium chloride solution to all three flasks and dilute to the mark with water.

To a second set of three 25-mL volumetric flasks add 10.0 mL (1.0 g) of sample stock solution. To two of the flasks add the specified amounts of potassium ion (K) and strontium ion (Sr) from the following table. Dilute the contents of the three flasks to the mark with water.

To a third set of three 25-mL volumetric flasks add 20.0 mL (2.0 g) of sample stock solution. To two of the flasks add the specified amounts of manganese ion (Mn) from the following table. Dilute the contents of the three flasks to the mark with water.

Analyze the solutions by means of a suitable atomic absorption spectrophotometer, using the conditions outlined in the following table. Calculate the metal content of the sample by the method of standard additions.

Element	Wavelength (nm)	Sample Wt (g)	Standard Added (mg)	Flame Type*	Background Correction
Ca	422.7	0.4	0.02; 0.04	N/A	No
Mn	279.5	4.0	0.02; 0.04	A/A	Yes
K	766.5	1.8	0.02; 0.04	A/A	No
Na	589.0	0.2	0.005; 0.01	A/A	No
Sr	460.7	2.0	0.05; 0.10	N/A	No

*A/A is air/acetylene; N/A is nitrous oxide/acetylene.

MERCURY DETERMINATION BY COLD VAPOR ATOMIC ABSORPTION SPECTROPHOTOMETRY

A sensitive determination of mercury is based upon its reduction in aqueous solution to the free element, vaporization of the latter into a stream of air or inert gas, and passage of the vapors through an optical cell. The cell is substituted for the burner and a high-intensity mercury electrodeless discharge lamp can be used to increase sensitivity. The apparatus for the reduction and vaporization is shown in Figure 2 (commercial instruments specifically designed for this determination are available). Care must be taken to guard against contamination of samples and solutions by mercury. Glassware should be washed with nitric acid and rinsed before use.

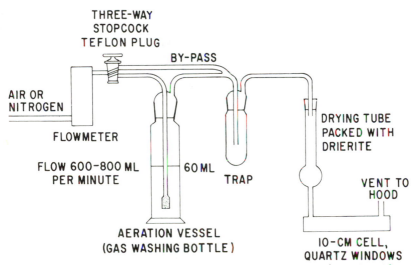

Figure 2. Mercury aeration apparatus. Connections are glass or poly(vinyl chloride).

Ultraviolet radiation from the lamp is directed through the cell and into the detector. The absorbance of mercury in the gas stream is measured and compared to a calibration curve prepared from known solutions. Standard additions can also be used.

Sample preparation is described in the test for each reagent. Reagent blanks should also be run as directed in the test. The solution of the sample, or an aliquot thereof, is transferred to the aeration vessel before addition of any reducing agent. If the sample has been treated with permanganate, the container is rinsed with the quantity of hydroxylamine hydrochloride solution specified, and the rinsings are also added to the aeration vessel with enough water to bring the level to a 60-mL calibration line. The instrument is adjusted to give a smooth base line with carrier gas flowing through the bypass and cell and with the radiation of the 253.7-nm mercury resonance line focused through the cell onto the detector. Stannous chloride solution as specified is introduced into the aeration vessel, and the valve is set to direct the carrier gas through the vessel and the cell. Bubbling is continued until the instrument returns to its zero point. Prior to running the next sample, the aeration vessel should be drained and rinsed. Periodically any deposit of stannic oxide should be removed.

Direct Electrometric Methods

pH POTENTIOMETRY

pH Range.   This pH requirement is intended to limit the amount of free acid or alkali that is allowed in a reagent-grade salt. Sodium chloride, for example, has a requirement that the pH of a 5% solution should be from 5.0 to 9.0 at 25°C. This requirement limits the free hydrochloric acid or sodium hydroxide to about 0.001%. The effect of excess acid or base on the pH of 5% solutions of salts of strong acids and bases is given in the following table.

Excess Acid / Base	0.01%, pH	0.001%, pH	0.0001%, pH
HCl	3.9	4.9	5.9
H_2SO_4	4.0	5.0	6.0
HNO_3	4.1	5.1	6.1
HBr	4.2	5.2	6.2
$HClO_4$	4.3	5.3	6.3
KOH	9.9	8.9	7.9
NaOH	10.1	9.1	8.1

The pH of the solution is determined by means of any reliable pH meter equipped with a glass electrode and a reference half cell, usually a saturated calomel electrode (SCE), while the solution is protected from absorption of atmospheric gases. The meter and electrodes are standardized at pH 4.00 at 25°C with 0.05 M potassium hydrogen phthalate ($KHC_8H_4O_4$) and at pH 9.18 at 25°C with 0.01 M NIST sodium borate decahydrate ($Na_2B_4O_7 \cdot 10H_2O$). Because the pH values are reported to only 0.1 pH unit, the meter and electrodes are considered to be in satisfactory working order if they show an error of less than 0.05 pH unit in the pH range of 4.00 to 9.18.

For more accurate measurement of a test solution, the meter and electrodes should be standardized against a buffer standard solution whose pH is close to the pH of the test solution.

pH of a 5% Solution at 25°C.   The solution to be tested is prepared by dissolving 5 g of the sample in 100 mL of carbon dioxide-free water* while protecting the solution from absorption of carbon dioxide from the atmosphere. The pH measurement is made as previously described.

Table II reports the pH of a 5% solution of the pure salts. This value was determined experimentally for each salt by dissolving 10 g of salt in approximately 200 mL of water, then making the solution slightly acid and titrating with standard alkali. Another similar solution was prepared, made slightly alkaline, and titrated with standard acid. Graphs were constructed for each titration by plotting pH vs. milliliters of titrating solution. The average of the two end points so determined is reported as the pH of a 5% solution containing no free acid or alkali.

pH of Other Concentrations.   Directions for the preparation of solutions at concentrations of other than 5% are given in individual specifications. The pH measurements are made as previously described.

pH of Buffer Standard Solutions.   For reagents that are suitable for use as pH standards, concentrations are stated in the individual specifications. They are expressed in the same units (molal) as the NIST standard reference materials to which they are compared. The

---

*Carbon dioxide-free water can be prepared by purging reagent water with carbon dioxide-free air (using a gas dispersion tube) or nitrogen for at least 15 min, or by boiling reagent water vigorously for at least 5 min and allowing it to cool while protected from absorption of carbon dioxide from the atmosphere.

Table II. pH of a 5% Solution of Pure Salts at 25°C

Salt	pH	Salt	pH
Ammonium Acetate	7.0	Potassium Chromate	9.3
Ammonium Bromide	4.9	Potassium Iodate	7.0
Ammonium Chloride	4.7	Potassium Iodide	7.0
Ammonium Nitrate	4.8	Potassium Nitrate	7.0
Ammonium Phosphate,		Potassium Phosphate,	
Monobasic	4.1	Monobasic	4.2
Ammonium Sulfate	5.2	Potassium Sodium Tartrate	
Ammonium Thiocyanate	4.9	Tetrahydrate	8.4
Barium Chloride		Potassium Sulfate	7.3
Dihydrate	6.9	Potassium Thiocyanate	7.0
Barium Nitrate	6.9	Sodium Acetate, Anhydrous	8.9
Calcium Chloride		Sodium Acetate Trihydrate	8.9
Dihydrate	6.6	Sodium Bromide	7.0
Calcium Nitrate		Sodium Chloride	7.0
Tetrahydrate	6.6	Sodium Molybdate Dihydrate	8.1
Lithium Perchlorate	7.0	Sodium Nitrate	7.0
Magnesium Nitrate		Sodium Phosphate, Dibasic,	
Hexahydrate	6.6	Heptahydrate	9.0
Magnesium Sulfate		Sodium Phosphate, Monobasic,	
Heptahydrate	6.8	Monohydrate	4.2
Manganese Chloride		Sodium Pyrophosphate	
Tetrahydrate	5.4	Decahydrate	10.4
Potassium Acetate	9.0	Sodium Sulfate	7.2
Potassium Bromate	7.0	Sodium Tartrate Dihydrate	8.4
Potassium Bromide	7.0	Sodium Thiosulfate	
Potassium Chloride	7.0	Pentahydrate	7.9
		Zinc Sulfate Heptahydrate	5.4

meter and electrodes must be capable of a precision of 0.005 pH unit at the specified pH for satisfactory results.

POLAROGRAPHIC ANALYSIS

Polarography offers a rapid and sensitive method of determining a number of ions or compounds that are electrolytically reducible, usually at a dropping mercury cathode, in a solution with appropriate electrical conductivity. The measurements depend upon obtaining the curve of current transported vs. applied potential as the latter is scanned in a negative direction. Typical apparatus consists of a reference half cell (usually a saturated calomel electrode), an electrolyte container or cell that can be deaerated by bubbling with nitrogen or argon, the dropping mercury electrode, a source of uniformly increasing potential, means of converting the resulting current to a measurable signal, and a recorder with suitable ranges of sensitivity down to 1 μA full scale. Various commercial instruments are available.

As the voltage applied to the cell rises, the residual current of the supporting electrolyte increases very gradually until the potential at the dropping electrode reaches the reduction potential of the most easily reduced species, ordinarily the substance under determination. At this point the current rises rapidly above the residual value to a limiting value dependent on the concentration of the electroactive species. The difference between the residual current and the limiting current comprises the diffusion current of the electroactive species. The resultant current–voltage curve, which is overall S-shaped, is called a polarographic curve or polarogram. The midpoint in the rise of the curve, one-half the distance between the residual current and the limiting current, is called the half-wave potential and is characteristic of the electroactive substance. Ultimately, as the applied voltage continues to rise, the current increases again as a result of the reduction of the supporting electrolyte.

Because the observed current is also a function of the size of the growing mercury drop at the cathode, the curve will show pronounced oscillations as the drops grow and fall. Most instruments incorporate means of damping out such oscillations. Reading of the current is simplified with the damped curve, but on some early-model instruments the potential readings may be shifted by up to 0.1 V from those observed without damping. A typical damped polarographic curve is shown in Figure 3. Calibration is performed with known solutions of the material to be determined.

Directions for individual reagent chemicals normally include appropriate means of treating the sample to convert it to a suitable electrolyte, free from oxygen and containing any needed pH buffer and maximum suppressor. Maxima are abnormally high current readings that may distort a polarogram—generally at the top of the step in the wave. Most maxima can be suppressed by gelatin, some indicators, or a surface-active agent. Use of too much suppressor, however, may lower the diffusion current of the material being determined. Maxima can be avoided if the method is sensitive and the concentration of the substance being determined is low.

Typical substances that can be determined polarographically include many metals (especially lead, zinc, and cadmium), some anions such as bromate and iodate, and several types of organic materials

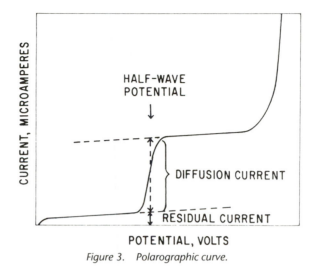

Figure 3.   Polarographic curve.

(aldehydes, ketones, peroxides, nitro-, and some unsaturated or halo-genated compounds).

A number of extensions of the original polarographic method have advanced the capabilities of polarography. For example, differential pulse polarography and square wave polarography, for which commercial instruments are widely available, permit both higher sensitivity and greater resolution of constituents. With the increased sensitivity, however, greater precautions against contamination and interfering materials are necessary.

Differential Pulse Polarography.   The detection limits in conventional dc polarography, typically 10^{-5} M, are set by the charging current resulting from the continuous growth of the mercury-drop electrode. One approach to improving the sensitivity of polarographic methods has been the development of pulse methods that reduce the contribution of charging current to the overall measured current. In pulse polarography, instead of applying a continuously increasing potential (as in dc polarography), a voltage pulse is applied to the mercury drop near the end of its life. The reason for this can be seen from Figure 4. The faradaic current, i_f (or that due to the reduction of the electroactive species), increases continuously during the life of the drop and is greatest at that moment just before the drop falls. The charging current (i_c), however, decreases steadily over the life of the drop. The ratio of faradaic to charging current (i.e., the sensitivity) can be optimized by sampling the current at the end of the drop life.

In differential pulse polarography, a voltage pulse, typically 50 ms, is applied to the electrode during the last portion of the drop's life

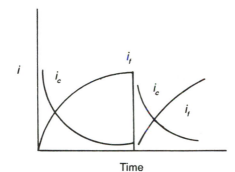

Figure 4. Comparison of faradaic current (i_f) and charging current (i_c) vs. time.

(which is controlled by a mechanical drop dislodger and is commonly 0.5, 1.0, or 2.0 s). When the voltage is first applied the charging current is very large, but it decays exponentially. The pulses have a constant amplitude of between 5 and 100 mV, and are superimposed on a slowly increasing voltage ramp, as shown in Figure 5. The current is measured twice: immediately preceding the pulse and near the end of the drop life. The overall response output to the recorder is the difference in the two currents (Δi) sampled. The plot of Δi as a function of voltage is peak-shaped, as shown in Figure 6. By measuring the difference in current before the pulse and toward the end of the pulse, the contribution of the charging current to the overall measured current is significantly reduced. In addition, the backgrounds are flat, as opposed to the sloping backgrounds found in dc and some other polarographic techniques. Because the polarograms are peak-shaped, the limiting currents are easy to determine in differential pulse polarography. The reduction in charging current results in detection limits as low as 10^{-8} M.

Instruments suitable for differential pulse polarography and other voltammetric techniques are commercially available from several manufacturers. Many of the procedures to be described were developed to use a Princeton Applied Research Model 174A instrument, which is no longer produced.* It has been succeeded by Model 384B, which employs an integral microprocessor to control the measurements, record the data, and compute the results. Several other vendors market computer-based instruments. The conditions originally specified have been modified to be applicable to one of the newer

---

*The value of the full-scale current range that is printed on the sensitivity control of this instrument is 10 times as large as the actual electrolysis current at the dropping mercury electrode when the instrument is used in the differential pulse mode.

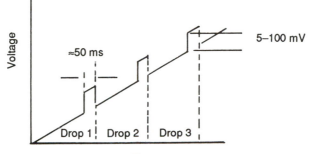

Figure 5.   Voltage pulse excitation for differential pulse polarography.

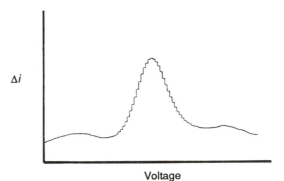

Figure 6.   Differential pulse polarogram.

Sargent-Welch Polarographs, Model 7001. Anyone using a different instrument may have to make changes suitable to its individual characteristics.

SPECIFIC PROCEDURES

Ammonium.   Transfer the specified amount of sample to each of two 25-mL volumetric flasks containing 5 mL of water. Neutralize, if necessary, with the appropriate quantity of ammonia-free 6 N sodium hydroxide (prepared by boiling and cooling 6 N sodium hydroxide solution). To one of the flasks add the indicated quantity of ammonium ion. To each add 5 mL of acetate buffer (to establish pH 4) and 5 mL of ammonia-free formaldehyde (*see* Reagents). Heat on a steam bath for 5 min with occasional shaking, cool, and dilute to volume with water. Transfer a suitable portion to the cell, deoxygenate for 5 min with nitrogen, and record the polarogram from -0.60 to -1.10 V vs. SCE. The following instrumental settings have proved to be satisfactory: drop time, 1 s; scan rate, 5 mV/s; sensitivity, 0.5 μA full scale; modulation amplitude, 50 mV. The peak for the ammonium–formaldehyde derivative occurs at about -0.85 V, but its position is dependent on pH.

The peak for the sample should not exceed one-half of the peak for the sample plus standard.

Reagents

1. ***Acetate Buffer.***   Dissolve 10.4 g of sodium acetate trihydrate, $CH_3COONa \cdot 3H_2O$, in 200 mL of water and add 32 mL of glacial acetic acid.

2. ***Ammonia-Free Formaldehyde.***   Add 50 g of a cation-exchange resin in hydrogen form (e.g., Dowex 50×8, 50–100 mesh) to a 1-lb bottle of 37% formaldehyde solution. Place the bottle in a shaker for 1 h, or let it stand for several hours and mix occasionally.

Bromate, Iodate.   Add to the polarographic cell an appropriate volume of solution prepared as described in the specification. Deoxygenate for 5 min with nitrogen and record the polarogram from -0.6 V to -1.8 V vs. SCE. The following conditions are satisfactory: drop time, 1 s; scan rate, 5 mV/s; sensitivity, 0.5 μA full scale; modulation amplitude, 25 mV. The peak occurs at about -1.25 V for iodate and -1.6 V for bromate. If the ionic strength is not sufficiently high, the bromate wave becomes more negative and the cathodic wave may interfere. In this case calcium chloride may be added, and a reagent blank correction may be applied.

The peak for the sample should not exceed one-half of the peak for the sample plus standard.

Carbonyl Compounds.   To the specified quantity of sample in each of two 25-mL volumetric flasks, add 10 mL of phosphate buffer (*see* Reagent) and 2 mL of 2% hydrazine sulfate solution. To one of the flasks add the indicated quantities of standard solutions. Dilute each to volume with water and mix. Promptly transfer a suitable portion to the cell, deoxygenate for 5 min with nitrogen, and record the polarogram from -0.6 V to -1.6 V vs. SCE. The following conditions are satisfactory: drop time, 1 s; scan rate, 5 mV/s; sensitivity, 0.5 μA full scale (*see* footnote on page 47); modulation amplitude, 50 mV.

The peak heights for the sample should not be greater than one-half of the peak heights for the sample plus standards. Approximate peak potentials for the hydrazones of known carbonyl compounds are as follows: formaldehyde, -1.0 V; acetaldehyde, -1.1 V; acetone, -1.3 V.

Reagent

Phosphate Buffer.   Dissolve 10.0 g each of monobasic sodium phosphate monohydrate ($NaH_2PO_4 \cdot H_2O$) and anhydrous dibasic sodium phosphate (Na_2HPO_4) in water and dilute with water to 500 mL.

Lead (and Cadmium) in Zinc Compounds.   Prepare a solution with the prescribed quantities of sample and hydrochloric acid in water to make 25.0 mL. Prepare a standard with the same volume of hydrochloric acid and the specified quantities of lead (and cadmium) in 25.0 mL. Place 10 mL of either solution in the polarographic cell, deoxygenate with nitrogen for 5 min, and record the polarogram from -0.25 to -0.80 V vs. SCE. The following conditions are satisfactory: drop time, 1 s; scan rate, 5 mV/s; sensitivity, 0.5 μA full scale; modulation amplitude, 50 mV. Read the peak heights for lead at approximately -0.44 V and those for cadmium at approximately -0.63 V.

The peak(s) for the sample should not be greater than the peak(s) for the standard.

Titrimetry: Electrometric End-Point Detection

POTENTIOMETRIC TITRATIONS

Potentiometric titrations are titrations in which the equivalence point is determined from the rate of change of the potential difference between two electrodes. The potential difference can be measured by any reliable millivoltmeter, including a pH meter that can be read either in pH units or millivolts. (These titrations can also be made by using commercially available automatic titrators, which either plot the complete titration on chart paper or act to close an electrically operated buret valve exactly at the equivalence point.) The electrode pair, which consists of an indicating electrode and a reference electrode, depends on the type of titration and is indicated in the individual specifications.

Relatively large increments of the titrant may be added until the equivalence point is approached, usually within 1 mL. In the vicinity of the end point, small equal increments (0.1 mL, for example) are added; after allowing sufficient time for the indicator electrode to reach a constant potential, the voltage and volume of titrant are recorded. The titration is continued for several increments of titrant beyond the end point. For acid–base titrations, the pH is recorded instead of the potential difference and is treated the same as potential difference for the determination of the end point.

**First Derivative Method for the Determination of the Equiva-
lence Point.** When the titration is symmetrical about the end point
(same rate of change of potential before and after end point), the
position of the end point may be found by a graphical plot of $\Delta E/\Delta V$
vs. V, where E is the potential and V is the volume of titrant. The
end point corresponds to the maximum value of $\Delta E/\Delta V$. (This is a
plot of the first derivative, dE/dV, of $E = f(V)$ vs. the volume of
titrant and is referred to as the first derivative method.)

**Second Derivative Method for the Determination of the Equiva-
lence Point.** The end point can be determined by the use of the
second derivative, d^2E/dV^2, of $E = f(V)$ and is the point where
d^2E/dV^2 becomes zero. This point can be determined mathematically, assuming there is no significant difference between the average
slope $\Delta E/\Delta V$ and the true slope dE/dV. The end point lies in the
increment where the ΔE is the greatest. The amount of titrant to be
added to the buret reading at the beginning of the interval is found by
multiplying the volume of the increment by the factor in which the
numerator is the last $+\Delta^2 E$ value, and the denominator is the sum of
the last $+\Delta^2 E$ value and the first $-\Delta^2 E$ value, disregarding the sign.
When the volume increments are identical, the method can be re-
duced to the use of the first and second differences. This simplification is employed in the following typical example:

V (mL)	E (mV)	ΔE	$\Delta^2 E$
16.00	504		
		6	
16.10	510		+4
		10	
16.20	520		+20
		30	
16.30	550		+60
		90	
16.40	640		-50
		40	
16.50	680		-30
		10	
16.60	690		

$$V(\text{e.p.}) = 16.30 + 0.10 \times \frac{60}{60 + 50} = 16.35 \text{ mL}$$

WATER BY THE KARL FISCHER METHOD

The water content of most of the organic reagents discussed in *Reagent Chemicals* is determined by the Karl Fischer method. This method involves the titration of the sample in methanol or any suitable solvent—for example, pyridine, formamide, or petroleum ether—with the Karl Fischer reagent. This reagent consists of iodine, sulfur dioxide, an amine, and a solvent in which the iodine and the sulfur dioxide are rapidly and quantitatively consumed by the water in the sample. The end point is detected either visually from the color change caused by free iodine or electrometrically. The latter method is preferred in most cases and necessary in the case of colored solutions.

Two different techniques can be used to add the iodine needed in the reaction. In the volumetric Karl Fischer titration the Karl Fischer reagent is added by means of a volumetric buret. With this technique it is necessary to standardize the reagent often, and the amount of reagent added must be measured accurately. This accuracy is especially important in determining small amounts of water. A small error in delivering this required amount of reagent can lead to a large error in the determined water content.

The development of commercial coulometric instrumentation has enabled the alternative coulometric titration method to be introduced into the analytical laboratory. Coulometry is an electrochemical process involving the generation of various materials in direct proportion to their equivalent weights. As applied to the determination of water by the Karl Fischer method, the iodine needed in the reaction is generated in situ (inside the actual titration vessel) from an iodide-containing solution. As in the volumetric Karl Fischer method, the excess iodine beyond the end point is usually determined by an electrometric technique. The amount of iodine generated for the reaction is determined accurately and related to the amount of water present in the sample. The coulometric Karl Fischer titration is a more sensitive procedure (i.e., it can determine smaller amounts of water). Because of this, exclusion of extraneous sources of moisture is absolutely necessary. No standardization of reagent is required because this is an absolute method depending only on electrochemical laws and the accurate measurement of electrical current. The method is generally recommended when the samples are liquid and have a low water content.

VOLUMETRIC PROCEDURE

The apparatus described here, and the Karl Fischer reagent described under Reagent Solutions, page 75, are suitable for the determination of water by this method.

There are, however, a number of commercial titration instruments and Karl Fischer reagents available. These instruments and reagents may be used in accord with their manufacturers' instructions.

Apparatus

Enclosed Titration System. The Karl Fischer reagent is highly sensitive to water, and exposure to atmospheric moisture must be avoided. The reagent is best stored in a reservoir connected to an automatic buret with all exits to the atmosphere protected with a suitable desiccant—for example, silica gel or anhydrous calcium chloride. The closed titration flask, with provision for insertion of platinum–platinum electrodes and insertion of the sample through a septum or removable closure, must be connected to the buret in such a way that atmospheric moisture cannot enter through the joint. Because of the necessity for a closed titration system, stirring is best accomplished magnetically.

Polarizing Unit for Use with the Electrometric End Point. A simple polarizing unit consists of a 1.5-V dry battery connected across a radio-type potentiometer of about 100 ohms. (One platinum electrode is connected to one end of the potentiometer; the other platinum electrode is connected through a microammeter to the movable control of the potentiometer.) By adjustment of the movable contact, the initial polarizing current can be set to the desired value of 100 microamperes through a 2000-ohm precision resistor. This current passes through the platinum–platinum electrodes, which are immersed in the solution to be titrated.

Reagents

Preparation of the Reagent. (Preparation should proceed according to Mitchell and Smith, *Aquametry*, 2nd ed., page 41, Wiley-Interscience, New York, 1980). *See* Reagent Solutions, Page 69.

Commercial Reagents. Stabilized Karl Fischer reagents for volumetric water determinations are available from several suppliers. Pyridine and pyridine-free reagents can be used for water determination. For ketones and aldehydes, specially formulated reagents are available and should be used.

Pyridine and pyridine-free reagents have defined water capacities, which should be considered when these reagents are used.

Standardization of Karl Fischer Reagents*

Method A.   Place 25 to 50 mL of methanol in the titration flask and titrate with Karl Fischer reagent to the electrometric end point. The deflection should be maintained for at least 30 s. This is a blank titration on the water contained in the methanol and the titration flask. Record the volume. Add 1 drop of water, accurately weighed, and again titrate with Karl Fischer reagent to the end point. Calculate the strength of the reagent in milligrams of water per milliliter of Karl Fischer reagent from the weight of the drop of water taken and the net volume of Karl Fischer reagent used.

Method B.   Use the same general procedure but add a weighed amount of sodium tartrate dihydrate** instead of the water and calculate the Karl Fischer titration factor (expressed in milligrams of water per milliliter) by the following:

$$\text{K.F. titration factor} = \frac{\text{wt of sodium tartrate (in mg)} \times 0.1566}{\text{mL of K.F. reagent}}$$

Sodium tartrate does not dissolve completely in methanol and thus can cause premature, transient end points near the true end point. Hence the titration must be performed slowly, and the suspension must be well stirred when near the final end point.

Procedure for Samples

Method 1.   Titrate 25 to 50 mL of methanol or other solvent, as specified, to an end point with the Karl Fischer reagent, as in the standardization. Unstopper the titration flask and rapidly add an accurately weighed portion of the solid or liquid sample to be tested. (For liquids, a weight buret or syringe is convenient or, when the density is known, a measured volume of sample may be introduced

---

*The Karl Fischer reagent, prepared as described under Reagent Solutions, gradually deteriorates. It should be standardized within an hour before use, or daily if it is in continuous use. The pyridine-free reagents are more stable, and titrant–solvent systems exist that are free from deterioration.

**Karl Fischer reagent, if used to assay more than 1% water, should not be standardized against sodium tartrate dihydrate because sodium tartrate dihydrate crystals have been shown to contain about 0.3% occluded water within the crystal structure. For trace water determinations (less than 1%) the error from occluded water is not significant.

with a syringe or pipet.) Quickly restopper the flask and titrate again with Karl Fischer reagent to an end point. The percentage of water in the sample is calculated from the known strength of the Karl Fischer reagent, the net volume used, and the sample weight.

COULOMETRIC PROCEDURE

Apparatus

Follow the instrument manufacturers' instructions for specific operation procedures.

Reagents

Coulometric pyridine and pyridine-free reagents have defined water capacities. Several suppliers offer solutions that are used for assuring the performance of coulometric instruments. These usually are nonhygroscopic mixtures of solvents with a known water content. Reagent water may be used if the small amount required is weighed accurately.

Procedure for Samples

Method 2.   Titrate the sample cell to dryness. Activate the sample titration procedure and add the specified amount of sample with a syringe (if liquid) or as a weighed solid. Determine the weight of the liquid sample by weighing the syringe before and after injection or by calculating from the volume injected and the known sample density. The instrument will indicate the end of the titration and the amount of water in the sample is automatically calculated.

Chromatography

Chromatography is an analytical technique used in quantitative determination of purity of most organic and an increasing number of inorganic chemical reagents. The broad scope of chromatography allows it to be used in the separation, identification, and assay of diverse chemical species, ranging from simple metal ions to compounds of complex molecular structure such as proteins.

Chromatography is a method of separation of individual components in a mixture achieved when a mobile phase is passed over a stationary phase. Differences in affinities of various substances for these phases result in their separation.

Chromatography can be divided into two main branches, depending on whether the mobile phase is a gas or a liquid. Gas chromatography is principally used for analysis of volatile thermally stable materials. Liquid chromatography is particularly useful for analysis of nonvolatile or thermally unstable organic substances. Ion chromatography, a technique in which anions and cations can be determined by using the principles of ion exchange, is a form of liquid chromatography.

GAS CHROMATOGRAPHY

Gas chromatography (GC) may be subdivided into gas–liquid and gas–solid chromatography. Gas–liquid chromatography is by far the most widely used form of gas chromatography. The heart of a GC system is the column, which is contained in an oven and operated in either the isothermal or the temperature-programmed mode. Until recently columns packed with a nonvolatile liquid phase coated on a porous solid support were extensively used. Capillary columns, constructed of fused silica onto which is bonded the liquid phase, are replacing these packed columns because of their greater resolving power and chemical inertness.

Conventional capillary GC uses long narrow-bore columns with an inside diameter (i.d.) from 220 to 320 μm, coated or bonded with a thin film of the liquid phase. This arrangement results in high resolution but low sample capacity. Special injection techniques and hardware have to be employed. The introduction of wide-bore capillary columns, which have an i.d. of typically 530 μm and a relatively thick film of the liquid phase (1 to 5 μm), allows a laboratory to use a standard packed column instrument and conditions, while gaining the advantages of capillary technology.

Direct flash vaporization or on-column injection of the sample with standard-gauge needles can be used with these columns. Flash vaporization is the predominant technique used with packed columns. When low-capacity (narrow-bore) capillary columns are used, the split mode of operation, in which part of the sample is vented from the injector, is normally required. The splitless mode using a 0.1-μL sample size is recommended for the assay of most reagents discussed in this monograph.

In conjunction with wide-bore columns, on-column injection can minimize sample degradation while increasing the reproducibility of results. Wide-bore columns can be used either in a high-resolution mode (carrier gas flow rates less than 10 mL/min) to obtain optimum resolution of sample components, or at higher flow rates of 10–30 mL/min, which will generate packed column-quality separations in a shorter time. Packed columns require a multitude of stationary phases to accomplish the typical separations performed in an analytical laboratory. The increased length of capillary columns allows for better separation of components so that as few as three columns of high, moderate, and low polarity can handle the majority of analytical requirements.

A wide variety of detectors are used to quantify and/or identify the components in the eluent from the column. Thermal conductivity (TCD) and flame ionization (FID) are examples of general detectors that respond to most organic compounds. Electron capture (ECD) and photoionization are examples of class-specific detectors used often in trace environmental analysis. Mass spectrometric type detectors are used to provide positive component identification. The detector output after amplification is used to produce the chromatogram, a plot of component response vs. elution time. Modern systems convert the analog output to digital form, which allows for further manipulation and interpretation of the data.

Results from a chromatogram, when used to determine the assay of a reagent, are often expressed as area percent. A response factor correction for each component is required for the most accurate results, especially when the sample components differ markedly in their detector response. An internal standard reduces error due to variations in injection quantities, column, and detector conditions. Use of a thermal conductivity detector minimizes the need to correct for response and is usually sufficient for determining reagent assay.

The use of control charts to aid the analyst in visualizing chromatographic variability is suggested. To certify that assay results are valid, use of a system suitability test as described later in this section is recommended. Tests of peak shape and asymmetry are not as critical when capillary rather than packed columns are employed.

LIQUID CHROMATOGRAPHY

Many reagent chemicals described in this monograph can be assayed by and/or tested for suitability for use in liquid chromatography (LC). LC is a technique in which the sample interacts with both the liquid mobile phase and the stationary phase to effect a separation. An LC system consists of a pump that delivers a liquid, usually a solvent for the sample, at a constant flow rate through a sample injector, a column, and a detector. The solvent is called the mobile phase. Typically, the flow rate is 1 to 10 mL/min for conventional LC. In isocratic operation the composition of the mobile phase is kept constant, while in gradient elution it is varied during the analysis. Gradient elution is required when the sample mixture contains components with a wide range of affinity for the stationary phase. In the isocratic mode the purity of the solvents is less critical, as the impurities are adsorbed–desorbed at a constant rate, whereas in gradient elution impurities may result in extraneous peaks and/or shifts in the baseline. This characteristic necessitates the incorporation of a gradient elution test in this monograph to verify the quality of a solvent when used in this most stringent LC system mode of operation.

In LC a dilute solution of the sample to be analyzed is introduced into the mobile phase via the sample injector and enters the column, where it is separated into its individual components. The columns are usually steel or glass densely packed with semirigid organic gels or rigid inorganic silica microspheres to which a variety of substrates can be chemically bonded and whose typical particle size is 3 to 10 μm.

Several mechanisms of separation are possible in LC. In adsorption chromatography separation is based on adsorption–desorption kinetics, whereas in partition chromatography the separation is based on partitioning of the components between the mobile and stationary phases. Ion exchange is the dominant mechanism in ion chromatography. In practice, a successful separation may involve a combination of separation mechanisms.

A commonly used term, coined by early chromatographers for describing separations dominated by adsorption–desorption, is "normal phase". In normal phase chromatography, the stationary phase is strongly polar (e.g., silica or aminopropyl) and the mobile phase is less polar (e.g., hexane). Polar components are thus retained on the column longer than less polar materials.

A second term of historical origin is "reverse phase". Reverse phase generally applies to separations dominated by partition chromatography, and the elution of components in a reverse phase separation are more or less reversed from the order that would be obtained in a normal phase separation. Reverse phase, the more widely used mode of chromatography, uses a nonpolar (hydrophobic) stationary phase, such as C-18 (octadecyl) chemically bonded to silica, while the mobile phase is a polar liquid such as acetonitrile–water. Hydrophobic (nonpolar) components are retained longer than hydrophilic (polar) components. The elution properties of the mobile phase can be adjusted to modify a separation by addition of appropriate ionic modifiers. These modifiers can be chosen for their ability to either suppress or enhance ion formation in the sample. When enhancement is chosen the separation mechanism may involve both ion exchange and partition.

The wide range of available stationary phases in combination with changes in mobile-phase composition makes LC a very flexible separation-assay technique. The back pressure on the pump is several hundred to several thousand psi, depending upon the particle size of the packing material, mobile-phase flow rate, and viscosity. The column eluent is continuously monitored by a sensitive detector chosen to respond either to the sample component alone (e.g., an ultraviolet (UV) photometer) or to a change in some physical property of the mobile phase due to the presence of the solute (e.g., a differential refractometer (RI)). Other detectors such as electrochemical, fluorescence, etc., are also used for more specialized applications. When photometric detectors are used, solvents are often specified in absorbance units at a specific wavelength. The response of the detector is related to the concentration or weight of the solute and is displayed on a recorder. A computer is usually interfaced to the LC system to control the method and to collect and analyze data.

RECOMMENDED CHROMATOGRAPHY PROCEDURE

Assay Methods

Procedures published in this monograph list the parameters and values established to give satisfactory results. These procedures should be considered adequate only for determining the constituents specified in the individual tests, and they may not necessarily separate other components in a given sample. An attempt is made to keep each procedure as simple as possible, while still retaining the desired sensitivity, because simplicity may lead to wider usage.

The list of parameters for GC includes such variables as column type, diameter and length of column, carrier-gas flow rate, detector type, and operating parameters. For LC the list includes diameter and length of column, packing type and particle size, mobile-phase composition and flow rate, detector type, and operating parameters. They represent one set of conditions that resulted in reproducible analyses. Because of differences among instruments, one or more of the stated parameters may require adjustment for any given instrument. Therefore these parameters, except for the column and type of detector, should be regarded as guides to aid the analyst in establishing the optimum conditions for a particular instrument. Relative retention times are given as an aid in peak identification. Exact reproduction of these times is not essential.

GAS CHROMATOGRAPHY

The volatile organic reagents in this monograph can be assayed by gas chromatography. A single set of instrument conditions shown here has been chosen and found to give satisfactory results for the assay of these reagents. Three columns of varying polarity can be used to achieve the desired separation: type I, low polarity, methyl silicone, 5 μm; type II, moderate polarity, mixed cyano, phenylmethyl silicone, 1.5 μm; and type III, high polarity, polyethylene glycol, 1 μm. The recommended column type and reagent-specific conditions can be found listed alphabetically under the individual reagents in Table III. The relative retentions of the various reagents vs. methanol (1.0) are listed in increasing order in Table IV.

Table III. GC Retention Time of Reagents by Alphabetical Order

Compounds	Column of Choice	Type I, Nonpolar		Type II, Med. Polar		Type III, High Polar	
		RT^a	RR^b	RT	RR	RT	RR
Acetaldehyde	I	2.9	0.97	2.2	0.89	2.8	0.44
Acetone	I	4.9	1.6	3.4	1.4	4.1	0.64
Acetonitrile	I	4.7	1.6	4.4	1.8	8.7	1.4
2-Aminoethanol	I	10.6	3.5	9.9	4.0	N/A	N/A
Aniline	I	19.5	6.5	17.3	7.0	21.9	3.4
Benzene	I	11.2	3.7	6.7	2.7	6.9	1.1
Benzyl Alcohol	I	20.6	6.8	18.6	7.5	23.5	3.7
2-Butanone	I	8.5	2.8	6.0	2.4	6.0	0.95
Butyl Acetate	I	15.3	5.1	11.7	4.7	10.6	1.7
Butyl Alcohol	I	11.1	3.7	9.0	3.6	12.3	1.9
t-Butyl Alcohol	I	6.3	2.1	4.1	1.7	6.5	1.0
Carbon Disulfide	I	7.4	2.5	3.0	1.2	3.0	0.47
Carbon Tetrachloride	I	11.2	3.7	6.0	2.4	5.3	0.84
Chlorobenzene	I	16.6	5.5	12.5	5.0	16.5	2.6
Chloroform	I	9.2	3.1	6.0	2.4	8.9	1.4
Cyclohexane	I	11.4	3.8	5.4	2.2	3.0	0.46
Cyclohexanone	I	17.5	5.8	14.9	6.0	15.4	2.4
1,2-Dichloroethane	I	10.3	3.4	7.4	3.0	10.0	1.6
Dichloromethane	I	6.2	2.0	3.5	1.4	6.4	1.0
Diethanolamine	I	20.6	6.9	20.0	8.0	N/A	N/A
N,N-Dimethylformamide	I	14.2	4.7	13.6	5.5	16.2	2.5
Dimethylsulfoxide	III	15.7	5.2	16.0	6.4	20.5	3.2
Dioxane	I	12.2	4.1	9.1	3.6	10.3	1.6
Ethanol	I	4.3	1.4	3.2	1.3	7.4	1.2
Ethyl Acetate	I	9.2	3.1	5.8	2.3	5.6	0.89
Ethyl Ether	I	5.6	1.9	2.7	1.1	2.4	0.38
Ethylbenzene	I	17.0	5.6	12.5	5.0	11.8	1.8
2-Furancarboxyaldehyde	I	15.8	5.2	14.2	5.7	17.9	2.8
Glycerol	II	19.8	6.6	19.7	7.9	N/A	N/A

Table III. Continued

Compounds	Column of Choice	Type I, Nonpolar		Type II, Med. Polar		Type III, High Polar	
		RT^a	RR^b	RT	RR	RT	RR
2-Hexane	I	8.6	2.9	3.2	1.3	2.3	0.36
n-Hexane	I	9.2	3.1	3.5	1.4	2.3	0.37
Isobutyl Alcohol	I	10.0	3.3	7.9	3.2	11.2	1.7
Isopentyl Alcohol	I	13.2	4.4	11.0	4.4	13.5	2.1
Isopropyl Alcohol	I	5.4	1.8	3.8	1.5	7.3	1.1
Isopropyl Ether	I	9.1	3.0	2.6	1.0	2.6	0.41
Methanol	I	3.0	1.0	2.5	1.0	6.4	1.0
2-Methoxyethanol	III	10.2	3.4	8.4	3.4	13.2	2.1
4-Methyl-2-Pentanone	I	13.3	4.4	10.2	4.1	8.9	1.4
Nitrobenzene	I	22.0	7.3	19.4	7.8	22.1	3.5
Nitromethane	III	7.2	2.4	7.0	2.8	12.2	1.9
1-Octanol	I	21.2	7.0	17.8	7.2	19.2	3.0
1-Pentanol	I	14.2	4.7	11.8	4.7	14.3	2.2
1,2-Propanediol	III	13.4	4.5	13.4	5.4	19.8	3.1
Propionic Acid	III	14.0	4.6	12.5	5.0	18.5	2.9
Pyridine	I	13.4	4.5	10.6	4.2	13.2	2.1
Quinoline	II	N/A	N/A	121.6	8.7	25.0	3.9
Tetrahydrofuran	I	9.8	3.3	5.8	2.3	5.1	0.80
Toluene	I	14.4	4.8	10.0	4.0	9.7	1.5
Trichloroethylene	I	12.3	4.1	7.8	3.1	8.4	1.3
2,2,4-Trimethylpentane	I	12.4	4.1	5.9	2.4	2.6	0.41
1,2-Xylene	I	17.8	5.9	13.3	5.4	12.8	2.0
1,3-Xylene	I	17.2	5.7	12.6	5.1	12.0	1.9
1,4-Xylene	I	16.9	5.6	12.4	5.0	11.6	1.8

[a]RT is retention time.
[b]RR is relative retention (methanol 1.0).

Procedure.

Column: 30-m × 530-μm i.d. fused silica capillary coated with a film that has been surface-bonded and cross-linked

Column Temperature: 40°C isothermal for 5 min, then programmed to 220°C at 10°C/min., hold 2 min.

Injector Temperature: 150–220°C

Detector Temperature: 250°C

Sample Size: 0.2 μL splitless

Carrier gas: Helium at 3 mL/min.

Detector: Thermal conductivity or flame ionization

Table IV. GC Retention Time of Reagents
by Chronological Order on Type I Column

Compounds	Column of Choice	Type I, Nonpolar RT^a	RR^b	Type II, Med. Polar RT	RR	Type III, High Polar RT	RR
Acetaldehyde	I	2.9	0.97	2.2	0.89	2.8	0.44
Methanol	I	3.0	1.0	2.5	1.0	6.4	1.0
Ethanol	I	4.3	1.4	3.2	1.3	7.4	1.2
Acetonitrile	I	4.7	1.6	4.4	1.8	8.7	1.4
Acetone	I	4.9	1.6	3.4	1.4	4.1	0.64
Isopropyl Alcohol	I	5.4	1.8	3.8	1.5	7.3	1.1
Ethyl Ether	I	5.6	1.9	2.7	1.1	2.4	0.38
Dichloromethane	I	6.2	2.0	3.5	1.4	6.4	1.0
t-Butyl Alcohol	I	6.3	2.1	4.1	1.7	6.5	1.0
Nitromethane	III	7.2	2.4	7.0	2.8	12.2	1.9
Carbon Disulfide	I	7.4	2.5	3.0	1.2	3.0	0.47
2-Butanone	I	8.5	2.8	6.0	2.4	6.0	0.95
2-Hexane	I	8.6	2.9	3.2	1.3	2.3	0.36
Isopropyl Ether	I	9.1	3.0	2.6	1.0	2.6	0.41
Chloroform	I	9.2	3.1	6.0	2.4	8.9	1.4
Ethyl Acetate	I	9.2	3.1	5.8	2.3	5.6	0.89
n-Hexane	I	9.2	3.1	3.5	1.4	2.3	0.37
Tetrahydrofuran	I	9.8	3.3	5.8	2.3	5.1	0.80
Isobutyl Alcohol	I	10.0	3.3	7.9	3.2	11.2	1.7
2-Methoxyethanol	III	10.2	3.4	8.4	3.4	13.2	2.1
1,2-Dichloroethane	I	10.3	3.4	7.4	3.0	10.0	1.6
2-Aminoethanol	I	10.6	3.5	9.9	4.0	N/A	N/A
Butyl Alcohol	I	11.1	3.7	9.0	3.6	12.3	1.9
Benzene	I	11.2	3.7	6.7	2.7	6.9	1.1
Carbon Tetrachloride	I	11.2	3.7	6.0	2.4	5.3	0.84
Cyclohexane	I	11.4	3.8	5.4	2.2	3.0	0.46
Dioxane	I	12.2	4.1	9.1	3.6	10.3	1.6
Trichloroethylene	I	12.3	4.1	7.8	3.1	8.4	1.3

LIQUID CHROMATOGRAPHY

Organic reagents not amenable to assay by gas chromatography and some inorganic reagents can be analyzed by liquid chromatography. Descriptions of specific procedures can be found under the individual reagents.

Solvent Suitability for Liquid Chromatography

Solvents used in LC are tested for suitability in gradient elution analysis. The objective is to assess the performance suitability of the

Table IV. Continued

Compounds	Column of Choice	Type I, Nonpolar		Type II, Med. Polar		Type III, High Polar	
		RT^a	RR^b	RT	RR	RT	RR
2,2,4-Trimethylpentane	I	12.4	4.1	5.9	2.4	2.6	0.41
Isopentyl Alcohol	I	13.2	4.4	11.0	4.4	13.5	2.1
4-Methyl-2-pentanone	I	13.3	4.4	10.2	4.1	8.9	1.4
1,2-Propanediol	III	13.4	4.5	13.4	5.4	19.8	3.1
Pyridine	I	13.4	4.5	10.6	4.2	13.2	2.1
Propionic Acid	III	14.0	4.6	12.5	5.0	18.5	2.9
1-Pentanol	I	14.2	4.7	11.8	4.7	14.3	2.2
N,N-Dimethylformamide	I	14.2	4.7	13.6	5.5	16.2	2.5
Toluene	I	14.4	4.8	10.0	4.0	9.7	1.5
Butyl Acetate	I	15.3	5.1	11.7	4.7	10.6	1.7
Dimethylsulfoxide	III	15.7	5.2	16.0	6.4	20.5	3.2
2-Furancarboxyaldehyde	I	15.8	5.2	14.2	5.7	17.9	2.8
Chlorobenzene	I	16.6	6.6	12.5	5.0	16.5	2.6
1,4-Xylene	I	16.9	5.6	12.4	5.0	11.6	1.8
Ethylbenzene	I	17.0	5.6	12.5	5.0	11.8	1.8
1,3-Xylene	I	17.2	5.7	12.6	5.1	12.0	1.9
Cyclohexanone	I	17.5	5.8	14.9	6.0	15.4	2.4
1,2-Xylene	I	17.8	5.9	13.3	5.4	12.8	2.0
Aniline	I	19.5	6.5	17.3	7.0	21.9	3.4
Glycerol	II	19.8	6.6	19.7	7.9	N/A	N/A
Benzyl Alcohol	I	20.6	6.8	18.6	7.5	23.5	3.7
Diethanolamine	I	20.6	6.9	20.0	8.0	N/A	N/A
1-Octanol	I	21.2	7.0	17.8	7.2	19.2	3.0
Nitrobenzene	I	22.0	7.3	19.4	7.8	22.1	3.5
Quinoline	II	N/A	N/A	21.6	8.7	25.0	3.9

[a]RT is retention time.
[b]RR is relative retention (methanol 1.0).

solvents under routine LC operating conditions. The operating parameters found under the specific reagent were selected to simulate a typical gradient run. The reverse gradient portion of the chromatogram is of little practical interest in actual sample analysis. Thus, only the peaks eluted in the "up-gradient" portion of the program are recorded (e.g., water to acetonitrile or water to methanol). Commercially available reversed-phase columns come in a variety of supports and stationary phases. Thus, parameters such as monomeric vs. polymeric phase, endcapped vs. nonendcapped, particle size, and %C loading are specified.

Procedure

Column: Octadecyl (C-18 polymeric phase), 250 × 4.6 mm i.d., 5μm, 12% C loading, endcapped

Mobile Phase: A. Solvent to be tested
B. LC reagent-grade water

Conditions: Detector: Ultraviolet at 254 nm
Sensitivity: 0.02 AUFS*

Gradient Elution:

1. Program the liquid chromatograph system to develop a solvent–water gradient as follows:

Time (min)	Flow (mL / min)	% A (Solvent)	% B (Water)
0	2.0	20	80
30	2.0	20	80
50	2.0	100	0
60	2.0	100	0

2. Start the program after achieving a stable baseline.
3. Disregard the results of the first program. Repeat the program and use these results in the Step 4 evaluation.
4. No peak should be greater than 25% of full scale (0.005 absorbance units).

Reagent-specific conditions can be found under the individual reagents.

Solvent Suitability for Extraction–Concentration

Trace-level analysis, such as for pollutants in the environment and toxic chemicals in the workplace, frequently require the use of solvent-extraction procedures to isolate the analyte. In some cases the solvent is then removed to concentrate the analyte for detection and quantification. It is critical that the solvent be of the highest purity, so that potential interferences during subsequent chromatographic analysis are minimized. For GC analysis, a pure solvent is usually characterized by a small or narrow solvent "front" and the absence of extraneous peaks under the temperature-programmed conditions commonly used for these analyses.

---

* AUFS is absorbance units full scale.

In this monograph, the purity of the solvent is specified as suitable for use with FID (universal) and ECD (class-specific) detectors. For analysis of solvents that are not compatible with the detector, as in the case of halogenated solvents and ECD, a solvent-exchange step is used prior to GC analysis. In LC analysis using gradient elution and UV detection, a pure solvent would not show extraneous peaks or cause unusual baseline drift when analyzed as a sample. In this monograph this suitability is specified by measuring the absorbance through a particular spectral region for each solvent. The solvent should give a smooth curve and contain no extraneous absorptions throughout the specified region.

Procedure: ECD and FID Suitability

Preconcentration.   Preconcentration of the solvent is required to reach the specification levels set for the FID and ECD detectors. Preconcentration can be accomplished with a conventional Kuderna–Danish evaporator concentrator or a vacuum rotary evaporator. The solvent should never be evaporated to dryness.

> *Procedure.*   Accurately transfer 500 mL of test solvent to a 1-L evaporator. Concentrate (100 ×) to approximately 5 mL. If solvent exchange is necessary, add two separate 50-mL portions and evaporate to 5 mL each time.

> *Analysis.*   Analyze the concentrated solvent and an external standard by using a gas chromatograph equipped with wide-bore capillary columns and FID and ECD detectors. The parameters cited here have given satisfactory results.

Instrument: GC–ECD

> **Column:**   30-m × 530-μm i.d. fused silica capillary, coated with 1.5-μm film of 5% diphenyl, 94% dimethyl, 1% vinyl polysiloxane that has been surface-bonded and cross-linked

> **Column Temperature:**   40 to 250°C at 15°C/min; hold at 250°C for 15 min

> **Injector Temperature:**   250°C

> **Detector Temperature:**   320°C

> **Carrier Gas:**   helium at 2–5 mL/min

Sample Size: 5 μL

Standard: 1 μg/L of heptachlor epoxide in hexane

Standard Size: 5 μL

Instrument: GC–FID

Column: 30-m × 530-μm i.d. fused silica capillary, coated with 1.5-μm film of 5% diphenyl, 94% dimethyl, 1% vinyl that has been surface-bonded and cross-linked

Column Temperature: 40 to 250°C at 15°C/min; hold at 250°C for 15 min

Injector Temperature: 250°C

Detector Temperature: 250°C

Carrier Gas: helium at 2–5 mL/min

Sample Size: 5 μL

Standard: 1 mg/L of 2-octanol in hexane

Standard Size: 5 μL

Results

The purity of the solvents is expressed in micrograms per liter for FID and nanograms per liter for ECD, measured against an external standard appropriate for the analysis being performed. In this monograph 2-octanol is used for FID and heptachlor epoxide is used for ECD. Measure the peak area–height for all peaks inside a window of 0.5 to 2 times the retention time of 2-octanol for GC–FID suitability. The sum of the peaks must be no greater than 10 μg/L, with no single peak greater than 5.0 μg/L. Measure the peaks for all peaks inside a window of 0.2 to 2 times the retention time of heptachlor epoxide for GC–ECD suitability. The sum of the peaks must be no greater than 10 ng/L, with no single peak greater than 5.0 ng/L. For pesticide residue analysis the standards might also be a series of pesticides, or for drinking water a series of halomethanes. Detailed procedures for such analysis can be found in many excellent references.

LC Suitability

Absorbance.   Measure the absorbance of the sample in a 1-cm cell against water (as the reference liquid) in a matched cell from 400 nm to the cutoff of the reagent solvent. The absorbance should not exceed 1.00 at the specified cutoff wavelength and not exceed the specified absorbances at each wavelength.

Calibration for GC and Assay Procedures

Because the response of the detector is not the same for equal concentrations of all components, the instrument must be calibrated to obtain concentrations in terms of weight percentages. Correction of area values to weight values is accomplished by the use of calibration factors (also referred to as response factors) that may be obtained in one of the following ways:

Standard Addition Technique.   The sample is analyzed by the prescribed procedure before and after the addition of a measured amount of the component of interest. The ratio of the area obtained for that component, before and after the addition, combined with the weight of added component, can be used to calculate the calibration factor.

Calibration Mixture.   A mixture is prepared to correspond as closely as possible in composition to the sample to be analyzed. (For the preparation of a calibration mixture it is desirable, but not always possible, to start with components of negligible impurity. When the impurity content cannot be ignored, it may be established by an independent means.) The mixture is analyzed by the prescribed procedure, and the area of each peak is measured. The calibration factor for each component is the ratio of its weight or concentration to its area. For an unidentified component, the calibration factor cannot be determined and is usually assumed to be the same as that of a known component in the calibration mixture.

System Suitability Tests for GC and LC Systems

To determine if a chromatographic system can provide acceptable, repeatable results, it can be subjected to a suitability test at the time of use. Underlying such a test is the concept that electronics, equipment, column, and standard or samples, coupled with sample handling, constitute a single system. Specific data are collected from repeated injections of the assay preparation or standard preparation. The results are compared with limiting values for key parameters such as peak symmetry, efficiency, precision, and resolution.

The most useful test parameter is the precision of replicate injections of the analytical reference solution, prepared as directed under the individual reagent. The precision of replicate injections is expressed as the relative standard deviation as follows:

$$S_R(\%) = (100/\overline{X})\left[\sum_{i=1}^{N} (X_i - \overline{X})^2/(N-1)\right]^{1/2}$$

where S_R is the relative standard deviation in percent, $\overline{X}$ is the mean of the set of N measurements, and X_i is an individual measurement. When an internal standard is used, the measurement X_i usually refers to the measurement of the relative area, A_S:

$$X_i = A_S = a_R/a_i$$

where a_R is the peak area of the substance being analyzed and a_i is the peak area of the internal standard.

The peak asymmetry value is useful and can be limited by specifying an asymmetry factor. The asymmetry factor, T, is defined as the ratio of the distance from the leading edge to the trailing edge divided by twice the distance from the leading edge to the peak perpendicular measured at 5% of the peak height (*see* Figure 7). Therefore,

$$T = AB/2\,AC$$

This asymmetry factor is unity for a symmetrical peak and increases in value as the asymmetry becomes more pronounced.

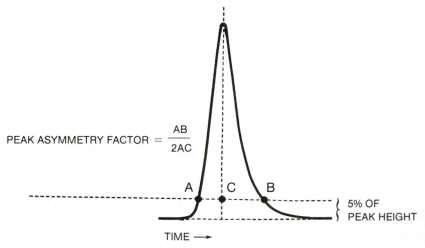

Figure 7.   Chromatogram showing peak asymmetry factor.

Resolution, R, can be specified to ensure separation of closely eluting components or to establish the general separation efficiency of the system. The resolution, R, can be calculated as follows:

$$R = 1.178 \, (t_2 - t_1) / (W_{(h/2)_1} + W_{(h/2)_2})$$

where t_2 and t_1 are the retention times of the two component peaks, and $W_{(h/2)_1}$ and $W_{(h/2)_2}$ are the half-height widths of the two peaks in units of time. An R value of 1.0 means that the resolution is 98% complete. This condition is sufficient, in most cases, for peak area calculations. An R value of 1.5 or greater represents baseline, or complete, separation of the peaks.

REAGENT SOLUTIONS: PREPARATION

Throughout the specifications the term "reagent solution" is used to designate a solution described in this section. The tests and limits in the specifications are based on the use of the solutions in the strength indicated in the descriptions.

Wherever the use of ammonium hydroxide or an acid is prescribed with no indication of strength or dilution, the reagent is to be used in full strength as described in its specification. Dilutions are indicated either by the percentage of some constituent or by the volumes of reagents and water mixed to prepare a dilute reagent. Dilute acid or ammonium hydroxide $(1 + X)$ means a dilute solution prepared by mixing 1 volume of the strong acid or ammonium hydroxide with X volumes of reagent water.

Unless otherwise indicated, the reagent solutions are prepared and diluted with reagent water by using standard Class A volumetric pipets and volumetric flasks.

Reagent Water

Throughout the specifications for reagent chemicals the term "water" means distilled water, or deionized water, which meets the requirements of Water, Reagent, page 777. However, for specific applications such as UV determinations, or liquid and ion chromatography, ASTM Type I reagent water (see ASTM D 1193) or water for which the suitability has been determined should be used. In tests for nitrogen compounds water should be "ammonia-free" or "nitrogen-free". For some tests freshly boiled water must be used to ensure freedom from material absorbed from the air, such as ammonia, carbon dioxide, or oxygen. For LC gradient elution analysis, water must pass the following suitability test.

Water (Suitable for LC Gradient Elution Analysis)

Analyze the sample by using the following gradient elution conditions. Use ACS grade or better acetonitrile.

Column: Octadecyl (C-18 polymeric phase), 250 × 4.6 mm i.d., 5 μm, 12% C loading, endcapped

Mobile Phase: A. Water to be tested
B. ACS reagent-grade acetonitrile

Conditions: Detector: Ultraviolet at 254 nm
Sensitivity: 0.02 AUFS

Gradient Elution:

1. Program the liquid chromatograph to develop a water–acetonitrile gradient as follows:

Time (min)	Flow (mL / min)	% A (Water)	% B (Acetonitrile)
0	2.0	100	0
20	2.0	100	0
40	2.0	0	100
50	2.0	0	100

2. Equilibrate the system at 100% water. Start the program after achieving a stable baseline. Repeat the gradient program and use these results to evaluate the water. No peak should be greater than 10% of full scale (0.002 absorbance units).

Reagent Solutions

Acetic Acid, 1 N.   Dilute 57.5 mL of glacial acetic acid with water to 1 L.

Alcohol.   "Alcohol" refers to the specification on page 319, Ethyl Alcohol. This is distinguished from the specification on page 614, Reagent Alcohol, which is a denatured form of ethyl alcohol.

Ammoniacal Buffer, pH 10. Dissolve 180 g of ammonium chloride in 750 mL of freshly opened ammonium hydroxide and dilute to 1 L with water. (The pH of a 1 + 10 dilution with water should be about 10.0.)

Ammonia–Cyanide Mixture, Lead-Free. Dissolve 20 g of potassium cyanide in 150 mL of ammonium hydroxide and dilute with water to 1 L. Remove lead by shaking the solution with small portions of dithizone extraction solution in chloroform until the dithizone retains its original green color; discard the extraction solution.

Ammonium Acetate–Acetic Acid Buffer Solution. Dissolve 77.1 g of ammonium acetate in water, add 57 mL of glacial acetic acid, and dilute with water to 1 L. (The pH of a 1 + 10 dilution with water should be about 4.5.)

Ammonium Citrate, Lead-Free. Dissolve 40 g of citric acid in 100 mL of water and make alkaline to phenol red with ammonium hydroxide. Remove lead by shaking the solution with small portions of dithizone extraction solution in chloroform until the dithizone solution retains its original green color; discard the extraction solution.

Ammonium Hydroxide, 10% NH_3. Dilute 400 mL of ammonium hydroxide with water to 1 L.

Ammonium Hydroxide, 2.5% NH_3. Dilute 100 mL of ammonium hydroxide with water to 1 L.

Ammonium Metavanadate. Dissolve 2.5 g of ammonium metavanadate, NH_4VO_3, in 500 mL of boiling water, cool, and add 20 mL of nitric acid. Cool, dilute with water to 1 L, and store in a polyethylene bottle.

Ammonium Molybdate. Dissolve 50 g of ammonium molybdate tetrahydrate, $(NH_4)_6Mo_7O_{24} \cdot 4H_2O$, in water and dilute with water to 1 L. Store in a polyethylene bottle.

Ammonium Molybdate–Nitric Acid Solution. Mix thoroughly 100 g of molybdic acid, 85%, with 240 mL of water and add 140 mL of ammonium hydroxide. Filter and add 60 mL of nitric acid. Cool and pour, with constant stirring, into a cool mixture of 400 mL of nitric acid and 960 mL of water. Add 0.1 g of ammonium phosphate dissolved in 10 mL of water, allow to stand 24 h, and filter through glass wool.

Ammonium Molybdate–Sulfuric Acid Solution.   Dissolve 50 g of ammonium molybdate tetrahydrate, $(NH_4)_6Mo_7O_{24} \cdot 4H_2O$, in about 500 mL of water, add 27.5 mL of sulfuric acid, and dilute with water to 1 L.

Ammonium Oxalate.   Dissolve 35 g of ammonium oxalate monohydrate, $(COONH_4)_2 \cdot H_2O$, in water and dilute with water to 1 L.

Ammonium Phosphate.   Dissolve 130 g of dibasic ammonium phosphate, $(NH_4)_2HPO_4$, in water and dilute with water to 1 L.

Ammonium Thiocyanate.   Dissolve 300 g of ammonium thiocyanate, NH_4SCN, in water and dilute with water to 1 L.

Barium Chloride.   Dissolve 120 g of barium chloride dihydrate, $BaCl_2 \cdot 2H_2O$, in water, filter if necessary, and dilute with water to 1 L.

Bromine Water.   A saturated aqueous solution of bromine. Add enough liquid bromine to water in a bottle so that when the mixture is shaken, undissolved bromine remains in a separate phase.

Bromphenol Blue Indicator.   Dissolve 0.10 g of the salt form of bromphenol blue in water and dilute with water to 100 mL (pH 3.0–4.6).

Bromthymol Blue Indicator.   Dissolve 0.10 g of bromthymol blue in 100 mL of dilute alcohol (1 + 1), and filter if necessary (pH 6.0–7.6).

Brucine Sulfate.   Dissolve 0.6 g of brucine sulfate in dilute sulfuric acid (2 + 1), previously cooled to room temperature, and dilute to 1 L with the dilute acid. The sulfuric acid should be nitrate-free acid prepared as follows: dilute the concentrated sulfuric acid (about 96% H_2SO_4) to about 80% H_2SO_4 by adding it to water, heat to dense fumes of sulfur trioxide, and cool. Repeat the dilution and fuming 3 or 4 times.

Chromotropic Acid.   Dissolve 1 g of chromotropic acid, 4,5-dihydroxy-2,7-naphthalenedisulfonic acid, in water and dilute with water to 100 mL.

Crystal Violet Indicator.   Dissolve 100 mg of crystal violet in 10 mL of glacial acetic acid.

5,5-Dimethyl-1,3-cyclohexanedione, 5% in Alcohol. Dissolve 5 g of 5,5-dimethyl-1,3-cyclohexanedione in alcohol and dilute with alcohol to 100 mL.

Dimethylglyoxime. Dissolve 1 g of dimethylglyoxime in alcohol and dilute with alcohol to 100 mL.

N, N-Dimethyl-p-phenylenediamine. Add 200 mL of sulfuric acid to approximately 700 mL of water. Cool to room temperature, add 1 g of N,N-dimethyl-p-phenylenediamine (p-aminodimethylaniline) monohydrochloride, and dilute with water to 1 L.

Diphenylamine. Dissolve 10 mg of colorless diphenylamine in 100 mL of sulfuric acid. In a separate beaker dissolve 2 g of ammonium chloride in 200 mL of water. Cool both solutions in an ice bath and cautiously add the sulfuric acid solution to the water solution, taking care to keep the resulting solution cold. The solution should be nearly colorless.

Dithizone Extraction Solutions.

Carbon Tetrachloride. Dissolve 30 mg of dithizone in 1 L of carbon tetrachloride. Store the solution in a refrigerator.

Chloroform. Dissolve 30 mg of dithizone in 1 L of chloroform plus 5 mL of alcohol. Store the solution in a refrigerator.

Dithizone Indicator Solution. Dissolve 25.6 mg of dithizone in 100 mL of alcohol. Store in a cold place and use within 2 months.

Dithizone Test Solutions.

Carbon Tetrachloride. Dissolve 10 mg of dithizone in 1 L of carbon tetrachloride.

Chloroform. Dissolve 10 mg of dithizone in 1 L of chloroform.

Keep the solutions in glass-stoppered lead-free bottles, protected from light and stored in a refrigerator.

Eosin Y. Dissolve 50 mg of eosin Y in 10 mL of water.

(Ethylenedinitrilo)tetraacetic Acid, Disodium Salt, 0.2 M. Dissolve 74.4 g of (ethylenedinitrilo)tetraacetic acid, disodium salt, dihydrate in 950 mL of water, and dilute with water to 1 L.

Ferric Ammonium Sulfate Indicator. Dissolve 80 g of clear crystals of ferric ammonium sulfate dodecahydrate, $FeNH_4(SO_4)_2 \cdot 12H_2O$, in water and dilute with water to 1 L. A few drops of sulfuric acid may be added, if necessary, to clear the solution.

Ferric Chloride. Dissolve 100 g of ferric chloride hexahydrate $FeCl_3 \cdot 6H_2O$, in dilute hydrochloric acid (1 + 99) and dilute with the dilute acid to 1 L.

Ferric Sulfate. Add 25 mL of sulfuric acid to 125 g of ferric sulfate n-hydrate, $Fe_2(SO_4)_3 \cdot nH_2O$. Add about 700 mL of water, digest on a steam bath to dissolve, cool to room temperature, and dilute with water to 1 L.

Ferroin Indicator, 0.025 M. Dissolve 0.70 g of ferrous sulfate heptahydrate and 1.5 g of 1,10-phenanthroline in 100 mL of water.

Hexamethylenetetramine Solution, Saturated. Dissolve about 20 g of hexamethylenetetramine in 100 mL of water.

Hydrazine Sulfate, 2%. Dissolve 2 g of hydrazine sulfate, $(NH_2)_2 \cdot H_2SO_4$, in water and dilute with water to 100 mL.

Hydrochloric Acid, 20% HCl. Dilute 470 mL of hydrochloric acid with water to 1 L.

Hydrochloric Acid, 10% HCl. Dilute 235 mL of hydrochloric acid with water to 1 L.

Hydrogen Sulfide Water. Saturate water with hydrogen sulfide gas. This solution must be freshly prepared.

Hydroxylamine Hydrochloride. Dissolve 10 g of hydroxylamine hydrochloride, $NH_2OH \cdot HCl$, in water and dilute with water to 100 mL.

Hydroxylamine Hydrochloride for Dithizone Test. Dissolve 20 g of hydroxylamine hydrochloride, $NH_2OH \cdot HCl$, in about 65 mL of water and add 0.10 to 0.15 mL of thymol blue indicator solution. Add ammonium hydroxide until a yellow color appears. Add 5 mL of a 4% solution of sodium diethyldithiocarbamate. Mix thoroughly and allow to stand for 5 min. Extract with successive portions of chloroform until no yellow color is developed in the chloroform layer when the extract is shaken with a dilute solution of a copper salt. Add hydrochloric acid until the indicator turns pink, and dilute with water to 100 mL.

Indigo Carmine. Dissolve 1 g of FD & C Blue No. 2, dried at 105°C, in a mixture of 800 mL of water and 100 mL of sulfuric acid, and dilute with water to 1 L. FD & C Blue No. 2, the disodium salt of 5,5'-indigotindisulfonic acid, $C_{16}H_8N_2O_2(SO_3Na)_2$, is one of the dyes that may be certified under the Federal Food, Drug, and Cosmetic Act as suitable for use in foods, drugs, and cosmetics.

Jones Reductor. The zinc is amalgamated by immersing it in a solution of mercuric chloride in hydrochloric acid. A 250-g portion of 20-mesh zinc is covered with reagent water in a 1-L suction flask, a solution containing 11 g of mercuric chloride ($HgCl_2$) in 100 mL of hydrochloric acid is poured into the flask, and the system is slowly mixed and shaken for about 2 min. The solution is poured off and the amalgam is washed thoroughly with hot tap water, then finally with reagent water. The column is charged with six 250-g portions.

> *Apparatus.* A dispensing buret, about 22 inches long and 2 inches in diameter, equipped with a glass stopcock and delivery tube, 6 mm wide × 3.5 inches long. The reductor is charged with an 8-inch column of 20-mesh amalgamated zinc (1500 g) and on top of this a 6-inch column of broken amalgamated zinc (about 750 g). The delivery tube is connected to a 1-L flask through a two-hole rubber stopper. One hole is used as an inlet; the other functions as an outlet for carbon dioxide gas.

Karl Fischer Reagent. Dissolve 762 g of iodine in 2420 mL of pyridine in a 9-L glass-stoppered bottle, and add 6 L of methanol. To prepare the active reagent, add 3 L of the foregoing stock to a 4-L bottle, and cool by placing the bottle in a slurry of chopped ice. Add carefully about 135 mL of liquid sulfur dioxide, collected in a calibrated cold trap, and stopper the bottle. Shake the mixture until it is homogeneous and set it aside for one or two days before use.

Lead Acetate. Dissolve 100 g of lead acetate trihydrate, $(CH_3COO)_2Pb \cdot 3H_2O$, in water. If necessary, add a few drops of acetic acid to clear the solution and dilute with water to 1 L.

Metalphthalein-Screened Indicator. Dissolve 0.18 g of metalphthalein and 0.02 g of Naphthol Green B (i.e., Acid Green 1; C.I. 10020) in water containing 0.5 mL of ammonium hydroxide and dilute to 100 mL with water.

4-(Methylamino)phenol Sulfate. Dissolve 2 g of 4-(methylamino)phenol sulfate in 100 mL of water. To 10 mL of this solution add 90 mL of water and 20 g of sodium bisulfite. Confirm the suitability of the reagent solution by the following test: Add 1 mL of this reagent solution to each of four solutions containing 25 mL of 0.5 N sulfuric acid and 1 mL of ammonium molybdate 5% reagent solution. Add 0.005 mg of phosphate ion (PO_4) to one of the solutions, 0.01 mg to a second, and 0.02 mg to a third.

Allow to stand at room temperature for 2 h. The solutions in the three tubes should show readily perceptible differences in blue color corresponding to the relative amounts of phosphate added, and the one to which 0.005 mg of phosphate was added should be perceptibly bluer than the blank.

Methyl Orange Indicator.   Dissolve 0.10 g of methyl orange in 100 mL of water (pH 3.2–4.4).

Methyl Red Indicator.   Dissolve 0.10 g of methyl red in 100 mL of alcohol. See page 480 for a description of the three forms of methyl red (pH 4.2–6.2).

Nessler Reagent.   Dissolve 143 g of sodium hydroxide in 700 mL of water. Dissolve 50 g of red mercuric iodide and 40 g of potassium iodide in 200 mL of water. Pour the iodide solution into the hydroxide solution and dilute with water to 1 L. Allow to settle and use the clear supernatant liquid. Nessler reagent prepared by any of the recognized methods may be used, but the reagent described reacts promptly and consistently.

Nitric Acid, 10% HNO$_3$.   Dilute 105 mL of nitric acid with water to 1 L.

Nitric Acid, 1% HNO$_3$.   Dilute 10.5 mL of nitric acid with water to 1 L.

Oxalic Acid.   Dissolve 40 g of oxalic acid dihydrate, $(COOH)_2 \cdot 2H_2O$, in water and dilute with water to 1 L.

1,10-Phenanthroline.   Dissolve 0.1 g of 1,10-phenanthroline monohydrate in 100 mL of water containing 0.1 mL of 10% hydrochloric acid reagent solution.

Phenoldisulfonic Acid.   Dissolve 50 g of phenol, C_6H_5OH, in 300 mL of sulfuric acid, add 150 mL of fuming sulfuric acid (15% SO_3), and heat at 100°C for 2 h.

Phenolphthalein Indicator.   Dissolve 1 g of phenolphthalein in 100 mL of alcohol (pH 8.0–10.0).

Phenol Red Indicator.   Dissolve 0.10 g of phenol red in 100 mL of alcohol and filter if necessary (pH 6.8–8.2).

Potassium Chromate.   Dissolve 10 g of potassium chromate, K_2CrO_4, in water and dilute with water to 100 mL.

Potassium Cyanide, Lead-Free.   Dissolve 10 g of potassium cyanide, KCN, in sufficient water to make 20 mL. Remove lead by shaking with portions of dithizone extraction solution. Part of the dithizone remains in the aqueous phase but can be removed, if desired, by shaking with chloroform. Dilute the potassium cyanide solution with water to 100 mL.

Potassium Dichromate.   Dissolve 10 g of potassium dichromate, $K_2Cr_2O_7$, in water and dilute with water to 100 mL.

Potassium Ferricyanide.   Dissolve 5 g of potassium ferricyanide, $K_3Fe(CN)_6$, in 100 mL of water. Prepare the solution at the time of use.

Potassium Ferrocyanide.   Dissolve 10 g of potassium ferrocyanide trihydrate, $K_4Fe(CN)_6 \cdot 3H_2O$, in 100 mL of water. Prepare the solution at the time of use.

Potassium Hydroxide, 0.5 N in Methanol.   Dissolve 34 to 36 g of potassium hydroxide, KOH, in 20 mL of water and dilute with methanol to 1 L. Allow the solution to stand in a stoppered bottle for 24 h. Decant the clear supernatant solution into a bottle provided with a tight-fitting stopper.

Potassium Iodide (for Mercury by Flameless Atomic Absorption). Dissolve 165 g of potassium iodide, KI, in water and dilute with water to 1 L.

Potassium Iodide (for Free Chlorine).   Dissolve 10 g of potassium iodide, KI, in water and dilute with water to 100 mL.

Silver Diethyldithiocarbamate.   Dissolve 1 g of silver diethyl-dithiocarbamate, $(C_2H_5)_2NCS_2Ag$, in 200 mL of freshly distilled pyridine.

Silver Nitrate.   Dissolve 17 g of silver nitrate, $AgNO_3$, in 1 L of water.

Sodium Citrate, 1 M.   Dissolve 294 g of sodium citrate dihydrate, $Na_3C_6H_5O_7 \cdot 2H_2O$, in 950 mL of water, and dilute with water to 1 L.

Sodium Cyanide.   Dissolve 10 g of sodium cyanide, NaCN, in water and dilute with water to 100 mL.

Sodium Diethyldithiocarbamate.   Dissolve 1.0 g of sodium diethyldithiocarbamate in water and dilute with water to 1 L.

Sodium Hydroxide.   Dissolve 100 g of sodium hydroxide, NaOH, in water and dilute with water to 1 L.

Stannous Chloride, 40%.   Dissolve 40 g of stannous chloride dihydrate, $SnCl_2 \cdot 2H_2O$, in 100 mL of hydrochloric acid. Store this solution in glass containers and use within 3 months.

Stannous Chloride, 2%.   Dissolve 2 g of stannous chloride dihydrate, $SnCl_2 \cdot 2H_2O$, in hydrochloric acid and dilute with hydrochloric acid to 100 mL.

Starch Indicator.   Mix 1 g of soluble starch with 10 mg of red mercuric iodide and enough cold water to make a thin paste, add 200 mL of boiling water, and boil for 1 min while stirring. Cool before use.

Sulfuric Acid, Chloride- and Nitrate-Free.   In a hood gently fume sulfuric acid in a crucible or dish for at least 30 min. Allow to cool and transfer to a tightly capped glass bottle for storage.

Sulfuric Acid, 25%.   Slowly add 170 mL of sulfuric acid to 750 mL of water, cool, and dilute with water to 1 L.

Sulfuric Acid, 10%.   Slowly add 60 mL of sulfuric acid to 750 mL of water, cool, and dilute with water to 1 L.

Thymol Blue Indicator.   Dissolve 0.10 g of thymol blue in 100 mL of alcohol (acid range, pH 1.2–2.8; alkaline range, pH 8.0–9.2).

Thymolphthalein Indicator.   Dissolve 0.10 g of thymolphthalein in 100 mL of alcohol (pH 8.8–10.5).

Titanium Tetrachloride.   Cool separately in small beakers surrounded by crushed ice 10 mL of 20% hydrochloric acid and 10 mL of water-white titanium tetrachloride. Add the tetrachloride dropwise to the chilled acid. Allow the mixture to stand at ice temperature until all of the yellow solid dissolves, then dilute the solution with 20% hydrochloric acid to 1 L.

Triton X-100, 0.20%. Dissolve 0.20 g of Triton X-100 (polyethylene glycol ether of isooctylphenol) in water, and dilute with water to 100 mL.

SOLID REAGENT MIXTURES

Eriochrome Black T Indicator. Grind 0.2 g of eriochrome black T to a fine powder with 20 g of potassium chloride.

Hydroxy Naphthol Blue Indicator. Use the mixture with sodium chloride commercially supplied (*see* specifications for this indicator, page 380).

Methylthymol Blue Indicator. Grind 0.2 g of methylthymol blue to a fine powder with 20 g of potassium nitrate.

Murexide Indicator. Grind 0.2 g of murexide to a fine powder with 20 g of potassium nitrate.

Xylenol Orange Indicator. Grind 0.2 g of xylenol orange (either free acid or sodium salt form) to a fine powder with 20 g of potassium nitrate. Alternatively, use a solution of 0.1 g of xylenol orange in 100 mL of alcohol.

STANDARD SOLUTIONS

Controls and Standards Preparation

Many of the tests for impurities require the comparison of the color or the turbidity produced under specified conditions by an impurity ion with the color or the turbidity produced under similar conditions by a known amount of the impurity. Three types of solutions are used to determine amounts of impurities in reagent chemicals: (1) the blank, (2) the standard, and (3) the control.

(1) Blank. A solution containing the quantities of solvents and reagents used in the test.

(2) Standard. A solution containing the quantities of solvents and reagents used in the test plus an added known amount of the impurity to be determined.

(3) **Control.** A solution containing the quantities of solvents and reagents used in the test plus an added known amount of the impurity to be determined and a known amount of the reagent chemical being tested. Some of the chemical being tested must be added to the solution because it is known that the chemical interferes with the test.

Throughout this book, the expressions "for the control" or "for the standard" add X mg of Y ion mean to add the proper aliquot of one of the following solutions. These solutions must be prepared at the time of use or checked often enough to make certain that the tests will not be vitiated by changes in strength of the standard solution during storage. Calibrated volumetric glassware must be used in preparing these solutions.

> NOTE. In place of standard solutions prepared as described below, commercially available atomic absorption standards, where applicable, may be used after proper dilution.

Acetaldehyde (0.1 mg of CH₃CHO in 1 mL). Prepare in water by appropriate dilution of an assayed acetaldehyde solution.

Acetone (0.1 mg of (CH₃)₂CO in 1 mL). Dilute 6.4 mL of acetone with water to 1 L. Dilute 2.0 mL of this solution with water to 100 mL. Prepare the solution at the time of use.

Aluminum (0.01 mg of Al in 1 mL). Dissolve 0.100 g of metallic aluminum, Al, in 10 mL of dilute hydrochloric acid (1 + 1) and dilute with water to 100 mL. To 10 mL of this solution add 25 mL of dilute hydrochloric acid (1 + 1) and dilute with water to 1 L.

Ammonium (0.01 mg of NH₄ in 1 mL). Dissolve 0.296 g of ammonium chloride, NH₄Cl, in water and dilute with water to 100 mL. Dilute 10 mL of this solution with water to 1 L.

Antimony (0.1 mg of Sb in 1 mL). Dissolve 0.2743 g of antimony potassium tartrate hemihydrate, $KSbOC_4H_4O_6 \cdot \frac{1}{2}H_2O$, in 100 mL of water. Add 100 mL of hydrochloric acid and dilute with water to 1 L.

Arsenic (0.001 mg of As in 1 mL). Dissolve 0.132 g of arsenic trioxide, As_2O_3, in 10 mL of 10% sodium hydroxide reagent solution, neutralize with 10% sulfuric acid reagent solution, add an excess of 10 mL, and dilute with water to 1 L. To 10 mL of this solution add 10 mL of 10% sulfuric acid reagent solution and dilute with water to 1 L.

Barium (0.1 mg of Ba in 1 mL). Dissolve 0.178 g of barium chloride dihydrate, $BaCl_2 \cdot 2H_2O$, in water and dilute with water to 1 L.

Bismuth (0.01 mg of Bi in 1 mL). Dissolve 0.232 g of bismuth trinitrate pentahydrate, $Bi(NO_3)_3 \cdot 5H_2O$, in 10 mL of dilute nitric acid (1 + 9) and dilute with water to 100 mL. Dilute 10 mL of this solution plus 10 mL of nitric acid with water to 1 L.

Bromate (0.1 mg of BrO_3 in 1 mL). Dissolve 0. 131 g of potassium bromate, $KBrO_3$, in water and dilute with water to 1 L.

Bromide (0.1 mg of Br in 1 mL). Dissolve 1.49 g of potassium bromide, KBr, in water and dilute with water to 100 mL. Dilute 10 mL of this solution with water to 1 L.

Cadmium (0.025 mg of Cd in 1 mL). Dissolve 0.10 g of cadmium chloride 2.5-hydrate, $CdCl_2 \cdot 2\frac{1}{2}H_2O$, in water, add 1 mL of hydrochloric acid, and dilute with water to 500 mL. To 25 mL of this solution add 1 mL of hydrochloric acid and dilute with water to 100 mL.

Calcium (0.1 mg of Ca in 1 mL). Dissolve 0.250 g of calcium carbonate, $CaCO_3$, in 20 mL of water plus 5 mL of 10% hydrochloric acid reagent solution and dilute with water to 1 L.

Calcium (0.01 mg of Ca in 1 mL). Dilute 100 mL of the preceding solution with water to 1 L.

Carbon (4 mg of C in 1 mL). Dissolve 8.5 g of potassium hydrogen phthalate, $HOCOC_6H_4COOK$, in water and dilute to 1 L.

Carbonate (0.3 mg of CO_3, 0.2 mg of CO_2, or 0.06 mg of C in 1 mL). Dissolve 0.53 g of sodium carbonate, Na_2CO_3, in carbon dioxide-free water, and dilute with carbon dioxide-free water to 1 L. Prepare the solution at the time of use.

Chloride (0.01 mg of Cl in 1 mL). Dissolve 0.165 g of sodium chloride, NaCl, in water and dilute with water to 100 mL. Dilute 10 mL of this solution with water to 1 L.

Chlorine (0.01 mg of Cl_2 in 1 mL). Dilute 1.0 mL of a 2.5% sodium hypochlorite, NaOCl, solution with water to 1 L. Correct for any assay not equal to 2.5%. Standardize the sodium hypochlorite solution before use by the following method: Pipet 3 mL into a tared, glass-stoppered flask. Weigh accurately and add 50 mL of water, 2 g of potassium iodide, and 10 mL of acetic acid. Titrate the liberated iodine with 0.1 N sodium thiosulfate, adding 3 mL of starch indicator solution near the end point. One milliliter of 0.1 N sodium thiosulfate corresponds to 3.723 mg of NaOCl.

Chlorobenzene (0.1 mg of C_6H_5Cl in 1 mL). Dilute 1.0 mL of chlorobenzene to 100 mL with 2,2,4-trimethylpentane. Dilute 9.0 mL of this solution to 100 mL with 2,2,4-trimethylpentane.

Chromate (0.01 mg of CrO_4 in 1 mL). Dissolve 1.268 g of potassium dichromate, $K_2Cr_2O_7$, in water and dilute with water to 1 L. Dilute 1.0 mL of this solution to 100 mL.

Copper Stock Solution (0.1 mg of Cu in 1 mL). Dissolve 0.393 g of cupric sulfate pentahydrate, $CuSO_4 \cdot 5H_2O$, in water and dilute with water to 1 L.

Copper Standard for Dithizone Test (0.001 mg of Cu in 1 mL). Dilute 1 mL of the copper stock solution with water to 100 mL. This dilute solution must be prepared immediately before use.

Ferric Iron (0.1 mg of Fe^{3+} in 1 mL). Dissolve 0.863 g of ferric ammonium sulfate dodecahydrate, $FeNH_4(SO_4)_2 \cdot 12H_2O$, in water and dilute with water to 1 L.

Fluoride Stock Solution (1 mg of F in 1 mL). Dissolve 2.21 g of sodium fluoride, NaF, previously dried at 110°C for 2 h, and dissolve with 200 mL of water in a 400-mL plastic beaker. Transfer to a 1-L volumetric flask, dilute with water to the mark, and store the solution in a plastic bottle.

Fluoride (0.005 mg in 1 mL). Transfer 5.0 mL of the fluoride stock solution to a 1-L plastic volumetric flask, dilute with water to the mark, and store in a plastic bottle. Prepare the solution at the time of use.

Formaldehyde (0.1 mg of HCHO in 1 mL). Dilute 2.7 g (2.7 mL) of 37% formaldehyde solution to 1 L with water (prepare freshly as needed). Dilute 10 mL of this solution to 100 mL.

Formaldehyde (0.01 mg of HCHO in 1 mL). Dilute 10 mL of the preceding solution with water to 100 mL. Prepare the solution at the time of use.

Hydrogen Peroxide (0.1 mg of H_2O_2 in 1 mL). Dilute 3.0 mL of 30% hydrogen peroxide with water to 100 mL. Dilute 1.0 mL of this solution with water to 100 mL. Prepare the solution at the time of use.

Iodate (0.10 mg of IO_3 in 1 mL). Dissolve 0.123 g of potassium iodate, KIO_3, in water and dilute to 1 L.

Iodate (0.01 mg of IO_3 in 1 mL). Dilute 10 mL of iodate solution (0.10 mg of IO_3 in 1 mL) to 100 mL.

Iodide (0.01 mg of I in 1 mL). Dissolve 0.131 g of potassum iodide, KI, in water and dilute with water to 100 mL. Dilute 10 mL of this solution with water to 1 L.

Iron (0.01 mg of Fe in 1 mL). Dissolve 0.702 g of ferrous ammonium sulfate hexahydrate, $Fe(NH_4)_2(SO_4)_2 \cdot 6H_2O$, in 10 mL of 10% sulfuric acid reagent solution and dilute with water to 100 mL. To 10 mL of this solution add 10 mL of 10% sulfuric acid reagent solution and dilute with water to 1 L.

Lead Stock Solution (0.1 mg of Pb in 1 mL). Dissolve 0.160 g of lead nitrate, $Pb(NO_3)_2$, in 100 mL of dilute nitric acid (1 + 99) and dilute with water to 1 L. The solution should be prepared and stored in containers free from lead. Its strength should be checked every few months to determine whether the lead content has changed by reaction with the container.

Lead Standard for Dithizone Test (0.001 mg of Pb in 1 mL). Dilute 1 mL of the lead stock solution to 100 mL with dilute nitric acid (1 + 99). This solution must be prepared immediately before use.

Lead Standard for Heavy Metals Test (0.01 mg of Pb in 1 mL). Dilute 10 mL of the lead stock solution to 100 mL with water. This dilute solution must be prepared at the time of use.

Magnesium (0.01 mg of Mg in 1 mL). Dissolve 1.014 g of clear crystals of magnesium sulfate heptahydrate, $MgSO_4 \cdot 7H_2O$, in water and dilute with water to 100 mL. Dilute 10 mL of this solution with water to 1 L.

Magnesium (0.002 mg of Mg in 1 mL). Dilute 100 mL of the preceding solution with water to 500 mL. Prepare the solution at the time of use.

Manganese (0.01 mg of Mn in 1 mL). Dissolve 3.08 g of manganese sulfate monohydrate, $MnSO_4 \cdot H_2O$, in water and dilute with water to 1 L. Dilute 10 mL of this solution with water to 1 L.

Mercury (0.05 mg of Hg in 1 mL). Dissolve 1.35 g of mercuric chloride, $HgCl_2$, in water, add 8 mL of hydrochloric acid, and dilute with water to 1 L. To 50 mL of this solution add 8 mL of hydrochloric acid and dilute with water to 1 L.

Methanol (0.1 mg of CH_3OH in 1 mL). Dilute 5.1 mL of methanol, CH_3OH, with water to 200 mL. Dilute 5.0 mL of this solution with water to 1 L. Prepare the solution at the time of use.

Molybdenum (0.01 mg of Mo in 1 mL). Dissolve 0.150 g of molybdenum trioxide, MoO_3, in 10 mL of dilute ammonium hydroxide $(1 + 9)$ and dilute with water to 100 mL. Dilute 16 mL of this solution with water to 1 L.

Nickel (0.01 mg of Ni in 1 mL). Dissolve 0.448 g of nickel sulfate hexahydrate, $NiSO_4 \cdot 6H_2O$, in water and dilute with water to 100 mL. Dilute 10 mL of this solution with water to 1 L.

Nitrate (0.01 mg of NO_3 in 1 mL). Dissolve 0.163 g of potassium nitrate, KNO_3, in water and dilute with water to 100 mL. Dilute 10 mL of this solution with water to 1 L.

Nitrilotriacetic Acid (10 mg of $(HOCOCH_2)_3N$ in 1 mL). Transfer 1.0 g of nitrilotriacetic acid, $(HOCOCH_2)_3N$, to a 100-mL volumetric flask. Dissolve in 10 mL of 10% potassium hydroxide solution, and dilute with water to volume.

Nitrite (0.001 mg in 1 mL). Dissolve 0.150 g of sodium nitrite, $NaNO_2$, in water and dilute with water to 100 mL. Dilute 0.1 mL of this solution with water to 100 mL. This solution must be prepared at the time of use.

Nitrogen (0.01 mg of N in 1 mL). Dissolve 0.382 g of ammonium chloride, NH_4Cl, in water and dilute with water to 100 mL. Dilute 10 mL of this solution with water to 1 L.

Phosphate (0.01 mg of PO_4 in 1 mL). Dissolve 0.143 g of monobasic potassium phosphate, KH_2PO_4, in water and dilute with water to 100 mL. Dilute 10 mL of this solution with water to 1 L.

Platinum–Cobalt Standard (APHA No. 500). *See* page 11.

Potassium (0.01 mg of K in 1 mL). Dissolve 0.191 g of potassium chloride, KCl, in water and dilute with water to 1 L. Dilute 100 mL of this solution with water to 1 L.

Silica (0.01 mg of SiO$_2$ in 1 mL). Dissolve 0.473 g of sodium silicate, Na$_2$SiO$_3 \cdot$9H$_2$O, in 100 mL of water in a platinum or polyethylene dish. Dilute 10 mL of this solution (polyethylene graduate) with water to 1 L in a polyethylene bottle.

Silver (0.1 mg of Ag in 1 mL). Dissolve 0.157 g of silver nitrate, AgNO$_3$, in water and dilute with water to 1 L.

Sodium (0.01 mg of Na in 1 mL). Dissolve 0.254 g of sodium chloride, NaCl, in water and dilute with water to 1 L. Dilute 100 mL of this solution with water to 1 L.

Strontium Stock Solution (1 mg of Sr in 1 mL). Dissolve 0.242 g of strontium nitrate, Sr(NO$_3$)$_2$, in water and dilute with water to 100 mL.

Strontium (0.1 mg of Sr in 1 mL). Dilute 10 mL of the strontium stock solution with water to 100 mL.

Strontium (0.01 mg of Sr in 1 mL). Dilute 10 mL of the strontium stock solution with water to 1 L.

Sulfate (0.01 mg of SO$_4$ in 1 mL). Dissolve 0.148 g of anhydrous sodium sulfate, Na$_2$SO$_4$, in water and dilute with water to 100 mL. Dilute 10 mL of this solution with water to 1 L.

Sulfide (0.01 mg of S in 1 mL). Dissolve 0.75 g of sodium sulfide nonahydrate, Na$_2$S$\cdot$9H$_2$O, in water and dilute with water to 100 mL. Dilute 10 mL of this solution with water to 1 L. This solution must be prepared immediately before use.

Tin (0.05 mg of Sn in 1 mL). Dissolve 0.100 g of metallic tin (Sn) in 10 mL of dilute hydrochloric acid (1 + 1) and dilute with water to 100 mL. Dilute 5 mL of this solution with dilute hydrochloric acid (1 + 9) to 100 mL. This solution must be prepared immediately before use.

Titanium (0.1 mg of Ti in 1 mL). Dissolve 0.739 g of potassium titanyl oxalate dihydrate, K$_2$TiO(C$_2$O$_4$)$_2 \cdot$2H$_2$O, in water and dilute with water to 1 L.

Zinc Stock Solution (0. 1 mg of Zn in 1 mL).   Dissolve 0.124 g of zinc oxide, ZnO, in 10 mL of dilute sulfuric acid $(1 + 9)$ and dilute with water to 1 L.

Zinc Standard for Dithizone Tests (0.001 mg of Zn in 1 mL). Dilute 1 mL of the zinc stock solution with water to 100 mL. This dilute solution must be prepared immediately before use.

VOLUMETRIC SOLUTIONS: PREPARATION AND STANDARDIZATION

General

Throughout the specifications the term "volumetric solution" is used to designate a solution prepared and standardized as described in this section.

Two background references that the analyst may find helpful are ASTM Standard Practice E 200–91, "Preparation, Standardization, and Storage of Standard Solutions for Chemical Analysis"* and the section on Volumetric Solutions in the United States Pharmacopeia–National Formulary.** Various textbooks on chemical analysis also deal with volumetric solutions and volumetric analysis, on both a theoretical and a practical basis.

The concentrations of volumetric solutions usually are expressed as normality (the number of gram equivalent weights in each liter of solution), although concentration sometimes is expressed as molarity (the number of gram molecular weights in each liter of solution).

The accuracy of many analytical procedures is dependent on the manner in which such solutions are prepared, standardized, and stored. The directions that follow describe the preparation of these solutions and their standardization against primary standards or against already-standardized solutions. Alternatively, in cases where suitable primary standard chemicals are available, volumetric solutions may be prepared by dissolving accurately weighed portions of

---

*Available from ASTM, 1916 Race St., Philadelphia, PA 19103.

**Available from the United States Pharmacopeial Convention, Inc., 12601 Twinbrook Parkway, Rockville, MD 20852.

these substances to make an accurately known volume of solution. All volumetric solutions, of course, must be thoroughly mixed as part of their preparation. Stronger or weaker solutions than those described are prepared and standardized in the same general manner, using proportional amounts of reagents and standards. Lower strengths frequently may be prepared by accurately diluting a stronger solution. However, when necessary because of the accuracy requirements of the analysis being performed, volumetric solutions prepared in this manner should be standardized before use against a primary standard or by comparison with an appropriate volumetric solution of known strength.

Certain reagents, such as arsenic trioxide, Reductometric Standard, are designated as Standards in this book. Certified Standards of a number of chemical compounds are available as Standard Reference Materials (SRMs)* from NIST.

It is desirable to standardize a solution in such a way that the end point is observed at the same pH value (or oxidation–reduction potential) anticipated during subsequent use. For example, a sodium hydroxide solution found to be 0.1000 N when standardized against potassium hydrogen phthalate to phenolphthalein (pH 8.5) may behave as though 0.1002 N when used for titrating a strong acid to methyl red (pH 5). Only a small part of this difference is due to the NaOH required to bring pure water from pH 5 to 8.5. In this case, the error can be avoided by titrating the strong acid to phenolphthalein (pH 8.5), the same indicator that was used in standardizing the sodium hydroxide.

Storage

Glass containers are suitable for the storage of most volumetric solutions. However, polyolefin containers are recommended for alkaline solutions. Volumetric solutions are stable for varying lengths of time, depending on their chemical nature and their concentration. Dilute solutions are likely to be less stable than those that are 0.1 N or stronger. If there is any doubt about the reliability of a solution, it should be restandardized at the time of use.

---

*Available from the Office of Standard Reference Materials, National Institute of Standards and Technology, Gaithersburg, MD 20899.

Temperature Considerations

If possible, volumetric solutions should be prepared, standardized, and used at 25°C. If a titration is carried out at a temperature different from that at which the solution was standardized, a temperature correction may be needed, depending on the accuracy requirement of the analysis being performed. If the temperature of use is higher than the temperature of standardization, the correction is to be subtracted from the standardization value; if the temperature of use is lower than the temperature of standardization, the correction is to be added. For guidance with respect to temperature corrections, consider that a 5°C temperature change will alter the normalities of solutions approximately as follows: For aqueous solutions 0.1 N or less, by 1 part in 1000; 0.5 N, by 1.2 parts in 1000; and 1 N, by 1.4 parts in 1000. For nonaqueous solutions, based on the expansion coefficients of the solvents, by 5.4 parts in 1000 for glacial acetic acid; 5.5 parts in 1000 for dioxane; and 6.2 parts in 1000 for methanol.

Errors caused by temperature differences, as well as by improper drainage of burets or pipets, can be avoided if measurements are made on a weight basis. With a sensitive direct-reading balance and the use of squeeze bottles, titrations by weight have become very practical and precise.

Correction Factors

It is not necessary that volumetric solutions have the identical normalities (or molarities) indicated in this section, or as specified in many of the standardization and assay calculations in this book. It is necessary, however, that the exact value be known, because if it is different from the nominal value specified, a correction factor will have to be applied. Thus, if a calculation is based on an exactly 0.1 N volumetric solution, and the solution actually used is 0.1008 N, its volume must be multiplied by a factor of 1.008 to obtain the corrected volume of 0.1 N solution to use in the calculation. Similarly, if the actual normality is 0.0984 N, the factor is 0.984.

Replication

Because the results obtained by titrations depend on the reliability of the volumetric solutions used, it is imperative that the latter be standardized at least in duplicate, with particular care, and preferably by experienced analysts. Duplicate standardizations for solutions from 0.1 N to 1 N should agree at least to within 2 parts in 1000.

Duplicates for 0.05 N solutions may differ by as much as 10 parts in 1000. If deviations are greater than indicated, the determinations should be repeated until the criteria are met.

Volumetric Solutions

Acetic Acid, 1 N. Add 61 g (58.1 mL) of glacial acetic acid to sufficient water to make 1 L, and mix thoroughly. Standardize as follows: Measure accurately 40 mL of the solution into a 250-mL conical flask, add 0.10 mL of phenolphthalein indicator solution, and titrate with 1 N sodium hydroxide volumetric solution to a permanent pink color.

$$N = \frac{mL\ NaOH \times N\ NaOH}{mL\ CH_3COOH\ solution}$$

Arsenious Acid, 0.1 N. Weigh accurately 4.946 g of arsenic triox-ide, Reductometric Standard, previously dried for 1 h at 105°C, and dissolve in 60 mL of 1 N sodium hydroxide in a 250-mL beaker, warming on the steam bath, if necessary. Neutralize with 59–60 mL of 1 N sulfuric acid, cool, transfer to a 1-L volumetric flask, dilute to volume, and mix thoroughly. The solution should have a pH between 5 and 9.

Bromine, 0.1 N, Solution I.   (Prepared from bromine.) Pipet 2.6 mL of bromine into a 1-L glass-stoppered volumetric flask containing 25 g of potassium bromide and 5 mL of hydrochloric acid dissolved in 500 mL of water. Stopper and swirl until the bromine is dissolved, dilute to volume, and mix thoroughly. Standardize as follows: Pipet 25 mL into a 500-mL iodine flask containing 100 mL of water and 2 g of potassium iodide, and titrate the liberated iodine with 0.1 N sodium thiosulfate volumetric solution. Add 3 mL of starch indicator solution near the end of the titration, and continue to a colorless end point.

$$N = \frac{mL\ Na_2S_2O_3 \times N\ Na_2S_2O_3}{mL\ Br_2\ solution}$$

Bromine, 0.1 N, Solution II. (Prepared as a potassium bromate–potassium bromide mixture.) Transfer 2.8 g of potassium bromate, plus 15 g of potassium bromide, to a 1-L volumetric flask, dilute to volume, and mix thoroughly. Standardize as follows: Measure accurately 40 mL of the solution into a 500-mL iodine flask, and dilute with 100 mL of water. Add 2 g of potassium iodide, stopper the flask, and swirl the contents carefully; then add 5 mL of hydrochloric acid, stopper the flask again, swirl vigorously, and allow to stand for 5 min. Titrate the liberated iodine with 0.1 N sodium thiosulfate volumetric solution. Add 3 mL of starch indicator solution near the end of the titration, and continue to a colorless end point.

$$N = \frac{mL\ Na_2S_2O_3 \times N\ Na_2S_2O_3}{mL\ Br_2\ solution}$$

Ceric Ammonium Sulfate, 0.1 N. Place 65 g of ceric ammonium sulfate dihydrate in a 1500-mL beaker, and slowly add 30 mL of sulfuric acid. Stir to a smooth paste, and cautiously add water in portions of 20 mL or less, with stirring, until the salt is dissolved. Dilute, if necessary, to approximately 500 mL, and mix thoroughly by stirring. Allow to stand overnight. If a residue has formed, or the solution is turbid, filter through a fine-porosity sintered-glass filter (do not filter through paper or similar material). Transfer the filtrate to a 1-L volumetric flask, dilute to volume, and mix thoroughly. Standardize as follows: Measure accurately 40 mL of 0.1 N arsenious acid volumetric solution into a 500-mL conical flask. Add 100 mL of water and 10 mL of 25% sulfuric acid solution, followed by 0.1 mL of 1,10-phenanthroline reagent solution and 0.2 mL of 0.1 N ferrous ammonium sulfate solution. Add 0.1 mL of osmium tetroxide solution,* and titrate slowly with the ceric solution to a change from pink to pale blue. Perform a blank titration on all of the reagents except the arsenious acid solution, and subtract the volume of ceric sulfate solution consumed in the blank titration from that consumed in the first titration.

$$N = \frac{mL\ As_2O_3 \times N\ As_2O_3}{mL\ Ce^{IV}\ solution\ (corrected\ for\ blank)}$$

Alternatively, the solution may be standardized against arsenic trioxide, Reductometric Standard: Weigh accurately into a 500-mL Erlenmeyer flask, 0.21 ± 0.01 g of arsenic trioxide, Reductometric Standard, previously dried for 1 h at 105°C. Add 15 mL of 1 N sodium hydroxide, warm gently, and swirl to hasten dissolution, being certain

---

*Caution: In a well-ventilated hood, dissolve 0.1 g of osmium tetroxide in 40 mL of 0.1 N H_2SO_4. Osmium tetroxide vapor is extremely hazardous.

that no particles of arsenic trioxide remain on the sides of the flask, and then neutralize with 15 mL of 1 N sulfuric acid. Add 110 mL of water, and proceed as directed, beginning with "add 10 mL of 25% sulfuric acid...", except substitute "arsenic trioxide" for "arsenious acid solution" in the sentence that calls for a blank titration.

$$N = \frac{g\ As_2O_3}{0.04946 \times mL\ Ce^{IV}\ solution\ (corrected\ for\ the\ blank)}$$

EDTA, 0.1 M.   Assay a sample of (ethylenedinitrilo)tetraacetic acid disodium salt dihydrate as described on page 331. Correcting for the assay value obtained, weigh exactly enough of the material to contain 37.22 g of $Na_2EDTA \cdot 2H_2O$. Dissolve in 500 mL of water in a 1-L volumetric flask, dilute to the mark, and mix thoroughly.

Ferrous Ammonium Sulfate, 0.1 N.   Dissolve 40 g of ferrous ammonium sulfate hexahydrate in 200 mL of 25% sulfuric acid, dilute to 1 L with water, and mix thoroughly. On each day of use, standardize as follows: Pipet 25 mL into a 250-mL conical flask, add 0.1 mL of 1,10-phenanthroline reagent solution, and titrate with 0.1 N ceric ammonium sulfate volumetric solution to the change from pink to pale blue.

$$N = \frac{mL\ Ce^{IV} \times N\ Ce^{IV}}{mL\ Fe^{II}\ solution}$$

Hydrochloric Acid, 1 N.   Dilute 85 mL of hydrochloric acid to 1 L, and mix thoroughly. Standardize as follows: Weigh accurately about 2.2 g of sodium carbonate, Alkalimetric Standard, that previously has been heated at 285°C for 2 h. Dissolve in 100 mL of water, and add 0.1 mL of methyl red indicator solution. Add the acid slowly from a buret, with stirring, until the solution becomes faintly pink. Heat the solution to boiling, cool, and again titrate until the solution becomes faintly pink. Repeat this procedure until the faint color is no longer discharged on further boiling.

$$N = \frac{g\ Na_2CO_3}{0.05300 \times mL\ HCl\ solution}$$

Iodine, 0.1 N. Place approximately 40 g of potassium iodide and 10 mL of water in a large glass-stoppered weighing bottle. Mix the contents by swirling; allow to come to room temperature, and then weigh accurately. Add approximately 12.7 g of assayed iodine (page 383), previously weighed on a rough balance, and reweigh the stoppered bottle and contents to obtain the exact weight of the iodine added. Mix and transfer quantitatively to a 1-L volumetric flask, dilute to the mark, and mix thoroughly. For approximate work, calculate the normality from the weight and assay value of the iodine taken (atomic weight = 126.90). For more exact work, standardize as follows: Measure accurately 40 mL of 0.1 N arsenious acid volumetric solution into a 250-mL conical flask. Add 2 g of sodium bicarbonate dissolved in 20 mL of water, and titrate with the iodine solution until within about 2 mL of the anticipated equivalence point. Then add 3 mL of starch indicator solution, and continue the titration to the first permanent blue tinge.

$$N = \frac{\text{mL As}_2\text{O}_3 \times \text{N As}_2\text{O}_3}{\text{mL I}_2 \text{ solution}}$$

Perchloric Acid in Glacial Acetic Acid, 0.1 N. Add slowly, with stirring, 8.5 mL of 70% perchloric acid* to a mixture of 500 mL of glacial acetic acid and 21 mL of acetic anhydride. Cool, dilute to 1 L with glacial acetic acid, and mix thoroughly. Determine the water content of the solution by titrating 30 g with Karl Fischer reagent (page 52). If the water content is greater than 0.05%, add a small amount of acetic anhydride, mix thoroughly, and again determine the water. Repeat until the water content is between 0.02% and 0.05%. In a similar manner, if the original water content is below 0.02%, add water to bring the water content into the 0.02–0.05% range. Allow the solution to stand for 24 h, and check the water content to be certain that it remains in this range. Standardize as follows: Weigh accurately about 0.7 g of potassium hydrogen phthalate, Acidimetric Standard, previously lightly crushed and dried for 2 h at 120°C. Dissolve in 50 mL of glacial acetic acid, add 0.1 mL of crystal violet indicator solution, and titrate with the perchloric acid solution until the violet color changes to blue-green. Perform a blank titration on 50 mL of glacial acetic acid in the same manner, and subtract the

---

*Caution. Perchloric acid in contact with some organic materials can form explosive mixtures. Use safety goggles and rubber gloves when preparing the solution. Rinse any glassware that has been in contact with perchloric acid before setting it aside.

volume of perchloric acid solution consumed in the blank titration from that consumed in the first titration.

$$N = \frac{g\ KHC_8H_4O_4}{0.20423 \times mL\ HClO_4\ \text{solution (corrected for the blank)}}$$

Potassium Bromate, 0.1 N.   Weigh accurately 2.784 g of potassium bromate, transfer to a 1-L volumetric flask, dilute to volume, and mix thoroughly. Standardize as follows: Measure accurately 40 mL of the solution into a 250-mL glass-stoppered conical flask, add 3 g of potassium iodide and 3 mL of hydrochloric acid, stopper the flask, and allow to stand for 5 min. Titrate the liberated iodine with the 0.1 N sodium thiosulfate volumetric solution. Add 3 mL of starch indicator solution near the end of the titration, and continue to a colorless end point.

$$N = \frac{mL\ Na_2S_2O_3 \times N\ Na_2S_2O_3}{mL\ KBrO_3\ \text{solution}}$$

Potassium Dichromate, 0.1 N.   Weigh exactly 4.903 g of potassium dichromate, NIST Standard Reference Material, previously dried for 4 h at 120°C, transfer to a 1-L volumetric flask, dilute to volume, and mix thoroughly.

Potassium Hydroxide, Methanolic, 0.1 N.   Dissolve 6.8 g of potassium hydroxide in a minimum amount of water (about 5 mL), dilute to 1 L with methanol, and mix thoroughly. Allow to stand in a container protected from carbon dioxide for 24 h. Carefully decant the clear solution, leaving any residue behind, into a suitable container, mix well, and store protected from carbon dioxide. Standardize as follows: Weigh accurately about 0.85 g of potassium hydrogen phthalate, Acidimetric Standard, previously lightly crushed and dried for 2 h at 120°C. Dissolve in 100 mL of carbon dioxide-free water, add 0.15 mL of phenolphthalein indicator solution, and titrate with the methanolic potassium hydroxide solution to a permanent faint pink color.

$$N = \frac{g\ KHC_8H_4O_4}{0.20423 \times mL\ \text{of methanolic KOH solution}}$$

Potassium Iodate, 0.05 M. Weigh exactly 10.700 g of potassium iodate, previously dried at 110°C to constant weight, transfer to a 1-L volumetric flask, dilute to volume, and mix thoroughly.

Potassium Permanganate, 0.1 N. Dissolve 3.3 g of potassium permanganate in 1 L of water in a 1500-mL conical flask, heat to boiling, and boil gently for 15 min. Allow to cool, stopper, and keep in the dark for 48 h. Then filter through a fine-porosity sintered-glass filter and place in a glass-stoppered amber bottle. Standardize as follows: Weigh accurately 0.24–0.28 g of sodium oxalate, NIST Standard Reference Material, dissolve in 250 mL of water containing 7 mL of sulfuric acid, and heat to 70°C. Titrate immediately with the permanganate solution to a pink color that persists for 15 s. The temperature at the end of the titration should not be less than 60°C.

$$N = \frac{g\ Na_2C_2O_4}{mL\ KMnO_4\ solution \times 0.06700}$$

Potassium Thiocyanate, 0.1 N. Weigh exactly 9.7 g of potassium thiocyanate, previously dried for 2 h at 110°C, transfer to a 1-L volumetric flask, dilute to volume, and mix thoroughly. Standardize as follows: Pipet 25 mL of 0.1 N silver nitrate volumetric solution into a 250-mL beaker and add 100 mL of water, 1 mL of nitric acid, and 2 mL of ferric ammonium sulfate indicator solution. Titrate with the thiocyanate solution, with agitation, to a permanent light pinkish-brown color of the supernatant solution.

$$N = \frac{mL\ AgNO_3 \times N\ AgNO_3}{mL\ KSCN\ solution}$$

Silver Nitrate, 0.1 N. Weigh exactly 16.987 g of silver nitrate, previously dried for 1 h at 100°C, transfer to a 1-L volumetric flask, dilute to volume, and mix thoroughly. Standardize as follows: Weigh accurately about 0.33 g of potassium chloride (NIST Standard Reference Material No. 999), previously dried for 2 h at 110°C, place in a 250-mL beaker, and dissolve in 50 mL of water and 1 mL of nitric acid. Pipet 50 mL of the silver nitrate solution into the beaker, slowly, and with agitation. Heat to boiling, cool in the dark, and filter, washing the precipitate thoroughly with 1% nitric acid. To the filtrate add 2 mL of ferric ammonium sulfate indicator solution, and titrate the excess silver nitrate with 0.1 N potassium thiocyanate volumetric solution, with agitation, to a permanent light pinkish-brown color of the supernatant solution.

$$N = \frac{g \; KCl \times 1000}{74.55(50 - A)}$$

where A = mL of 0.1 N potassium thiocyanate volumetric solution.

Sodium Hydroxide, 1 N.   Dissolve 162 g of sodium hydroxide in 150 mL of carbon dioxide-free water, cool the solution to room temperature, and filter through a suitable filter, such as hardened filter paper. (Note: A reagent-grade 50% sodium hydroxide solution may be used in place of the solution prepared in this way.) Prepare a 1 N solution by diluting 54.5 mL of the concentrated solution to 1 L with carbon dioxide-free water. Mix thoroughly, and store in a tight polyolefin container. Standardize as follows: Weigh accurately about 8.5 g of potassium hydrogen phthalate, Acidimetric Standard, previously lightly crushed and dried for 2 h at 120°C. Dissolve in 100 mL of carbon dioxide-free water, add 0.15 mL of phenolphthalein indicator solution, and titrate with the sodium hydroxide solution to a permanent faint pink color.

$$N = \frac{g \; KHC_8H_4O_4}{0.20423 \times mL \; NaOH \; solution}$$

Sodium Methoxide in Methanol, 0.01 N.   Dilute 20.0 mL of sodium methoxide in methanol, 0.5 M methanolic solution, with methanol to 1 L, and mix thoroughly. Standardize as follows: Measure accurately 40 mL of 0.01 N sulfuric acid volumetric solution into a suitable container, add 0.10 mL of thymol blue indicator solution, and titrate with the sodium methoxide in methanol solution to a blue end point.

$$N = \frac{mL \; H_2SO_4 \times N \; H_2SO_4}{mL \; CH_3ONa \; solution}$$

Sodium Nitrite, 0.1 N.   Assay a sample of sodium nitrite as described on page 685. Correcting for the assay value obtained, weigh exactly enough of the material to contain 6.900 g of $NaNO_2$. Dissolve in 100 mL of water in a 1-L volumetric flask, dilute to volume, and mix thoroughly.

Sodium Thiosulfate, 0.1 N.   Dissolve 25 g of sodium thiosulfate pentahydrate and 200 mg of sodium carbonate in 1 L of recently boiled and cooled water, and mix thoroughly. Allow to stand for 24 h, and then standardize as follows: Measure accurately 40 mL of 0.1 N iodine volumetric solution that has been freshly standardized against 0.1 N arsenious acid volumetric solution (page 89), into 75 mL of water in a 250-mL glass-stoppered Erlenmeyer flask. Titrate with the thiosulfate solution to a light yellow color, add 3 mL of starch indicator solution, and continue to a colorless end point.

$$N = \frac{mL\ I_2 \times N\ I_2}{mL\ Na_2S_2O_3\ solution}$$

Sulfuric Acid, 1 N.   Add slowly (caution), with stirring, 30 mL of sulfuric acid to about 500 mL of water. Allow the mixture to cool to room temperature, dilute to 1 L, and mix thoroughly. Standardize against sodium carbonate, Alkalimetric Standard, as directed for hydrochloric acid, 1 N.

Zinc Chloride, 0.1 M.   Weigh accurately 6.537 g of zinc and dissolve in 80 mL of 10% hydrochloric acid. Warm if necessary to complete dissolution, cool, dilute with water to volume in a 1-L volumetric flask, and mix thoroughly.

REGISTERED TRADEMARKS

The tests in Reagent Chemicals employ some products having registered trademarks. These are listed below with the corporate owners.

Amberlite (Rohm and Haas Corp.)
Carbowax (Union Carbide Chemical and Plastics Inc.)
Chromosorb (Johns-Mansville Corp.)
Dowex (Dow Chemical Co.)
Drierite (W. Hammond Drierite Co.)
Eriochrome (Ciba-Geigy Corp.)
Flexol (Union Carbide Chemicals and Plastics Inc.)
GasChrom (Altex Associates, Inc.)
Igepal (GAF Corp.)
Porapak (Waters Associates, Inc.)
Regisil (Regis Chemical Co.)
Supelcoport (Supelco, Inc.)
Teflon (E.I. du Pont de Nemours & Co., Inc.)
Tenax (Enka Research Institute)
Triton (Union Carbide Chemicals and Plastics Inc.)

SPECIFICATIONS FOR REAGENT CHEMICALS

REQUIREMENTS AND TESTS

Acetaldehyde
Ethanal

CH_3CHO Formula Wt 44.05

CAS Number 75 –07 –0

REQUIREMENTS

Assay (GC) . . . . $\geq$ 99.5% CH_3CHO

MAXIMUM ALLOWABLE

Residue after evaporation 0.005%
Titrable acid . . . . 0.008 meq / g

TESTS

ASSAY. Analyze the sample by gas chromatography by using the general parameters cited on page 56. The following specific conditions are also required.

Column: Type I: methyl silicone

Detector: Flame ionization

Measure the area under all peaks and calculate the area percent for acetaldehyde.

RESIDUE AFTER EVAPORATION.   (page 14). Evaporate 40 g (51 mL) to dryness in a tared platinum dish on the steam bath in a well-ventilated hood, and dry the residue at 105°C for 30 min.

TITRABLE ACID. Chill 25 g (32 mL) of sample in a graduated cylinder in an ice bath. To a 250-mL Erlenmeyer flask add about 25 mL of deionized water and 75 g of deionized ice. Add 0.5 mL of phenolphthalein indicator solution and titrate with 0.1 N sodium hydroxide to the first perceptible pink color that persists for 15 s. Working in a fume hood, add the chilled sample and titrate immediately to the same faint pink color. Not more than 2.0 mL of 0.1 N sodium hydroxide should be required.

Acetic Acid, Glacial

CH_3COOH Formula Wt 60.05

CAS Number 64-19-7

REQUIREMENTS

Assay . . . . $\geq$ 99.7%
CH_3COOH

MAXIMUM ALLOWABLE

Color (APHA). 10
Dilution test . . . . Passes test
Residue after evaporation 0.001%
Acetic anhydride [$(CH_3CO)_2O$]. 0.01%
Chloride (Cl) . . . . 1 ppm
Sulfate (SO_4). . . . . 1 ppm
Heavy metals (as Pb) . . . . 0.5 ppm
Iron (Fe) . . . . 0.2 ppm
Substances reducing dichromate . . . . Passes test
Substances reducing permanganate . . . . Passes test
Titrable base . . . . 0.0004 meq / g

TESTS

ASSAY.   (Page 20). The freezing point should not be below 16.0°C.

COLOR (APHA).   (Page 17).

DILUTION TEST.   Dilute 1 volume of the acid with 3 volumes of water and allow to stand for 1 h. The solution should be as clear as an equal volume of water.

RESIDUE AFTER EVAPORATION.   (Page 14). Evaporate 100 g (95 mL) to dryness in a tared dish on the steam bath and dry the residue at 105°C for 30 min.

ACETIC ANHYDRIDE.   Place 52.5 g (50 mL) of sample in a 250-mL titration flask. In a second flask place 50 mL of glacial acetic acid known to be free from acetic anhydride. Into each flask pipet 10 mL of 1.0% solution of 4,4'-methylenedianiline (4,4'-diaminodiphenyl-methane) in glacial acetic acid and add 0.10 mL of 1.0% solution of crystal violet in glacial acetic acid. Titrate each solution with a 0.1 N solution of perchloric acid in glacial acetic acid to a green end point. Subtract the volume for the titration of the sample from the volume for the other titration. One milliliter of 0.1 N perchloric acid corresponds to 0.0194% $(CH_3CO)_2O$ for a 50-mL sample.

Reagents

Glacial Acetic Acid.   Glacial acetic acid suspected of containing anhydride may be purified by adding 0.50 mL of water per 100 mL and digesting overnight in a glass-stoppered flask, slightly ajar, on low heat. If this acid is used for comparison in the test described, 0.25 mL of water should be added to the flask containing the test sample just before the end point because water affects the indicator change slightly.

4,4'-Methylenedianiline.   Dissolve 2.50 g of 4,4'-methylene-dianiline (colorless or only slightly colored) in glacial acetic acid to make 250 mL. Protect the solution from light.

CHLORIDE.   Dilute 10 g (9.5 mL) with 10 mL of water and add 1 mL of silver nitrate reagent solution. Prepare a standard containing 0.01 mg of chloride ion (Cl) in 20 mL of water and add 1 mL of silver nitrate reagent solution. Evaporate the solutions to dryness on the steam bath. Dissolve the residues with 0.5 mL of ammonium hydroxide, dilute with 20 mL of water, and add 1.5 mL of nitric acid. Any turbidity in the solution of the sample should not exceed that of the standard.

SULFATE. (Page 32, Procedure B). Transfer 20 g (19 mL) to a reaction tube, add 1 mL of 0.1 N potassium permanganate, and evaporate to dryness on a hot plate at low setting. Add 1 mL of hydrochloric acid to the residue, digest on a hot plate until dissolved, and continue with the procedure.

HEAVY METALS. (Page 26, Method 1). To 40 g (38 mL) add about 10 mg of sodium carbonate, evaporate to dryness on the steam bath, dissolve the residue in about 20 mL of water, and dilute with water to 25 mL.

IRON. (Page 28, Method 1). To 50 g (48 mL) add 10 mg of sodium carbonate and evaporate to dryness. Dissolve the residue in 2 mL of hydrochloric acid, dilute with water to 50 mL, and use the solution without further acidification.

SUBSTANCES REDUCING DICHROMATE. To 10 mL add 1.0 mL of 0.1 N potassium dichromate and cautiously add 10 mL of sulfuric acid. Cool the solution to room temperature and allow to stand for 30 min. While the solution is swirled, dilute slowly and cautiously with 50 mL of water, cool, and add 1 mL of freshly prepared potassium iodide reagent solution. Titrate the liberated iodine with 0.1 N thiosulfate, using starch as the indicator. Not more than 0.40 mL of the 0.1 N potassium dichromate should be consumed (not less than 0.60 mL of the 0.1 N thiosulfate should be required). Correct for a complete blank.

SUBSTANCES REDUCING PERMANGANATE. Add 40 mL of the sample to 10 mL of water. Cool to 15°C, add 0.30 mL of 0.1 N potassium permanganate, and allow to stand at 15°C for 10 min. The pink color should not be entirely discharged.

TITRABLE BASE. To 25 mL add 0.10 mL of a solution of 1 g of methyl violet (or crystal violet) in 100 mL of glacial acetic acid. The color should be violet. Titrate the solution with 0.1 N perchloric acid in glacial acetic acid to a green color. Not more than 0.10 mL should be required.

Acetic Anhydride

$(CH_3CO)_2O$

$$CH_3-\overset{\overset{\textstyle O}{\|}}{C}-O-\overset{\overset{\textstyle O}{\|}}{C}-CH_3$$

Formula Wt 102.09

CAS Number 108–24–7

REQUIREMENTS

Assay . . . . $\geq$ 97.0%
$(CH_3CO)_2O$

MAXIMUM ALLOWABLE

Residue after evaporation 0.003%
Chloride (Cl) . . . . 5 ppm
Phosphate (PO_4) 0.001%
Sulfate (SO_4) . . . . 5 ppm
Heavy metals (as Pb) . . . . 2 ppm
Iron (Fe) . . . . 5 ppm
Substances reducing permanganate . . . . Passes test

TESTS

ASSAY. (By acid–base titration). Carefully transfer 50.0 mL of morpholine methanolic solution into each of two 250-mL glass-stoppered flasks. Weigh accurately the first flask for the sample. Immediately add 2.0 mL of the acetic anhydride from a pipet, reweigh the sample flask, and swirl to effect dissolution. Reserve the second flask for the determination of the quantity of morpholine mixed with the sample. Allow the flasks to stand at room temperature for 5 min. Add 0.35–0.40 mL of bis-4-azo-brilliant yellow indicator. Titrate each solution with 0.5 N hydrochloric acid methanolic solution to the end point at which the reddish orange color changes to purple. Calculate the assay value as follows:

$$\%(CH_3CO)_2O = 10.209\frac{(V_2 - V_1)N}{W}$$

where
V_1 = volume, in mL, of 0.5 N HCl added to the first flask
V_2 = volume, in mL, of 0.5 N HCl added to the second flask
N = normality of HCl methanolic solution
W = weight, in grams, of sample

Reagents

Morpholine Methanolic Solution, 0.5 N.   Transfer 44 mL of redistilled morpholine to a 1-L reagent bottle and dilute to 1 L with reagent methanol. To facilitate removal of aliquots, fit the bottle with a two-hole rubber stopper and insert a 50-mL pipet through one hole so that the tip dips below the surface of the liquid. Through the other hole insert a short piece of glass tubing to which is attached a rubber atomizer bulb.

Bis-4-azo Brilliant Yellow Indicator Solution.   Weigh 0.050 g of 4,4'-bis(4-amino-1-naphthylazo)-2,2'-stilbenedisulfonic acid and 0.01 g of brilliant yellow into a 60-mL vial. Pipet 1.5 mL of 0.1 N sodium hydroxide into the vial, stir well, add 3.5 mL of water, and stir again. Transfer the mixture to a storage bottle, rinsing the vial with 45 mL of methanol. Add the rinsings to the bottle, cap, and shake to mix thoroughly. For details, see *Anal. Chem.* **1975**, *47*, 2057.

Hydrochloric Acid Methanolic Solution, 0.5 N.   Transfer 84 mL of 6 N hydrochloric acid to a 1-L volumetric flask and dilute to volume with reagent methanol. Standardize daily against 0.5 N sodium hydroxide by using phenolphthalein indicator.

RESIDUE AFTER EVAPORATION.   (Page 14). Evaporate 50 g (46 mL) in a tared dish on the steam bath, and dry the residue at 105°C for 30 min.

Sample Solution A.   Dilute 40 g (37 mL) of the sample with water to 200 mL (1 mL = 0.2 g).

CHLORIDE.   (Page 25). Use 10 mL of sample solution A (2-g sample).

PHOSPHATE.   (Page 30, Method 1). Evaporate 10 mL of sample solution A (2-g sample) to dryness on the steam bath, and dissolve the residue in approximately 0.5 N sulfuric acid.

SULFATE.   (Page 32, Procedure A, Method 3). Use 50 mL of sample solution A (10-g sample).

HEAVY METALS.   (Page 26). To 50 mL of sample solution A (10-g sample) add about 10 mg of sodium carbonate, evaporate to dryness on the steam bath, dissolve the residue in about 20 mL of water, and dilute with water to 25 mL.

IRON.   (Page 28, Method 1). To 10 mL of sample solution A (2-g sample) add 10 mg of sodium carbonate and evaporate to dryness.

Dissolve the residue in 2 mL of hydrochloric acid, dilute with water to 50 mL, and use the solution without further acidification.

SUBSTANCES REDUCING PERMANGANATE. To 10 mL of sample solution A (2-g sample) add 0.4 mL of 0.1 N potassium permanganate and allow to stand for 5 min. The pink color should not be entirely discharged.

Acetone
2-Propanone

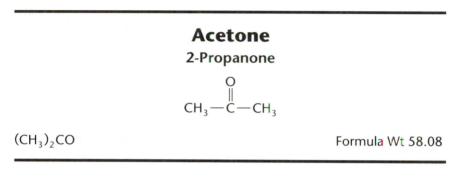

$(CH_3)_2CO$ Formula Wt 58.08

CAS Number 67–64–1

Suitable for use in ultraviolet spectrophotometry or general use. Product labeling shall designate the one or more of those uses for which suitability is represented on the basis of meeting the relevant requirements and tests. The ultraviolet spectrophotometry requirements include all of the requirements for general use.

REQUIREMENTS

General Use

Assay . . . . ≥ 99.5% $(CH_3)_2CO$

MAXIMUM ALLOWABLE

Color (APHA). 10
Residue after evaporation 0.001%
Solubility in water . . . . Passes test
Titrable acid . . . . 0.0003 meq / g
Titrable base . . . . 0.0006 meq / g
Aldehyde (as HCHO) 0.002%
Isopropyl alcohol 0.05%
Methanol 0.05%
Substances reducing permanganate . . . . Passes test
Water 0.5%

Specific Use

ULTRAVIOLET SPECTROPHOTOMETRY

Wavelength (nm)	Absorbance (AU)
400	0.01
350	0.02
340	0.10
330	1.00

TESTS

ASSAY, ISOPROPYL ALCOHOL, AND METHANOL. Analyze the sample by gas chromatography by using the general parameters cited on page 56. The following specific conditions are also required.

   Column: Type I, methyl silicone

Measure the area under all peaks and calculate the area percent for acetone, isopropyl alcohol, and methanol. Correct for water content.

COLOR (APHA).   (Page 17).

RESIDUE AFTER EVAPORATION.   (Page 14). Evaporate 100 g (125 mL) to dryness in a tared dish on the steam bath and dry the residue at 105°C for 30 min.

SOLUBILITY IN WATER.   Mix 30 g (38 mL) with 38 mL of carbon dioxide-free water. The solution should remain clear for 30 min. Reserve this solution for the test for titrable acid.

TITRABLE ACID.   Add 0.10 mL of phenolphthalein indicator solution to the solution prepared in the preceding test. Not more than 1.0 mL of 0.01 N sodium hydroxide should be required to produce a pink color.

TITRABLE BASE.   Mix 18 g (23 mL) with 25 mL of water and add 0.05 mL of methyl red indicator solution. Not more than 1.0 mL of 0.01 N hydrochloric acid should be required to produce a red color.

ALDEHYDE.   Dilute 2 g (2.5 mL) with water to 10 mL. Prepare a standard containing 0.04 mg of formaldehyde in 10 mL of water and a complete reagent blank in 10 mL of water alone. To each add 0.15 mL of a freshly prepared 5% solution of 5,5-dimethyl-1,3-cyclohexanedione in alcohol. Evaporate each on the steam bath until the acetone is volatilized. Dilute each with water to 10 mL and cool quickly in an ice bath while stirring vigorously. Any turbidity in the solution of the sample should not exceed that in the standard. If any turbidity develops in the reagent blank, the 5,5-dimethyl-1,3-cyclohexanedione should be discarded and the entire test rerun with a fresh solution.

SUBSTANCES REDUCING PERMANGANATE. Add 0.05 mL of 0.1 N potassium permanganate to 10 mL of the sample and allow to stand for 15 min at 25°C. The pink color should not be entirely discharged, when compared with an equal volume of water.

WATER. (Page 54, Method 1). Use 25 mL (20 g) of the sample and a methanol-free system to prevent ketal formation with liberation of water.

ULTRAVIOLET SPECTROPHOTOMETRY. Use the procedure on page 67 to determine the absorbance.

Acetonitrile

CH_3CN Formula Wt 41.05

CAS Number 75–05–8

Suitable for use in high-performance liquid chromatography, extraction–concentration analysis, ultraviolet spectrophotometry, or general use. Product labeling shall designate the one or more of these uses for which suitability is represented on the basis of meeting the relevant requirements and tests. The ultraviolet spectrophotometry and liquid chromatography suitability requirements include all of the requirements for general use. The extraction–concentration suitability requirements include only the general use requirement for color.

REQUIREMENTS

General Use

Appearance. . . . . Clear
Assay (CH_3CN). . . . . ≥ 99.5%

MAXIMUM ALLOWABLE

Color (APHA). 10
Residue after evaporation 0.005%
Titrable acid . . . . 8 μeq / g
Titrable base . . . . 0.6 μeq / g
Water 0.3%

Specific Use

ULTRAVIOLET SPECTROPHOTOMETRY

Wavelength (nm) . . . . Absorbance (AU)
254 . . . . 0.01
220 . . . . 0.05
190 . . . . 1.00

LIQUID CHROMATOGRAPHY SUITABILITY

Absorbance . . . . Passes Test
Gradient elution . . . . Passes Test

EXTRACTION–CONCENTRATION SUITABLITY

Absorbance . . . . Passes Test
GC–FID . . . . Passes Test
GC–ECD . . . . Passes Test

TESTS

ASSAY.   Analyze the sample by gas chromatography using the general parameters cited on page 56. The following specific conditions are also required.

Column: Type I, methyl silicone

Measure the area under all peaks and calculate the acetonitrile content in area percent. Correct for water content.

COLOR (APHA).   (Page 17).

RESIDUE AFTER EVAPORATION.   (Page 14). Evaporate 39 g (50 mL) to dryness in a tared platinum dish on the steam bath, in a well-ventilated hood, and dry the residue at 105°C for 30 min.

TITRABLE ACID.   Mix 10 g (13 mL) with 13 mL of carbon dioxide-free water. Add 0.10 mL of phenolphthalein indicator solution. Not more than 8.0 mL of 0.01 N sodium hydroxide solution should be required in the titration.

TITRABLE BASE.   To 78 g (100 mL) in a 250-mL conical flask, add 0.30 mL of 0.05% solution of bromcresol green in alcohol and 0.10 mL of methyl red indicator solution. Titrate with 0.01 N hydrochloric acid to a light orange-pink end point. Not more than 4.5 mL of 0.01 N hydrochloric acid should be required.

WATER.   (Page 54, Method 1). Use 10 mL (7.8 g) of the sample.

ULTRAVIOLET SPECTROPHOTOMETRY.   Use the procedure on page 67 to determine the absorbance.

LIQUID CHROMATOGRAPHY SUITABILITY.   Analyze the sample by using the gradient elution procedure cited on page 64. Use the procedure on page 67 to determine the absorbance.

EXTRACTION–CONCENTRATION SUITABILITY.   Analyze the sample by using the general procedure cited on page 65. In the sample preparation step exchange the sample with ACS grade 2,2,4-trimethylpentane suitable for extraction–concentration. Use the procedure on page 67 to determine the absorbance.

Acetyl Chloride
Ethanoic Acid Chloride

CH_3COCl Formula Wt 78.50

CAS Number 75–36–5

REQUIREMENTS

Assay (GC) . . . . $\geq$ 98.5% CH_3COCl

MAXIMUM ALLOWABLE

Color (APHA). 20
Residue after evaporation 0.005%
Insoluble matter 0.0025%
Phosphate (PO_4) 0.002%
Heavy metals (as Pb) . . . . 5 ppm
Iron (Fe) . . . . 5 ppm

TESTS

ASSAY. Analyze the sample by gas chromatography using the general parameters cited on page 56. The following specific conditions are also required.

Column: Type I, methyl silicone

Detector: Flame ionization

Column Temperature: Initial 75°C for 5 min, then increased at a rate of 10°C per minute to a final temperature of 200°C.

Approximate Retention Times (min): Acetaldehyde, 2.5; acetyl chloride, 3.5; acetic acid, 4.5. Measure the area under all peaks and calculate the area percent for acetyl chloride.

COLOR (APHA). (Page 17).

RESIDUE AFTER EVAPORATION. (Page 14). Evaporate 20 g (18 mL) to dryness in a tared dish on the steam bath and dry the residue at 105°C for 30 min.

INSOLUBLE MATTER. (Page 14). Add 40 g (36 mL) cautiously, with stirring, to 150 mL of water and dilute to 200 mL with water. The solution is clear and colorless. Continue with the procedure as described on page 14. Retain the filtrate separate from the washings for the tests for phosphate, heavy metals, and iron.

PHOSPHATE. (Page 30, Method 1). To 5 mL of the solution from the test for insoluble matter (1.0-g sample) add 5 mg of anhydrous sodium carbonate and evaporate to dryness. Take up the residue in 20 mL of 0.5 N H_2SO_4. For the standard, use 0.02 mg of PO_4 and 5 mg of Na_2CO_3, and treat exactly as the sample.

HEAVY METALS. (Page 27, Method 1). Use 20 mL (4-g sample) of the solution retained from the test for insoluble matter.

IRON. (Page 28, Method 1). Use 10 mL (2.0-g sample) of the solution retained from the test for insoluble matter.

Aluminum

Al   Atomic Wt 26.98

CAS Number 7429–90–5

REQUIREMENTS

MAXIMUM ALLOWABLE

Insoluble in dilute hydrochloric acid 0.05%
Silicon (Si). 0.1%
Nitrogen compounds (as N) 0.001%
Copper (Cu) 0.02%
Iron (Fe) 0.1%
Manganese (Mn) 0.002%
Titanium (Ti) 0.03%

TESTS

INSOLUBLE IN DILUTE HYDROCHLORIC ACID. Suspend 10 g in 100 mL of water, heat to 80°C, and add dropwise 100 mL of hydrochloric acid to dissolve the aluminum. Filter through a tared filtering crucible. Retain the filtrate, without washings, for sample solution A. Wash thoroughly, rejecting the washings, and dry at 105°C.

> *Sample Solution A for the Determination of Iron, Manganese, and Titanium.* Dilute the filtrate, without washings, obtained in the test for insoluble in dilute hydrochloric acid with water to 250 mL (1 mL = 0.04 g).

SILICON. *NOTE.* The molybdenum blue test for silicon is extremely sensitive; therefore, all solutions that are more alkaline than pH 4 should be handled in containers other than glass or porcelain. Platinum dishes and plastic graduated cylinders are recommended. The 10% solution of sodium hydroxide is prepared and measured in plastic.

Dissolve 0.25 g in 10 mL of 10% sodium hydroxide solution. Add the aluminum to the sodium hydroxide solution in small portions, particularly if the sample is finely divided. When the vigorous reaction has subsided, digest the covered container on the steam bath until dissolution is complete. Uncover and evaporate to a moist residue. Dissolve the residue in 10 mL of water, and add 0.05–0.10 mL of 30% hydrogen peroxide. Boil the solution gently to decompose the excess peroxide, cool, and dilute with water to 100 mL. Adjust the acidity of the solution to pH 2 (external indicator) by adding dilute sulfuric acid (1 + 1). Digest on the steam bath until any precipitate is dissolved and the solution is clear. If necessary, add more of the dilute sulfuric acid (1 + 1). When dissolution is complete, cool and dilute with water to 250 mL. Prepare a blank solution by treating 10 mL of the 10% sodium hydroxide solution exactly as described above, but omit the 0.25-g sample. Dilute 5.0 mL of the sample solution with water to 80 mL. For the standard add 0.01 mg of SiO_2 (0.005 mg of Si) to 5 mL of the blank solution and dilute with water to 80 mL. To each solution add 5.0 mL of a freshly prepared 10% solution of ammonium molybdate. Adjust the pH of each solution to 1.7–1.9 (using a pH meter) with dilute hydrochloric acid (1 + 9) or silica-free ammonium hydroxide. Heat the solutions just to boiling, cool, add 20 mL of hydrochloric acid, and dilute with water to 110 mL. Transfer the solutions to separatory funnels, add 50 mL of butyl alcohol, and shake vigorously. Allow the layers to separate and draw off and discard the aqueous phase. Wash the butyl alcohol three times with 20-mL portions of dilute hydrochloric acid (1 + 99), discarding each aqueous phase. Add 0.5 mL of a freshly prepared 2% solution of stannous chloride in hydrochloric acid and shake. If the butyl alcohol is turbid, wash it with 10 mL of dilute hydrochloric acid (1 + 99). Any blue color in the butyl alcohol from the solution of the sample should not exceed that in the butyl alcohol from the standard.

NITROGEN COMPOUNDS.   (Page 30). Use 2.5 g dissolved in 60 mL of freshly boiled 10% sodium hydroxide. For the standard use 0.5 g of aluminum and 0.02 mg of nitrogen (N).

COPPER.   Dissolve 0.25 g in 10 mL of dilute hydrochloric acid (1 + 1) by digesting on the steam bath. Cool and add 1 mL of 30% hydrogen peroxide. Let stand for 5 min without heating, then boil for 12 min. Add about 150 mL of water, adjust the pH to 1.0–1.5 (using a pH meter) with 1 N sodium hydroxide (about 1 mL will be required), and dilute with water to 200 mL. To 20 mL of this solution in a separatory funnel add 5 mL of standard dithizone solution in carbon tetrachloride. Shake for 2 min, and draw off the carbon tetrachloride layer. Any pink color in the carbon tetrachloride extract should not exceed that produced by a standard containing 0.005 mg of copper ion (Cu) in 20 mL of 0.05 N hydrochloric acid treated exactly like the solution of the sample.

IRON. (Page 28, Method 1). Dilute 5.0 mL of sample solution A (0.2-g sample) with water to 100 mL, and use 5.0 mL of this dilution.

MANGANESE. To 12.5 mL of sample solution A (0.5-g sample) add 20 mL of nitric acid and 10 mL of sulfuric acid. Prepare a standard containing 0.01 mg of manganese ion (Mn) in 7.5 mL of water plus 5 mL of hydrochloric acid, 20 mL of nitric acid, and 10 mL of sulfuric acid. Evaporate each solution to dense fumes of sulfur trioxide, cool, and cautiously add 50 mL of water, 20 mL of nitric acid, and 5 mL of phosphoric acid. Dilute with water to 100 mL and boil gently for 5 min. Cool, add 0.25 g of potassium periodate to each, and again boil gently for 5 min. Cool and dilute with water to 100 mL. Any pink color in the solution of the sample should not exceed that in the standard.

TITANIUM. To 25 mL of sample solution A (1-g sample) add 3 mL of phosphoric acid and 0.5 mL of 30% hydrogen peroxide. Any color should not exceed that produced by 0.3 mg of titanium ion (Ti) in an equal volume of solution containing the quantities of reagents used in the test.

Aluminum Ammonium Sulfate Dodecahydrate
Ammonium Alum

$AlNH_4(SO_4)_2 \cdot 12H_2O$ Formula Wt 453.33

CAS Number 7784–26–1

REQUIREMENTS

Assay 98.0–102.0%
$AlNH_4(SO_4)_2 \cdot 12H_2O$

MAXIMUM ALLOWABLE

Insoluble matter 0.005%
Chloride (Cl) 0.001%
Calcium (Ca) 0.05%
Heavy metals (as Pb) 0.001%
Iron (Fe) 0.001%
Potassium (K) 0.05%
Sodium (Na) 0.01%

TESTS

ASSAY. (By complexometry). Weigh accurately 0.8 g and transfer to a 400-mL beaker. Moisten with 1 mL of glacial acetic acid, add 50 mL of water, 40.0 mL of standard 0.1 M EDTA, and 20 mL of ammonium acetate–acetic acid buffer solution. Warm on a steam bath until solution is complete, and boil gently for 5 min. Cool, add 50 mL of alcohol and 2 mL of dithizone indicator solution, and titrate with standard 0.1 M zinc chloride to a bright rose-pink color. Perform a blank titration of 40.0 mL of standard 0.1 M EDTA, using the same quantities of reagents as for the assay. One milliliter of 0.1 M EDTA consumed by the sample corresponds to 0.04533 g of $AlNH_4(SO_4)_2 \cdot 12H_2O$.

INSOLUBLE MATTER. (Page 14). Use 10 g dissolved in 150 mL of water containing 5 mL of 25% sulfuric acid.

CHLORIDE. (Page 25). Use 1.0 g.

HEAVY METALS. (Page 26, Method 1). Dissolve 2.0 g in 20 mL of water.

CALCIUM, IRON, POTASSIUM, AND SODIUM. Determine the calcium, iron, potassium, and sodium content by the flame atomic absorption spectrophotometric method described on page 35.

> ***Sample Stock Solution.*** Dissolve 10.0 g in 100 mL of water (1 mL = 0.1 g).

Analyze the solutions by means of a suitable atomic absorption spectrophotometer using the conditions outlined in the following table. Calculate the metal content of the sample by the method of standard additions.

Element	Wavelength (nm)	Sample Wt (g)	Standard Added (mg)	Flame Type*	Background Correction
Ca	422.7	0.2	0.05; 0.10	N/A	No
Fe	248.3	5.0	0.025; 0.05	A/A	Yes
K	766.5	0.1	0.025; 0.05	A/A	No
Na	589.0	0.2	0.01; 0.02	A/A	No

*A/A is air/acetylene; N/A is nitrous oxide/acetylene.

Aluminum Nitrate Nonahydrate

$Al(NO_3)_3 \cdot 9H_2O$ Formula Wt 375.13

CAS Number 7784−27−2

REQUIREMENTS

Assay . . . . 98.0−102.0%
$Al(NO_3)_3 \cdot 9H_2O$

MAXIMUM ALLOWABLE

Insoluble matter 0.005%
Chloride (Cl) 0.001%
Sulfate (SO_4) 0.005%
Substances not precipitated by
ammonium hydroxide (as SO_4) 0.05%
Heavy metals (as Pb) 0.001%
Iron (Fe) 0.002%

TESTS

ASSAY. (By indirect complexometric titration of Al^{III}). Weigh accurately 5 g and transfer to a 250-mL volumetric flask. Dissolve in about 75 mL of water, add 5 mL of hydrochloric acid, dilute to volume with water, and mix. Place a 25.0-mL aliquot of this solution in a 250-mL beaker, add 50 mL of water, and mix. While stirring, add 40.0 mL of standard 0.1 M EDTA and 20 mL of ammonium acetate−acetic acid buffer solution. Heat the solution to near boiling for 5 min and cool. (The pH should be 4.0−5.0 and should be maintained in that range by adding more buffer during the titration if necessary.) Add 50 mL of alcohol and 2 mL of dithizone indicator solution and titrate the excess of EDTA with 0.1 M zinc sulfate (28.75 g of zinc sulfate heptahydrate diluted to 1 L with water) until the color changes from a green-violet to a rose-pink. Similarly, titrate 40.0 mL of the standard 0.1 M EDTA with 0.1 M zinc sulfate, substituting 25 mL of water for the sample solution. Let V_s and V_b be the volumes of the 0.1 M zinc sulfate required in the two titrations; let V_e be the volume in milliliters and M_e be the molarity of the standard EDTA solution; and let W be the weight of sample, in grams, corresponding to the aliquot taken.

$$\%Al(NO_3)_3 \cdot 9H_2O = [1 - (V_s/V_b)](V_e M_e)(37.51/W)$$

INSOLUBLE MATTER. (Page 14). Dissolve 20 g in 150 mL of water and filter immediately without digesting.

CHLORIDE. (Page 25). Use 1.0 g.

SULFATE. Dissolve 4.0 g in 10 mL of hydrochloric acid and evaporate to dryness on a hot plate at low setting. Take up the residue in 10 mL of hydrochloric acid and again evaporate to dryness. Dissolve the residue in 100 mL of water, add a slight excess of ammonium hydroxide, and boil the solution until the odor of ammonia is entirely gone. Dilute with water to 200 mL, filter, and evaporate 50 mL of the filtrate to 10 mL. Retain the remainder of the filtrate for the substances not precipitated by the ammonium hydroxide test. To the evaporated filtrate add 1 mL of dilute hydrochloric acid (1 + 19) and 1 mL of barium chloride reagent solution. Any turbidity should not exceed that produced by 0.05 mg of sulfate ion (SO_4) in an equal volume of solution containing 1 mL of dilute hydrochloric acid (1 + 19) and 1 mL of barium chloride reagent solution. Compare 10 min after adding the barium chloride to the sample and standard solutions.

SUBSTANCES NOT PRECIPITATED BY AMMONIUM HYDROXIDE. To 100 mL of the filtrate retained from the test for sulfate, add 0.5 mL of sulfuric acid and evaporate the solution to dryness. Ignite the residue at 800 ± 25°C for 15 min.

HEAVY METALS. (Page 26, Method 1). Dissolve 4.0 g in 40 mL of water, and dilute with water to 48 mL. Use 36 mL to prepare the sample solution, and use the remaining 12 mL to prepare the control solution.

IRON. (Page 28, Method 1). Dissolve 1.0 g in 50 mL of water, and use 25 mL of this solution.

Aluminum Potassium Sulfate Dodecahydrate

$AlK(SO_4)_2 \cdot 12H_2O$ Formula Wt 474.39

CAS Number 7784–24–9

REQUIREMENTS

Assay 98.0–102.0%
$$AlK(SO_4)_2 \cdot 12H_2O$$

MAXIMUM ALLOWABLE

Insoluble matter 0.005%
Chloride (Cl) . . . . 5 ppm
Ammonium (NH$_4$) 0.005%
Arsenic (As) . . . . 2 ppm
Heavy metals (as Pb) 0.001%
Iron (Fe) 0.001%
Sodium (Na) 0.02%

TESTS

ASSAY.   (By indirect complexometric titration of Al^{III}). Weigh accurately 10 g and transfer to a 250-mL volumetric flask. Dissolve in about 75 mL of water, add 5 mL of hydrochloric acid, dilute to volume with water, and mix. Place a 25.0-mL aliquot of this solution in a 250-mL beaker, add 50 mL of water, and mix. While stirring, add 40.0 mL of standard 0.1 M EDTA and 20 mL of ammonium acetate–acetic acid buffer solution. Heat the solution to near boiling for 5 min and cool. (The pH should be 4.0–5.0 and should be a maintained in that range by adding more buffer during the titration if necessary.) Add 50 mL of alcohol and 2 mL of dithizone indicator solution and titrate the excess of EDTA with 0.1 M zinc sulfate (28.75 g of zinc sulfate heptahydrate diluted to 1 L with water) until the color changes from a green-violet to a rose-pink. Similarly, titrate 40.0 mL of the standard 0.1 M EDTA with 0.1 M zinc sulfate, substituting 25 mL of water for the sample solution. Let V_s and V_b be the volumes of the 0.1 M zinc sulfate required in the two titrations; let V_e be the volume in milliliters and M_e be the molarity of the standard EDTA solution; and let W be the weight of sample, in grams, corresponding to the aliquot taken.

$$\%AlK(SO_4)_2 \cdot 12H_2O = [1 - (V_s/V_b)](V_e M_e)(47.44/W)$$

INSOLUBLE MATTER.   (Page 14). Use 20 g dissolved in 150 mL of hot water.

CHLORIDE.   (Page 25). Use 2.0 g dissolved in warm water.

AMMONIUM.   Dissolve 1.0 g in 100 mL of ammonia-free water. To 20 mL of this solution add 10% sodium hydroxide reagent solution until the precipitate first formed is redissolved. Dilute with water to 50 mL and add 2 mL of Nessler reagent. Any color should not exceed that produced by 0.01 mg of ammonium ion (NH_4) in an equal volume of solution containing the quantities of reagents used in the test.

ARSENIC. (Page 23). Use 1.5 g. For the standard use 0.003 mg of arsenic (As).

HEAVY METALS. (Page 26, Method 1). Dissolve 5.0 g in 40 mL of water, and dilute with water to 50 mL. Use 30 mL to prepare the sample solution, and use 10 mL to prepare the control solution.

IRON. (Page 28, Method 1). Use 1.0 g.

SODIUM. Determine the sodium content by the flame atomic absorption spectrophotometric method described on page 35.

> *Sample Stock Solution.* Dissolve 1.0 g of sample in a 100-mL volumetric flask and dilute to the mark with water (1 mL = 0.01 g).

Analyze the solutions by means of a suitable atomic absorption spectrophotometer, using the conditions outlined in the following table. Calculate the metal content of the sample by the method of standard additions.

Element	Wavelength (nm)	Sample Wt (g)	Standard Added (mg)	Flame Type*	Background Correction
Na	589.0	0.05	0.01; 0.02	A/A	No

*A/A is air/acetylene.

Aluminum Sulfate, Hydrated

$Al_2(SO_4)_3 \cdot (14-18)H_2O$

CAS Number 7784−31−8 (for 18-hydrate)

> NOTE. This salt is available in degrees of hydration ranging from 14 to 18 molecules of water.

REQUIREMENTS

Assay (as the labeled hydrate) 98.0–102.0%

MAXIMUM ALLOWABLE

Insoluble matter 0.01%
Chloride (Cl) 0.005%
Arsenic (As) . . . . 0.5 ppm
Substances not precipitated by
    ammonium hydroxide (as SO_4) 0.2%
Heavy metals (as Pb) 0.001%
Iron (Fe) 0.002%

TESTS

ASSAY. (By indirect complexometric titration of Al^{III}). Weigh accurately 8 g and transfer to a 250-mL volumetric flask. Dissolve in about 75 mL of water, add 5 mL of hydrochloric acid, dilute to volume with water, and mix. Place a 25.0-mL aliquot of this solution in a 250-mL beaker, add 50 mL of water, mix, and then add 40.0 mL of standard 0.1 M EDTA and 20 mL of ammonium acetate–acetic acid buffer solution. Heat the solution to near boiling for 5 min and cool. (The pH should be 4.0–5.0 and should be maintained in that range by adding more buffer during the titration if necessary.) Add 50 mL of alcohol and 2 mL of dithizone indicator solution and titrate the excess of EDTA with 0.1 M zinc sulfate (28.75 g of zinc sulfate heptahydrate diluted to 1 L with water) until the color changes from a green-violet to a rose-pink. Similarly, titrate 40 mL of the standard 0.1 M EDTA with 0.1 M zinc sulfate, substituting 25 mL of water for the sample. Let V_s and V_b be the volumes of the 0.1 M zinc sulfate required in the two titrations; let V_e be the volume in milliliters and M_e be the molarity of the standard EDTA solution; and let W be the weight of sample, in grams, corresponding to the aliquot taken.

$$\%Al_2(SO_4)_3 \cdot n\,H_2O = [1 - (V_s/V_b)](V_e M_e)(K/W)$$

where K has values of 29.72, 31.52, and 33.32 for n values of 14, 16, and 18, respectively.

INSOLUBLE MATTER. (Page 14). Use 20 g dissolved in 150 mL of water containing 5 mL of 25% sulfuric acid.

CHLORIDE. (Page 25). Use 0.20 g.

ARSENIC. (Page 23). Use 6.0 g, dissolved in 50 mL of water, and add 30 mL of dilute sulfuric acid (1 + 4). For the standard use 0.003 mg of arsenic (As).

SUBSTANCES NOT PRECIPITATED BY AMMONIUM HYDROXIDE. Dissolve 2.0 g in 130 mL of boiling water, add 0.15 mL of methyl red indicator solution, then add ammonium hydroxide until the solution is alkaline. Heat to boiling, boil for 5 min, and dilute with water to 150 mL. Filter, transfer 75 mL of the filtrate to a tared evaporating dish, and evaporate to dryness. Ignite the residue at 800 ± 25°C for 15 min.

HEAVY METALS. (Page 26, Method 1). Dissolve 4.0 g in 40 mL of water, and dilute with water to 48 mL. Use 36 mL to prepare the sample solution, and use the remaining 12 mL to prepare the control solution.

IRON. (Page 28, Method 1). Dissolve 1.0 g in 40 mL of water; use 20 mL of this solution, and only 1 mL of hydrochloric acid.

2-Aminoethanol
Monoethanolamine

$HOCH_2CH_2NH_2$ Formula Wt 61.08

CAS Number 141−43−5

REQUIREMENTS

Assay 98.0–100.5%
$$HOCH_2CH_2NH_2$$

MAXIMUM ALLOWABLE

Color (APHA). 15
Water. 0.30%
Iron (Fe) . . . . 5 ppm
Heavy metals (as Pb) . . . . 5 ppm

TESTS

ASSAY.   Analyze the sample by gas chromatography, using the general parameters cited on page 56. The following specific conditions are also required.

Column: Type I, methyl silicone

Measure the area under all peaks and calculate the 2-aminoethanol content in area percent. Correct for water content.

COLOR (APHA).   (Page 17).

WATER.   (Page 55, Method 2). Use 0.10 mL (0.1 g) of the sample.

IRON.   (Page 28, Method 1). Use 2.0 g (2.0 mL) in 40 mL of water and 5 mL of hydrochloric acid.

HEAVY METALS.   (Page 26, Method 1). Use 4.0 g (4.0 mL) in 30 mL of water and 6 mL of hydrochloric acid.

4-Amino-3-hydroxy-1-naphthalene Sulfonic Acid
1-Amino-2-naphthol-4-sulfonic Acid

$H_2N(HO)C_{10}H_5SO_3H$ Formula Wt 239.25

CAS Number 116−63−2

REQUIREMENTS

Assay . . . . $\geq$ 98.0%
$H_2N(HO)C_{10}H_5SO_3H$

MAXIMUM ALLOWABLE

Solubility in sodium carbonate. . . . . Passes test
Residue after ignition 0.1%
Sulfate (SO_4) 0.2%
Sensitivity to phosphate . . . . Passes test

TESTS

ASSAY. (By acid–base titrimetry). Weigh accurately about 0.5 g and transfer to a beaker. Add 100 mL of dimethyl sulfoxide and, while stirring, add 25 mL of water in small portions. Insert the electrodes of a suitable pH meter, and titrate with 0.1 N sodium hydroxide, establishing the end point potentiometrically. Correct for a blank determination. One milliliter of 0.1 N sodium hydroxide corresponds to 0.02392 g of $H_2N(HO)C_{10}H_5SO_3H$.

SOLUBILITY IN SODIUM CARBONATE. Dissolve 0.1 g in 3 mL of sodium carbonate reagent solution, and dilute with water to 10 mL. The solution should be clear and dissolution complete, or nearly so.

RESIDUE AFTER IGNITION. (Page 15). Ignite 1.0 g.

SULFATE. Transfer 0.5 g to a beaker, add 25 mL of water and 0.10 mL of hydrochloric acid, and heat on a hot plate at low setting for 10 min. Cool, dilute with water to 50 mL, and filter. To 10 mL of the clear filtrate add 1 mL of 1 N hydrochloric acid and 2 mL of barium chloride reagent solution. Any turbidity should not exceed that produced by 0.2 mg of sulfate ion (SO_4) in an equal volume of solution containing the quantities of reagents used in the test. Compare 10 min after adding the barium chloride to the sample and standard solutions.

SENSITIVITY TO PHOSPHATE. Dissolve 0.10 g in a mixture of 50 mL of water and 10 g of sodium bisulfite, warm if necessary to complete dissolution, and filter. Add 1.0 mL of this solution to a test solution prepared as follows: To 20 mL of water add 0.02 mg of phosphate ion (PO_4) and 2 mL of 25% sulfuric acid reagent solution, mix, and then add 1.0 mL of ammonium molybdate reagent solution. A distinct blue color should be produced in the test solution in 5 min.

Ammonium Acetate

CH_3COONH_4 Formula Wt 77.08

CAS Number 631–61–8

REQUIREMENTS

Assay . . . . ≥ 97%
pH of a 5% solution . . . . 6.7–7.3 at 25°C

MAXIMUM ALLOWABLE

Insoluble matter 0.005%
Residue after ignition 0.01%
Chloride (Cl) . . . . 5 ppm
Nitrate (NO_3) 0.001%
Sulfate (SO_4) 0.001%
Heavy metals (as Pb) . . . . 5 ppm
Iron (Fe) . . . . 5 ppm

TESTS

ASSAY. (By acid–base titrimetry). Weigh accurately about 3 g of sample and dissolve in 30 mL of water. Add 50 mL of 30% formaldehyde solution (1 + 1), already neutralized to thymol blue (0.15 mL) with 0.1 N sodium hydroxide. Let stand for 30 min and titrate with 1 N sodium hydroxide using thymol blue as the indicator. One milliliter of 1 N sodium hydroxide corresponds to 0.07708 g of CH_3COONH_4.

pH OF A 5% SOLUTION. (Page 43). The pH should be between 6.7 and 7.3 at 25°C.

INSOLUBLE MATTER. (Page 14). Use 20 g dissolved in 200 mL of water.

RESIDUE AFTER IGNITION. (Page 15). Ignite 10 g.

CHLORIDE. (Page 25). Use 2.0 g.

NITRATE. (Page 29). For sample solution A use 1.0 g. For control solution B use 1.0 g and 1 mL of standard nitrate solution.

Sample Solution A and Blank Solution B for the Determination of Sulfate, Heavy Metals, and Iron.

Sample Solution A. Dissolve 15 g in 15 mL of water and add about 10 mg of sodium carbonate, 2 mL of hydrogen peroxide, 15 mL of hydrochloric acid, and 25 mL of nitric acid.

Blank Solution B.   Prepare a solution containing the quantities of reagents used in sample solution A, omitting only the sample.

Digest each in covered beakers on a hot plate at low setting until reaction ceases, uncover, and evaporate to dryness. Dissolve each in 3 mL of 1 N acetic acid plus 15 mL of water, filter through a small filter, and dilute with water to 30 mL (1 mL = 0.5 g).

SULFATE.   (Page 32, Procedure A, Method 1). Use 10 mL of sample solution A (5-g sample).

HEAVY METALS.   (Page 26, Method 1). Dilute 8.0 mL of sample solution A (4-g sample) with water to 25 mL. Use 8.0 mL of blank solution B to prepare the standard solution.

IRON.   (Page 28, Method 1). Use 4.0 mL of sample solution A (2-g sample). Prepare the standard from 4.0 mL of blank solution B.

Ammonium Bromide

NH_4Br Formula Wt 97.94

CAS Number 12124–97–9

REQUIREMENTS

Assay . . . . $\geq$ 99.0% NH_4Br
pH of a 5% solution . . . . 4.5–6.0 at 25°C

MAXIMUM ALLOWABLE

Insoluble matter 0.005%
Residue after ignition 0.01%
Bromate (BrO_3) 0.002%
Chloride (Cl) 0.2%
Iodide (I) . . . . Passes test
                                                                     (limit about
                                                                     0.005%)
Sulfate (SO_4) 0.005%
Barium (Ba) 0.002%
Heavy metals (as Pb) . . . . 5 ppm
Iron (Fe) . . . . 5 ppm

TESTS

ASSAY. (By indirect argentimetric titration of bromide). Weigh accurately about 0.4 g of sample in 50 mL of water. Add 10 mL of dilute nitric acid and 50.0 mL of 0.1 N silver nitrate, and titrate the excess silver nitrate with 0.1 N ammonium thiocyanate, using 0.5 mL of ferric sulfate as an indicator. One milliliter of 0.1 N silver nitrate consumed corresponds to 0.009794 g of NH_4Br.

pH OF A 5% SOLUTION. (Page 43). The pH should be 4.5–6.0 at 25°C.

INSOLUBLE MATTER. (Page 14). Use 20 g dissolved in 150 mL of water.

RESIDUE AFTER IGNITION. (Page 15). Ignite 20 g.

BROMATE. (Page 49). Determine bromate by differential pulse polarography. Use 5.0 g of sample in 25 mL of solution. For the standard add 0.10 mg of bromate (BrO_3).

CHLORIDE. (Page 25). Dissolve 0.50 g in 15 mL of dilute nitric acid (1 + 2) in a small flask. Add 3 mL of 30% hydrogen peroxide and digest on the steam bath until the solution is colorless. Wash down the sides of the flask with a little water, digest for an additional 15 min, cool, and dilute with water to 200 mL. Dilute 2.0 mL of the solution with water to 20 mL.

IODIDE. Dissolve 5.0 g in 20 mL of water. Add 1 mL of chloroform, 0.15 mL of 10% ferric chloride reagent solution, and 0.25 mL of 10% sulfuric acid, and shake. No violet tint should be produced in the chloroform.

SULFATE. (Page 32, Procedure A, Method 1). Allow 30 min for the turbidity to form.

BARIUM. For the sample dissolve 6.0 g in 15 mL of water. For the control dissolve 1.0 g in 15 mL of water and add 0.1 mg of barium ion (Ba). To each solution add 5 mL of glacial acetic acid, 5 mL of 30% hydrogen peroxide, and 1 mL of hydrochloric acid. Digest in a covered beaker on the steam bath until reaction ceases, uncover, and evaporate to dryness. Dissolve each residue in 15 mL of water, filter if necessary, and dilute with water to 23 mL. Add 2 mL of 10% potassium dichromate reagent solution, then add ammonium hydroxide until the orange color is just dissipated and the yellow color persists. Add 25 mL of methanol, stir vigorously, and allow to stand for 10 min. Any turbidity in the solution of the sample should not exceed that in the control.

HEAVY METALS. (Page 26, Method 1). Dissolve 6.0 g in about 20 mL of water, and dilute with water to 30 mL. Use 25 mL to prepare the sample solution, and use the remaining 5.0 mL to prepare the control solution.

IRON. (Page 28, Method 1). Use 2.0 g.

Ammonium Carbonate

CAS Number 506–87–6

> *NOTE.* This product is a mixture of variable proportions of ammonium bicarbonate and ammonium carbamate.

REQUIREMENTS

Assay . . . . $\geq$ 30.0% NH_3

MAXIMUM ALLOWABLE

Insoluble matter 0.005%
Nonvolatile matter 0.01%
Chloride (Cl) . . . . 5 ppm
Sulfur compounds (as SO_4) 0.002%
Heavy metals (as Pb) . . . . 5 ppm
Iron (Fe) . . . . 5 ppm

TESTS

ASSAY. (By acid–base titrimetry). Tare a glass-stoppered flask containing about 25 mL of water. Add 2.0–2.5 g of the sample and weigh accurately. Add slowly 50.0 mL of 1 N hydrochloric acid and titrate the excess with 1 N sodium hydroxide, using methyl orange indicator. One milliliter of 1 N hydrochloric acid corresponds to 0.01703 g of NH_3.

INSOLUBLE MATTER.   (Page 14). Use 20 g dissolved in 100 mL of water.

NONVOLATILE MATTER.   To 20 g in a tared dish add 10 mL of water, volatilize on the steam bath, and dry for 1 hour at 105°C. Retain the residue for the test for heavy metals.

CHLORIDE.   (Page 25). Dissolve 2.0 g in 25 mL of hot water, add about 10 mg of sodium carbonate, and evaporate to dryness on the steam bath. Dissolve the residue in 20 mL of water.

SULFUR COMPOUNDS.   Dissolve 2.0 g in 20 mL of water, add about 10 mg of sodium carbonate, and evaporate to dryness. Dissolve the residue in a slight excess of hydrochloric acid, add 2 mL of bromine water, and again evaporate to dryness. Dissolve the residue in 4 mL of water plus 1 mL of dilute hydrochloric acid (1 + 19). Filter through a small filter, wash with two 2-mL portions of water, dilute with water to 10 mL, and add 1 mL of barium chloride reagent solution. Any turbidity should not exceed that produced by 0.04 mg of sulfate ion (SO_4) in an equal volume of solution containing the quantities of reagents used in the test. Compare 10 min after adding the barium chloride to the sample and standard solutions.

HEAVY METALS.   (Page 26, Method 1). Dissolve the residue from the test for nonvolatile matter in 5 mL of 1 N acetic acid, and dilute with water to 50 mL. Dilute 10 mL of this solution with water to 25 mL.

IRON.   (Page 28, Method 1). To 2.0 g add 5 mL of water and evaporate on the steam bath. Dissolve the residue in 2 mL of hydrochloric acid, add 20 mL of water, and filter if necessary. Use this solution without further acidification.

Ammonium Chloride

NH$_4$Cl   Formula Wt 53.49

CAS Number 12125-02-9

REQUIREMENTS

Assay . . . . ≥ 99.5%
pH of a 5% solution . . . . 4.5-5.5 at 25°C

MAXIMUM ALLOWABLE

Insoluble matter 0.005%
Residue after ignition 0.01%
Phosphate (PO$_4$) . . . . 2 ppm
Sulfate (SO$_4$) 0.002%
Calcium and magnesium precipitate 0.002%
Heavy metals (as Pb) . . . . 5 ppm
Iron (Fe) . . . . 2 ppm

TESTS

ASSAY. (By argentimetric titration). Dissolve about 200 mg of ammonium chloride sample, accurately weighed, in 20 mL of water. Add about 10 mL of glacial acetic acid, 100 mL of methanol, and 0.25 mL of eosin Y indicator. With stirring, titrate with 0.1 N silver nitrate to a pink end point. One milliliter of 0.1 N silver nitrate corresponds to 0.005358 g of NH$_4$Cl.

pH OF A 5% SOLUTION. (Page 43). The pH should be 4.5-5.5 at 25°C.

INSOLUBLE MATTER. (Page 14). Use 20 g dissolved in 200 mL of water.

RESIDUE AFTER IGNITION. (Page 15). Ignite 10 g. Retain the residue for the test for calcium and magnesium precipitate.

Sample Solution A and Blank Solution B for the Determination of Phosphate, Sulfate, Heavy Metals, and Iron.

Sample Solution A.   Dissolve 25 g in 75 mL of water and add about 10 mg of sodium carbonate, 10 mL of hydrogen peroxide, and 50 mL of nitric acid.

Blank Solution B.   To 75 mL of water add about 10 mg of sodium carbonate, 10 mL of hydrogen peroxide, and 50 mL of nitric acid.

Digest sample solution A and blank solution B in covered beakers on the steam bath until reaction ceases, uncover, and evaporate to dryness. Dissolve the two residues in separate 5-mL portions of 1 N acetic acid and 50 mL of water, filter if necessary, and dilute with water to 100 mL (1 mL of sample solution A is 0.25 g).

PHOSPHATE.   (Page 30, Method 1). Evaporate 40 mL of sample solution A (10-g sample) to dryness on the steam bath. Dissolve the residue in 25 mL of approximately 0.5 N sulfuric acid. Use 40 mL of blank solution B to prepare the standard solution.

SULFATE.   (Page 32, Procedure A, Method 1). Use 10 mL of sample solution A (2.5-g sample). Use 10 mL of blank solution B to prepare the standard solution.

CALCIUM AND MAGNESIUM PRECIPITATE.   Warm the residue obtained in the test for residue after ignition with 1 mL of hydrochloric acid and 3 mL of water. Add 2 mL of ammonium hydroxide, filter, and wash with a few milliliters of water. Add to the filtrate 2 mL of ammonium oxalate reagent solution and 2 mL of ammonium phosphate reagent solution, and allow to stand overnight. Filter, wash with a solution containing 2.5% of ammonium hydroxide and 0.1% of ammonium oxalate, and ignite.

HEAVY METALS.   (Page 26, Method 1). Dilute 16 mL of sample solution A (4-g sample) with water to 25 mL. Use 16 mL of blank solution B to prepare the standard solution.

IRON.   (Page 28, Method 1). Use 20 mL of sample solution A (5-g sample). Prepare the standard with 20 mL of blank solution B.

Ammonium Citrate, Dibasic
Citric Acid, Diammonium Salt
2-Hydroxy-1,2,3-propanetricarboxylic Acid, Diammonium Salt

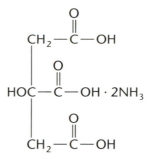

$(NH_4)_2HC_6H_5O_7$ Formula Wt 226.19

CAS Number 3012–65–5

REQUIREMENTS

Assay 98.0–103.0%
$(NH_4)_2HC_6H_5O_7$

MAXIMUM ALLOWABLE

Insoluble matter 0.005%
Residue after ignition 0.01%
Chloride (Cl) 0.001%
Oxalate (C_2O_4) . . . . Passes test
(limit about 0.05%)
Phosphate (PO_4) . . . . 5 ppm
Sulfur compounds (as SO_4) 0.005%
Heavy metals (as Pb) . . . . 5 ppm
Iron (Fe) 0.001%

TESTS

ASSAY.   (By acid–base titration based on ammonium content). Weigh accurately 3 g and dissolve in 40 mL of water. Dilute 20 mL of formaldehyde solution with 20 mL of water, add 0.15 mL of phenolphthalein indicator solution, and neutralize with 0.1 N sodium hydroxide. Add the formaldehyde solution to the solution of the sample, mix, and allow to stand 30 min. Add 0.15 mL of phenolphthalein indicator solution and titrate with 1 N sodium hydroxide to a pink color that persists for 5 min. One milliliter of 1 N sodium hydroxide corresponds to 0.07540 g of $(NH_4)_2HC_6H_5O_7$.

INSOLUBLE MATTER. (Page 14). Use 20 g dissolved in 200 mL of water.

RESIDUE AFTER IGNITION. (Page 15). Ignite 10 g.

CHLORIDE. (Page 25). Use 1.0 g.

OXALATE. Dissolve 5 g in 25 mL of water, and add 3 mL of glacial acetic acid and 2 mL of 10% solution of calcium acetate. No turbidity or precipitate should appear after standing 4 h.

PHOSPHATE. (Page 30, Method 1). Mix 4.0 g with 0.5 g of magnesium nitrate in a platinum dish and ignite. Dissolve the residue in 5 mL of water, add 5 mL of nitric acid, and evaporate to dryness. Dissolve the residue in 25 mL of approximately 0.5 N sulfuric acid and continue as described.

SULFUR COMPOUNDS. To 1.0 g add 1 mL of hydrochloric acid and 3 mL of nitric acid. Prepare a standard containing 0.05 mg of sulfate ion (SO_4), 1 mL of hydrochloric acid, and 3 mL of nitric acid. Digest each in a covered beaker on a hot plate at low setting until the reaction ceases, remove the covers, and evaporate to dryness. Add 0.2–0.5 mg of ammonium vanadate and 10 mL of nitric acid and digest in covered beakers on the hot plate until reactions cease. Remove the covers and evaporate to dryness. Add 10 mL more of nitric acid and repeat the digestions and evaporations. Add 5 mL of dilute hydrochloric acid (1 + 1) and evaporate to dryness. Dissolve the residues in 4 mL of water plus 1 mL of dilute hydrochloric acid (1 + 19), and filter through a small filter. Wash with two 2-mL portions of water, dilute with water to 10 mL, and add 1 mL of barium chloride reagent solution to each. Any turbidity in the solution of the sample should not exceed that in the standard. Compare 10 min after adding the barium chloride to the sample and standard solutions.

HEAVY METALS. (Page 26, Method 1). Dissolve 6.0 g in about 20 mL of water, add 5 mL of dilute hydrochloric acid (1 + 1), and dilute with water to 30 mL. Use 25 mL to prepare the sample solution, and use the remaining 5.0 mL to prepare the control solution.

IRON. (Page 28, Method 1). Use 1.0 g.

Ammonium Dichromate

$(NH_4)_2Cr_2O_7$ Formula Wt 252.07

CAS Number 7789–09–5

> *NOTE.* This reagent may be stabilized by the addition of 0.5% to 3.0% water.

REQUIREMENTS

Assay (dried basis) . . . . ≥ 99.5%
$(NH_4)_2Cr_2O_7$

MAXIMUM ALLOWABLE

Insoluble matter and ammonium
  hydroxide precipitate 0.005%
Loss on drying at 105°C. 3.0%
Chloride (Cl) 0.005%
Sulfate (SO$_4$) 0.01%
Calcium (Ca) 0.002%
Iron (Fe) 0.002%
Sodium (Na) 0.005%

TESTS

ASSAY. (By iodometry). Weigh accurately 1.0 g, previously dried to constant weight at 105°C, transfer to a 100-mL volumetric flask, dissolve in water, dilute to volume, and mix well. Pipet a 10.0-mL aliquot to a glass-stoppered flask, add 50 mL of water, 10 mL of 10% sulfuric acid reagent solution, and 3 g of potassium iodide. Mix, stopper, and allow to stand in the dark for 10 min. Titrate the liberated iodine with 0.1 N sodium thiosulfate, adding 3 mL of starch indicator solution near the end point, which is a greenish-blue color. One milliliter of 0.1 N sodium thiosulfate is equivalent to 0.004201 g of $(NH_4)_2Cr_2O_7$.

INSOLUBLE MATTER AND AMMONIUM HYDROXIDE PRECIPITATE. Dissolve 20 g in 200 mL of water, add 2 mL of ammonium hydroxide, heat to boiling, and digest in a covered beaker on the steam bath for 1 hour. Filter, wash thoroughly, and ignite.

LOSS ON DRYING AT 105°C. Weigh accurately 2.0 g and dry to constant weight at 105°C.

CHLORIDE. (Page 25). Dissolve 0.2 g in 10 mL of water, filter if necessary through a small chloride-free filter, and add 1 mL of ammonium hydroxide and 1 mL of silver nitrate reagent solution. Prepare a standard containing 0.01 mg of chloride ion (Cl) in 10 mL of water, and add 1 mL of ammonium hydroxide and 1 mL of silver nitrate reagent solution. Add 2 mL of nitric acid to each. The comparison is best made by the general method for chloride in colored solutions, page 25.

SULFATE. Dissolve 10 g in 250 mL of water, filter if necessary, and heat to boiling. Add 25 mL of a solution containing 1 g of barium chloride and 2 mL of hydrochloric acid per 100 mL of solution. Digest in a covered beaker on a hot plate at low setting for 2 hours and allow to stand overnight. If a precipitate is formed, filter, wash thoroughly, and ignite. Fuse the residue with 1 g of sodium carbonate. Extract the fused mass with water and filter off the insoluble residue. Add 5 mL of hydrochloric acid to the filtrate, dilute with water to about 200 mL, heat to boiling, and add 10 mL of alcohol. Digest in a covered beaker on the hot plate until reduction of chromate is complete, as indicated by the change to a clear green or colorless solution. Neutralize the solution with ammonium hydroxide and add 2 mL of hydrochloric acid. Heat to boiling and add 10 mL of barium chloride reagent solution. Digest in a covered beaker on a hot plate at low setting for 2 hours and allow to stand overnight. Filter, wash thoroughly, and ignite. The weight of the precipitate should not exceed the weight obtained in a complete blank test by more than 0.0024 g. If the original precipitate of barium sulfate weighs less than the requirement permits, the fusion with sodium carbonate is not necessary.

CALCIUM, IRON, AND SODIUM. Determine the calcium, iron, and sodium content by the flame atomic absorption spectrophotometric method described on page 35.

> *Sample Stock Solution.* Dissolve 20 g in water and dilute with water to 100 mL (1 mL = 0.2 g).

Analyze the solutions by means of a suitable atomic absorption spectrophotometer, using the conditions outlined in the following table. Calculate the metal content of the sample by the method of standard additions.

Element	Wavelength (nm)	Sample Wt (g)	Standard Added (mg)	Flame Type*	Background Correction
Ca	422.7	2.0	0.02; 0.04	N/A	No
Fe	248.3	4.0	0.04; 0.08	A/A	Yes
Na	589.0	0.2	0.005; 0.01	A/A	No

*A/A is air/acetylene; N/A is nitrous oxide/acetylene.

Ammonium Fluoride

NH_4F Formula Wt 37.04

CAS Number 12125–01–8

REQUIREMENTS

Assay . . . . $\geq$ 98.0% NH_4F

MAXIMUM ALLOWABLE

Insoluble matter 0.005%
Residue after ignition 0.01%
Chloride (Cl) 0.001%
Sulfate (SO_4) 0.005%
Heavy metals (as Pb) . . . . 5 ppm
Iron (Fe) . . . . 5 ppm

TESTS

ASSAY. (By acidimetric titration of fluoride content). Weigh accurately about 1.5 g of sample and dissolve in a plastic container with 50 mL of water. Add 50 mL of 30% formaldehyde solution (1 + 1), already neutralized with 0.1 N sodium hydroxide, and let stand for 30 min. Titrate with 1 N sodium hydroxide, using 0.15 mL of thymol blue indicator. One milliliter of 1 N sodium hydroxide corresponds to 0.03704 g of NH_4F.

INSOLUBLE MATTER. (Page 14). Use 20 g dissolved in 200 mL of hot water.

RESIDUE AFTER IGNITION. (Page 15). Ignite 10 g in a tared platinum crucible or dish.

CHLORIDE.   Dissolve 1.0 g in a mixture of 10 mL of water and 1 mL of nitric acid in a platinum dish. In a test tube or small beaker mix 20 mL of water and 1 mL of silver nitrate reagent solution, add this solution to the sample solution, and mix. Any turbidity should not exceed that produced by 0.01 mg of chloride ion (Cl) in an equal volume of solution containing the quantities of reagents used in the test.

SULFATE.   (Page 32, Procedure A, Method 2). Use 10 mL of hydrochloric acid in a platinum dish, and perform four evaporations.

HEAVY METALS.   (Page 26, Method 1). Dissolve 5.0 g in about 25 mL of water, add 2 g of sodium acetate, and dilute with water to 30 mL. Adjust the pH with short-range pH paper. Use 1.0 g of sample and 2 g of sodium acetate to prepare the control solution.

IRON.   (Page 28, Method 1). Treat 2.0 g in a platinum dish with 10 mL of dilute hydrochloric acid $(1 + 1)$, and evaporate on the steam bath to dryness. Repeat the evaporation with a second portion of the dilute acid. Warm the residue with 2 mL of hydrochloric acid, dilute with water to 50 mL, and use this solution without further acidification. Prepare the standard from the residue remaining from the evaporation of 10 mL of hydrochloric acid.

Ammonium Hydroxide
Aqueous Ammonia

NH_4OH Formula Wt 35.05

CAS Number 1336–21–6

REQUIREMENTS

Appearance...............................	Colorless and free from suspended matter or sediment
Assay (as NH_3)...........................	28.0–30.0%

MAXIMUM ALLOWABLE

Residue after ignition 0.002%
Carbon dioxide (CO_2) 0.002%
Chloride (Cl) . . . . 0.5 ppm
Phosphate (PO_4) . . . . 2 ppm
Total sulfur (as SO_4) . . . . 2 ppm
Heavy metals (as Pb) . . . . 0.5 ppm
Iron (Fe) . . . . 0.2 ppm
Substances reducing permanganate . . . . Passes test

TESTS

APPEARANCE. Mix the material in the original container, pour 10 mL into a test tube (20 × 150 mm), and compare with distilled water in a similar tube. The liquids should be equally clear and free from suspended matter; looking across the columns by means of transmitted light should reveal no apparent difference in color between the two liquids.

ASSAY. (By acidimetry). Tare a small glass-stoppered Erlenmeyer flask containing 15 mL of water. Using a measuring pipet, without suction, introduce barely under the surface about 2 mL of the ammonium hydroxide, stopper, and weigh accurately. Titrate with 1 N hydrochloric acid, using methyl red as indicator. One milliliter of 1 N acid is equivalent to 0.01703 g of NH_3.

RESIDUE AFTER IGNITION. (Page 15). Evaporate 50 g (56 mL) and ignite.

CARBON DIOXIDE. Dilute 10 g (11 mL) of the sample with 10 mL of water free from carbon dioxide, and add 5 mL of clear saturated barium hydroxide solution. Any turbidity should not be greater than is produced when the same quantity of barium hydroxide solution is added to 21 mL of water (free from carbon dioxide) containing 0.3 mg of carbonate standard (0.2 mg of carbon dioxide).

CHLORIDE. (Page 25). To 20 g (22 mL) add about 10 mg of sodium carbonate and evaporate to dryness. Dissolve the residue in 20 mL of dilute nitric acid (1 + 19).

PHOSPHATE. (Page 30, Method 1). Evaporate 10 g (11 mL) to dryness on the steam bath. Dissolve the residue in 25 mL of approximately 0.5 N sulfuric acid and continue as described.

TOTAL SULFUR. To 30 g (33 mL) add about 10 mg of sodium carbonate and evaporate to about 5 mL. Add 1 mL of bromine water and evaporate to dryness. Dissolve the residue in 4 mL of water and 1 mL of hydrochloric acid (1 + 19), and evaporate again to dryness. Dissolve the residue in 4 mL of water plus 1 mL of dilute hydrochloric acid (1 + 19), filter through a small filter, wash with two 2-mL portions of water, and dilute with water to 10 mL. Add 1 mL of barium chloride reagent solution. Any turbidity should not exceed that produced by 0.06 mg of sulfate ion (SO_4) in an equal volume of solution containing the quantities of reagents used in the test. Compare 10 min after adding the barium chloride to the sample and standard solutions.

HEAVY METALS. (Page 26, Method 1). To 40 g (44 mL) add about 10 mg of sodium carbonate, evaporate to dryness on the steam bath, dissolve the residue in about 20 mL of water, and dilute with water to 25 mL.

IRON. (Page 28, Method 1). To 50 g (56 mL) add about 10 mg of sodium carbonate, and evaporate to dryness on the steam bath. Dissolve the residue in 3 mL of hydrochloric acid, dilute with water to 50 mL, and use this solution without further acidification.

SUBSTANCES REDUCING PERMANGANATE. Dilute 3 mL of the sample with 5 mL of water and add 50 mL of 10% sulfuric acid reagent solution. Add 0.05 mL of 0.1 N potassium permanganate, heat to boiling, and keep at this temperature for 5 min. The pink color should not be entirely discharged.

Ammonium Iodide

NH_4I Formula Wt 144.94

CAS Number 12027−06−4

NOTE: This compound may turn yellow on storage.

REQUIREMENTS

Assay . . . . ≥ 99.0% NH_4I

MAXIMUM ALLOWABLE

Insoluble matter 0.005%
Residue after ignition 0.05%
Chloride and bromide (as Cl). 0.005%
Phosphate (PO_4) 0.001%
Sulfate (SO_4). 0.05%
Barium (Ba). 0.002%
Heavy metals (as Pb) 0.001%
Iron (Fe) . . . . 5 ppm

TESTS

ASSAY. (By oxidation–reduction titration of iodide). Weigh to the nearest 0.1 mg about 0.3 g of sample, and dissolve with about 20 mL of water in a 250-mL glass-stoppered titration flask. Add 30 mL of hydrochloric acid and 5 mL of chloroform. Cool, if necessary, and titrate with 0.05 M potassium iodate solution until the iodine color disappears from the aqueous layer. Stopper, shake vigorously for 30 s, and continue the titration, shaking vigorously after each addition of the iodate until the iodine color in the chloroform is discharged. One milliliter of 0.05 M potassium iodate corresponds to 0.007247 g of NH_4I.

INSOLUBLE MATTER. (Page 14). Use 20 g dissolved in 200 mL of water.

RESIDUE AFTER IGNITION. (Page 15). Ignite 2.0 g.

CHLORIDE AND BROMIDE. Dissolve 1.0 g in 100 mL of water in a distilling flask. Add 1 mL of hydrogen peroxide and 1 mL of phosphoric acid, heat to boiling, and boil gently until all the iodine is expelled and the solution is colorless. Cool, wash down the sides of the flask, and add 0.5 mL of hydrogen peroxide. If an iodine color develops, boil until the solution is colorless and for 10 min longer. If no color develops, boil for 10 min, filter if necessary through a chloride-free filter, and dilute with water to 100 mL. Dilute 20 mL with water to 23 mL, and add 1 mL of nitric acid and 1 mL of silver nitrate reagent solution. Any turbidity should not exceed that produced by 0.01 mg of chloride ion (Cl) in an equal volume of solution containing the quantities of nitric acid and silver nitrate used in the test.

Sample Solution A and Blank Solution B for the Determination of Phosphate, Sulfate, Barium, Heavy Metals, and Iron.

Sample Solution A. Dissolve 20 g in 40 mL of water in a 600-mL beaker and add about 10 mg of sodium carbonate. Add 16 mL of hydrochloric acid and 32 mL of nitric acid, cover with a watch glass, and warm on the steam bath. When the rapid evolution of iodine ceases, add an additional 16 mL of nitric acid and 24 mL of hydrochloric acid. Digest in the covered beaker until the bubbling ceases, remove the watch glass, and evaporate to dryness.

Blank Solution B. Evaporate to dryness the quantities of acids and sodium carbonate used to prepare sample solution A.

Dissolve the residues in separate 20-mL portions of water, filter if necessary, and dilute each with water to 100 mL (1 mL of sample solution A = 0.2 g).

PHOSPHATE. (Page 30, Method 1). Evaporate 10 mL of sample solution A (2-g sample) to dryness on the steam bath. Dissolve the residue in 25 mL of approximately 0.5 N sulfuric acid. Use 10 mL of blank solution B to prepare the standard solution.

SULFATE. (Page 32, Procedure A, Method 1). Use 0.5 mL of sample solution A. Use 0.5 mL of blank solution B to prepare the standard solution.

BARIUM. To 20 mL of sample solution A (4-g sample) add 5 mL of a 1% solution of potassium sulfate. For the standard add 0.08 mg of barium ion (Ba) and 5 mL of a 1% solution of potassium sulfate to 20 mL of blank solution B. Any turbidity in the solution of the sample should not exceed that in the standard. Compare 10 min after adding the potassium sulfate to the sample and standard solutions.

HEAVY METALS. (Page 26, Method 1). Dilute 10 mL of sample solution A (2-g sample) with water to 25 mL. Use 10 mL of blank solution B to prepare the standard solution.

IRON. (Page 28, Method 1). Use 10 mL of sample solution A (2-g sample). Prepare the standard with 10 mL of blank solution B.

Ammonium Metavanadate

NH_4VO_3 Formula Wt 116.98

CAS Number 7803–55–6

REQUIREMENTS

Assay . . . . $\geq$ 99.0% NH_4VO_3

MAXIMUM ALLOWABLE

Solubility in ammonium hydroxide . . . . Passes test
Carbonate (CO_3) . . . . Passes test
                                                                     (about 0.3%)
Chloride (Cl) 0.2%
Sulfate (SO_4) 0.05%

TESTS

ASSAY.   (By oxidation–reduction titration of vanadium). Weigh accurately about 0.4 g, dissolve in 50 mL of warm water, and add 1 mL of sulfuric acid and 30 mL of sulfurous acid. Boil gently until the sulfur dioxide is expelled, then boil for 5 min longer. Cool, dilute with water to 100 mL, and titrate with 0.1 N potassium permanganate. One milliliter of 0.1 N potassium permanganate corresponds to 0.01170 g of NH_4VO_3.

SOLUBILITY IN AMMONIUM HYDROXIDE.   Dissolve 5.0 g in a mixture of ammonium hydroxide and 250 mL of hot water, and heat the solution to boiling. The solution should be clear and free from insoluble matter.

CARBONATE.   To 0.5 g add 1 mL of water and 2 mL of 10% hydrochloric acid reagent solution. No effervescence should be produced.

CHLORIDE.   (Page 25). Dissolve 0.5 g in 50 mL of hot water, add 2 mL of nitric acid, and let stand for 1 h. Filter through a chloride-free filter, wash with a few milliliters of water, and dilute with water to 100 mL; use 1.0 mL.

SULFATE.   (Page 32, Procedure 1, Method A). Dissolve 0.5 g in 40 mL of hot water, add 2 mL of 10% hydrochloric acid reagent solution and 1.5 g of hydroxylamine hydrochloride, and heat at 60°C for 5 min. Filter through a suitable filter paper, cool, and dilute with water to 100 mL. To 20 mL add 1 mL of dilute hydrochloric acid (1 + 19), and continue as directed.

Ammonium Molybdate Tetrahydrate
Ammonium Heptamolybdate Tetrahydrate

$(NH_4)_6Mo_7O_{24} \cdot 4H_2O$ Formula Wt 1235.86

CAS Number 12027–67–7

REQUIREMENTS

Assay (as MoO_3) 81.0–83.0%

MAXIMUM ALLOWABLE

Insoluble matter 0.005%
Chloride (Cl) 0.002%
Nitrate (NO_3) . . . . Passes test (limit
                                                          about 0.003%)
Arsenate, phosphate, and silicate (as SiO_2) 0.001%
Phosphate (PO_4) . . . . 5 ppm
Sulfate (SO_4) 0.02%
Heavy metals (as Pb) 0.001%
Magnesium and other alkaline earths 0.02%

TESTS

ASSAY. (Molybdenum content by gravimetry). Weigh accurately 1 g and dissolve in 10 mL of water plus 1 mL of ammonium hydroxide. Transfer to a 250-mL flask, dilute with water to volume, and mix thoroughly. To 50.0 mL of the solution (filtered if necessary) in a 600-mL beaker add 250 mL of water, 20 g of ammonium chloride, 15 mL of hydrochloric acid, and 0.15 mL of methyl orange indicator solution. Heat nearly to boiling and add 18 mL of 10% lead acetate solution. To the hot solution add, slowly and with constant stirring, a saturated ammonium acetate solution until the color turns yellow, and then add an excess of 15 mL of the acetate solution. Digest in the covered beaker on the hot plate below the boiling temperature until the precipitate has settled (from 30 min to 1 h). Filter through a tared Gooch crucible or a porous porcelain crucible, wash 7 or 8 times with an aqueous solution containing 10 mL of nitric acid and 100 mL of saturated ammonium acetate solution in a liter, and finally wash 3 times with hot water. Ignite to constant weight in a muffle furnace at 560–625°C. The weight of the lead molybdate times 0.3921 corresponds to the weight of MoO_3.

INSOLUBLE MATTER. (Page 14). Use 20 g dissolved in 200 mL of water.

CHLORIDE.   (Page 25). Use 0.5 g.

NITRATE.   Dissolve 1 g in 10 mL of water containing 5 mg of sodium chloride. Add 0.10 mL of indigo carmine solution and 10 mL of sulfuric acid. The blue color should not be completely discharged in 5 min.

ARSENATE, PHOSPHATE, AND SILICATE.   Dissolve 2.5 g in 70 mL of water. For the control, dissolve 0.5 g in 70 mL of water and add 0.02 mg of silica (SiO_2). These solutions are prepared in containers other than glass to avoid excessive silica contamination. Adjust the pH to between 3 and 4 (pH paper) with dilute hydrochloric acid (1 + 9), then transfer to glass containers and treat each solution as follows: add 1–2 mL of bromine water and adjust the pH to between 1.7 and 1.9 (using a pH meter) with dilute hydrochloric acid (1 + 9). Heat just to boiling, but do not boil, and cool to room temperature. Dilute with water to 90 mL, add 10 mL of hydrochloric acid, and transfer to a separatory funnel. Add 1 mL of butyl alcohol and 30 mL of 4-methyl-2-pentanone, shake vigorously, and allow the phases to separate. Draw off and discard the aqueous phase and wash the ketone phase three times with 10-mL portions of dilute hydrochloric acid (1 + 99), discarding each aqueous phase. To the washed ketone phase add 10 mL of dilute hydrochloric acid (1 + 99) to which has just been added 0.2 mL of a freshly prepared 2% solution of stannous chloride in hydrochloric acid. Any blue color in the solution of the sample should not exceed that in the control.

PHOSPHATE. (Page 30, Method 3). Proceed as described.

SULFATE.   (Page 32, Procedure A, Method 2). Use 5 mL of nitric acid.

HEAVY METALS. Dissolve 2.0 g in about 20 mL of water, add 10 mL of 10% sodium hydroxide reagent solution and 2 mL of ammonium hydroxide, and dilute with water to 40 mL. For the control, add 0.01 mg of lead ion (Pb) to 10 mL of the solution, adjust the pH to 3–4, and dilute with water to 40 mL. For the sample also, adjust the pH to 3–4, and dilute the remaining 30-mL portion with water to 40 mL. Add 10 mL of freshly prepared hydrogen sulfide water to each. Any color in the solution of the sample should not exceed that in the control.

MAGNESIUM AND OTHER ALKALINE EARTHS. Dissolve 5.0 g in 50 mL of water, filter if not clear, and add 0.5 g of sodium carbonate and 25 mL of 10% sodium hydroxide reagent solution. Boil the solution gently for 5 min, cool, filter, wash with 2.5% ammonia solution, and ignite at $800 \pm 25°C$ for 30 min.

Ammonium Nitrate

NH_4NO_3 Formula Wt 80.04

CAS Number 6484–52–2

REQUIREMENTS

Assay . . . . $\geq$ 98.0% NH_4NO_3
pH of a 5% solution . . . . 4.5–6.0 at 25°C

MAXIMUM ALLOWABLE

Insoluble matter 0.005%
Residue after ignition 0.01%
Chloride (Cl) . . . . 5 ppm
Nitrite (NO_2) . . . . Passes test (limit
                                                     about 5 ppm)
Phosphate (PO_4) . . . . 5 ppm
Sulfate (SO_4) 0.002%
Heavy metals (as Pb) . . . . 5 ppm
Iron (Fe) . . . . 2 ppm

TESTS

ASSAY. (By alkalimetry for ammonium). Weigh, to the nearest 0.1 mg, 3 g of sample and dissolve in 50 mL of water in a 500-mL Erlenmeyer flask. Add exactly 50.0 mL of 1 N sodium hydroxide, place a filter funnel loosely in the neck of the flask, and boil until all the ammonia is expelled (about 10–15 min) as determined with litmus paper. Cool, add 0.15 mL of thymol blue indicator solution (0.1% in alcohol). Titrate the excess sodium hydroxide with 1 N sulfuric acid. One milliliter of 1 N sodium hydroxide corresponds to 0.08004 g of NH_4NO_3.

INSOLUBLE MATTER. (Page 14). Use 20 g dissolved in 200 mL of water.

RESIDUE AFTER IGNITION. (Page 15). Ignite 10 g.

pH OF A 5% SOLUTION. (Page 43). The pH should be 4.5–6.0 at 25°C.

CHLORIDE. (Page 25). Use 2.0 g.

NITRITE. Dissolve 1.0 g in 10 mL of water, and add 1 mL of 10% sulfuric acid and 1 mL of colorless 0.5% 1,3-phenylenediamine hydrochloride solution. No yellowish or brownish color should be produced in 5 min.

> NOTE. 1,3-Phenylenediamine hydrochloride solution can be decolorized by treating with a little activated carbon and filtering.

PHOSPHATE. (Page 30, Method 1). Dissolve 4.0 g in 25 mL of approximately 0.5 N sulfuric acid.

Sample Solution A and Blank Solution B for the Determination of Sulfate, Heavy Metals, and Iron.

Sample Solution A. Dissolve 20 g in 20 mL of water, add 10 mg of sodium carbonate, 5 mL of 30% hydrogen peroxide, 20 mL of hydrochloric acid, and 20 mL of nitric acid.

Blank Solution B. Prepare a solution containing the quantities of reagents used in sample solution A.

Digest each in covered beakers on a hot plate at low setting until reaction ceases, uncover, and evaporate to dryness. Dissolve the residues in 5 mL of 1 N acetic acid and 50 mL of water, filter if necessary, and dilute to 100 mL with water (1 mL = 0.2 g).

SULFATE.   (Page 32, Procedure A, Method 1). Use 12.5 mL of sample solution A (2.5-g sample). Use 12.5 mL of blank solution B to prepare the standard solution.

HEAVY METALS.   (Page 26, Method 1). Dilute 20 mL of sample solution A (4-g sample) with water to 25 mL. Use 20 mL of blank solution B to prepare the standard solution.

IRON.   (Page 28, Method 1). Use 25 mL of sample solution A (5-g sample). Prepare the standard solution with 25 mL of blank solution B.

Ammonium Oxalate Monohydrate
Ethanedioic Acid Diammonium Salt Monohydrate

$(COONH_4)_2 \cdot H_2O$ Formula Wt 142.11

CAS Number 6009−70−7

REQUIREMENTS

Assay . . . . 99.0−101.0%
$(COONH_4)_2 \cdot H_2O$

MAXIMUM ALLOWABLE

Insoluble matter 0.005%
Residue after ignition 0.02%
Chloride (Cl) 0.002%
Sulfate (SO_4) 0.002%
Heavy metals (as Pb) . . . . 5 ppm
Iron (Fe) . . . . 2 ppm

TESTS

ASSAY. (By oxidation−reduction titration of oxalate). Transfer 5 g of sample, accurately weighed, to a 500-mL volumetric flask. Dissolve it in water, dilute with water to volume, and mix. Pipet 25.0 mL of the solution into 70 mL of water and 5 mL of sulfuric acid in a beaker. Titrate slowly with 0. 1 N potassium permanganate until about 25 mL has been added, then heat the mixture to about 70°C, and complete the titration. One milliliter of 0.1 N potassium permanganate corresponds to 0.007106 g of $(COONH_4)_2 \cdot H_2O$.

INSOLUBLE MATTER.   (Page 14). Use 20 g dissolved in 400 mL of hot water.

RESIDUE AFTER IGNITION.   (Page 15). Ignite 5.0 g.

CHLORIDE.   Dissolve 2.0 g in water plus 10 mL of nitric acid, filter if necessary through a chloride-free filter, and dilute with water to 100 mL. To 25 mL of the solution add 1 mL of silver nitrate reagent solution. Any turbidity should not exceed that produced by 0.01 mg of chloride ion (Cl), in an equal volume of solution containing the quantities of reagents used in the test.

Sample Solution A and Blank Solution B for the Determination of Sulfate, Heavy Metals, and Iron.

Sample Solution A.   Digest 20 g of sample plus 10 mg of sodium carbonate with 40 mL of nitric acid plus 35 mL of hydrochloric acid in a covered beaker on a hot plate at low setting until no more bubbles of gas are evolved. Remove the cover and evaporate until a small amount of crystals form in the beaker. Add 10 mL of 30% hydrogen peroxide, cover the beaker, and digest on the hot plate until reaction ceases. Add an additional 10 mL of 30% hydrogen peroxide, re-cover the beaker, digest on a hot plate at low setting until reaction ceases, remove the cover, and evaporate to dryness. Add 5 mL of dilute hydrochloric acid (1 + 1), cover, digest on the hot plate for 15 min, remove the cover, and evaporate to dryness. Dissolve the residue in about 50 mL of water, filter if necessary, and dilute with water to 100 mL (1 mL = 0.2 g).

Blank Solution B.   Prepare a similar solution containing the quantities of reagents used in preparing sample solution A and treated in the identical manner.

SULFATE. (Page 32, Procedure A, Method 1). Use 12.5 mL of sample solution A (2.5-g sample). Use 12.5 mL of blank solution B to prepare the standard solution.

HEAVY METALS.   (Page 26, Method 1). Dilute 20 mL of sample solution A (4-g sample) with water to 25 mL. Use 20 mL of blank solution B to prepare the standard solution.

IRON. (Page 28, Method 1). Use 25 mL of sample solution A (5-g sample). Prepare the standard solution from 25 mL of blank solution B.

Ammonium Peroxydisulfate
Ammonium Persulfate

$(NH_4)_2S_2O_8$ Formula Wt 228.19

CAS Number 7727−54−0

> *NOTE.* Because of inherent instability, this reagent may be expected to decrease in strength and to increase in acidity during storage. After storage for some time the reagent may fail to meet the specified requirements for assay and acidity. Store in a dry, cool place.

REQUIREMENTS

Assay . . . . $\geq$ 98.0%
$(NH_4)_2S_2O_8$

MAXIMUM ALLOWABLE

Insoluble matter 0.005%
Residue after ignition 0.05%
Titrable free acid . . . . 0.04 meq / g
Chloride and chlorate (as Cl) 0.001%
Heavy metals (as Pb) 0.005%
Iron (Fe) 0.001%
Manganese (Mn) . . . . 0.5 ppm

TESTS

ASSAY. (By oxidation−reduction titration of persulfate). Weigh accurately 0.5 g and add it to 25.0 mL of standard ferrous sulfate solution (0.25 N) in a glass-stoppered flask. Stopper the flask, allow it to stand for 1 h with frequent shaking, and titrate the excess ferrous sulfate with 0.1 N potassium permanganate. One milliliter of 0.1 N potassium permanganate corresponds to 0.01141 g of $(NH_4)_2S_2O_8$.

Standard Ferrous Sulfate Solution. Dissolve 7 g of clear crystals of ferrous sulfate heptahydrate, $FeSO_4 \cdot 7H_2O$, in 80 mL of freshly boiled and cooled water, and add dilute sulfuric acid (1 + 1) to make 100 mL. Standardize the solution with 0.1 N potassium permanganate. The solution must be prepared and standardized at the time of use.

INSOLUBLE MATTER. (Page 14). Use 20 g dissolved in 200 mL of water.

RESIDUE AFTER IGNITION.   (Page 15). Ignite 5.0 g.

TITRABLE FREE ACID.   Dissolve 10 g in 100 mL of water and add 0.01 mL of methyl red indicator solution. If a red color is produced, not more than 4.0 mL of 0.1 N sodium hydroxide should be required to discharge it.

CHLORIDE AND CHLORATE.   (Page 25). Mix 1.0 g with 1 g of sodium carbonate and heat until no more gas is evolved. Dissolve the residue in 20 mL of water, and neutralize with nitric acid.

HEAVY METALS. (Page 26, Method 1). Gently ignite 2.0 g in a porcelain or silica crucible or dish (not in platinum), and to the residue add 1 mL of hydrochloric acid, 1 mL of nitric acid, and about 10 mg of sodium carbonate. Evaporate to dryness on the steam bath, dissolve the residue in about 20 mL of water, and dilute with water to 50 mL. Dilute 10 mL of the solution with water to 25 mL.

IRON.   (Page 28, Method 1). To 1.0 g add 5 mL of water and 10 mL of hydrochloric acid, and evaporate to dryness. Dissolve the residue in 2 mL of hydrochloric acid, dilute with water to 50 mL, and use this solution without further acidification. Prepare the standard solution from the residue remaining from evaporation of 10 mL of hydrochloric acid.

MANGANESE.   Gently ignite 20 g (not in platinum) until the sample is decomposed, and finally ignite at 600°C until the sample is volatilized. Dissolve the residue by boiling for 5 min with 35 mL of water plus 10 mL of nitric acid, 5 mL of sulfuric acid, and 5 mL of phosphoric acid, and cool. Prepare a standard containing 0.01 mg of manganese ion (Mn) in an equal volume of solution containing the quantities of reagents used to dissolve the residue. To each solution add 0.25 g of potassium periodate, boil gently for 5 min, and cool. Any pink color in the solution of the sample should not exceed that in the standard.

Ammonium Phosphate, Dibasic
Diammonium Hydrogen Phosphate

$(NH_4)_2HPO_4$ Formula Wt 132.06

CAS Number 7783–28–0

REQUIREMENTS

Assay . . . . $\geq$ 98.0%
$(NH_4)_2HPO_4$
pH of a 5% solution . . . . 7.7–8.1 at 25°C

MAXIMUM ALLOWABLE

Insoluble matter 0.005%
Ammonium hydroxide precipitate 0.005%
Chloride (Cl) 0.001%
Nitrate (NO_3) 0.003%
Sulfur compounds (as SO_4) 0.004%
Arsenic (As) . . . . 2 ppm
Heavy metals (as Pb) 0.001%
Iron (Fe) 0.001%
Potassium (K) 0.005%
Sodium (Na) 0.005%

TESTS

ASSAY. (By acid–base titration). Weigh accurately about 0.5 g and dissolve in 50 mL of water. Titrate with 0.1 N hydrochloric acid to the potentiometric end point. One milliliter of 0.1 N hydrochloric acid corresponds to 0.01321 g of $(NH_4)_2HPO_4$.

pH OF A 5% SOLUTION. (Page 43). The pH should be 7.7–8.1 at 25°C.

INSOLUBLE MATTER. (Page 14). Use 20 g dissolved in 200 mL of water. Save the filtrate separate from the washings for the test for ammonium hydroxide precipitate.

AMMONIUM HYDROXIDE PRECIPITATE. To the filtrate from the test for insoluble matter add successively enough hydrochloric acid to make the solution acid to methyl red and enough ammonium hydroxide to make the solution again alkaline to methyl red. Heat to boiling, boil gently for 5 min, and cool. Filter, wash, and ignite the precipitate.

CHLORIDE.   (Page 25). Use 1.0 g of sample and 3 mL of nitric acid.

NITRATE.   (Page 29). For sample solution A use 0.50 g. For control solution B use 0.50 g and 1.5 mL of standard nitrate solution.

SULFUR COMPOUNDS.   Dissolve 10 g in about 80 mL of water, add 5 mL of bromine water, and heat to boiling. Add 12 mL of hydrochloric acid, heat to boiling again, and add 5 mL of barium chloride reagent solution. Digest in a covered beaker on a hot plate at low setting for 2 h and allow to stand overnight. If a precipitate is formed, filter, wash thoroughly, and ignite. Correct for the weight obtained in a complete blank test.

ARSENIC.   (Page 23). Use 1.5 g. For the standard use 0.003 mg of arsenic (As).

HEAVY METALS.   (Page 26, Method 1). Dissolve 4.0 g in 10 mL of water, add 15 mL of 2 N hydrochloric acid, and dilute with water to 32 mL. Use 24 mL to prepare the sample solution, and use the remaining 8.0 mL to prepare the control solution.

IRON.   (Page 28, Method 2). Use 1.0 g.

POTASSIUM AND SODIUM.   Determine the potassium and sodium content by the flame atomic absorption spectrophotometric method described on page 35.

> *Sample Stock Solution.*   Dissolve 10.0 g of sample with water in a 100-mL volumetric flask and dilute to the mark with water (0.1 g/mL).

Analyze the solutions by means of a suitable atomic absorption spectrophotometer, using the conditions outlined in the following table. Calculate the metal content of the sample by the method of standard additions.

Element	Wavelength (nm)	Sample Wt (g)	Standard Added (mg)	Flame Type*	Background Correction
K	766.5	1.0	0.05; 0.10	A/A	No
Na	589.0	0.1	0.005; 0.01	A/A	No

*A/A is air/acetylene.

Ammonium Phosphate, Monobasic

Ammonium Dihydrogen Phosphate

$NH_4H_2PO_4$ Formula Wt 115.03

CAS Number 7722–76–1

REQUIREMENTS

Assay . . . . ≥ 98.0%
                                                                 $NH_4H_2PO_4$
pH of a 5% solution . . . . 3.8–4.4 at 25°C

MAXIMUM ALLOWABLE

Insoluble matter 0.005%
Ammonium hydroxide precipitate 0.005%
Chloride (Cl) . . . . 5 ppm
Nitrate (NO_3) 0.001%
Sulfur compounds (as SO_4) 0.005%
Arsenic (As) . . . . 0.5 ppm
Heavy metals (as Pb) . . . . 5 ppm
Iron (Fe) 0.001%
Potassium (K) 0.005%
Sodium (Na) 0.005%

TESTS

ASSAY.   (By acid–base titration). Weigh accurately about 2 g and dissolve in 50 mL of water. Add exactly 5.0 mL of 1 N hydrochloric acid and stir until the sample is completely dissolved. Titrate the excess acid, stirring constantly, with 1 N sodium hydroxide to the inflection point occurring at about pH 4.0, as measured with a standardized pH meter and electrode system.

Calculate A, the volume of 1 N hydrochloric acid consumed by the sample. Continue the titration with 1 N sodium hydroxide to the inflection point occurring at about pH 8. Calculate B, the volume of 1 N sodium hydroxide required in the titration between the two inflection points.

CALCULATION

$A = (5.0 \times N \text{ of HCl}) - (V_1 \times N \text{ of NaOH})$
    where V_1 is mL of NaOH needed to reach the first inflection
    point

$B = (V_2 - V_1)(N \text{ of NaOH})$
  where V_2 is mL of NaOH needed to reach the second inflection
  point from the beginning of the titration

If A is a negative integer, there is a slight excess of acidity. Calculate
the assay as follows:

$$\%NH_4H_2PO_4 = (B + A) \times 11.503/\text{sample wt, g}$$

pH OF A 5% SOLUTION.   (Page 43). The pH should be 3.8–4.4 at
25°C.

INSOLUBLE MATTER.   (Page 14). Use 20 g dissolved in 200 mL of
water. Save the filtrate separate from the washings for the test for
ammonium hydroxide precipitate.

AMMONIUM HYDROXIDE PRECIPITATE.   To the filtrate from the
test for insoluble matter add enough ammonium hydroxide to make
the solution alkaline to methyl red. Heat to boiling, boil gently for 5
min, and cool. Filter, wash, and ignite the precipitate.

CHLORIDE.   (Page 25). Use 2.0 g of sample and 3 mL of nitric acid.

NITRATE.   (Page 29). For sample solution A use 1.5 g. For control
solution B use 1.5 g and 1.5 mL of standard nitrate solution.

SULFUR COMPOUNDS.   Dissolve 10 g in 100 mL of water, add 6 mL
of hydrogen peroxide, and boil the solution for a few min. Add 6.5 mL
of hydrochloric acid, boil, add 5 mL of barium chloride reagent
solution, digest in a covered beaker on the steam bath for 2 h, and
allow to stand overnight. If a precipitate is formed, filter, wash
thoroughly, and ignite. Correct for the weight obtained in a complete
blank test.

ARSENIC.   (Page 23). Use 6.0 g. For the standard use 0.003 mg of
arsenic (As).

HEAVY METALS.   (Page 26, Method 1). Dissolve 6.0 g in 30 mL of
water. Use 25 mL to prepare the sample solution, and use the
remaining 5.0 mL to prepare the control solution.

IRON.   (Page 28, Method 2). Use 1.0 g. Add 3 mL of dilute ammonium hydroxide (1 + 4) to both the sample and the standard solutions.

POTASSIUM AND SODIUM. Determine the potassium and sodium content by the flame atomic absorption spectrophotometric method described on page 35.

> *Sample Stock Solution.* Dissolve 10.0 g of sample with water in a 100-mL volumetric flask and dilute to the mark with water (0.1 g/mL).

Analyze the solutions by means of a suitable atomic absorption spectrophotometer, using the conditions outlined in the following table. Calculate the metal content of the sample by the method of standard additions.

Element	Wavelength (nm)	Sample Wt (g)	Standard Added (mg)	Flame Type*	Background Correction
K	766.5	1.0	0.05; 0.10	A/A	No
Na	589.0	0.1	0.005; 0.01	A/A	No

*A/A is air/acetylene.

Ammonium Sulfamate

$NH_4OSO_2NH_2$ Formula Wt 114.13

CAS Number 7773–06–0

REQUIREMENTS

Assay . . . . ≥ 98.0%
                                                                      $NH_4OSO_2NH_2$
Melting point . . . . Within a range of
                                                                      2.0°C, including
                                                                      133.0°C

MAXIMUM ALLOWABLE

Insoluble matter 0.02%
Residue after ignition 0.10%
Heavy metals (as Pb) . . . . 5 ppm

TESTS

ASSAY. (By oxidation–reduction titration). Weigh accurately about 0.35 g and transfer to a 250-mL flask. Add 75 mL of water and 5 mL of sulfuric acid, swirl to dissolve the sample, and titrate slowly with 0.1 M sodium nitrite, shaking the flask vigorously from time to time. Titrate dropwise near the end point, shaking after each addition, until a blue color is produced immediately when a glass rod dipped into the titrated solution is streaked on starch–iodide test paper. One milliliter of 0.1 M sodium nitrite corresponds to 0.01141 g of $NH_4OSO_2NH_2$.

INSOLUBLE MATTER. (Page 14). Use 5.0 g dissolved in 100 mL of water.

RESIDUE AFTER IGNITION. (Page 15). Ignite 1.0 g.

HEAVY METALS. (Page 26, Method 1). Dissolve 6.0 g in 20 mL of water, and dilute with water to 30 mL. Use 25 mL to prepare the sample solution, and use the remaining 5.0 mL to prepare the control solution.

Ammonium Sulfate

$(NH_4)_2SO_4$

Formula Wt 132.14

CAS Number 7783–20–2

REQUIREMENTS

Assay	≥ 99.0% $(NH_4)_2SO_4$
pH of a 5% solution	5.0–6.0 at 25°C

MAXIMUM ALLOWABLE

Insoluble matter	0.005%
Residue after ignition	0.005%
Chloride (Cl)	5 ppm
Nitrate (NO_3)	0.001%
Phosphate (PO_4)	5 ppm
Heavy metals (as Pb)	5 ppm
Iron (Fe)	5 ppm

TESTS

ASSAY. (By acidimetry). Weigh, to the nearest 0.1 mg, 2 g of sample and dissolve with 40 mL of water in a 250-mL glass-stoppered flask. Add 20 mL of formaldehyde, previously mixed with 20 mL of water, and neutralize with 0.1 N sodium hydroxide solution, using 0.15 mL of phenolphthalein indicator solution. Mix, allow to stand for 30 min, and then titrate with 1 N sodium hydroxide solution to a pink color that persists for 5 min. One milliliter of 1 N sodium hydroxide corresponds to 0.06606 g of $(NH_4)_2SO_4$.

pH OF A 5% SOLUTION.   (Page 43). The pH should be 5.0–6.0 at 25°C.

INSOLUBLE MATTER.   (Page 14). Use 20 g dissolved in 200 mL of water.

RESIDUE AFTER IGNITION.   (Page 15). Ignite 20 g.

CHLORIDE.   (Page 25). Use 2.0 g.

NITRATE.   (Page 29). For sample solution A use 1.0 g. For control solution B use 1.0 g and 1 mL of standard nitrate solution.

PHOSPHATE.   (Page 30, Method 1). Dissolve 4.0 g in 25 mL of 0.5 N sulfuric acid.

HEAVY METALS.   (Page 26, Method 1). Dissolve 6.0 g in 20 mL of water, and dilute with water to 30 mL. Use 25 mL to prepare the sample solution, and use the remaining 5.0 mL to prepare the control solution.

IRON.   (Page 28, Method 1). Use 2.0 g.

Ammonium Thiocyanate

NH_4SCN Formula Wt 76.12

CAS Number 1762–95–4

REQUIREMENTS

Assay . . . . $\geq$ 97.5% NH_4SCN
Appearance. . . . . Colorless or white
                                                                      crystals
pH of a 5% solution . . . . 4.5–6.0 at 25°C

MAXIMUM ALLOWABLE

Insoluble matter 0.005%
Residue after ignition 0.025%
Chloride (Cl) 0.005%
Sulfate (SO_4) 0.005%
Heavy metals (as Pb) . . . . 5 ppm
Iron (Fe) . . . . 3 ppm
Iodine-consuming substances . . . . 0.004 meq / g

TESTS

ASSAY. (By argentimetric titration of thiocyanate content). Weigh, to the nearest 0.1 mg, 7 g of sample. Dissolve with 100 mL of water in a 1-L volumetric flask, dilute with water to the mark, and mix well. Transfer a 50.0-mL aliquot to a 250-mL glass-stoppered iodine-type flask, add 5 mL of 5 N nitric acid, then add with agitation exactly 50.0 mL of 0.1 N silver nitrate solution, and shake vigorously. Add 2 mL of ferric ammonium sulfate indicator solution, and chill in an ice bath to approximately 10°C or lower. Titrate the excess silver nitrate with 0.1 N ammonium thiocyanate solution. Near the end point, shake after the addition of each drop.

$$\% \ NH_4SCN = \frac{[(50.0 \ \text{mL} \times N \ \text{of AgNO}_3) - (\text{mL} \times N \ \text{of NH}_4SCN)](152.23)}{\text{sample wt, g}}$$

pH OF A 5% SOLUTION. (Page 43). The pH should be 4.5–6.0 at 25°C.

INSOLUBLE MATTER. (Page 14). Use 20 g dissolved in 150 mL of water.

RESIDUE AFTER IGNITION. (Page 15). Ignite 3.3 g. Retain the residue for the test for iron.

CHLORIDE. (Page 25). Dissolve 1.0 g in 20 mL of water in a small flask. Add 10 mL of 25% sulfuric acid reagent solution and 7 mL of 30% hydrogen peroxide. Evaporate to 20 mL by boiling in a well-ventilated hood, add 17 mL of water, and evaporate again. Repeat until all the cyanide has been volatilized. Cool, filter if necessary through a chloride-free filter, and dilute with water to 100 mL. Use 20 mL of this solution.

SULFATE. (Page 32, Procedure A, Method 1). Use 1.0-g sample, and allow 30 min for the turbidity to form.

HEAVY METALS. (Page 26). Dissolve 6.0 g in 20 mL of water, and dilute with water to 30 mL. Use 25 mL to prepare the sample solution, and use the remaining 5.0 mL to prepare the control solution.

IRON. (Page 28, Method 1). To the residue after ignition add 3 mL of dilute hydrochloric acid (1 + 1), cover with a watch glass, and digest on the steam bath for 15–20 min. Remove the watch glass and evaporate to dryness. Dissolve the residue in 2 mL of hydrochloric acid, filter if necessary, and dilute with water to 50 mL. Use the solution without further acidification.

IODINE-CONSUMING SUBSTANCES. Dissolve 5.0 g in 50 mL of water plus 1.7 mL of 10% sulfuric acid reagent solution. Add 1 g of potassium iodide and 2 mL of starch indicator solution, and titrate with 0.01 N iodine solution. Not more than 2.0 mL of the iodine solution should be required.

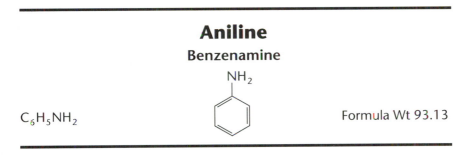

Aniline
Benzenamine

$C_6H_5NH_2$

Formula Wt 93.13

CAS Number 62−53−3

NOTE. This product darkens to a reddish-brown color on storage.

REQUIREMENTS

Assay . . . . $\geq 99.0\%$ $C_6H_5NH_2$

MAXIMUM ALLOWABLE

Color (APHA). 250
Residue after ignition 0.005%
Chlorobenzene (C_6H_5Cl). 0.01%
Hydrocarbons . . . . Passes test
Nitrobenzene ($C_6H_5NO_2$) . . . . Passes test (limit
about 0.001%)

TESTS

ASSAY AND CHLOROBENZENE. Analyze the sample by gas chromatography, using the general parameters cited on page 56. The following specific conditions are also required.

Column: Type I, methyl silicone

Measure the area under all peaks and calculate the area percent for aniline and chlorobenzene. Correct for water content.

COLOR (APHA). (Page 17).

RESIDUE AFTER IGNITION. (Page 15). Evaporate 20 g (20 mL) to dryness in the hood and ignite.

HYDROCARBONS. Mix 5 mL with 10 mL of hydrochloric acid. The solution should be clear after dilution with 15 mL of cold water.

NITROBENZENE

Sample Solution A. Place 10 mL of methanol in a 25-mL glass-stoppered graduated cylinder.

Control Solution B. Place 10 mL of methanol in a 25-mL glass-stoppered graduated cylinder and add 1 mL of a methanol solution containing 0.10 mg of nitrobenzene per mL.

Immerse the cylinders in a beaker of water at or below room temperature, and add 10 mL of the aniline and 2.5 mL of hydrochloric acid to each. Mix well and bring to room temperature.

Transfer a portion of sample solution A to a polarographic cell and deaerate with nitrogen or hydrogen. Record the polarogram from

−0.2 to −0.7 V vs. SCE with a current sensitivity of 0.02 μA/mm. Repeat this procedure with control solution B. The diffusion current in sample solution A should not exceed the difference in diffusion current between control solution B and sample solution A.

Anthrone
9(10*H*)-Anthracenone

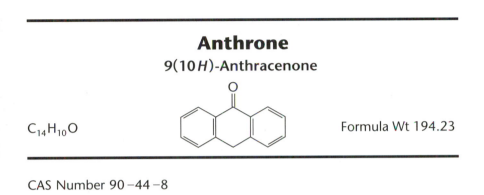

$C_{14}H_{10}O$ Formula Wt 194.23

CAS Number 90−44−8

REQUIREMENTS

Appearance. . . . . Off-white to light
                                                              yellow crystals
Melting point . . . . No more than a 3°
                                                              range, including
                                                              156°C
Sensitivity to carbohydrates . . . . Passes test

MAXIMUM ALLOWABLE

Absorbance of reagent solution . . . . Passes test
Solubility in ethyl acetate . . . . Passes test

TESTS

ABSORBANCE OF REAGENT SOLUTION.   Measure the absorbance of the reagent solution (*see* the test for sensitivity to carbohydrates) in a spectrophotometer, using 1.00-cm cells, with sulfuric acid as reference. Record the absorbance at 425 nm and 620 nm, respectively. The absorbance should not exceed 0.75 at 425 nm and 0.045 at 620 nm.

SOLUBILITY IN ETHYL ACETATE.   Dissolve 1 g in 50 mL of freshly distilled ethyl acetate in a dry 50-mL glass-stoppered graduate. The mixture should not be heated, dissolution should be complete, and the resulting solution should be clear.

SENSITIVITY TO CARBOHYDRATES

Reagent Solution. Slowly add 132 mL of sulfuric acid to 68 mL of ice-cold water in an ice bath. Allow the mixture to cool to room temperature, add 0.100 g of the sample, and heat if necessary to dissolve.

Glucose Standard. (1,000 μg in 1 mL). Dissolve 1.00 g of glucose in water, and dilute with water to 100 mL. Pipet 10.0 mL of this solution (1 mL = 10 mg) into a 100-mL volumetric flask, and dilute to volume with water.

Mark four 25-mL volumetric flasks as 0 μg (blank), 250 μg, 500 μg, and 750 μg. Add the glucose standard and water as specified in the following table, and dilute to volume with the reagent solution. Place all flasks in a vigorously boiling water bath for 7.0 min, allow the flasks to cool to room temperature, and determine the absorbance of each at 620 nm against the blank in 1.00-cm cells. A plot of the absorbance against concentration should be linear, and the absorbance of the 750-μg flask should not be less than 0.90.

Flask (μg)	Glucose Standard (mL)	Water (mL)
0	0	0.75
250	0.25	0.50
500	0.50	0.25
750	0.75	0

Antimony Trichloride
Antimony(III) Chloride

$SbCl_3$ Formula Wt 228.12

CAS Number 10025–91–9

REQUIREMENTS

Assay . . . . $\geq$ 99.0% $SbCl_3$

MAXIMUM ALLOWABLE

Insoluble in chloroform. 0.05%
Sulfate (SO_4) 0.005%
Arsenic (As) 0.02%
Copper (Cu) 0.001%
Iron (Fe) 0.002%
Lead (Pb) 0.005%
Substances not precipitated by
  hydrogen sulfide (as sulfates) 0.1%

TESTS

ASSAY. (By oxidation–reduction titration). Weigh accurately 0.5 g and, in a glass-stoppered Erlenmeyer flask, dissolve in 5 mL of 10% hydrochloric acid reagent solution. When dissolution is complete, add a solution of 4.0 g of potassium sodium tartrate tetrahydrate in 30 mL of water. Stopper and swirl. Add 50 mL of a cold saturated solution of sodium bicarbonate, and titrate immediately with 0.1 N iodine, adding 3 mL of starch indicator solution near the end of the titration. One milliliter of 0.1 N iodine corresponds to 0.01140 g of $SbCl_3$.

INSOLUBLE IN CHLOROFORM. Dissolve 5.0 g in 25 mL of chloroform, filter through a tared filtering crucible, wash the crucible with several portions of chloroform, and dry at 105°C.

> NOTE. Weigh quickly to avoid moisture absorption and the formation of insoluble oxychlorides. The use of anhydrous, ethanol-free chloroform is recommended.

SULFATE. Dissolve 10 g in the minimum volume of hydrochloric acid required to achieve complete dissolution, dilute with water to 75 mL, neutralize with ammonium hydroxide, and filter. Add 2 mL of hydrochloric acid to the filtrate, dilute with water to 100 mL, and heat to boiling. Add 10 mL of barium chloride reagent solution, digest in a covered beaker on the hot plate at low setting for 2 h, and allow to stand overnight. If a precipitate is formed, filter, wash thoroughly, and ignite. Correct for the weight obtained in a complete blank test.

ARSENIC. (Page 23). Dissolve 1.0 g in 5 mL of hydrochloric acid, pour this solution into a solution of 2 g of stannous chloride dihydrate ($SnCl_2 \cdot 2H_2O$) in 2 mL of hydrochloric acid, and allow to stand overnight. If no precipitate is visible, the arsenic content is less than the limit and the test need not be completed. If a precipitate has formed, filter it on a glass-fiber filter. Wash the beaker and the precipitate with two 5-mL portions of hydrochloric acid and then with water, passing the washings through the filter. Discard the filtrate. Dissolve the precipitate by passing a mixture of 5 mL of hydrochloric acid and 0.15 mL of bromine through the filter, and wash the filter with 5 mL of hydrochloric acid and then with water. Collect the filtrate and washings in the beaker in which the precipitation was made, warm the solution on the steam bath to remove the excess bromine, and dilute with water to 1 L. Dilute 15 mL of this solution with water to 35 mL. For the standard use 0.003 mg of arsenic (As).

COPPER.   Heat 1.0 g in a 50-mL beaker on a hot plate in a well-ventilated hood until the sample has been completely volatilized. (Note. A temperature of at least 230°C is required.) Dissolve the residue in 0.25 to 0.30 mL of dilute hydrochloric acid (1 + 1) and evaporate to dryness. Repeat the addition of hydrochloric acid and the evaporation to dryness. For the standard add 0.50 mL of dilute hydrochloric acid (1 + 1) to a solution containing 0.01 mg of copper ion (Cu), evaporate to dryness, dissolve the residue in 0.50 mL of dilute hydrochloric acid (1 + 1), and evaporate again to dryness. Dissolve the residues separately in 0.50 mL of 10% hydrochloric acid reagent solution and add a few milliliters of water to each. Transfer the solutions to Nessler tubes and dilute with water to 10 mL. Add to each tube 5 mL of ammonium citrate reagent solution, 1.0 mL of 10% ammonium hydroxide reagent solution, and 1 mL of 0.1% sodium diethyldithiocarbamate reagent solution, and mix well. Add 5 mL of isopentyl alcohol to each, shake for 1 min, and allow the layers to separate. Add 5 mL of ethyl ether and swirl gently for 1 min. Any yellow color in the upper layer from the solution of the sample should not exceed that in the upper layer from the standard.

Sample Solution A and Blank Solution B for the Determination of Iron and Lead.

Sample Solution A.   Dissolve 5.0 g in 15 mL of nitric acid, heat on the steam bath, and finally evaporate to dryness. Heat the residue gently with a flame to expel any remaining traces of acid or moisture, cool, and digest for 10 min with 5 mL of warm nitric acid. Add 20 mL of water, filter, wash thoroughly, and dilute with water to 50 mL (1 mL = 0.1 g.).

Blank Solution B.   Treat 5 mg of sodium carbonate in the same manner described for sample solution A.

IRON.   (Page 28, Method 1). Evaporate 5 mL of sample solution A (0.5-g sample) to dryness on the steam bath. To the residue add 2 mL of hydrochloric acid and 20 mL of water, and heat until dissolution is complete. Use this solution without further acidification. Prepare the standard solution from 5 mL of blank solution B, treated exactly as the sample solution.

LEAD.   For the sample use 20 mL of sample solution A (2-g sample). For the standard add 0.1 mg of lead ion (Pb) to 20 mL of blank solution B. Add 0.1 g of potassium sulfate to the solutions and evaporate to dryness on the steam bath. Digest each residue with a hot solution of 10 g of ammonium acetate in 10 mL of water, filter, and cool to room temperature. Add 0.5 mL of potassium chromate reagent solution to each filtrate and let stand for 30 min. Any turbidity in the solution of the sample should not exceed that in the standard.

SUBSTANCES NOT PRECIPITATED BY HYDROGEN SULFIDE.   Dissolve 2.0 g in 5 mL of hydrochloric acid, dilute with water to 100 mL, and pass a stream of hydrogen sulfide through the solution to precipitate the antimony completely. Allow the precipitate to settle and decant the supernatant liquid through a retentive filter, but do not wash. To 50 mL of the clear filtrate add 0.5 mL of sulfuric acid and evaporate to dryness. Finally, ignite at $800 \pm 25°C$ for 15 min.

Arsenic Trioxide
Arsenic(III) Oxide
Reductometric Standard

As_2O_3 Formula Wt 197.84

CAS Number 1327–53–3

REQUIREMENTS

Assay 99.95–100.05%
As_2O_3

MAXIMUM ALLOWABLE

Residue after ignition 0.02%
Insoluble in dilute hydrochloric acid 0.01%
Chloride (Cl) 0.005%
Sulfide (S) . . . . Passes test (limit
                                                          about 0.001%)
Antimony (Sb) 0.05%
Lead (Pb) 0.002%
Iron (Fe) . . . . 5 ppm

TESTS

ASSAY. (By oxidation–reduction titration of arsenic). This assay is based on a direct titration with NIST Standard Reference Material Potassium Dichromate. Accurately weighed portions of the arsenic trioxide are reacted with accurately weighed portions of the NIST potassium dichromate of such size as to provide a small excess of the latter. The excess dichromate is determined by titration with standardized ferrous ammonium sulfate solution, using a potentiometrically determined end point.

> *CAUTION.* Due to the small tolerance in assay limits for this reductometric standard, extreme care must be observed in the weighing, transferring, and titrating operations in the following procedure. Strict adherence to the specified sample weights and final titration volumes is absolutely necessary. It is recommended that the titrations be run at least in duplicate. Duplicate values for the assay of this material should agree within 2 parts in 5000 to be acceptable for averaging.

Procedure. Place 2.00-g portions of the arsenic trioxide in a weighing bottle, dry for 1 h at 105°C, and cool for 2 h in a desiccator. Weigh accurately to within 0.1 mg and transfer to 400-mL beakers. Weigh each bottle again and determine by difference to the nearest 0.1 mg the weight of each portion of arsenic trioxide.

Weigh accurately to within 0.1 mg an amount of NIST potassium dichromate equivalent to a slight excess of the weight of arsenic trioxide taken, and transfer it to a 50-mL beaker. Dissolve the arsenic trioxide in 20 mL of 20% sodium hydroxide solution, add 100 mL of water and 10 mL of dilute sulfuric acid (1 + 1), and stir. Transfer the potassium dichromate to the solution, using water to loosen any crystals adhering to the beakers. Stir until all of the potassium dichromate is dissolved, and let stand for 10 min.

Add 20 mL of dilute sulfuric acid (1 + 1), dilute with water to 300 mL, and stir. Titrate the excess dichromate with standardized 0.02 N

ferrous ammonium sulfate solution, using a platinum indicator electrode and a calomel reference electrode for the measurement of the
potential difference in millivolts. The end point of the titration is
determined potentiometrically, using the second derivative method
(page 51). A correction for "blank" oxidants or reductants must be
applied by titrating an accurately weighed 40-mg portion of the NIST
potassium dichromate with the standardized ferrous ammonium sulfate in an equal volume of solution containing the quantities of
reagents used in this assay. The "blank" is equal to the calculated
milliequivalents of $K_2Cr_2O_7$ minus the calculated milliequivalents of
$Fe(NH_4)_2(SO_4)_2 \cdot 6H_2O$. A positive "blank" correction results from
other oxidants and a negative correction from other reductants.

CALCULATIONS

$$M_1 = \left[\frac{W_1}{49.0307} - VN + B \right] \times 1.00032F$$

where
M_1 = number of milliequivalents of $K_2Cr_2O_7$ equivalent to As_2O_3
W_1 = weight of $K_2Cr_2O_7$ in milligrams
V = volume of $Fe(NH_4)_2(SO_4)_2$ solution in milliliters
N = concentration of $Fe(NH_4)_2(SO_4)_2$ solution in equivalents
per liter
B = blank correction in milliequivalents (may be either positive
or negative)
F = assay value of the NIST standard in percent divided by 100
(The equivalent formula weight of $K_2Cr_2O_7$ is 49.0307, and the
conversion factor for air to vacuum weight of $K_2Cr_2O_7$ is 1.00032.)
Calculate the assay value of the arsenic trioxide from

$$A = \%As_2O_3 = \frac{(49.4603)(M_1)}{(1.00017)W_2} 100$$

(W_2 = weight of As_2O_3 in milligrams; 1.00017 is the conversion factor
for air to vacuum weight of As_2O_3.)

*Preparation and Standardization of Ferrous Ammonium
Sulfate Solution, 0.02 N.* Weigh 8.0 g of $Fe(NH_4)_2(SO_4)_2 \cdot$
$6H_2O$ and transfer to a 1-L flask. Add 200 mL of dilute sulfuric
acid (1 + 19) and, after the salt is dissolved, dilute to the liter
mark with the dilute acid. Mix thoroughly.

Transfer a 40-mg portion of the potassium dichromate, accurately
weighed to within 0.1 mg, to a 400-mL beaker. Add 300 mL of dilute
sulfuric acid (1 + 19) and stir to dissolve the salt. Titrate the solution

with the 0.02 N ferrous ammonium sulfate, using a potentiometrically determined end point. The concentration of the ferrous ammonium sulfate, in equivalents per liter (N), is calculated from the
following equation:

$$N = \left[\frac{W}{49.03}\right]\left[\frac{1000}{V}\right]$$

where
   N = concentration of ferrous ammonium sulfate in equivalents
      per liter
   W = weight of $K_2Cr_2O_7$ in grams
   V = volume of $Fe(NH_4)_2(SO_4)_2$ solution in milliliters

RESIDUE AFTER IGNITION.   (Page 15). Ignite 5.0 g in a tared platinum dish. Retain the residue for the test for iron.

INSOLUBLE IN DILUTE HYDROCHLORIC ACID.   To 10 g add 90 mL
of dilute hydrochloric acid (3 + 7) and 10 mL of hydrogen peroxide.
Cover with a watch glass and allow to stand until the sample goes
into solution, heating if necessary. Heat to boiling, digest in a covered
beaker on the steam bath for 1 h, and filter through a tared filtering
crucible. Retain the filtrate for the antimony and lead test, wash the
insoluble matter thoroughly, and dry at 105°C.

CHLORIDE.   (Page 25). Dissolve 1.0 g in 10 mL of dilute ammonium
hydroxide (1 + 2) with the aid of gentle heating, and dilute with
water to 100 mL. Neutralize 20 mL with nitric acid.

SULFIDE.   Dissolve 1.0 g in 10 mL of 10% sodium hydroxide reagent
solution, and add 0.05 mL of lead acetate solution (about 10%). The
color should be the same as that of an equal volume of sodium
hydroxide solution to which only the lead acetate is added.

ANTIMONY AND LEAD.   Determine the antimony and lead by the
flame atomic absorption spectrophotometric method described on
page 35.

   Sample Stock Solution.   Dilute the filtrate retained from the
   insoluble in hydrochloric acid test with water to the mark in a
   200-mL volumetric flask (0.05 g/mL).

Analyze the solutions by means of a suitable atomic absorption
spectrophotometer, using the conditions outlined in the following

table. Calculate the metal content of the sample by the method of standard additions.

Element	Wavelength (nm)	Sample Wt (g)	Standard Added (mg)	Flame Type*	Background Correction
Sb	217.6	1.0	0.5; 1.0	A/A	Yes
Pb	217.0	1.0	0.05; 0.10	A/A	Yes

*A/A is air/acetylene.

IRON. (Page 28, Method 1). To the residue from the test for residue after ignition, add 3 mL of dilute hydrochloric acid (1 + 1), and warm. Add 5 mL of hydrochloric acid, dilute with water to 125 mL, and use 50 mL of this solution without further acidification.

Ascorbic Acid

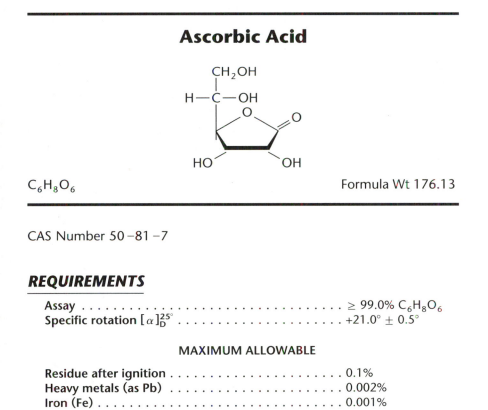

$C_6H_8O_6$ Formula Wt 176.13

CAS Number 50–81–7

REQUIREMENTS

Assay . . . . $\geq$ 99.0% $C_6H_8O_6$
Specific rotation $[\alpha]_D^{25°}$ . . . . +21.0° ± 0.5°

MAXIMUM ALLOWABLE

Residue after ignition 0.1%
Heavy metals (as Pb) 0.002%
Iron (Fe) 0.001%

TESTS

ASSAY. (By oxidation–reduction titration). Weigh accurately 0.4 g and in a conical flask dissolve in a mixture of 100 mL of water (made oxygen-free by bubbling nitrogen gas through it) and 25 mL of 10% sulfuric acid. Swirl and titrate immediately with 0.1 N iodine, adding 3 mL of starch indicator near the end of the titration. One milliliter of 0.1 N iodine corresponds to 0.008806 g of $C_6H_8O_6$.

SPECIFIC ROTATION. Weigh accurately about 10 g, dissolve in 90 mL of oxygen-free water in a 100-mL volumetric flask, and dilute with water to volume. Adjust the temperature of the solution to 25°C. Observe the optical rotation in a polarimeter at 25°C using sodium light, and calculate the specific rotation.

RESIDUE AFTER IGNITION. (Page 15). Ignite 1.0 g. Moisten the char with 1 mL of sulfuric acid. Retain the residue for the test for iron.

HEAVY METALS. (Page 27, Method 2). Use 1.0 g.

IRON. (Page 28, Method 1). To the residue from the test for Residue after ignition add 3 mL of dilute hydrochloric acid (1 + 1) and 0.10 mL of nitric acid, cover with a watch glass, and digest on the steam bath for 15–20 min. Remove the watch glass and evaporate to dryness. Dissolve the residue in a mixture of 2 mL of hydrochloric acid and 10 mL of water, dilute with water to 50 mL, and use without further acidification.

Aurin Tricarboxylic Acid, [tri]Ammonium Salt

5-[(3-Carboxy-4-hydroxyphenyl)(3-carboxy-4-oxo-2,5-cyclohexadien-1-ylidene)methyl]-2-hydroxybenzoic Acid, Triammonium Salt

Aluminon

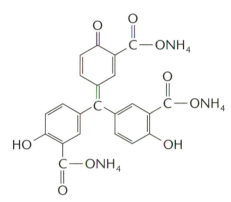

$(HOC_6H_3COONH_4)_2C{:}C_6H_3(COONH_4){:}O$ Formula Wt 473.44

CAS Number 569–58–4

REQUIREMENTS

Sensitivity to aluminum . . . . Passes test

MAXIMUM ALLOWABLE

Insoluble matter 0.1%
Residue after ignition 0.2%

TESTS

INSOLUBLE MATTER.   (Page 14). Use 1.0 g dissolved in 100 mL of water plus 0.5 mL of dilute ammonium hydroxide (10% NH_3).

RESIDUE AFTER IGNITION.   (Page 15). Ignite 0.50 g.

SENSITIVITY TO ALUMINUM.   Dissolve 25 mg in 20 mL of water and 0.1 mL of dilute ammonium hydroxide (10% NH_3). Add 0.1 mL of this solution and 0.1 mL of acetic acid to 10 mL of a solution containing 0.001 mg of aluminum ion (Al). A distinct pink color should appear within 15 min.

Barium Acetate

$(CH_3COO)_2Ba$ Formula Wt 255.42

CAS Number 543−80−6

REQUIREMENTS

Assay . . . . 99.0−102.0%
$(CH_3COO)_2Ba$

MAXIMUM ALLOWABLE

Insoluble matter 0.01%
Chloride (Cl) 0.001%
Oxidizing substances (as NO_3) 0.005%
Substances not precipitated by sulfuric acid 0.1%
Calcium (Ca) 0.05%
Heavy metals (as Pb) . . . . 5 ppm
Iron (Fe) 0.001%
Strontium (Sr) 0.2%

TESTS

ASSAY. (By complexometry for barium). Weigh accurately 0.28 g, transfer to a 400-mL beaker, and dissolve in 100 mL of carbon dioxide-free water. Add 100 mL of ethanol, 10 mL of ammonium hydroxide, and 3.0 mL of metalphthalein-screened indicator solution. Titrate immediately with standard 0.1 M EDTA to a color change from magenta to grey-green. One milliliter of 0.1 M EDTA corresponds to 0.02554 g of $(CH_3COO)_2Ba$.

INSOLUBLE MATTER. (Page 14). Use 10 g dissolved in 100 mL of water.

CHLORIDE. (Page 25). Use 1.0 g.

OXIDIZING SUBSTANCES. Place 0.10 g in a dry beaker. Cool the beaker thoroughly in an ice bath and add 22 mL of sulfuric acid that has been cooled to ice-bath temperature. Allow the mixture to warm to room temperature and swirl the beaker at intervals to effect gentle dissolution with slow evolution of the acetic acid vapor. When dissolution is complete, add 3 mL of diphenylamine reagent solution and digest on the steam bath for 90 min. Prepare a standard by evaporating to dryness a solution containing 0.005 mg of nitrate (0.5 mL of the nitrate standard solution) and 0.01 g of sodium carbonate. Treat the residue exactly like the sample. Any color produced in the solution of the sample should not exceed that in the standard.

SUBSTANCES NOT PRECIPITATED BY SULFURIC ACID. Dissolve 5.0 g in 150 mL of water. Add 1 mL of hydrochloric acid and heat to boiling. Add 25 mL of 10% sulfuric acid reagent solution, cool, dilute with water to 250 mL, and allow to stand overnight. Decant through a filter and evaporate 100 mL of the filtrate to dryness in a tared dish. Ignite gently to remove the excess acids, and finally ignite at $800 \pm 25°C$ for 15 min.

CALCIUM AND STRONTIUM. Determine calcium and strontium by the atomic absorption spectrophotometric method described on page 35.

> *Sample Stock Solution.* Dissolve 1.0 g of sample in sufficient water and add 5 mL of nitric acid in a 100-mL volumetric flask. Dilute to the mark with water (0.01 g/mL).

Analyze the solutions by means of a suitable atomic absorption spectrophotometer, using the conditions in the following table. Calculate the metal content of the sample by the method of standard additions.

Element	Wavelength (nm)	Sample Wt (g)	Standard Added (mg)	Flame Type*	Background Correction
Ca	422.7	0.04	0.02; 0.04	N/A	No
Sr	460.7	0.05	0.05; 0.10	N/A	No

*N/A is nitrous oxide/acetylene.

HEAVY METALS. (Page 26, Method 1). Dissolve 6.0 g in about 15 mL of water, add 8 mL of dilute hydrochloric acid $(1 + 1)$, and dilute with water to 30 mL. Use 25 mL to prepare the sample solution, and use the remaining 5.0 mL to prepare the control solution.

IRON. Dissolve 1.0 g in 40 mL of water. For the standard add 0.01 mg of iron (Fe) to 40 mL of water. Add 2 mL of hydrochloric acid to each, dilute with water to 50 mL, and add 0.10 mL of 0.1 N potassium permanganate. Allow to stand for 5 min and add 3 mL of ammonium thiocyanate reagent solution. Any red color in the solution of the sample should not exceed that in the standard.

Barium Carbonate

$BaCO_3$ Formula Wt 197.34

CAS Number 513–77–9

REQUIREMENTS

Assay 99.0–101.0%
$BaCO_3$

MAXIMUM ALLOWABLE

Insoluble in dilute hydrochloric acid 0.015%
Chloride (Cl) 0.002%
Water-soluble titrable base . . . . 0.002 meq / g
Oxidizing substances (as NO_3) 0.005%
Sulfide (S) 0.001%
Substances not precipitated by sulfuric acid 0.25%
Calcium (Ca) 0.05%
Heavy metals (as Pb) 0.001%
Iron (Fe) 0.002%
Strontium (Sr) 0.7%

TESTS

ASSAY. (By acid–base titrimetry). Weigh to the nearest 0.1 mg 4 g of sample, transfer to a 250-mL beaker, add 50 mL of water, and mix. Cautiously add exactly 50.0 mL of 1 N hydrochloric acid, boil, and cool. Add a few drops of bromphenol blue indicator (0.1% in water), and titrate the excess acid with 1 N sodium hydroxide to the blue end point. One milliliter of 1 N hydrochloric acid corresponds to 0.0987 g of $BaCO_3$.

INSOLUBLE IN DILUTE HYDROCHLORIC ACID. Cautiously dissolve 10 g in 100 mL of dilute hydrochloric acid (1 + 9) and dilute with water to 200 mL. Heat the solution to boiling and digest in a covered beaker on the steam bath for 1 hour. Filter through a tared filtering crucible, wash thoroughly, and dry at 105°C.

CHLORIDE. To 1.0 g add 20 mL of water and add dropwise, with stirring, about 2–3 mL of nitric acid. Filter if necessary through a chloride-free filter, wash with a little hot water, and dilute with water to 50 mL. To 25 mL of the solution add 1 mL of silver nitrate reagent solution. Any turbidity should not exceed that produced by 0.01 mg of chloride ion (Cl) in an equal volume of solution containing the quantities of reagents used in the test.

WATER-SOLUBLE TITRABLE BASE. Shake 5.0 g for 5 min with 50 mL of carbon dioxide-free water, cool, and filter. To 25 mL of the filtrate add 0.10 mL of phenolphthalein indicator solution. If a pink color is produced, it should be discharged by 0.50 mL of 0.01 N hydrochloric acid.

OXIDIZING SUBSTANCES. Place 0.10 g in a dry beaker. Cool the beaker thoroughly in an ice bath and add 22 mL of sulfuric acid that has been cooled to ice-bath temperature. Allow the mixture to warm to room temperature and swirl the beaker at intervals to effect gentle dissolution with slow evolution of carbon dioxide. When dissolution is complete, add 3 mL of diphenylamine reagent solution and digest on the steam bath for 90 min. Prepare a standard by evaporating to dryness a solution containing 0.005 mg of nitrate (0.5 mL of the nitrate standard solution) and 0.01 g of sodium carbonate. Treat the residue exactly like the sample. Any color produced in the solution of the sample should not exceed that in the standard.

SULFIDE. Dissolve 1.0 g in 10 mL of dilute acetic acid (1 + 4). As soon as dissolution is complete, add 1 mL of silver nitrate reagent solution. Any color after 5 min should not exceed that produced by 0.01 mg of sulfide ion (S) in an equal volume of solution containing the quantities of reagents used in the test.

SUBSTANCES NOT PRECIPITATED BY SULFURIC ACID. Cautiously dissolve 2.0 g in 30 mL of dilute hydrochloric acid (1 + 9) and dilute with water to 80 mL. Heat the solution to boiling, add 15 mL of 2 N sulfuric acid, cool, dilute with water to 100 mL, mix well, and allow to stand overnight. Decant the solution through a dry filter paper and evaporate 50 mL of the filtrate to dryness in a tared dish. Ignite carefully to remove the excess acids, and finally ignite at 800 ± 25°C for 15 min.

CALCIUM AND STRONTIUM.   Determine the calcium and sodium by the flame atomic absorption spectrophotometric method described on page 35.

> ***Sample Stock Solution.***   Cautiously dissolve 1.0 g of sample in 40 mL of water and sufficient hydrochloric acid to effect solution. Dilute with water to the mark in a 100-mL volumetric flask (0.01 g/mL) To each of three flasks add 2 mL of a 2.5% potassium solution.

Analyze the solutions by means of a suitable atomic absorption spectrophotometer, using the conditions outlined in the following table. Calculate the metal content of the sample by the method of standard additions.

Element	Wavelength (nm)	Sample Wt (g)	Standard Added (mg)	Flame Type *	Background Correction
Ca	422.7	0.05	0.025; 0.05	N/A	Yes
Sr	460.7	0.01	0.035; 0.07	N/A	Yes

*N/A is nitrous oxide/acetylene.

HEAVY METALS.   (Page 26, Method 1). Cautiously dissolve 5.0 g in 30 mL of dilute hydrochloric acid (1 + 4), and evaporate to dryness on the steam bath. Dissolve the residue in about 20 mL of water, and dilute with water to 25 mL. Dilute 15 mL with water to 25 mL, and use to prepare the sample solution. Use 5.0 mL of the remaining solution to prepare the control solution.

IRON.   Cautiously dissolve 1.0 g in 15 mL of dilute hydrochloric acid (1 + 2). For the standard add 0.02 mg of iron (Fe) to 15 mL of dilute hydrochloric acid (1 + 2). Evaporate both solutions to dryness on the steam bath. Dissolve each residue in about 20 mL of water, add 4 mL of hydrochloric acid, and dilute with water to 100 mL. To 50 mL of each solution add 0.10 mL of 0.1 N potassium permanganate, allow to stand for 5 min, and add 3 mL of ammonium thiocyanate reagent solution. Any red color in the solution of the sample should not exceed that in the standard.

Barium Chloride Dihydrate

$BaCl_2 \cdot 2H_2O$ Formula Wt 244.26

CAS Number 10326 – 27 – 9

REQUIREMENTS

Loss on drying at 150°C. 14.0 – 16.0%
pH of a 5% solution . . . . 5.2 – 8.2 at 25°C

MAXIMUM ALLOWABLE

Insoluble matter 0.005%
Oxidizing substances (as NO_3). 0.005%
Substances not precipitated by sulfuric acid 0.05%
Calcium (Ca) 0.05%
Heavy metals (as Pb) . . . . 5 ppm
Iron (Fe) . . . . 2 ppm
Strontium (Sr) 0.1%

TESTS

LOSS ON DRYING AT 150°C.   Weigh accurately about 1.0 g, and dry at 150°C to constant weight. The theoretical loss for $BaCl_2 \cdot 2H_2O$ is 14.75%.

pH OF A 5% SOLUTION.   (Page 43). The pH should be 5.2–8.2 at 25°C.

INSOLUBLE MATTER.   (Page 14). Use 20 g dissolved in 200 mL of water. The solution should be clear and colorless.

OXIDIZING SUBSTANCES.   Place 0.10 g in a dry beaker. Cool the beaker thoroughly in an ice bath and add 22 mL of sulfuric acid that has been cooled to ice-bath temperature. Allow the mixture to warm to room temperature and swirl the beaker at intervals to effect slow dissolution with gentle evolution of hydrogen chloride. When dissolution is complete, add 3 mL of diphenylamine reagent solution and digest on the steam bath for 90 min. Prepare a standard by evaporating to dryness a solution containing 0.005 mg of nitrate (0.5 mL of the standard nitrate solution) and 0.01 g of sodium carbonate. Treat the residue exactly like the sample. Any color produced in the solution of the sample should not exceed that in the standard.

SUBSTANCES NOT PRECIPITATED BY SULFURIC ACID. Dissolve 5.0 g in about 150 mL of water. Add 1 mL of hydrochloric acid, heat to boiling, and add 50 mL of 1 N sulfuric acid. Cool the solution, dilute with water to 250 mL, and allow to stand overnight. Decant through a dry filter paper and evaporate 100 mL of the filtrate to dryness in a tared dish. Ignite gently to volatilize the excess acids, and finally ignite at 800 ± 25°C for 15 min.

CALCIUM AND STRONTIUM. Determine the calcium and strontium by the flame atomic absorption spectrophotometric method described on page 35.

> *Sample Stock Solution.* Dissolve 1.0 g of sample in sufficient water and add 5 mL of nitric acid in a 100-mL volumetric flask. Dilute to the mark with water (0.01 g/mL).

Analyze the solutions by means of a suitable atomic absorption spectrophotometer, using the conditions in the following table. Calculate the metal content of the sample by the method of standard additions.

Element	Wavelength (nm)	Sample Wt (g)	Standard Added (mg)	Flame Type*	Background Correction
Ca	422.7	0.04	0.02; 0.04	N/A	No
Sr	460.7	0.05	0.05; 0.10	N/A	No

*N/A is nitrous oxide/acetylene.

HEAVY METALS. (Page 26, Method 1). Dissolve 6.0 g in about 20 mL of water, and dilute with water to 30 mL. Use 25 mL to prepare the sample solution, and use the remaining 5.0 mL to prepare the control solution.

IRON. Dissolve 5.0 g in 40 mL of water plus 2 mL of hydrochloric acid. Prepare a standard containing 0.01 mg of iron (Fe) in 40 mL of water plus 2 mL of hydrochloric acid. Add 0.10 mL of 0.1 N potassium permanganate to each, dilute with water to 50 mL, and allow to stand for 5 min. Add 3 mL of ammonium thiocyanate reagent solution to each. Any red color in the solution of the sample should not exceed that in the standard.

Barium Hydroxide Octahydrate

$Ba(OH)_2 \cdot 8H_2O$ Formula Wt 315.46

CAS Number 12230−71−6

REQUIREMENTS

Assay . . . . $\geq$ 98.0%
$Ba(OH)_2 \cdot 8H_2O$

MAXIMUM ALLOWABLE

Carbonate (as $BaCO_3$) 2.0%
Insoluble in dilute hydrochloric acid 0.01%
Chloride (Cl) 0.001%
Sulfide (S) . . . . Passes test (limit
about 0.001%)
Substances not precipitated by sulfuric acid 0.2%
Calcium (Ca) 0.05%
Heavy metals (as Pb) . . . . 5 ppm
Iron (Fe) 0.001%
Strontium (Sr) 0.8%

TESTS

ASSAY. (By acid−base titrimetry). Weigh accurately 4−5 g of clear crystals and dissolve in about 200 mL of carbon dioxide-free water. Add 0.10 mL of phenolphthalein indicator solution and titrate with 1 N hydrochloric acid. Reserve this solution for the determination of carbonate. Each milliliter of 1 N hydrochloric acid corresponds to 0.1577 g of $Ba(OH)_2 \cdot 8H_2O$.

CARBONATE. To the solution reserved from the test for assay add 5.00 mL of 1 N hydrochloric acid, heat to boiling, boil gently to expel all of the carbon dioxide, and cool. Add 0.10 mL of methyl orange indicator and titrate the excess acid with 1 N sodium hydroxide. Each milliliter of 1 N hydrochloric acid consumed corresponds to 0.09867 g of $BaCO_3$.

INSOLUBLE IN DILUTE HYDROCHLORIC ACID. Dissolve 10 g in 100 mL of dilute hydrochloric acid (1 + 9). Heat the solution to boiling, then digest in a covered beaker on the steam bath for 1 h. Filter through a tared filtering crucible, wash thoroughly, and dry at 105°C.

CHLORIDE. Dissolve 2.0 g in 25 mL of water, add 0.05 mL of phenolphthalein indicator solution, and neutralize with nitric acid. If the solution is not perfectly clear, filter it through a chloride-free filter and dilute the filtrate with water to 50 mL. To 25 mL of the solution add 1 mL of nitric acid and 1 mL of silver nitrate reagent solution. Any turbidity should not exceed that produced by 0.01 mg of chloride ion (Cl) in an equal volume of solution containing the quantities of reagents used in the test.

SULFIDE. Dissolve 1.0 g in 8 mL of warm water and add 0.25 mL of alkaline lead solution (prepared by adding 10% sodium hydroxide solution to a 10% lead acetate solution until the precipitate is redissolved) and 2 mL of glacial acetic acid. No darkening should occur.

SUBSTANCES NOT PRECIPITATED BY SULFURIC ACID. Dissolve 5.0 g in 150 mL of water plus 5 mL of hydrochloric acid. Heat the solution to boiling, and add 25 mL of 10% sulfuric acid reagent solution. Allow the solution to cool, dilute with water to 250 mL, and allow to stand overnight. Decant the solution through a dry filter and evaporate 100 mL to dryness in a tared dish. Heat gently to volatilize the excess acids, and finally ignite at $800 \pm 25°C$ for 15 min.

CALCIUM AND STRONTIUM. Determine the calcium and strontium by the flame atomic absorption spectrophotometric method described on page 35.

> **Sample Stock Solution.** Dissolve 1.0 g of sample in water in a 100-mL volumetric flask, add 5 mL of nitric acid, and dilute to the mark with water (0.01 g/mL).

Analyze the solutions by means of a suitable atomic absorption spectrophotometer, using the conditions outlined in the following table. Calculate the metal content of the sample by the method of standard additions.

Element	Wavelength (nm)	Sample Wt (g)	Standard Added (mg)	Flame Type*	Background Correction
Ca	422.7	0.04	0.02; 0.04	N/A	No
Sr	460.7	0.01	0.04; 0.08	N/A	No

*N/A is nitrous oxide/acetylene.

HEAVY METALS. (Page 26, Method 1). To 5.0 g in a 150-mL beaker, add 25 mL of water, mix, and cautiously add 10 mL of hydrochloric acid. Evaporate to dryness on the steam bath, dissolve the residue in about 2 mL of water, and dilute with water to 25 mL. For the control solution treat 1.0 g of the sample and 0.02 mg of lead ion (Pb) exactly as the 5.0 g of sample.

IRON. Dissolve 1.0 g in 40 mL of water. For the standard add 0.01 mg of iron (Fe) to 40 mL of water. Add 2.5 mL of hydrochloric acid to each, dilute with water to 50 mL, and add 0.10 mL of 0.1 N potassium permanganate solution. Allow to stand for 5 min and add 3 mL of ammonium thiocyanate reagent solution. Any red color in the solution of the sample should not exceed that in the standard.

Barium Nitrate

$Ba(NO_3)_2$ Formula Wt 261.35

CAS Number 10022–31–8

REQUIREMENTS

Assay	$\geq 99.0\%\ Ba(NO_3)_2$
pH of a 5% solution	5.0–8.0 at 25°C

MAXIMUM ALLOWABLE

Insoluble matter	0.01%
Chloride (Cl)	5 ppm
Substances not precipitated by sulfuric acid	0.05%
Calcium (Ca)	0.05%
Heavy metals (as Pb)	5 ppm
Iron (Fe)	2 ppm
Strontium (Sr)	0.1%

TESTS

ASSAY. (By complexometry for barium). Weigh accurately 0.2 g, transfer to a 400-mL beaker, and dissolve in 100 mL of freshly boiled (carbon dioxide-free) water. Add 100 mL of ethanol, 10 mL of ammonium hydroxide, and 0.3 mL of metalphthalein-screened indicator solution. Titrate immediately with standard 0.1 M EDTA to a color change from magenta to grey-green. One milliliter of 0.1 M EDTA corresponds to 0.02614 g of $Ba(NO_3)_2$.

pH OF A 5% SOLUTION.   (Page 43). The pH should be 5.0–8.0 at 25°C.

INSOLUBLE MATTER.   (Page 14). Use 10 g dissolved in 150 mL of hot water.

CHLORIDE.   Dissolve 2.0 g in 30 mL of warm water, filter if necessary through a chloride-free filter, and add 0.10 mL of nitric acid and 1 mL of silver nitrate reagent solution. Any turbidity should not exceed that produced by 0.01 mg of chloride ion (Cl) in an equal volume of solution containing the quantities of reagents used in the test.

SUBSTANCES NOT PRECIPITATED BY SULFURIC ACID.   Dissolve 5.0 g in about 150 mL of water plus 1 mL of hydrochloric acid. Heat the solution to boiling, add 25 mL of approximately 2 N sulfuric acid, cool, dilute with water to 250 mL, and allow to stand overnight. Decant the solution through a dry filter and evaporate 100 mL of the filtrate to dryness in a tared porcelain or silica dish. Heat gently to volatilize the excess acid, and finally ignite at 800 ± 25°C for 15 min.

CALCIUM AND STRONTIUM.   Determine the calcium and strontium by the flame atomic absorption spectrophotometric method described on page 35.

> *Sample Stock Solution.*   Dissolve 1.0 g in water in a 100-mL volumetric flask, add 5 mL of nitric acid, and dilute to the mark with water (0.01 g/mL).

Analyze the solutions by means of a suitable atomic absorption spectrophotometer, using the conditions in the following table. Calculate the metal content of the sample by the method of standard additions.

Element	Wavelength (nm)	Sample Wt (g)	Standard Added (mg)	Flame Type *	Background Correction
Ca	422.7	0.1	0.05; 0.10	N/A	No
Sr	460.7	0.1	0.10; 0.20	N/A	No

*N/A is nitrous oxide/acetylene.

HEAVY METALS. Dissolve 4.0 g in about 50 mL of water and dilute with water to 60 mL. For the control add 0.01 mg of lead ion (Pb) to 15 mL of the solution and dilute with water to 45 mL. For the sample use the remaining 45-mL portion. Adjust the pH of the control and sample solutions to between 3 and 4 (using a pH meter) with 1 N acetic acid or ammonium hydroxide (10% NH_3), dilute with water to 48 mL, and mix. Add 10 mL of freshly prepared hydrogen sulfide water to each and mix. Any color in the solution of the sample should not exceed that in the control.

IRON. Dissolve 5.0 g in 45 mL of hot water plus 2 mL of hydrochloric acid and allow to cool. Prepare a standard containing 0.01 mg of iron (Fe) in 45 mL of water plus 2 mL of hydrochloric acid. Add 0.10 mL of 0.1 N potassium permanganate to each, allow to stand for 5 min, and add 3 mL of ammonium thiocyanate reagent solution. Any red color in the solution of the sample should not exceed that in the standard.

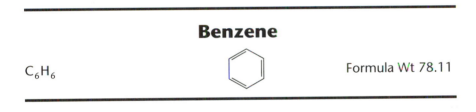

Benzene

C_6H_6 Formula Wt 78.11

CAS Number 71–43–2

Suitable for use in ultraviolet spectrophotometry or general use. Product labeling shall designate one or both of these uses for which suitability is represented on the basis of meeting the relevant requirements and tests. The ultraviolet spectrophotometry requirements include all the requirements for general use.

REQUIREMENTS

General Use

Assay . . . . $\geq$ 99.0% C_6H_6

MAXIMUM ALLOWABLE

Color (APHA)............................... 10
Residue after evaporation 0.001%
Substances darkened by sulfuric acid .... Passes test
Thiophene (limit about 1 ppm) .... Passes test
Sulfur compounds (as S) 0.005%
Water (H_2O) 0.05%

Specific Use

ULTRAVIOLET SPECTROPHOTOMETRY

Wavelength (nm)	Absorbance (AU)
380 to 400	0.01
350	0.02
330	0.04
300	0.10
290	0.30
280	1.00

TESTS

ASSAY. Analyze the sample by gas chromatography using the general parameters cited on page 35. The following specific conditions are also required.

Column: Type I, methyl silicone

Measure the area under all peaks and calculate the benzene content in area percent. Correct for water content.

COLOR (APHA). (Page 17).

RESIDUE AFTER EVAPORATION. (Page 14). Evaporate 100 g (115 mL) to dryness in a tared dish on the steam bath and dry the residue at 105°C for 30 min.

SUBSTANCES DARKENED BY SULFURIC ACID. Shake 25 mL with 15 mL of sulfuric acid for 15–20 s and allow to separate. Neither the benzene nor the acid should be darkened.

THIOPHENE. Add 5.0 mL of freshly prepared isatin reagent solution to a dry, clean 50-mL porcelain crucible. Carefully overlay with 5.0 mL of the sample, allow to stand undisturbed for 1 h, and compare with a blank containing 5.0 mL of isatin reagent solution in a 50-mL porcelain crucible. No bluish-green color should be present.

Solution A.   Dissolve 0.25 g of isatin in 25 mL of sulfuric acid.

Solution B.   Dissolve 0.25 g of ferric chloride in 1 mL of water. Dilute to 50 mL with sulfuric acid and allow the evolution of gas to cease.

To 2.5 mL of solution A add 5.0 mL of solution B and dilute to 100 mL with sulfuric acid.

SULFUR COMPOUNDS.   Place 30 mL of 0.5 N potassium hydroxide in methanol in a conical flask, add 5.2 g (6.0 mL) of the sample, and boil the mixture gently for 30 min under a reflux condenser, avoiding the use of a rubber stopper or connection. Detach the condenser, dilute with 50 mL of water, and heat on a hot plate at low setting until the benzene and methanol are evaporated. Add 50 mL of bromine water and heat for 15 min longer. Transfer the solution to a beaker, neutralize with dilute hydrochloric acid (1 + 3), add an excess of 1 mL of the acid, and concentrate to about 50 mL. Filter if necessary, heat the filtrate to boiling, add 5 mL of barium chloride reagent solution, digest in a covered beaker on a hot plate at low setting for 2 hours, and allow to stand overnight. If a precipitate is formed, filter, wash thoroughly, and ignite. Correct for the weight obtained in a complete blank test.

WATER.   (Page 55, Method 2). Use 100 μL (88 mg) of the sample.

ULTRAVIOLET SPECTROPHOTOMETRY.   Use the procedure on page 67 to determine the absorbance.

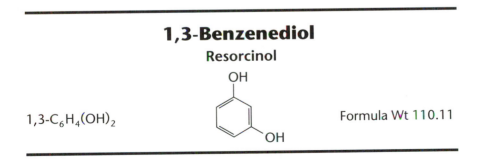

1,3-Benzenediol
Resorcinol

1,3-C$_6$H$_4$(OH)$_2$

Formula Wt 110.11

CAS Number 108–46–3

REQUIREMENTS

Assay 99.0–100.5%
$C_6H_4(OH)_2$
Melting range . . . . between 110.0
and 112.0°C

MAXIMUM ALLOWABLE

Insoluble matter 0.005%
Residue after ignition 0.01%
Titrable acid . . . . 0.004 meq / g
Phenols. 0.001%

TESTS

ASSAY. (By bromination). Weigh accurately 1.5 g, transfer to a 500-mL volumetric flask, dissolve in water, and dilute with water to volume. Stopper and mix thoroughly. Place a 25-mL aliquot in a 500-mL iodine flask. Add 50.0 mL of 0.1 N bromine, 50 mL of water, and 5 mL of hydrochloric acid. Stopper, shake, and allow to stand in the dark for 15 min, shaking occasionally. Add 5 mL of potassium iodide reagent solution (16.5 g/100 mL), taking care to prevent loss of bromine. Stopper and shake. Rinse the stopper, neck, and inner walls of the flask with about 25 mL of water. Titrate the liberated iodine with 0.1 N sodium thiosulfate, adding 3 mL of starch indicator solution near the end of the titration. Correct for a complete blank. One milliliter of 0.1 N bromine corresponds to 0.001835 g of $C_6H_4(OH)_2$

INSOLUBLE MATTER. (Page 14). Use 20 g dissolved in 150 mL of water.

RESIDUE AFTER IGNITION. (Page 15). Ignite 10 g in a tared crucible or dish.

TITRABLE ACID. Dissolve 1 g in 50 mL of water, add 0.15 mL of methyl orange indicator solution, and neutralize with 0.02 N sodium hydroxide. Add 5.0 g of sample to the neutralized solution, and titrate with 0.02 N sodium hydroxide. Not more than 1.0 mL of 0.02 N sodium hydroxide should be consumed in the titration.

PHENOLS. Dissolve 1.0 g in 20 mL of water. The solution should be perfectly clear. Warm the solution and note the odor. Any odor should not exceed that produced by warming a solution of 0.01 mg of phenol (C_6H_5OH) in 20 mL of water.

Benzoic Acid

C_6H_5COOH Formula Wt 122.12

CAS Number 65–85–0

REQUIREMENTS

Assay . . . . $\geq$ 99.5%
                                               C_6H_5COOH
Freezing point . . . . 122–123°C

MAXIMUM ALLOWABLE

Residue after ignition 0.005%
Insoluble in methanol 0.005%
Chlorine compounds (as Cl) 0.005%
Sulfur compounds (as S) 0.002%
Heavy metals (as Pb) . . . . 5 ppm
Substances reducing permanganate . . . . Passes test

TESTS

ASSAY. (By acid–base titrimetry). Dissolve about 0.5 g of the sample, accurately weighed, in 25 mL of 50% alcohol previously neutralized with 0.1 N sodium hydroxide, add 0.15 mL of phenolphthalein indicator solution, and titrate with 0.1 N sodium hydroxide. One milliliter of 0.1 N sodium hydroxide corresponds to 0.01221 g of C_6H_5COOH.

FREEZING POINT. Place 12–15 g in test tube (20 × 150 mm), and heat gently to melt the benzoic acid. Insert an accurate thermometer in the test tube so that the bulb of the thermometer is centrally located, and heat gently to about 130°C. Place the tube containing the melted sample in a larger test tube (38 × 200 mm) and keep it centered by means of corks. Allow the sample to cool slowly. If crystallization does not start when the temperature is about 120°C, stir gently with the thermometer to start freezing, and read the thermometer every half minute. The temperature that remains constant for 2–3 min is the freezing point.

RESIDUE AFTER IGNITION. (Page 15). Heat 20 g in a platinum dish at a temperature sufficient to volatilize the acid slowly, but do not char or ignite it. When nearly all the acid has been volatilized, cool, and continue as described.

INSOLUBLE IN METHANOL. Dissolve 20 g in 200 mL of methanol and digest under complete reflux for 30 min. Filter through a tared filtering crucible, wash thoroughly with methanol, and dry at 105°C.

CHLORINE COMPOUNDS. (Page 25). Mix 1.0 g with 0.5 g of sodium carbonate and add 10–15 mL of water. Evaporate on the steam bath and ignite until the mass is thoroughly charred, avoiding an unduly high temperature. Extract the fusion with 20 mL of water and 5.5 mL of nitric acid. Filter through a chloride-free filter, wash with two 10-mL portions of water, and dilute with water to 50 mL. Dilute 10 mL with water to 20 mL.

SULFUR COMPOUNDS. Mix 2.0 g with 1.0 g of sodium carbonate and add in small portions 15 mL of water. Evaporate and thoroughly ignite, protected from sulfur in the flame. Treat the residue with 20 mL of water and 2 mL of 30% hydrogen peroxide and heat on a hot plate at low setting for 15 min. Add 5 mL of hydrochloric acid and evaporate to dryness on the hot plate. Dissolve the residue in 10 mL of water, filter, wash with two 5-mL portions of water, and dilute with water to 25 mL. Add to this solution 0.5 mL of 1 N hydrochloric acid and 2 mL of barium chloride reagent solution. Any turbidity should not exceed that in a standard prepared as follows: treat 1 g of sodium carbonate with 2 mL of 30% hydrogen peroxide and 5 mL of hydrochloric acid and evaporate to dryness on a hot plate at low setting. Dissolve the residue and 0.12 mg of sulfate ion (SO_4) in sufficient water to make 25 mL and add 0.5 mL of 1 N hydrochloric acid and 2 mL of barium chloride reagent solution. Compare 10 min after adding the barium chloride to the sample and standard solutions.

HEAVY METALS. (Page 26, Method 1). Volatilize 4.0 g over a low flame, cool, and add 5 mL of nitric acid and about 10 mg of sodium carbonate. Evaporate to dryness on the steam bath, dissolve the residue in 20 mL of water, and dilute with water to 25 mL. For the control add 0.02 mg of lead ion (Pb) to 5 mL of nitric acid and about 10 mg of sodium carbonate, evaporate to dryness on the steam bath, dissolve the residue in 20 mL of water, and dilute with water to 25 mL.

SUBSTANCES REDUCING PERMANGANATE. Dissolve 1.0 g in 100 mL of water containing 1 mL of sulfuric acid. Heat to 85°C (steam bath temperature) and add 0.50 mL of 0.1 N potassium permanganate solution. The pink color should not be entirely discharged in 5 min.

Benzoyl Chloride

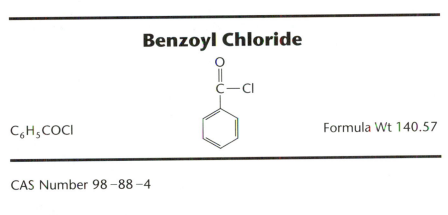

C_6H_5COCl

Formula Wt 140.57

CAS Number 98–88–4

REQUIREMENTS

Assay 98.0–100.5%
C_6H_5COCl
Freezing point . . . . −2.0 to 0.0°C

MAXIMUM ALLOWABLE

Residue after ignition 0.005%
Phosphorus compounds (as P) 0.002%
Heavy metals (as Pb) 0.001%
Iron (Fe) 0.001%

TESTS

ASSAY. (By acid–base titrimetry). Weigh accurately about 2 mL in a glass-stoppered flask, add 50.0 mL of 1 N sodium hydroxide, and stopper the flask. Allow to stand, with frequent agitation, until the sample has dissolved. Add 0.15 mL of phenolphthalein indicator solution and titrate the excess sodium hydroxide with 1 N hydrochloric acid. One milliliter of 1 N sodium hydroxide corresponds to 0.07028 g of C_6H_5COCl.

FREEZING POINT. Place 15 mL in a test tube (20 × 150 mm) in which is centered an accurate thermometer. The sample tube is centered by corks in an outer tube about 38 × 200 mm. Cool the whole apparatus, without stirring, in a bath of shaved or crushed ice mixed with sufficient salt and water to provide a temperature of $-5°C$. When the temperature is about $-2°C$, stir to start the freezing and read the thermometer every 30 s. The temperature that remains constant for 1–2 min is the freezing point.

RESIDUE AFTER IGNITION.   (Page 15). Evaporate 40 g (33 mL) to dryness in a tared dish, adding sufficient sulfuric acid to wet the sample. Add to the carbonized mass 2 mL of nitric acid and 0.25 mL of sulfuric acid, and heat cautiously until white fumes of sulfur trioxide are no longer evolved. Ignite at 500–600°C until the carbon is burned off. Retain the residue to prepare sample solution A.

PHOSPHORUS COMPOUNDS.   To 1 g (0.85 mL) add 7 mL of water and 3 mL of nitric acid. Heat the mixture to boiling, boil for 2 min, cool, dilute with water to 20 mL, and filter. To 10 mL of the filtrate add 5 mL of 10% sulfuric acid reagent solution, and evaporate to fumes of sulfur trioxide. Cool, dilute with water to 25 mL, add 1 mL of ammonium molybdate reagent solution and 1 mL of 4-(methyl-amino)phenol sulfate reagent solution, and allow to stand for 2 h at room temperature. Any blue color should not exceed that produced by 0.03 mg of phosphate ion (PO_4) in an equal volume of solution containing the quantities of reagents used in the test.

> *Sample Solution A.*   Add 10 mL of 10% hydrochloric acid reagent solution to the residue obtained in the test for residue after ignition and evaporate to dryness on the steam bath. Dissolve the residue in 0.5 mL of hydrochloric acid and 25 mL of hot water, filter if necessary, and dilute with water to 200 mL (1 mL = 0.2 g).

HEAVY METALS.   (Page 26, Method 1). Use 10 mL of sample solution A (2-g sample).

IRON.   (Page 28, Method 1). Use 5.0 mL of sample solution A (1-g sample).

Benzyl Alcohol
Benzenemethanol

$$CH_2OH$$

$C_6H_5CH_2OH$ Formula Wt 108.14

CAS Number 100–51–6

REQUIREMENTS

Assay (GC) . . . . $\geq$ 99.0%
$C_6H_5CH_2OH$

MAXIMUM ALLOWABLE

Color (APHA). 20
Residue after ignition 0.005%
Acetophenone ($C_6H_5COCH_3$) 0.02%
Benzaldehyde (C_6H_5CHO). 0.01%

TESTS

ASSAY, ACETOPHENONE, AND BENZALDEHYDE. Analyze the sample by gas chromatography, using the parameters cited on page 56. The following specific conditions are also required.

Column: Type III (polyethylene glycol)

Detector: Flame ionization

Approximate Retention Times (min.): Benzaldehyde, 18.6; acetophenone, 20.4; benzyl alcohol, 23.2

Approximate Relative Retention: Benzaldehyde, 2.9; acetophenone, 3.2; benzyl alcohol, 3.6

Measure the area under all peaks and calculate the area percent for benzyl alcohol, acetophenone, and benzaldehyde.

COLOR (APHA). (Page 17).

RESIDUE AFTER IGNITION. (Page 15). Use 25 g (24 mL).

Bismuth Nitrate Pentahydrate
Bismuth(III) Nitrate Pentahydrate

$Bi(NO_3)_3 \cdot 5H_2O$ Formula Wt 485.07

CAS Number 10035–06–0

REQUIREMENTS

Assay . . . . ≥ 98.0%
$$Bi(NO_3)_3 \cdot 5H_2O$$

MAXIMUM ALLOWABLE

Insoluble matter 0.005%
Chloride (Cl) 0.001%
Sulfate (SO_4) 0.005%
Arsenic (As) 0.001%
Copper (Cu) 0.002%
Iron (Fe) 0.001%
Lead (Pb) 0.002%
Silver (Ag) 0.001%
Substances not precipitated by hydrogen sulfide
   (as sulfates) 0.1%

TESTS

ASSAY. (By complexometric titration for bismuth). Weigh accurately 1.8 g and transfer to a 400-mL beaker. Dissolve in 5 mL of nitric acid and add about 250 mL of water. Add about 50 mg of ascorbic acid and 50 mg of xylenol orange indicator mixture. Titrate with standard 0.1 M EDTA to canary yellow. (The solution color changes from purple to salmon pink just before the canary yellow end point). One milliliter of 0.1 M EDTA corresponds to 0.04851 g of $Bi(NO_3)_3 \cdot 5H_2O$.

INSOLUBLE MATTER. (Page 14). Use 20 g dissolved in 100 mL of dilute nitric acid (1 + 4).

CHLORIDE. (Page 25). Use 1.0 g.

SULFATE. Dissolve 3.0 g in 5 mL of warm hydrochloric acid, dilute with water to 100 mL, and nearly neutralize with 10% ammonium hydroxide. Filter, and wash the residue and the filter with hot water to 150 mL. Transfer 100.0 mL to an evaporating dish, evaporate to about 15 mL, add 0.5 g of sodium carbonate, evaporate to dryness, and ignite gently. Cool, add 15 mL of hot water, neutralize with 10% hydrochloric acid, filter, and wash with water to 20 mL. Add to the combined filtrate and washings 1 mL of 1 N hydrochloric acid and 2 mL of barium chloride reagent solution, and allow to stand for 10 min. Any turbidity should not exceed that of a standard prepared as follows: To 0.5 g of sodium carbonate add 0.1 mg of sulfate ion (SO_4) and 2 mL of hydrochloric acid, and evaporate to dryness on the steam bath. Dissolve the residue in 20 mL of water, and add 1 mL of 1 N hydrochloric acid and 2 mL of barium chloride reagent solution.

ARSENIC. (Page 23). Mix 0.5 g with 5 mL of water, and cautiously add 2 mL of sulfuric acid. Heat the mixture until white fumes of sulfur trioxide are evolved copiously. Cool, cautiously add 10 mL of water, and evaporate to strong fumes of sulfur trioxide. Cool, cautiously add a second 10 mL of water, and again evaporate to strong fumes of sulfur trioxide. Cool, and cautiously wash the solution into a generator flask with sufficient water to make the volume 35 mL. The solution should show not more than 0.005 mg of arsenic (As).

COPPER, IRON, AND LEAD. Determine the copper, iron, and lead by the flame atomic absorption spectrophotometric method described on page 35.

> *Sample Stock Solution A.* Dissolve 10.0 g of sample in 20 mL of nitric acid and 20 mL of water. Transfer to a 100-mL volumetric flask and dilute to the mark with water (0.1 g/mL).

Analyze the solutions by means of a suitable atomic absorption spectrophotometer, using the conditions in the following table. Calculate the metal content of the sample by the method of standard additions.

Element	Wavelength (nm)	Sample Wt (g)	Standard Added (mg)	Flame Type*	Background Correction
Cu	324.7	1.0	0.02; 0.04	A/A	Yes
Fe	248.3	2.0	0.02; 0.04	A/A	Yes
Pb	217.0	2.0	0.04; 0.08	A/A	Yes

*A/A is air/acetylene.

SILVER. Dissolve 1.0 g in a mixture of 5 mL of nitric acid and 15 mL of water, and add 0.05 mL of hydrochloric acid. Any turbidity should not exceed that of a standard containing 0.01 mg of silver ion (Ag).

SUBSTANCES NOT PRECIPITATED BY HYDROGEN SULFIDE. Dissolve 2.0 g in 3 mL of warm nitric acid, and dilute with water to 100 mL. Pass hydrogen sulfide through the solution until all of the bismuth is precipitated, and filter until the filtrate is clear. Transfer 50.0 mL of the clear filtrate to an evaporating dish, add 0.25 mL of sulfuric acid, and evaporate to dryness. Ignite to constant weight, cool, and weigh.

Boric Acid

H_3BO_3 Formula Wt 61.83

CAS Number 10043–35–3

REQUIREMENTS

Assay . . . . $\geq$ 99.5% H_3BO_3

MAXIMUM ALLOWABLE

Insoluble in methanol 0.005%
Nonvolatile with methanol 0.05%
Chloride (Cl) 0.001%
Phosphate (PO_4) 0.001%
Sulfate (SO_4) 0.01%
Arsenic (As) . . . . 1 ppm
Calcium (Ca) 0.005%
Heavy metals (as Pb) 0.001%
Iron (Fe) 0.001%

TESTS

ASSAY. (By acid–base titrimetry). Weigh, to the nearest 0.1 mg, 2.5 g of sample and transfer to a 250-mL glass-stoppered flask. Dissolve in a mixture of 50 mL of warm water and 75 mL of glycerol, previously neutralized to a pH of 9.0 with 0.1 N sodium hydroxide. Titrate with 1 N sodium hydroxide to a pH of 9.0. One milliliter of 1 N sodium hydroxide corresponds to 0.06183 g of H_3BO_3.

INSOLUBLE IN METHANOL. Heat 20 g with 200 mL of methanol under complete reflux until the acid is dissolved, then reflux for 30 min. Filter through a tared filtering crucible, wash thoroughly with hot methanol, and dry at 105°C.

NONVOLATILE WITH METHANOL.   To 2.0 g of the powdered acid in a platinum dish, add 25 mL of methanol and 0.5 mL of hydrochloric acid and evaporate to dryness. Add 15 mL of methanol and 0.3 mL of hydrochloric acid and repeat the evaporation. Add to the residue 0.10 or 0.15 mL of sulfuric acid and ignite at 800 ± 25°C for 15 min.

CHLORIDE.   (Page 25). Use 1.0 g dissolved in 20 mL of warm water.

Sample Solution A for the Determination of Phosphate, Sulfate, Heavy Metals, and Iron.   To 10 g add 10 mg of sodium carbonate, 100 mL of methanol, and 5 mL of hydrochloric acid. Digest in a covered beaker on the steam bath to effect dissolution, then uncover and evaporate to dryness. Add 75 mL of methanol and 5 mL of hydrochloric acid and repeat the digestion and evaporation. Dissolve the residue in 5 mL of 1 N acetic acid and digest in a covered beaker on the steam bath for 15 min. Filter and dilute with water to 100 mL (1 mL = 0.1 g).

PHOSPHATE.   (Page 30, Method 1). Add 0.5 mL of nitric acid to 20 mL of sample solution A (2-g sample), and evaporate to dryness. Dissolve the residue in 25 mL of approximately 0.5 N sulfuric acid and continue as described.

SULFATE.   (Page 32, Procedure A, Method 1). Use 5.0 mL of sample solution A (0.5-g sample).

ARSENIC.   (Page 23). Use 3.0 g. For the standard use 0.003 mg of arsenic (As).

CALCIUM.   Determine the calcium by the flame atomic absorption spectrophotometric method described on page 35.

Sample Stock Solution.   To 20 g add 20 mg of sodium carbonate, 200 mL of methanol, and 10 mL of hydrochloric acid. Digest in a covered beaker on the steam bath to effect solution, then uncover and evaporate to dryness. Add 75 mL of methanol and 5 mL of hydrochloric acid, and repeat the digestion and evaporation. Dissolve the residue in 5 mL of 1 N acetic acid, and digest in a covered beaker on the steam bath for 15 min. Filter and dilute with water to volume in a 200-mL volumetric flask (0.1 g/mL).

Analyze the solutions by means of a suitable atomic absorption spectrophotometer, using the conditions outlined in the following

table. Calculate the metal content of the sample by the method of standard additions.

Element	Wavelength (nm)	Sample Wt (g)	Standard Added (mg)	Flame Type*	Background Correction
Ca	422.7	1.0	0.05; 0.10	N/A	No

*N/A is nitrous oxide/acetylene.

HEAVY METALS. (Page 26, Method 1). Dilute 20 mL of sample solution A (2-g sample) with water to 25 mL.

IRON. (Page 28, Method 1). Use 10 mL of sample solution A (1-g sample).

Bromcresol Green

3',3",5',5"-Tetrabromo-*m*-cresolsulfonphthalein

4,4'-(3*H*-2,1-Benzoxathiol-3-ylidene)bis-[(2,6-dibromo-3-methyl)phenol] *S*,*S*-Dioxide

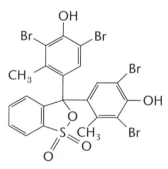

$C_{21}H_{14}Br_4O_5S$ Formula Wt 698.02

CAS Numbers 76−60−8; 62625−32−5 (sodium salt)

NOTE. This specification applies both to the sultone form and to the salt form of this indicator.

REQUIREMENTS

Clarity of solution . . . . Passes test
Visual transition interval . . . . pH 3.8 (yellow) to
pH 5.4 (blue)

TESTS

CLARITY OF SOLUTION.   If the indicator is the acid form, dissolve
0.1 g in 100 mL of alcohol. If the indicator is a salt form, dissolve 0.1 g
in 100 mL of water. Not more than a faint trace of turbidity or
insoluble matter should remain. Reserve the solution for the test for
visual transition interval.

VISUAL TRANSITION INTERVAL.   Dissolve 1 g of potassium chloride
in 100 mL of water. Adjust the pH of the solution to 3.8 (using a pH
meter as described on page 43) with 0.01 N hydrochloric acid. Add
0.10–0.30 mL of the 0.1% solution reserved from the test for insoluble
matter. The color of the solution should be yellow. Titrate the solution
with 0.01 N sodium hydroxide to a pH of 4.5 (using the pH meter).
The color of the solution should be green. Continue the titration to a
pH of 5.4. The color of the solution should be blue. Not more than 2.0
mL of 0.01 N sodium hydroxide should be consumed in the entire
titration.

Bromine

Br_2 Formula Wt 159.808

CAS Number 7726–95–6

REQUIREMENTS

Assay . . . . $\geq$ 99.5% Br_2

MAXIMUM ALLOWABLE

Residue after evaporation 0.005%
Chlorine (Cl) 0.05%
Iodine (I) 0.001%
Organic bromine compounds . . . . Passes test
Sulfur compounds (as S) 0.001%
Heavy metals (as Pb) . . . . 2 ppm
Nickel (Ni). . . . . 5 ppm

TESTS

ASSAY. (By iodometric titration). Weigh a 250-mL volumetric flask containing 15 g of potassium iodide in 45 mL of water. Add 1 mL of sample. Stopper promptly, mix by shaking, and reweigh. Dilute to volume with water and mix thoroughly. Place a 25.0-mL aliquot of this solution in a conical flask and titrate the liberated iodine with 0.1 N sodium thiosulfate, adding 3 mL of starch indicator solution near the end of the titration. One milliliter of 0.1 N sodium thiosulfate corresponds to 0.007990 g of Br_2.

RESIDUE AFTER EVAPORATION. (Page 14). Evaporate 31 g (10 mL) to dryness from a tared container on the steam bath under a hood and dry the residue at 105°C for 30 min. Reserve the residue for the test for heavy metals.

CHLORINE. To each of two 250-mL wide-mouth conical flasks add 0.5 mL of dilute sulfuric acid (1 + 4), 5 mL of potassium bromide solution (1.50 g of KBr per liter), and 35 mL of water. For the sample add 3 g (1.0 mL) of the bromine to one of the flasks. For the standard add 1.5 mg of chloride ion (3.2 mg of KCl) to the other. Digest each on the steam bath until the sample solution is colorless, to each add 4 mL of potassium peroxydisulfate solution (1.0 g per 100 mL), and wash down the sides of the flasks with a little water. Digest again on the steam bath for 20 min, or until all odor of free bromine is gone from both flasks. Cool and dilute each with water to 100 mL. Transfer 1.0 mL of each solution into separate 50-mL beakers and add 22 mL of water, 1 mL of nitric acid, and 1 mL of silver nitrate reagent solution. Any turbidity in the solution of the sample should not exceed that in the standard.

IODINE. To each of two 250-mL wide-mouth conical flasks, add 1.0 mL of 10% potassium chloride solution, 50 mL of water, and a few silicon carbide boiling chips. For the sample add 10.5 g (3.5 mL) of the bromine to one of the flasks. For the control add 0.3 mL of the bromine and 0.1 mg of iodide ion (0.13 mg of KI) to the other flask. Boil cautiously in a hood to remove the excess bromine, adding water as required to maintain a volume of not less than 50 mL. When no trace of yellow color remains in either flask, cool, and add 5 mL of 10% potassium iodide solution and 5 mL of dilute sulfuric acid (1 + 4) to each. Any color in the solution of the sample should not exceed that in the control.

ORGANIC BROMINE COMPOUNDS. Use matching infrared cells consisting of rock salt windows separated by 1/2-inch Teflon spacers. Fill one of the cells with bromine from a pipet inserted through an opening drilled in the plastic. Scan the sample from 2.5 to 15 μm, using the empty matching cell for reference. The absorbance of the sample should not exceed 0.10 in the region from 2.5 to 12 μm (except for possible water bands at 2.6 to 2.9 μm and 6.3 μm) or 0.20 in the region from 12 to 15 μm.

SULFUR COMPOUNDS. To 8 g (2.6 mL) add 5 mL of water and evaporate to dryness on a hot plate at low setting. To the residue add 5 mL of dilute hydrochloric acid (1 + 19), filter if necessary, dilute to 50 mL with water, and add 1 mL of barium chloride reagent solution to 10 mL of the solution. Any turbidity should not exceed that produced by 0.05 mg of sulfate ion (SO_4) in an equal volume of solution containing the quantities of reagents used in the test. Compare 10 min after adding the barium chloride to the sample and standard solutions. Reserve the remaining solution for the test for nickel.

HEAVY METALS. For the sample use the residue reserved from the test for residue after evaporation. For the standard use a solution contining 0.06 mg of lead ion (Pb). To each add 3 mL of nitric acid, 1 mL of water, and about 10 mg of sodium carbonate. Evaporate to dryness on the steam bath, dissolve the residues in about 20 mL of water, and dilute with water to 25 mL. Adjust the pH of the standard and sample solutions to between 3 and 4 (using a pH meter) with 1 N acetic acid or ammonium hydroxide (10% NH_3), dilute with water to 40 mL, and mix. Add 10 mL of freshly prepared hydrogen sulfide water to each and mix. Any color in the solution of the sample should not exceed that in the standard.

NICKEL. To 25 mL of the solution reserved from the test for sulfur compounds, add 5 mL of bromine water. Stir and add ammonium hydroxide (1 + 1) until the bromine color is discharged. Add 5 mL of

1% dimethylglyoxime solution in ethyl alcohol and 5 mL of 10% sodium hydroxide reagent solution. Any red color should not exceed that produced by 0.02 mg of nickel ion (Ni) in an equal volume of solution containing the quantities of reagents used in the test. Compare 10 min after adding the dimethylglyoxime to the sample and standard solutions.

Bromphenol Blue

3′,3″,5′,5″-Tetrabromophenolsulfonphthalein

4,4′-(3 H-2,1-Benzoxathiol-3-ylidene)bis-[2,6-dibromophenol] S, S-Dioxide

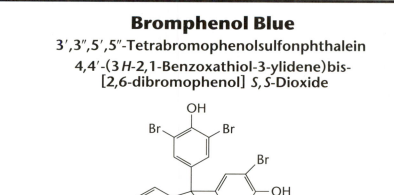

$C_{19}H_{10}Br_4O_5S$ Formula Wt 669.97

CAS Numbers 115–39–9; 62625–28–9 (sodium salt)

> *NOTE.* This specification applies both to the sultone form and to the salt form of this indicator.

REQUIREMENTS

Clarity of solution . . . . Passes test
Visual transition interval . . . . pH 3.0 (yellow) to
                                                               pH 4.6 (blue)

TESTS

CLARITY OF SOLUTION.   If the indicator is the acid form, dissolve 0.1 g in 100 mL of alcohol. If the indicator is the salt form, dissolve 0.1 g in 100 mL of water. Not more than a faint trace of turbidity or insoluble matter should remain. Reserve the solution for the test for visual transition interval.

VISUAL TRANSITION INTERVAL.   Dissolve 1 g of potassium chloride in 100 mL of water. Adjust the pH of the solution to 3.0 (using a pH meter as described on page 43) with 0.01 N hydrochloric acid. Add 0.10–0.30 mL of the 0.1% solution of the indicator reserved from the test for insoluble matter. The color of the solution should be yellow with a slight greenish hue. Titrate the solution with 0.01 N sodium hydroxide to a pH of 3.4 (using the pH meter). The color of the solution should be green. Continue the titration to pH 4.6. The color of the solution should be blue. Not more than 13.0 mL of 0.01 N sodium hydroxide should be consumed in the entire titration.

Bromthymol Blue

3′,3″-Dibromothymolsulfonphthalein

4,4′-(3 *H*-2,1-Benzoxathiol-3-ylidene)bis-[2-bromo-3-methyl-6-(1-methylethyl)phenol] *S,S*-Dioxide

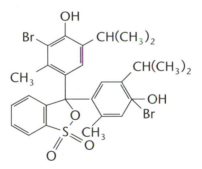

$C_{27}H_{28}Br_2O_5S$ Formula Wt 624.39

CAS Numbers 76–59–5; 34722–90–2 (sodium salt)

NOTE. This specification applies both to the sultone form and to the salt form of this indicator.

REQUIREMENTS

Clarity of solution . . . . Passes test
Visual transition interval . . . . pH 6.0 (yellow) to
pH 7.6 (blue)

TESTS

CLARITY OF SOLUTION.   If the indicator is the acid form, dissolve 0.1 g in 100 mL of alcohol. If the indicator is a salt form, dissolve 0.1 g in 100 mL of water. Not more than a faint trace of turbidity or insoluble matter should remain. Reserve the solution for the test for visual transition interval.

VISUAL TRANSITION INTERVAL.   Dissolve 1 g of potassium chloride in 100 mL of water. Adjust the pH of the solution to 6.0 (using a pH meter as described on page 43) with 0.01 N hydrochloric acid or sodium hydroxide. Add 0.10–0.30 mL of the 0.1% solution reserved from the test for insoluble matter. The color of the solution should be yellow, with not more than a faint trace of green color. Titrate the solution with 0.01 N sodium hydroxide to a pH of 6.7 (using a pH meter). The color of the solution should be green. Continue the titration to pH 7.6. The color of the solution should be blue. Not more than 0.30 mL of 0.01 N sodium hydroxide should be consumed in the entire titration.

Brucine Sulfate Heptahydrate
2,3-Dimethoxystrychnidin-10-one Sulfate Heptahydrate

$(C_{23}H_{26}N_2O_4)_2 \cdot H_2SO_4 \cdot 7H_2O$ Formula Wt 1013.13

CAS Number 5787–00–8

REQUIREMENTS

Sensitivity to nitrate . . . . Passes test

MAXIMUM ALLOWABLE

Clarity of solution . . . . Passes test
Loss on drying at 105°C. 13.0%
Residue after ignition 0.1%

TESTS

CLARITY OF SOLUTION. Dissolve 1 g in 100 mL of water, and digest in a covered beaker on the steam bath for 1 h. The solution should be as colorless and clear as an equal volume of water.

LOSS ON DRYING AT 105°C. Weigh accurately about 1 g and dry on a tared dish for 6 h at 105°C.

RESIDUE AFTER IGNITION. (Page 15). Ignite 1.0 g. Omit the addition of sulfuric acid.

SENSITIVITY TO NITRATE

Test Solutions 1, 2, 3, and 4. Prepare a sample solution by dissolving 0.6 g in dilute sulfuric acid (2 + 1), as described under Reagent Solutions, Brucine Sulfate, page 72. Place 50 mL of this sample solution in each of four test tubes, and add nitrate ion (NO_3) to each as follows: No. 1, none; No. 2, 0.01 mg; No. 3, 0.02 mg; and No. 4, 0.03 mg.

Heat the four test tubes in a preheated (boiling) water bath for 10 min. Cool rapidly in an ice bath to room temperature. Set a spectrophotometer at 410 nm and, using 1-cm cells, adjust the instrument to read 0 absorbance with solution 1 in the light path. Determine the absorbance for solutions 2, 3, and 4 at this adjustment, using similar cells. The absorbances for solutions 2, 3, and 4 should not be less than 0.025, 0.050, and 0.075, respectively, and the plot of absorbances versus nitrate concentrations should be linear.

2-Butanone
Methyl Ethyl Ketone

$$CH_3COCH_2CH_3 \qquad CH_3CH_2 - \overset{\overset{\displaystyle O}{\|}}{C} - CH_3 \qquad \text{Formula Wt 72.11}$$

CAS Number 78–93–3

REQUIREMENTS

Assay . . . . ≥ 99.0%
$$CH_3COCH_2CH_3$$

MAXIMUM ALLOWABLE

Color (APHA). 15
Residue after evaporation 0.0025%
Titrable acid . . . . 0.0005 meq / g
Water. 0.20%

TESTS

ASSAY.   Analyze the sample by gas chromatography, using the general parameters cited on page 56. The following specific conditions are also required.

   Column: Type I, methyl silicone

Measure the area under all peaks and calculate the 2-butanone content in area percent. Correct for water content.

COLOR (APHA). (Page 17).

RESIDUE AFTER EVAPORATION.   (Page 14). Use 80 g (100 mL) in a tared beaker on a hot water bath. When the liquid has evaporated, complete the drying at 105°C to constant weight after cooling in a desiccator.

TITRABLE ACID.   Mix 40 g (50 mL) with 50 mL of carbon dioxide-free water, and add 0.15 mL of phenolphthalein indicator solution. Not more than 2.0 mL of 0.01 N sodium hydroxide should be required to produce a pink color.

WATER. (Page 54, Method 1). Use 25 mL (20 g) of the sample and a methanol-free system to prevent ketal formation with liberation of water.

n-Butyl Acetate

$CH_3COO(CH_2)_3CH_3$ Formula Wt 116.16

CAS Number 123–86–4

REQUIREMENTS

Assay (GC) . . . . $\geq$ 99.5%
$CH_3COO(CH_2)_3CH_3$

MAXIMUM ALLOWABLE

Color (APHA). 10
Residue after evaporation 0.001%
Titrable acid . . . . 0.0016 meq / g
Substances darkened by sulfuric acid . . . . Passes test
Water (H$_2$O) 0.1%
n-Butyl alcohol (C$_4$H$_9$OH) 0.2%
n-Butyl formate (HCOOC$_4$H$_9$) 0.1%
n-Butyl propionate (C$_2$H$_5$COOC$_4$H$_9$) 0.1%

TESTS

ASSAY, *n*-BUTYL ALCOHOL, *n*-BUTYL FORMATE, AND *n*-BUTYL PROPIONATE. Analyze the sample by gas chromatography, using the general parameters cited on page 56. The following specific conditions are also required.

Column: Type I (methyl silicone).

Detector: Flame ionization.

Approximate Retention Times (min.): *n*-Butyl acetate, 15.3; *n*-butyl alcohol, 11.1; *n*-butyl formate, 13.2; *n*-butyl propionate, 18.2.

Approximate Relative Retention: *n*-Butyl acetate, 5.3; *n*-butyl alcohol, 3.7; *n*-butyl formate, 4.4; *n*-butyl propionate, 6.1.

Measure the area under all peaks and calculate the area percent for *n*-butyl acetate, *n*-butyl alcohol, *n*-butyl formate, and *n*-butyl propionate. Correct for the water content.

COLOR (APHA). (Page 17).

RESIDUE AFTER EVAPORATION. (Page 14). Evaporate 100 g (114 mL) to dryness in a tared dish on the steam bath and dry the residue at 105°C for 30 min.

TITRABLE ACID. Neutralize 25 mL of reagent alcohol, to which 0.05 mL of bromthymol blue indicator solution has been added, with 0.01 N sodium hydroxide to a blue-green color. Add 25 g (28 mL) of the sample, mix, and titrate with 0.01 N sodium hydroxide to the same color. Not more than 4.0 mL of 0.01 N sodium hydroxide should be required.

SUBSTANCES DARKENED BY SULFURIC ACID. Cool 5 mL of the sample to below 10°C. Cautiously add, dropwise, 5 mL of $95.0 \pm 0.5\%$ sulfuric acid. The color of the solution should not exceed that of a color standard composed of 9 volumes of water plus one volume of a color standard containing 3.25 g of $CoCl_2 \cdot 6H_2O$, 2.25 g of $FeCl_3 \cdot 6H_2O$, and 3 mL of hydrochloric acid per 100 mL of solution.

WATER. (Page 54, Method 1). Use 25 g (28 mL) of the sample.

Butyl Alcohol
1-Butanol

$CH_3(CH_2)_2CH_2OH$ Formula Wt 74.12

CAS Number 71−36−3

REQUIREMENTS

Assay . . . . ≥ 99.4%
$$CH_3(CH_2)_2CH_2OH$$

MAXIMUM ALLOWABLE

Color (APHA). 10
Residue after evaporation 0.005%
Titrable acid . . . . 0.0008 meq / g
Aldehydes . . . . Passes test
Butyl ether 0.2%
Water (H_2O) 0.1%

TESTS

ASSAY AND BUTYL ETHER.   Analyze the sample by gas chromatography using the general parameters cited on page 56. The following specific conditions are also required.

   Column: Type I, methyl silicone

Measure the areas under all peaks and calculate the butyl alcohol and butyl ether content in area percent. Correct for water content.

COLOR (APHA).   (Page 17).

RESIDUE AFTER EVAPORATION.   (Page 14). Evaporate 20 g (25 mL) to dryness in a tared platinum dish on the steam bath and dry the residue at 105°C for 30 min.

TITRABLE ACID.   To 60 g (74 mL) in a glass-stoppered flask add 0.10 mL of phenolphthalein indicator solution, and titrate with 0.01 N alcoholic potassium hydroxide to a pink end point that persists for at least 15 s. Not more than 5.0 mL of 0.01 N potassium hydroxide should be consumed.

ALDEHYDES.   Transfer 10 mL of ammoniacal silver nitrate reagent solution to a test tube, add 10 mL of the sample, and mix thoroughly. Allow the mixture to stand in a dark place for 30 min. No color should be produced, but a slight precipitate may form at the interface of the two layers.

   Ammoniacal Silver Nitrate Reagent Solution.   To a 5% solution of silver nitrate in water add dropwise a 10% solution of ammonium hydroxide until the precipitate which first forms is almost, but not entirely, dissolved. Filter the solution and store the filtrate in a dark-colored bottle. Do not store the reagent for long periods, since it may form explosive compounds on standing. Prepare a fresh quantity for each series of determinations.

WATER.   (Page 54, Method 1). Use 30 mL (24 g) of the sample.

tert-Butyl Alcohol
2-Methyl-2-propanol

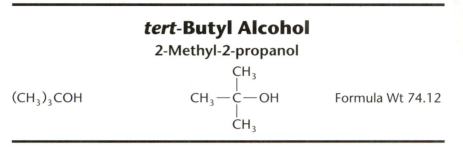

$(CH_3)_3COH$ Formula Wt 74.12

CAS Number 75 – 65 – 0

> *NOTE.* Since the melting point of this reagent is close to 25°C, the sample may be warmed to about 30°C for testing.

REQUIREMENTS

Assay . . . . $\geq$ 99.0%
$(CH_3)_3COH$

MAXIMUM ALLOWABLE

Color (APHA). 20
Residue after evaporation 0.003%
Titrable acid . . . . 0.001 meq / g
Water (H_2O) 0.1%

TESTS

ASSAY.   Analyze the sample by gas chromatography, using the parameters cited on page 56. The following specific conditions are also required.

    Column: Type I, methyl silicone

Measure the areas under all peaks and calculate the *tert*-butyl alcohol content in area percent. Correct for water content.

COLOR (APHA). (Page 17).

RESIDUE AFTER EVAPORATION.   (Page 14). Evaporate 60 g (77 mL) of sample to dryness in a tared evaporating dish on a steam bath and dry the residue at 105°C for 30 min.

TITRABLE ACID.   To 35 g (45 mL) in a glass-stoppered Erlenmeyer flask, add 0.15 mL of phenolphthalein indicator solution and shake for 1 min. No pink color should appear. Titrate the solution with

0.01 N alcoholic potassium hydroxide to a faint pink color that persists for at least 15 s. Not more than 3.5 mL should be required.

WATER. (Page 54, Method 1). Use 45 mL (35 g) of the sample.

Cadmium Chloride, Anhydrous

$CdCl_2$ Formula Wt 183.32

CAS Number 10108−64−2

REQUIREMENTS

Assay . . . . $\geq$ 99.0% $CdCl_2$

MAXIMUM ALLOWABLE

Insoluble matter 0.01%
Nitrate and nitrite (as NO_3). 0.003%
Sulfate (SO_4) 0.01%
Ammonium (NH_4) 0.01%
Copper (Cu) 0.001%
Iron (Fe) 0.001%
Lead (Pb) 0.005%
Substances not precipitated by
    hydrogen sulfide (as sulfates) 0.3%
Zinc (Zn) 0.1%

TESTS

ASSAY. (By complexometric titration of cadmium). Weigh accurately about 0.4 g of sample and dissolve in about 200 mL of water. Add 15 mL of a saturated aqueous hexamethylenetetramine solution and 50 mg of xylenol orange indicator mixture. Titrate with standard 0.1 M EDTA to a color change of red-purple to lemon-yellow. One milliliter of 0.1 M EDTA corresponds to 0.01832 g of $CdCl_2$.

INSOLUBLE MATTER. (Page 14). Use 10 g dissolved in 150 mL of water. Retain the filtrate and washings for the test for sulfate.

NITRATE AND NITRITE

>*Sample Solution A.*   Dissolve 0.30 g in a 50-mL centrifuge tube in 3 mL of water by heating in a boiling-water bath. Add 20 mL of nitrate-free sulfuric acid (2 + 1) and cool in an ice bath. Centrifuge for 10 min, decant the solution into a 200-mm test tube, and dilute to 50 mL with brucine sulfate reagent solution.

>*Control Solution B.*   Dissolve 0.30 g in a 50-mL centrifuge tube in 2 mL of water and 1 mL of the standard nitrate solution containing 0.01 mg of nitrate ion (NO_3) per mL by heating in a boiling-water bath. Add 20 mL of nitrate-free sulfuric acid (2 + 1) and cool in an ice bath. Centrifuge for 10 min, decant the solution into a 200-mm test tube, and dilute to 50 mL with brucine sulfate reagent solution.

Continue with the procedure described on page 29, starting with the preparation of blank solution C.

SULFATE.   (Page 32, Procedure A, Method 1). Use 1/20 of the filtrate from the test for insoluble matter. Allow 30 min for the turbidity to form.

AMMONIUM.   Dissolve 1.0 g in 80 mL of water and add with stirring 20 mL of freshly boiled 10% sodium hydroxide reagent solution. After allowing the precipitate to settle, dilute 10 mL of the clear supernatant liquid with water to 50 mL, and add 2 mL of Nessler reagent. Any color should not exceed that produced by 0.01 mg of ammonium ion (NH_4) in an equal volume of solution containing 4 mL of the sodium hydroxide solution and 2 mL of the Nessler reagent.

COPPER, LEAD, AND ZINC.   Determine the copper, lead, and zinc by the flame atomic absorption spectrophotometric method described on page 35.

>*Sample Stock Solution.*   Dissolve 20.0 g of sample in a 200-mL volumetric flask, add 5 mL of nitric acid, and dilute to the mark with water (0.1 g/mL).

Analyze the solutions by means of a suitable atomic absorption spectrophotometer, using the conditions outlined in the following table. Calculate the metal content of the sample by the method of standard additions.

Element	Wavelength (nm)	Sample Wt (g)	Standard Added (mg)	Flame Type*	Background Correction
Cu	324.7	2.0	0.02; 0.04	A/A	Yes
Pb	217.0	2.0	0.10; 0.20	A/A	Yes
Zn	213.9	0.05	0.025; 0.05	A/A	Yes

*A/A is air/acetylene.

IRON. (Page 28, Method 1). Use 1.0 g.

SUBSTANCES NOT PRECIPITATED BY HYDROGEN SULFIDE. Dissolve 2.0 g in 150 mL of water containing 15 mL of approximately 1 N sulfuric acid. Heat to boiling and pass a rapid stream of hydrogen sulfide through the solution as it cools to room temperature. Filter the solution, but do not wash the precipitate. Add 0.25 mL of sulfuric acid to 75 mL of the clear filtrate and evaporate to dryness in a tared dish. Finally, ignite at $800 \pm 25°C$ for 15 min.

Cadmium Chloride, Crystals

$CdCl_2 \cdot 2\frac{1}{2}H_2O$ Formula Wt 228.35

CAS Number 7790−78−5

REQUIREMENTS

Assay (as $CdCl_2$) 79.5–81.0%

MAXIMUM ALLOWABLE

Insoluble matter 0.005%
Nitrate and nitrite (as NO_3) 0.003%
Sulfate (SO_4) 0.005%
Ammonium (NH_4) 0.005%
Copper (Cu) . . . . 5 ppm
Iron (Fe) . . . . 5 ppm
Lead (Pb) 0.005%
Substances not precipitated by
    hydrogen sulfide (as sulfates) 0.2%
Zinc (Zn) 0.1%

TESTS

ASSAY. (By complexometric titration of cadmium). Weigh accurately about 0.8 g of sample and dissolve in about 200 mL of water. Add 15 mL of a saturated aqueous hexamethylenetetramine solution and 50 mg of xylenol orange indicator mixture. Titrate with standard 0.1 M EDTA to a color change of red-purple to lemon-yellow. One milliliter of 0.1 M EDTA corresponds to 0.01832 g of $CdCl_2$.

INSOLUBLE MATTER. (Page 14). Use 20 g dissolved in 150 mL of water. Reserve the filtrate for the test for sulfate.

NITRATE AND NITRITE

> *Sample Solution A.* Dissolve 0.30 g in a 50-mL centrifuge tube in 3 mL of water by heating in a boiling-water bath. Add 20 mL of nitrate-free sulfuric acid $(2 + 1)$ and cool in an ice bath. Centrifuge for 10 min, decant the solution into a 200-mm test tube, and dilute to 50 mL with brucine sulfate reagent solution.

> *Control Solution B.* Dissolve 0.30 g in a 50-mL centrifuge tube in 2 mL of water and 1 mL of the standard nitrate solution containing 0.01 mg of nitrate ion (NO_3) per mL by heating in a boiling-water bath. Add 20 mL of nitrate-free sulfuric acid $(2 + 1)$ and cool in an ice bath. Centrifuge for 10 min, decant the solution into a 200-mm test tube, and dilute to 50 mL with brucine sulfate reagent solution.

Continue with the procedure described on page 29, starting with the preparation of blank solution C.

SULFATE. (Page 32, Procedure A, Method 1). Use 1/20 of the filtrate from the test for insoluble matter. Allow 30 min for the turbidity to form.

AMMONIUM. Dissolve 1.0 g in 80 mL of water and add with stirring 20 mL of freshly boiled 10% sodium hydroxide reagent solution. After allowing the precipitate to settle, dilute 20 mL of the clear supernatant liquid with water to 50 mL, and add 2 mL of Nessler reagent. Any color should not exceed that produced by 0.01 mg of ammonium ion (NH_4) in an equal volume of solution containing 4 mL of the sodium hydroxide solution and 2 mL of the Nessler reagent.

COPPER, LEAD, AND ZINC. Determine the copper, lead, and zinc by the flame atomic absorption spectrophotometric method described on page 35.

Sample Stock Solution.   Dissolve 10.0 g of sample in a 100-mL volumetric flask, add 5 mL of nitric acid, and dilute to the mark with water (0.1 g/mL).

Analyze the solutions by means of a suitable atomic absorption spectrophotometer, using the conditions outlined in the following table. Calculate the metal content of the sample by the method of standard additions.

Element	Wavelength (nm)	Sample Wt (g)	Standard Added (mg)	Flame Type*	Background Correction
Cu	324.7	2.0	0.01; 0.02	A/A	Yes
Pb	217.0	2.0	0.01; 0.20	A/A	Yes
Zn	213.9	0.05	0.025; 0.05	A/A	Yes

*A/A is air/acetylene.

IRON.   (Page 28, Method 1). Use 2.0 g.

SUBSTANCES NOT PRECIPITATED BY HYDROGEN SULFIDE.   Dissolve 2.0 g in 150 mL of water containing 15 mL of approximately 1 N sulfuric acid. Heat to boiling and pass a rapid stream of hydrogen sulfide through the solution as it cools to room temperature. Filter the solution, but do not wash the precipitate. Add 0.25 mL of sulfuric acid to 75 mL of the clear filtrate and evaporate to dryness in a tared dish. Finally, ignite at $800 \pm 25°C$ for 15 min.

Cadmium Sulfate

Cadmium Sulfate, Anhydrous

$CdSO_4$ Formula Wt 208.47

CAS Number 10124–36–4

REQUIREMENTS

Assay . . . . $\geq$ 99.0% $CdSO_4$

MAXIMUM ALLOWABLE

Insoluble Matter 0.005%
Loss on drying at 150°C. 1.0%
Chloride (Cl) 0.001%
Nitrate and nitrite (as NO$_3$). 0.003%
Arsenic (As). . . . . 2 ppm
Copper (Cu) 0.002%
Iron (Fe) 0.001%
Lead (Pb) 0.003%
Substances not precipitated by
   hydrogen sulfide (as sulfates) 0.15%
Zinc (Zn) 0.1%

TESTS

ASSAY. (By complexometric titration of cadmium). Weigh accurately 0.9 g, transfer to a 400-mL beaker, and dissolve in 200 mL of water. Add 15 mL of a saturated aqueous hexamethylenetetramine solution and 50 mg of xylenol orange indicator mixture. Titrate with standard 0.1 M EDTA to a color change of red-purple to lemon-yellow. One milliliter of 0.1 M EDTA corresponds to 0.02085 g of $CdSO_4$.

INSOLUBLE MATTER. (Page 14). Use 20 g dissolved in 150 mL of water.

LOSS ON DRYING AT 150°C. Weigh accurately about 1 g and dry to constant weight at 150°C.

CHLORIDE. (Page 25). Use 1.0 g.

NITRATE AND NITRITE. (Page 29). For sample solution A use 0.50 g. For control solution B use 0.50 g and 1.5 mL of standard nitrate solution.

ARSENIC. (Page 23). Dissolve 1.0 g in 35 mL of water, add 2 g of granulated zinc (20 mesh), and allow to stand for a few minutes until the cadmium is precipitated, stirring occasionally to break up the mass of spent zinc. The suspension shows not more than 0.002 mg of arsenic (As), using the following modifications in amounts and order of addition of the reagents: Add 3 g more of the granulated zinc (20 mesh), 2 mL of potassium iodide solution (16.5% KI in water), and 0.5 mL of stannous chloride solution (40% $SnCl_2 \cdot H_2O$ in concentrated HCl). Then add 20 mL of dilute sulfuric acid (1 + 4) and immediately connect the generator flask to the scrubber–absorber assembly.

COPPER, LEAD, AND ZINC. Determine the copper, lead, and zinc by the flame atomic absorption spectrophotometric method described on page 35.

> **Sample Stock Solution A.** Dissolve 20.0 g of sample with water in a 100-mL volumetric flask, and dilute to the mark with water (0.2 g/mL).

> **Sample Stock Solution B.** Transfer 5.0 mL of sample stock solution A to a 200-mL volumetric flask, and dilute to the mark with water (0.01 g/mL).

Analyze the solutions by means of a suitable atomic absorption spectrophotometer, using the conditions outlined in the following table. Calculate the metal content of the sample by the method of standard additions.

Element	Wavelength (nm)	Sample Wt (g)	Standard Added (mg)	Flame Type*	Background Correction
Cu	324.7	2.0	0.04; 0.08	A/A	Yes
Pb	217.0	2.0	0.04; 0.08	A/A	Yes
Zn	213.9	0.05	0.025; 0.05	A/A	Yes

*A/A is air/acetylene.

IRON. (Page 28, Method 1). Dissolve 1.0 g in 30 mL of water, add 5 mL of hydrochloric acid, and dilute with water to 50 mL. Use this solution without further acidification.

SUBSTANCES NOT PRECIPITATED BY HYDROGEN SULFIDE. Dissolve 2.0 g in 150 mL of water containing 15 mL of approximately 1 N sulfuric acid. Heat to boiling and pass a rapid stream of hydrogen sulfide through the solution as it cools to room temperature. Filter the solution, but do not wash the precipitate. Add 0.25 mL of sulfuric acid to 75 mL of the clear filtrate and evaporate to dryness in a tared dish. Finally, ignite at $800 \pm 25°C$ for 15 min.

Cadmium Sulfate $\frac{8}{3}$-Hydrate

Cadmium Sulfate, Crystals

$CdSO_4 \cdot \frac{8}{3}H_2O$ Formula Wt 256.52

CAS Number 7790–84–3

REQUIREMENTS

Assay 98.0–102.0%
$CdSO_4 \cdot \frac{8}{3}H_2O$

MAXIMUM ALLOWABLE

Insoluble matter 0.005%
Chloride (Cl) 0.001%
Nitrate and nitrite (as NO_3). 0.003%
Arsenic (As). . . . . 2 ppm
Copper (Cu) 0.002%
Iron (Fe) 0.001%
Lead (Pb) 0.003%
Substances not precipitated by
  hydrogen sulfide (as sulfates) 0.1%
Zinc (Zn). 0.1%

TESTS

ASSAY. (By complexometric titration of cadmium). Weigh accurately 10 g, and transfer to a 250-mL volumetric flask. Dissolve in about 100 mL of water, dilute to volume with water, and mix thoroughly. Pipet 25 mL of this solution into a 400-mL beaker and add about 150 mL of water. Add 15 mL of a saturated aqueous hexamethylenetetramine solution and 50 mg of xylenol orange indicator mixture. Titrate with standard 0.1 M EDTA to a color change of red-purple to lemon-yellow. One milliliter of 0.1 M EDTA corresponds to 0.02565 g of $CdSO_4 \cdot \frac{8}{3}H_2O$.

INSOLUBLE MATTER. (Page 14). Use 20 g dissolved in 150 mL of water.

CHLORIDE. (Page 25). Use 1.0 g.

NITRATE AND NITRITE. (Page 29). For sample solution A use 0.50 g. For control solution B use 0.50 g and 1.5 mL of standard nitrate solution.

ARSENIC. (Page 23). Dissolve 1.0 g in 35 mL of water, add 2 g of granulated zinc (20 mesh), and allow to stand for a few minutes until the cadmium is precipitated, stirring occasionally to break up the mass of spent zinc. The suspension shows not more than 0.002 mg of arsenic (As), using the following modifications in amounts and order of addition of the reagents: Add 3 g more of the granulated zinc (20 mesh), 2 mL of potassium iodide solution (16.5% KI in water), and 0.5 mL of stannous chloride solution (40% $SnCl_2 \cdot 2H_2O$ in concentrated HCl). Then add 20 mL of dilute sulfuric acid (1 + 4) and immediately connect the generator flask to the scrubber–absorber assembly.

COPPER, LEAD, AND ZINC. Determine the copper, lead, and zinc by the flame atomic absorption spectrophotometric method described on page 35.

> **Sample Stock Solution A.** Dissolve 20.0 g of sample with water in a 100-mL volumetric flask, and dilute to the mark with water (0.2 g/mL).

> **Sample Stock Solution B.** Transfer 5.0 mL of sample stock solution A to a 200-mL volumetric flask, and dilute to the mark with water (0.01 g/mL).

Analyze the solutions by means of a suitable atomic absorption spectrophotometer, using the conditions outlined in the following table. Calculate the metal content of the sample by the method of standard additions.

Element	Wavelength (nm)	Sample Wt (g)	Standard Added (mg)	Flame Type*	Background Correction
Cu	324.7	2.0	0.04; 0.08	A/A	Yes
Pb	217.0	2.0	0.04; 0.08	A/A	Yes
Zn	213.9	0.05	0.025; 0.05	A/A	Yes

*A/A is air/acetylene.

IRON. (Page 28, Method 1). Dissolve 1.0 g in 30 mL of water, add 5 mL of hydrochloric acid, and dilute with water to 50 mL. Use this solution without further acidification.

SUBSTANCES NOT PRECIPITATED BY HYDROGEN SULFIDE. Dissolve 2.0 g in 150 mL of water containing 15 mL of approximately 1 N sulfuric acid. Heat to boiling and pass a rapid stream of hydrogen sulfide through the solution as it cools to room temperature. Filter the solution but do not wash the precipitate. Add 0.25 mL of sulfuric acid to 75 mL of the clear filtrate and evaporate to dryness in a tared dish. Finally, ignite at 800 ± 25°C for 15 min.

Calcium Acetate Monohydrate

$Ca(CH_3COO)_2 \cdot H_2O$ Formula Wt 176.18

CAS Number 5743−26−0

REQUIREMENTS

Assay . . . . ≥ 99.0%
$Ca(CH_3COO)_2 \cdot H_2O$

MAXIMUM ALLOWABLE

Insoluble matter 0.005%
Alkalinity. . . . . Passes test
Titrable acid . . . . 0.035 meq / g
Chloride (Cl) 0.001%
Sulfate (SO_4) 0.01%
Barium (Ba). 0.01%
Heavy metals (as Pb) 0.005%
Iron (Fe) 0.001%
Magnesium and alkali salts (as sulfates) 0.3%

TESTS

ASSAY. (By complexometric titration for calcium). Weigh accurately
0.70 g, transfer to a 250-mL beaker, and dissolve in 150 mL of water.
While stirring, add from a 50-mL buret about 30 mL of standard
0.1 M EDTA. Add 10% sodium hydroxide solution until the pH is
above 12 (using a pH meter). Add about 300 mg of hydroxy naphthol blue indicator mixture and continue the titration with the EDTA
solution to a blue color. One milliliter of 0.1 M EDTA corresponds to
0.01762 g of $Ca(CH_3COO)_2 \cdot H_2O$.

INSOLUBLE MATTER. (Page 14). Use 20 g dissolved in 200 mL of
water.

ALKALINITY AND TITRABLE ACID. Dissolve 2.0 g in 25 mL of
carbon dioxide-free water, and add 0.10 mL of phenolphthalein indicator solution. The sample solution should show no pink color due to
excess alkalinity. Titrate the solution with 0.10 N sodium hydroxide
until a pink color is produced after shaking. Not more than 0.70 mL
of 0.10 N sodium hydroxide should be required in the titration.

CHLORIDE.   (Page 25). Dissolve 1.0 g in 20 mL of water.

SULFATE.   (Page 32, Procedure A, Method 1). Use 0.5 g.

BARIUM.   Dissolve 2.0 g in 15 mL of water, add 2 drops of glacial acetic acid, filter, and add to the filtrate 0.3 mL of potassium dichromate solution (1 in 10). Compare the turbidity to a standard containing 0.2 mg of barium ion treated exactly as the sample.

HEAVY METALS.   (Page 26, Method 1). Dissolve 0.4 g in 20 mL of water.

IRON.   (Page 28, Method 1). Dissolve 1.0 g in 20 mL of water.

MAGNESIUM AND ALKALI SALTS.   Dissolve 2.0 g in 100 mL of water, add 5 mL of hydrochloric acid and 0.15 mL of methyl red indicator solution, and heat to boiling. Add slowly, with stirring, 100 mL of warm ammonium oxalate reagent solution, and heat to between 70° and 80°C. Add ammonium hydroxide (10% NH_3) dropwise until the solution is just alkaline to methyl red, dilute with water to 250 mL, and allow to stand without further heating for 1 hour. Filter, add 0.5 mL of sulfuric acid to 125 mL of the filtrate, and evaporate to about 30 mL. Cool, add 25 mL of nitric acid, and evaporate to dryness on the steam bath. Dissolve the residue in a few milliliters of water, filter through a small filter paper, and wash. Transfer the filtrate and washings to a small tared dish, evaporate to dryness, and heat on a hot plate to remove all of the sulfuric acid. Finally, ignite at 800 ± 25°C for 15 min. Correct for the weight obtained in a complete blank test.

Calcium Carbonate

$CaCO_3$ Formula Wt 100.09

CAS Number 471−34−1

REQUIREMENTS

Assay (dried basis) . . . . $\geq$ 99.0% $CaCO_3$

MAXIMUM ALLOWABLE

Insoluble in dilute hydrochloric acid 0.01%
Ammonium hydroxide precipitate 0.01%
Water-soluble titrable base . . . . 0.002 meq / g
Chloride (Cl) 0.001%
Fluoride (F) 0.0015%
Oxidizing substances (as NO_3). 0.005%
Sulfate (SO_4) 0.01%
Ammonium (NH_4) 0.003%
Barium (Ba). 0.005%
Heavy metals (as Pb) 0.001%
Iron (Fe) 0.003%
Magnesium (Mg) 0.02%
Potassium (K) 0.01%
Sodium (Na) 0.1%
Strontium (Sr) 0.1%

TESTS

ASSAY. (By complexometric titration of calcium). Weigh accurately 0.4 g, previously dried at 300°C for 4 h, and transfer to a 400-mL beaker. Cover the beaker with a watch glass. Add 2 mL of dilute hydrochloric acid from a pipet placed between the lip of the beaker and the watch glass. Swirl the beaker to aid dissolution. With water, wash down the inner wall of the beaker, the outer surface of the pipet, and the watch glass. Dilute to about 100 mL with water. While stirring, add from a 50-mL buret about 30 mL of standard 0.1 M EDTA. Adjust the solution to above pH 12 (pH meter with suitable high-pH glass electrode) with 10% sodium hydroxide solution. Add 300 mg of hydroxy naphthol blue indicator mixture and continue the titration immediately with the EDTA solution to a blue color. One milliliter of 0.1 M EDTA corresponds to 0.01001 g of $CaCO_3$.

INSOLUBLE IN DILUTE HYDROCHLORIC ACID. Add 10 g to about 100 mL of water, swirl, slowly and carefully add 20 mL of hydrochloric acid, and dilute with water to 150 mL. Heat the solution to boiling, boil gently to expel the carbon dioxide, and digest in a covered beaker on the steam bath for 1 h. Filter through a tared filtering crucible (reserve the filtrate for the determination of ammonium hydroxide precipitate), wash thoroughly, and dry at 105°C.

AMMONIUM HYDROXIDE PRECIPITATE. Add 0.10 mL of methyl red indicator solution to the filtrate reserved from the test for insoluble in dilute hydrochloric acid and make slightly alkaline with ammonium hydroxide. Heat the solution to boiling and boil gently for 5 min to coagulate the precipitate. If necessary, add more ammonium hydroxide to maintain the alkalinity throughout this operation. Filter through a small filter paper and wash with a small amount of hot water. Redissolve the precipitate from the filter paper with hot dilute hydrochloric acid (1 + 3) and wash the filter paper free of acid. Heat the filtrate plus washings, which should amount to about 25 mL, to boiling and boil gently for 1 to 2 min. Add 0.05 mL of methyl red indicator solution, make slightly alkaline with ammonium hydroxide, and again boil gently to coagulate the precipitate. Filter through the same small filter paper, wash thoroughly, and ignite in a tared crucible.

WATER-SOLUBLE TITRABLE BASE. Suspend 3 g in 30 mL of warm water in a small stoppered flask. Shake for 10 min, cool, and filter. To 20 mL of the filtrate add 0.05 mL of phenolphthalein indicator solution. If a red color is produced, it should be discharged by 0.50 mL of 0.01 N hydrochloric acid.

CHLORIDE. (Page 25). Dissolve 1.0 g in 10 mL of water plus 2 mL of nitric acid, filter if necessary through a small chloride-free filter, wash with hot water, and dilute with water to 20 mL.

FLUORIDE. Dissolve 1.0 g in 50 mL of water and 20 mL of 1 N hydrochloric acid in a 250-mL beaker. For the standards add 1.0, 2.0, 3.0, 5.0, 10.0, and 15.0 mL of standard fluoride solution to 250-mL glass beakers containing 50 mL of water and 5 mL of 1 N hydrochloric acid. Boil the sample and six standard solutions for a few seconds, cool rapidly, and transfer to plastic beakers. To each solution add 10 mL of 1 M sodium citrate reagent solution and 10 mL of 0.2 M disodium (ethylenedinitrilo)tetraacetate reagent solution and mix. If necessary, adjust the pH to 5.5 ± 0.1 with dilute hydrochloric acid or dilute sodium hydroxide. Transfer to 100-mL volumetric flasks, dilute to the mark with water, mix, and pour into plastic beakers. For each solution measure immediately the potential of a fluoride electrode versus a reference electrode using a pH meter with an expanded scale or an appropriate potentiometer. Plot a calibration curve on 2-cycle semilogarithmic paper with micrograms of fluoride per 100 mL of solution on the logarithmic scale. Determine the concentration of fluoride from the calibration curve.

OXIDIZING SUBSTANCES. Place 0.10 g in a dry beaker. Cool the beaker thoroughly in an ice bath and add 22 mL of sulfuric acid that has been cooled to ice-bath temperature. Allow the mixture to warm to room temperature and swirl the beaker at intervals to effect gentle dissolution with slow evolution of carbon dioxide. When dissolution is complete, add 3 mL of diphenylamine reagent solution and digest on the steam bath for 90 min. Prepare a standard by evaporating to dryness a solution containing 0.005 mg of nitrate (0.5 mL of the nitrate standard solution) and 0.01 g of sodium carbonate. Treat the residue exactly like the sample. Any color produced in the solution of the sample should not exceed that in the standard.

SULFATE.   (Page 32, Procedure B). Cautiously dissolve 2.0 g in 5 mL of water, 3.5 mL of hydrochloric acid, and 0.05 mL of bromine water. Dilute with water to 10 mL and mix. Transfer 2.0 mL to a reaction tube, and continue with the procedure.

> ***Sample Solution A for the Determination of Ammonium, Heavy Metals, and Iron.***   Cautiously dissolve 20 g in 100 mL of dilute hydrochloric acid (1 + 1) and evaporate on the steam bath to dryness or a moist residue. Dissolve the residue in about 100 mL of water, filter, and dilute with water to 200 mL (1 mL = 0.1 g).

AMMONIUM.   Dilute 10 mL (1-g sample) of sample solution A with water to 80 mL, add 20 mL of 10% sodium hydroxide reagent solution, stopper, mix well, and allow to stand for 1 h. Decant 50 mL through a filtering crucible that has been washed with 10% sodium hydroxide solution, and add to the filtrate 2 mL of Nessler reagent. Any color should not exceed that produced by 0.015 mg of ammonium ion (NH_4) in an equal volume of solution containing the quantities of reagents used in the test.

BARIUM.   For the sample dissolve 3.0 g in 15 mL of water and sufficient nitric acid (about 3.5 mL) to effect dissolution. For the control dissolve 1.0 g in 15 mL of water and sufficient nitric acid (about 1.5 mL) to effect dissolution and add 0.1 mg of barium ion (Ba). Heat each solution to boiling, boil gently for 5 min, cool, filter if necessary, and dilute with water to 23 mL. Add 2 mL of potassium dichromate reagent solution and neutralize with ammonium hydroxide until the orange color is just dissipated and the yellow color persists. To each solution slowly add, with constant stirring, 25 mL of methanol. Any turbidity in the solution of the sample should not exceed that in the control.

HEAVY METALS.   (Page 26, Method 1).   Use 30 mL (3-g sample) of sample solution A to prepare the sample solution, and use 10 mL (1-g sample) of sample solution A to prepare the control solution.

IRON.   (Page 28, Method 1). Use 3.3 mL of sample solution A (0.33-g sample).

MAGNESIUM, POTASSIUM, SODIUM, AND STRONTIUM.   Determine the magnesium, potassium, sodium, and strontium by the flame atomic absorption spectrophotometric method described on page 35.

>   ***Sample Stock Solution A.***   Cautiously dissolve 5.0 g of sample in 10 mL of nitric acid (1 + 1). Transfer to a 100-mL volumetric flask and dilute to the mark with water (0.05 g/mL).

>   ***Sample Stock Solution B.***   Transfer 10.0 mL (0.5 g) to a 100-mL volumetric flask, and dilute to the mark with water (0.005 g/mL).

Analyze the solutions by means of a suitable atomic absorption spectrophotometer, using the conditions outlined in the following table. Calculate the metal content of the sample by the method of standard additions.

Element	Wavelength (nm)	Sample Wt (g)	Standard Added (mg)	Flame Type*	Background Correction
Mg	285.2	0.1	0.01; 0.02	A/A	Yes
K	766.5	0.5	0.05; 0.10	A/A	No
Na	589.0	0.01	0.01; 0.02	A/A	No
Sr	460.7	0.1	0.1; 0.2	N/A	No

*A/A is air/acetylene; N/A is nitrous oxide/acetylene.

Calcium Carbonate, Low in Alkalies

$CaCO_3$ Formula Wt 100.09

CAS Number 471−34−1

REQUIREMENTS

Assay (dried basis) . . . . ≥ 99.0% $CaCO_3$

MAXIMUM ALLOWABLE

Insoluble in dilute hydrochloric acid 0.01%
Ammonium hydroxide precipitate 0.01%
Chloride (Cl) 0.001%
Fluoride (F) 0.0015%
Sulfate (SO$_4$) 0.005%
Barium (Ba) 0.01%
Heavy metals (as Pb) 0.001%
Iron (Fe) 0.002%
Magnesium (Mg) 0.01%
Potassium (K) 0.01%
Sodium (Na) 0.01%
Strontium (Sr) 0.1%

TESTS

ASSAY. (By complexometric titration of calcium). Weigh accurately 0.4 g, previously dried at 300°C for 4 h, and transfer to a 400-mL beaker. Cover the beaker with a watch glass. Add 2 mL of dilute hydrochloric acid from a pipet placed between the lip of the beaker and the watch glass. Swirl the beaker to aid dissolution. With water, wash down the inner wall of the beaker, the outer surface of the pipet, and the watch glass. Dilute to about 100 mL with water. While stirring, add from a 50-mL buret about 30 mL of standard 0.1 M EDTA. Adjust the solution to above pH 12 (pH meter with suitable high-pH glass electrode) with 10% sodium hydroxide solution. Add 300 mg of hydroxy naphthol blue indicator mixture and continue the titration immediately with the EDTA solution to a blue color. One milliliter of 0.1 M EDTA corresponds to 0.01001 g of CaCO$_3$.

INSOLUBLE IN DILUTE HYDROCHLORIC ACID. Add 20 g to about 200 mL of water, swirl, slowly and carefully add 40 mL of hydrochloric acid, and dilute with water to 300 mL. Heat the solution to boiling, boil gently to expel the carbon dioxide, and digest in a covered beaker on the steam bath for 1 h. Filter through a tared filtering crucible (reserve the filtrate for the determination of ammonium hydroxide precipitate), wash thoroughly, and dry at 105°C.

AMMONIUM HYDROXIDE PRECIPITATE. Add 0.20 mL of methyl red indicator solution to the filtrate reserved from the test for insoluble in dilute hydrochloric acid and make slightly alkaline with ammonium hydroxide. Heat the solution to boiling and boil gently for 5 min to coagulate the precipitate. If necessary, add more ammonium hydroxide to maintain the alkalinity throughout the operation. Filter through a small filter paper and wash with a small amount of hot water. (Reserve this filtrate for the determination of sulfate.) Redissolve the precipitate from the filter paper with hot dilute hydrochloric acid (1 + 3) and wash the filter paper free of acid. Heat the filtrate plus washings, which should amount to about 25 mL, to boiling and boil gently for 1 to 2 min. Add 0.05 mL of methyl red indicator solution, make slightly alkaline with ammonium hydroxide, and again boil gently to coagulate the precipitate. Filter through the same small filter paper, wash thoroughly, and ignite in a tared crucible.

CHLORIDE.   (Page 25). Dissolve 1.0 g in 10 mL of water plus 2 mL of nitric acid, filter if necessary through a small chloride-free filter, wash with hot water, and dilute with water to 20 mL.

FLUORIDE.   Dissolve 1.0 g in 50 mL of water and 25 mL of 1 N hydrochloric acid in a 250-mL beaker. For the standards add 1.0, 2.0, 3.0, 5.0, 10.0, and 15.0 mL of standard fluoride solution to 250-mL glass beakers containing 50 mL of water and 5 mL of 1 N hydrochloric acid. Boil the sample and six standard solutions for a few seconds, cool rapidly, and transfer to plastic beakers. To each solution add 20 mL of 1 M sodium citrate reagent solution and 10 mL of 0.2 M disodium (ethylenedinitrilo)tetraacetate reagent solution and mix. If necessary, adjust the pH to 5.5 $\pm$ 0.1 with dilute hydrochloric acid or dilute sodium hydroxide. Transfer to 100-mL volumetric flasks, dilute to the mark with water, mix, and pour into plastic beakers. For each solution measure immediately the potential of a fluoride electrode versus a reference electrode (Orion Model Nos. 94–09 and 90–01 are satisfactory) using a pH meter with an expanded scale or an appropriate potentiometer. Plot a calibration curve on 2-cycle semilogarithmic paper with micrograms of fluoride per 100 mL of solution on the logarithmic scale. Determine the concentration of fluoride from the calibration curve.

SULFATE.   (Page 32, Procedure A, Method 1). Use 1/20 of the filtrate from the test for ammonium hydroxide precipitate. Neutralize with hydrochloric acid.

BARIUM.   For the sample dissolve 2.0 g in 15 mL of water and sufficient nitric acid (about 2.5 mL) to effect dissolution. For the control dissolve 1.0 g in 15 mL of water and sufficient nitric acid (about 1.5 mL) to effect dissolution and add 0.1 mg of barium ion (Ba). Heat each to boiling, boil gently for 5 min, cool, filter if necessary, and dilute with water to 23 mL. Add 2 mL of potassium dichromate reagent solution and neutralize with ammonium hydroxide until the orange color is just dissipated and the yellow color persists. To each solution slowly add, with constant stirring, 25 mL of methanol. Any turbidity in the solution of the sample should not exceed that in the control.

> ***Sample Solution A for the Determination of Heavy Metals and Iron.*** Cautiously dissolve 20 g in 100 mL of dilute hydrochloric acid (1 + 1) and evaporate on the steam bath to dryness or a moist residue. Dissolve the residue in about 100 mL of water, filter, and dilute with water to 200 mL (1 mL = 0.1 g).

HEAVY METALS. (Page 26, Method 1). Use 30 mL (3-g sample) of sample solution A to prepare the sample solution, and use 10 mL of sample solution A (1-g sample) to prepare the control solution.

IRON. (Page 28, Method 1). Use 5.0 mL of sample solution A (0.5-g sample).

MAGNESIUM, POTASSIUM, SODIUM, AND STRONTIUM. Determine the magnesium, potassium, sodium, and strontium by the flame atomic absorption spectrophotometric method described on page 35.

> ***Sample Stock Solution.*** Cautiously dissolve 5.0 g of sample in 10 mL of nitric acid (1 + 1). Transfer to a 100-mL volumetric flask, and dilute to the mark with water (0.05 g/mL).

Analyze the solutions by means of a suitable atomic absorption spectrophotometer, using the conditions outlined in the following table. Calculate the metal content of the sample by the method of standard additions.

Element	Wavelength (nm)	Sample Wt (g)	Standard Added (mg)	Flame Type*	Background Correction
Mg	285.2	0.1	0.01; 0.02	A/A	Yes
K	766.5	0.5	0.05; 0.10	A/A	No
Na	589.0	0.1	0.01; 0.02	A/A	No
Sr	460.7	0.1	0.1; 0.2	N/A	No

*A/A is air/acetylene; N/A is nitrous oxide/acetylene.

Calcium Carbonate, Chelometric Standard

$CaCO_3$ Formula Wt 100.09

CAS Number 471–34–1

REQUIREMENTS

Assay 99.95–100.05%
$CaCO_3$

MAXIMUM ALLOWABLE

Insoluble in dilute hydrochloric acid 0.01%
Ammonium hydroxide precipitate 0.01%
Chloride (Cl) 0.001%
Fluoride (F) 0.0015%
Sulfate (SO_4) 0.005%
Barium (Ba) 0.01%
Heavy metals (as Pb) 0.001%
Iron (Fe) 0.002%
Magnesium (Mg) 0.01%
Potassium (K) 0.01%
Sodium (Na) 0.01%
Strontium (Sr) 0.1%

TESTS

ASSAY.   (By complexometric titration for calcium). This method is a comparative procedure in which the calcium carbonate to be assayed is compared with NIST Standard Reference Material Calcium Carbonate (SRM 915). A minimum of two samples and two standards should be run according to this procedure.

Procedure.   Dry the sample and NIST standard at 110°C for 2 h. Weigh (to the nearest 0.1 mg) 0.4000 ± 0.0040 g of calcium carbonate and place in a 400-mL beaker with a magnetic stirring bar and 50 mL of water. Cover with a watch glass and add 12 mL of 10% hydrochloric acid with the pipet tip under the watch glass. Wash down the watch glass and the sides of the beaker. All of the sample must be dissolved.

Weigh (to the nearest 0.1 mg) 1.4400 ± 0.0100 g of (ethylenedinitrilo)-tetraacetic acid disodium salt (dihydrate) (NOTE 1). Add the EDTA directly to the calcium solution, dilute to about 125 mL, and add 15 mL of 5 M potassium hydroxide. This addition should dissolve all the EDTA and leave a clear solution. Add 0.05–0.10 mL of 0.02% calcein indicator solution and titrate the excess calcium with 0.0200 M EDTA under UV illumination in a dark room or box (NOTE 2). The end point is extremely sharp from a bright yellow-green fluorescence to black.

$$\%CaCO_3 = \frac{[(\text{Wt EDTA}) + (0.37226 \times \text{mL} \times M)]/(\text{Wt sample } CaCO_3)}{[(\text{Wt EDTA}) + (0.37226 \times \text{mL} \times M)]/(\text{Wt NIST } CaCO_3 \times \text{NIST assay factor})}$$

NOTE 1. This method is independent of the quality of the disodium EDTA used, but it is recommended that reagent-quality material be selected, then crushed, and mixed thoroughly prior to use. Do not dry the EDTA because water loss will change the proportions needed for the test. A disodium dihydrogen EDTA dihydrate having an assay of 99.9–100.1% is commercially available. ACS Reagent is 99–101% and suitable for this test.

NOTE 2. The UV lamp should have dark blue glass (little visible light) and should be placed at a right angle to the viewer, shielded from the viewer's eyes.

NOTE 3. Buoyancy corrections cancel.

Reagents

Potassium Hydroxide, 5 M.   Dissolve 56 g of KOH in water and dilute with water to 200 mL.

Calcein Indicator Solution, 0.02%.   Dissolve 20 mg of calcein in 10 mL of water by addition of a few drops of 5 M potassium hydroxide and dilute with water to 100 mL. This solution is not stable at room temperature for more than 1 day. Frozen solutions are stable, so the indicator solution may be split into portions in plastic bottles and frozen, and a fresh bottle may be thawed as needed.

EDTA, 0.0200 M.   Dissolve 7.446 g of reagent-quality (ethylenedinitrilo)tetraacetic acid disodium salt dihydrate in water and dilute with water to 1 L. Standardize against a calcium solution prepared by dissolving 2.002 g (weighed to the nearest 0.1 mg) of NIST calcium carbonate in water by addition of 6 mL of concentrated hydrochloric acid, followed by dilution to exactly 1 L. The molarity of this solution should be calculated from the weight taken. The standardization of a 25-mL aliquot of the calcium solution will require 10 mL of 5 M potassium hydroxide and 0.05–0.10 mL of calcein indicator solution, as in the procedure cited.

Insoluble in dilute hydrochloric acid and other tests except assay are the same as for calcium carbonate, low in alkalies, page 219.

Calcium Chloride Desiccant

$CaCl_2$ Formula Wt 110.98

CAS Number 10043–52–4

REQUIREMENTS

Assay . . . . $\geq$ 96.0% $CaCl_2$

MAXIMUM ALLOWABLE

Titrable base . . . . 0.006 meq / g
Magnesium and alkali salts (as sulfates) 2.0%

TESTS

ASSAY. (By argentimetric titration of chloride). Weigh accurately 2 g, dissolve in water in a 200-mL volumetric flask, and dilute to volume with water. Transfer 25.0 mL to a second 200-mL volumetric flask, and dilute with 100 mL of dilute nitric acid (1 + 49). Add, slowly and with constant agitation, 50.0 mL of 0.1 N silver nitrate, dilute to volume with water, and mix well. Filter through a dry filter into a dry flask, rejecting the first 30 mL of the filtrate. To 100.0 mL of the resultant filtrate add 2 mL of ferric ammonium sulfate indicator solution, and titrate the excess silver nitrate with 0.1 N thiocyanate solution. One milliliter of 0.1 N silver nitrate corresponds to 0.005549 g of $CaCl_2$.

TITRABLE BASE. Dissolve 5.0 g in 50 mL of water and add 0.10 mL of phenolphthalein indicator solution. If any pink color is produced, it should be discharged by not more than 3.0 mL of 0.01 N hydrochloric acid.

MAGNESIUM AND ALKALI SALTS. To 1.0 g in 100 mL of water add 5 mL of hydrochloric acid and a few drops of methyl red indicator solution and heat to boiling. Add dropwise with stirring 100 mL of warm ammonium oxalate reagent solution. Heat the solution between 70 and 80°C and add dropwise 10% ammonium hydroxide until the solution is just alkaline to methyl red. Dilute with water to 250 mL and allow the solution to stand without further heating for 1 h. Filter, and to 125 mL of the filtrate add 0.5 mL of sulfuric acid, evaporate to about 30 mL, and cool. Add 25 mL of nitric acid and evaporate to dryness on the steam bath. Dissolve the residue in a few milliliters of water, filter through a small filter paper, and wash. Transfer the filtrate and washings to a small tared dish. Evaporate to dryness, ignite gently to volatilize the excess acids and salts, and finally ignite at 800 ± 25°C for 15 min.

Calcium Chloride Dihydrate

$CaCl_2 \cdot 2H_2O$ Formula Wt 147.01

CAS Number 10035−04−8

REQUIREMENTS

Assay (as $CaCl_2$). . . . . 74.0−78.0%
pH of a 5% solution . . . . 4.5−8.5 at 25°C

MAXIMUM ALLOWABLE

Insoluble and ammonium hydroxide precipitate . . . 0.01%
Oxidizing substances (as NO_3). 0.003%
Sulfate (SO_4). 0.01%
Ammonium (NH_4) 0.005%
Barium (Ba). 0.005%
Heavy metals (as Pb) . . . . 5 ppm
Iron (Fe) 0.001%
Magnesium (Mg). 0.005%
Potassium (K) 0.01%
Sodium (Na) 0.02%
Strontium (Sr) 0.1%

TESTS

ASSAY. (By argentometric titration of chloride). Weigh accurately 2 g, dissolve in water in a 200-mL volumetric flask, and dilute to volume with water. Transfer 25.0 mL to a second 200-mL volumetric flask, and dilute with 100 mL of dilute nitric acid (1 + 49). Add, slowly and with constant agitation, 50.0 mL of 0.1 N silver nitrate, dilute to volume with water, and mix well. Filter through a dry filter into a dry flask, rejecting the first 30 mL of the filtrate. To 100.0 mL of the resultant filtrate add 2 mL of ferric ammonium sulfate indicator solution, and titrate the excess silver nitrate with 0.1 N thiocyanate solution. One milliliter of 0.1 N silver nitrate corresponds to 0.005549 g of $CaCl_2$.

pH OF A 5% SOLUTION. (Page 43). The pH should be 4.5–8.5 at 25°C.

INSOLUBLE AND AMMONIUM HYDROXIDE PRECIPITATE. Dissolve 10 g in 100 mL of water and heat to boiling. Add 0.10 mL of methyl red indicator solution, make slightly alkaline with ammonium hydroxide (free from carbonate), and boil for 5 min. Filter through a small paper and wash with a little hot water. Redissolve the precipitate with hot dilute hydrochloric acid (1 + 3), wash the paper free of acid, boil the solution (which should amount to about 30 mL) for 1–2 min, add 0.05 mL of methyl red indicator solution, make slightly alkaline with ammonium hydroxide, and again boil gently to coagulate the precipitate. Filter through the same paper, wash thoroughly, and ignite.

OXIDIZING SUBSTANCES. Place 0.20 g in a dry beaker. Cool the beaker thoroughly in an ice bath and add 22 mL of sulfuric acid that has been cooled to ice-bath temperature. Allow the mixture to warm to room temperature and swirl the beaker at intervals to effect gentle dissolution with slow evolution of hydrogen chloride. When dissolution is complete, add 3 mL of diphenylamine reagent solution and digest on the steam bath for 90 min. Prepare a standard by evaporating to dryness a solution containing 0.006 mg of nitrate (0.6 mL of the nitrate standard solution) and 0.01 g of sodium carbonate. Treat the residue exactly like the sample. Any color produced in the solution of the sample should not exceed that in the standard.

SULFATE. (Page 32, Procedure B). Dissolve 10 g in 10 mL of water in a 25-mL glass-stoppered cylinder. Cool to room temperature, dilute to 20 mL with water, and transfer 1.0 mL to a reaction tube. Continue with the procedure.

Sample Solution A for Determining Ammonium, Barium, Heavy Metals, and Iron. Dissolve 50 g in about 200 mL of water, filter if necessary, and dilute with water to 250 mL in a volumetric flask (1 mL = 0.2 g).

AMMONIUM. Dilute 10 mL (2-g sample) of sample solution A to 45 mL and add 15 mL of 10% sodium hydroxide reagent solution. Filter through a filtering crucible previously washed with 10% sodium hydroxide reagent solution. Dilute 6 mL of the filtrate to 50 mL and add 2 mL of Nessler reagent. Any color should not exceed that produced by 0.01 mg of ammonium ion (NH_4) in an equal volume of solution containing 1.5 mL of 10% sodium hydroxide reagent solution and 2 mL of Nessler reagent.

BARIUM. For the sample add 2 g of sodium acetate and 0.05 mL of glacial acetic acid to 15 mL (3-g sample) of sample solution A. For the control add 2 g of sodium acetate, 0.05 mL of glacial acetic acid, and 0.1 mg of barium ion to 5 mL (1-g sample) of sample solution A and dilute with water to 15 mL. To each solution add 2 mL of potassium dichromate reagent solution and allow to stand for 15 min. Any turbidity in the solution of the sample should not exceed that in the control.

HEAVY METALS. (Page 26, Method 1). Use 30 mL (6-g sample) of sample solution A to prepare the sample solution, and use 10 mL (2-g sample) of sample solution A to prepare the control solution.

IRON. (Page 28, Method 1). Use 5.0 mL (1-g sample) of sample solution A.

MAGNESIUM, POTASSIUM, SODIUM, AND STRONTIUM. Determine the magnesium, potassium, sodium, and strontium by the flame atomic absorption spectrophotometric method described on page 35.

Sample Stock Solution. Dissolve 2.0 g of sample in 50 mL of water and 5 mL of nitric acid in a 200-mL volumetric flask, and dilute to the mark with water (0.01 g/mL).

Analyze the solutions by means of a suitable atomic absorption spectrophotometer, using the conditions outlined in the following table. Calculate the metal content of the sample by the method of standard additions.

Element	Wavelength (nm)	Sample Wt (g)	Standard Added (mg)	Flame Type*	Background Correction
Mg	285.2	0.1	0.005; 0.01	A/A	Yes
K	766.5	0.2	0.02; 0.04	A/A	No
Na	589.0	0.05	0.01; 0.02	A/A	No
Sr	460.7	0.05	0.05; 0.10	N/A	No

*A/A is air/acetylene; N/A is nitrous oxide/acetylene.

Calcium Hydroxide

$Ca(OH)_2$ Formula Wt 74.09

CAS Number 1305–62–0

REQUIREMENTS

Assay . . . . $\geq$ 95.0% $Ca(OH)_2$
and $\geq$ 3.0% $CaCO_3$

MAXIMUM ALLOWABLE

Insoluble in hydrochloric acid 0.03%
Chloride (Cl) 0.03%
Sulfur compounds (as SO_4) 0.1%
Heavy metals (as Pb) 0.003%
Iron (Fe) 0.05%
Magnesium and alkali salts (as sulfates) 1.0%

TESTS

ASSAY. (By acid–base titrimetry). Weigh accurately 1.4 g of sample, mix with 100 mL of carbon dioxide-free water, and add 0.15 mL of phenolphthalein indicator solution. Titrate with 1.0 N hydrochloric acid to the disappearance of the pink color (end point $= A$ mL). (The sample is initially suspended in the solution, but dissolves as the acid is added.) Add 0.20 mL of methyl orange indicator solution, and continue the titration to a pinkish-orange color (end point = B mL).

$$\%Ca(OH)_2 = \frac{A \times N \times 3.705}{Wt\ sample\ (g)}$$

$$\%CaCO_3 = \frac{(B - A) \times N \times 5.005}{Wt\ sample\ (g)}$$

INSOLUBLE IN HYDROCHLORIC ACID.   Add 5.0 g to 100 mL of water, add 25 mL of hydrochloric acid, heat the solution to boiling, and digest in a covered beaker on the steam bath for 1 h. Filter through a tared filtering crucible, wash thoroughly, and dry at 105°C.

CHLORIDE.   (Page 25). Dissolve 0.17 g in 10 mL of water plus 1 mL of nitric acid, dilute with water to 50 mL, and dilute 10 mL of this solution with water to 20 mL.

SULFUR COMPOUNDS.   Mix 4.0 g with 100 mL of water, add 5 mL of bromine water, and digest on a hot plate at low setting for 10 min. Add 6 mL of hydrochloric acid, boil to expel the excess bromine, filter, and wash the filter with hot water. Heat the solution to boiling, add 5 mL of barium chloride reagent solution, digest in a covered beaker on a hot plate at low setting for 2 h, and allow to stand overnight. If a precipitate is formed, filter, wash thoroughly, and ignite. Correct for the weight obtained in a complete blank test.

> ***Sample Solution A for the Determination of Heavy Metals and Iron.***   To 3.0 g add 40 mL of water, mix, cautiously add 10 mL of hydrochloric acid and 3 mL of nitric acid, and evaporate on the steam bath to dryness. Take up the residue with 1 mL of 10% hydrochloric acid and 30 mL of hot water, filter, and wash with a few mL of water. Add 0.10 mL of phenolphthalein indicator solution and sufficient 1 N sodium hydroxide to produce a pink color, then add sufficient 1 N hydrochloric acid to discharge the pink color, and dilute with water to 60 mL (10 mL = 0.5 g).

HEAVY METALS.   (Page 26, Method 1). Use 20 mL (1-g sample) of sample solution A to prepare the sample solution, and use 6.7 mL (0.33-g sample) of sample solution A to prepare the control solution.

IRON.   (Page 28, Method 1). Dilute 4.0 mL of sample solution A (0.2-g sample) with water to 100 mL, and use 10 mL of this dilution.

MAGNESIUM AND ALKALI SALTS. Mix 1.0 g with 30 mL of water and add, cautiously and with stirring, 5 mL of hydrochloric acid. Heat the solution to boiling, boil for 2–3 min, and dilute with water to 50 mL. To the hot solution add 85 mL of ammonium oxalate reagent solution, 0.15 mL of methyl red indicator solution, and sufficient ammonium hydroxide to neutralize. Cool, dilute with water to 150 mL, and allow to stand for 4 h or overnight. Filter, evaporate 75 mL of the clear filtrate to near dryness, add 0.50 mL of sulfuric acid, evaporate to dryness, and ignite at $800 \pm 25°C$ for 15 min.

Calcium Nitrate Tetrahydrate

$Ca(NO_3)_2 \cdot 4H_2O$ Formula Wt 236.15

CAS Number 13477–34–4

REQUIREMENTS

Assay	99.0–103.0% $Ca(NO_3)_2 \cdot 4H_2O$
pH of a 5% solution	5.0–7.0 at 25°C

MAXIMUM ALLOWABLE

Insoluble matter and ammonium hydroxide precipitate	0.005%
Chloride (Cl)	0.005%
Sulfate (SO_4)	0.002%
Barium (Ba)	0.005%
Heavy metals (as Pb)	5 ppm
Iron (Fe)	5 ppm
Magnesium and alkali salts (as sulfates)	0.2%

TESTS

ASSAY. (By complexometric titration of calcium). Weigh accurately about 0.90 g, transfer to a 250-mL beaker, and dissolve in 150 mL of water. While stirring, add from a 50-mL buret about 30 mL of standard 0.1 M EDTA. Then adjust the solution to above pH 12 (pH meter with suitable high-pH glass electrode) with 10% sodium hydroxide solution. Add about 300 mg of hydroxy naphthol blue indicator mixture and continue the titration immediately with the EDTA solution to a blue color. One milliliter of 0.1 M EDTA corresponds to 0.02362 g of $Ca(NO_3)_2 \cdot 4H_2O$.

pH OF A 5% SOLUTION.   (Page 43). The pH should be 5.0–7.0 at 25°C.

INSOLUBLE MATTER AND AMMONIUM HYDROXIDE PRECIPITATE. Dissolve 20 g in 100 mL of water, heat to boiling, render slightly alkaline with ammonium hydroxide, and boil for 5 min. If a precipitate is formed, filter through a small filter paper, and wash with hot water. Dissolve the precipitate in 20 mL of hot 10% hydrochloric acid, boil for 1–2 min, and make slightly alkaline with ammonium hydroxide. Filter through the same filter paper, wash thoroughly, and ignite gently in a tared crucible.

CHLORIDE.   (Page 25). Dissolve 1.0 g in 50 mL of water, and dilute 10 mL of the solution with water to 20 mL.

SULFATE.   (Page 32, Procedure A, Method 2). Use 6 mL of hydrochloric acid.

BARIUM.   For the sample dissolve 2.0 g in 15 mL of water, add 2 g of sodium acetate and 0.05 mL of glacial acetic acid, and filter if necessary. For the standard, add 0.1 mg of barium ion to 15 mL of water, then add 2 g of sodium acetate and 0.05 mL of glacial acetic acid. Add to each solution 2 mL of 10% potassium dichromate reagent solution, mix, and let stand for 15 min. Any turbidity in the solution of the sample should not exceed that in the standard.

HEAVY METALS.   (Page 26, Method 1). Dissolve 6.0 g in about 20 mL of water, and dilute with water to 30 mL. Use 25 mL to prepare the sample solution, and use the remaining 5.0 mL to prepare the control solution.

IRON.   (Page 28, Method 1). Use 2.0 g.

MAGNESIUM AND ALKALI SALTS.   Dissolve 2.0 g in 100 mL of water, add 5 mL of hydrochloric acid and 0.15 mL of methyl red indicator solution, and heat to boiling. Add slowly, with stirring, 100 mL of warm ammonium oxalate reagent solution, and heat to between 70 and 80°C. Add ammonium hydroxide (10% NH_3) dropwise until the solution is just alkaline to methyl red, dilute with water to 250 mL, and allow to stand without further heating for 1 h. Filter, add 0.5 mL of sulfuric acid to 125 mL of the filtrate, and evaporate to about 30 mL. Cool, add 25 mL of nitric acid, and evaporate to dryness on the steam bath. Dissolve the residue in a few mL of water, filter through a small filter paper, and wash. Transfer the filtrate and washings to a small tared dish, evaporate to dryness, and bake on a hot plate to remove all of the sulfuric acid. Finally, ignite at 800 $\pm$ 25°C for 15 min. Correct for the weight obtained in a complete blank test.

Calcium Sulfate Dihydrate

$CaSO_4 \cdot 2H_2O$ Formula Wt 172.17

CAS Number 10101 −41 −4

REQUIREMENTS

Assay . . . . 98.0 –102.0%
$CaSO_4 \cdot 2H_2O$

MAXIMUM ALLOWABLE

Insoluble in dilute hydrochloric acid 0.02%
Chloride (Cl) 0.005%
Nitrate (NO_3) . . . . Passes test (limit about 0.005%)
Carbonate (CO_3) . . . . Passes test
Heavy metals (as Pb) 0.002%
Iron (Fe) 0.001%
Magnesium and alkali salts (as sulfates) 0.3%

TESTS

ASSAY. (By complexometric titration for calcium). Dissolve 0.30 g, accurately weighed, in 100 mL of water and 5 mL of hydrochloric acid; boil if necessary to effect dissolution. Cool and stir, preferably with a magnetic stirrer, while adding the following reagents in the cited order: 0.5 mL of 2,2′,2″-nitrilotriethanol (i.e., triethanolamine), 300 mg of hydroxy naphthol blue indicator mixture, and (from a 50-mL buret) about 30 mL of standard 0.05 M EDTA. Now add 10% sodium hydroxide solution until the initial red color changes to clear blue; then continue the addition dropwise until the pH is above 12 (pH meter with suitable high-pH glass electrode). Continue the titration immediately with the EDTA solution to a blue color. One milliliter of 0.05 M EDTA corresponds to 0.008608 g of $CaSO_4 \cdot 2H_2O$.

INSOLUBLE IN DILUTE HYDROCHLORIC ACID.   Dissolve 4.0 g in a mixture of 100 mL of water and 20 mL of hydrochloric acid, heating on a steam bath if necessary until dissolution is complete. Filter through a tared filtering crucible, wash thoroughly with hot 10% hydrochloric acid, followed by water, and dry at 105°C.

CHLORIDE.   (Page 25). Dissolve 0.20 g in a mixture of 10 mL of water and 7 mL of nitric acid, warming if necessary to complete dissolution. Filter if necessary through a chloride-free filter, and dilute with water to 20 mL.

NITRATE.   Mix 1.0 g with 9 mL of water and 1 mL of 0.5% sodium chloride solution. Add 0.10 mL of indigo carmine reagent solution, mix, and add 10 mL of sulfuric acid. The blue color should not be completely discharged in 10 min.

CARBONATE.   Mix 1 g with 5 mL of water and slowly add 2 mL of hydrochloric acid. The evolution of a gas should not be observed.

HEAVY METALS.   (Page 26, Method 1). Add 2.0 g to a mixture of 25 mL of water and 5 mL of hydrochloric acid, heat the solution to boiling, and nearly neutralize with ammonium hydroxide. Filter, wash with water to about 40 mL, complete the neutralization with ammonium hydroxide (10% NH_3), and dilute with water to 50 mL. Use 25 mL for the sample solution. To prepare the standard use 25 mL of the following solution: add 0.04 mg of lead ion (Pb) to 5 mL of hydrochloric acid, neutralize with ammonium hydroxide, and dilute with water to 50 mL.

IRON.   (Page 28, Method 1). Boil 1.0 g with 6 mL of hydrochloric acid and 40 mL of water, cool, filter, and wash with water to 50 mL. Test the solution without further acidification.

MAGNESIUM AND ALKALI SALTS.   Dissolve 2.0 g in 100 mL of water, add 5 mL of hydrochloric acid and 0.15 mL of methyl red indicator solution, and heat to boiling. Add slowly, with stirring, 100 mL of warm ammonium oxalate reagent solution, and heat to between 70 and 80°C. Add ammonium hydroxide (10% NH_3) dropwise until the solution is just alkaline to methyl red, dilute with water to 250 mL, and allow to stand without further heating for 1 h. Filter, add 0.5 mL of sulfuric acid to 125 mL of the filtrate, and evaporate to about 30 mL. Cool, add 25 mL of nitric acid, and evaporate to dryness on the steam bath. Dissolve the residue in a few milliliters of water, filter through a small filter paper, and wash. Transfer the filtrate and washings to a small tared dish, evaporate to dryness, and bake on a hot plate to remove all of the sulfuric acid. Finally, ignite at 800 ± 25°C for 15 min. Correct for the weight obtained in a complete blank test.

Carbon Disulfide

CS_2 Formula Wt 76.13

CAS Number 75 –15 –0

> NOTE. Carbon disulfide should be supplied and stored in amber glass containers and protected from direct sunlight.

REQUIREMENTS

Assay . . . . ≥ 99.9% CS_2

MAXIMUM ALLOWABLE

Color (APHA). 10
Residue after evaporation 0.002%
Hydrogen sulfide (H_2S). Passes test (limit
                                                             about 1.5 ppm)
Sulfur dioxide (SO_2). . . . . Passes test (limit
                                                             about 2.5 ppm)
Water (H_2O) 0.05%

TESTS

ASSAY.   Analyze the sample by gas chromatography, using the general parameters cited on page 56. The following specific conditions are also required.

   Column: Type I, methyl silicone

Measure the area under all peaks and calculate the carbon disulfide content in area percent. Correct for water content.

COLOR (APHA).   (Page 17).

RESIDUE AFTER EVAPORATION.   (Page 14). Evaporate 50 g (40 mL) to dryness in a tared dish at 50 to 60°C, and heat at 60°C for 1 h.

HYDROGEN SULFIDE AND SULFUR DIOXIDE.   Shake vigorously 25 g (20 mL) with 0.2 mL of 0.01 N iodine for 15 s in a glass-stoppered cylinder. The pink color should not disappear.

WATER.   (Page 55, Method 2). Use 20 μL (25 mg) of the sample.

Carbon Tetrachloride
Tetrachloromethane

CCl_4 Formula Wt 153.82

CAS Number 56 – 23 – 5

Suitable for use in ultraviolet spectrophotometry or general use. Product labeling shall designate the one or more of these uses for which suitability is represented on the basis of meeting the relevant requirements and tests. The ultraviolet spectrophotometry requirements include all of the requirements for general use.

REQUIREMENTS

General Use

Assay . . . . $\geq$ 99.9% CCl_4

MAXIMUM ALLOWABLE

Color (APHA).	10
Residue after evaporation	0.001%
Water-soluble titrable acid	0.0005 meq / g
Free chlorine (Cl)	Passes test
Sulfur compounds (as S)	Passes test (limit about 0.005%)
Iodine-consuming substances	Passes test
Substances darkened by sulfuric acid	Passes test
Suitability for use in dithizone tests	Passes test

Specific Use

ULTRAVIOLET SPECTROPHOTOMETRY

Wavelength (nm)	Absorbance (AU)
330 to 400	0.01
300	0.02
290	0.05
280	0.10
270	0.35
265	1.00

TESTS

ASSAY. Analyze the sample by gas chromatography, using the general parameters cited on page 56. The following conditions are also required.

Column: Type I, methyl silicone

Measure the area under all peaks and calculate the carbon tetrachloride content in area percent.

COLOR (APHA). (Page 17).

RESIDUE AFTER EVAPORATION. (Page 14). Evaporate 100 g (63 mL) to dryness in a tared dish on the steam bath and dry the residue at 105°C for 30 min.

WATER-SOLUBLE TITRABLE ACID. Shake 20 g (13 mL) with 20 mL of carbon dioxide-free water for 5 min, separate, and discard the carbon tetrachloride. To 10 mL of the aqueous layer add 0.05 mL of phenolphthalein indicator solution and 0.50 mL of 0.01 N sodium hydroxide. A pink color should be produced.

FREE CHLORINE. Shake 10 mL for 2 min with 10 mL of water to which 0.10 mL of 10% potassium iodide reagent solution has been added, and allow to separate. The lower layer should not show a violet tint.

SULFUR COMPOUNDS. To 3.0 mL in a conical flask add 30 mL of approximately 0.5 N potassium hydroxide in methanol and boil the mixture gently for 30 min under a reflux condenser. Detach the condenser, dilute with 50 mL of water, and heat on a hot plate at low setting until the carbon tetrachloride and methanol are evaporated. Add 50 mL of saturated bromine water and heat for 15 min longer.

Transfer the solution to a beaker, neutralize with dilute hydrochloric acid (1 + 4); if alkaline, add an excess of 1 mL of hydrochloric acid, and concentrate to a volume of 100 mL. Filter if necessary, heat the filtrate to boiling, add 5 mL of barium chloride reagent solution, digest in a covered beaker on a hot plate at low setting for 2 h, and allow to stand overnight. If a precipitate is formed, filter, wash thoroughly, and ignite. Correct for the weight obtained in a complete blank test.

IODINE-CONSUMING SUBSTANCES.   To 25 mL add 0.05 mL of 0.1 N iodine solution, shake well, and allow to stand for 30 min. The violet color of the iodine should not be entirely discharged.

SUBSTANCES DARKENED BY SULFURIC ACID.   To 40 mL in a glass-stoppered separatory funnel that has been rinsed with sulfuric acid add 5 mL of sulfuric acid and shake vigorously for 5 min. Allow to separate completely and transfer the acid to a comparison tube. The acid should have no more color than a color standard of the following composition: 0.1 mL of cobaltous chloride solution (5.95 g of $CoCl_2 \cdot 6H_2O$ and 2.5 mL of hydrochloric acid in 100 mL), 0.4 mL of ferric chloride solution (4.50 g of $FeCl_3 \cdot 6H_2O$ and 2.5 mL of hydrochloric acid in 100 mL), 0.1 mL of cupric sulfate solution (6.24 g of $CuSO_4 \cdot 5H_2O$ and 2.5 mL of hydrochloric acid in 100 mL), and 4.4 mL of water.

SUITABILITY FOR USE IN DITHIZONE TESTS.

Solution A.   Dissolve 4 mg of dithizone in 200 mL of the sample of carbon tetrachloride.

Alkaline Extractions. To 50 mL of a 5% solution of sodium hydroxide in a 200-mL glass-stoppered flask add 0.025 mg of cadmium ion (Cd) and 25 mL of solution A, shake, and allow to stand for 10 min. The color after standing 10 min should be identical in hue and intensity with that in a similar, freshly prepared solution.

Acid Extractions. Dilute 25 mL of solution A with the carbon tetrachloride to 100 mL and transfer 25 mL of this diluted solution to each of two 200-mL glass-stoppered flasks. Add 25 mL of 0.1 N hydrochloric acid to one flask and 25 mL of water to the other flask and shake each intermittently for 10 min. The green color in the carbon tetrachloride layers should be identical. Add 0.5 mg of mercury ion (Hg) to the flask to which 0.1 N hydrochloric acid was added, shake, and allow to stand for 10 min. The orange color should be identical in hue and intensity with that in a similar, freshly prepared solution that is allowed to stand for only 1 min instead of 10 min. Since mercury dithizonate is slightly light-sensitive, both flasks should be shaken for 15 s before final comparison is made.

> *NOTE.* Dithizone is extremely sensitive to various metal ions. There-fore, the distilled water, 0.1 N hydrochloric acid, and the 5% solution of sodium hydroxide used must be free of metal ions that react with dithizone. In case of doubt, the solutions must be freed of such metals before use by shaking with portions of chloroform–dithizone solution until the color of the dithizone remains unchanged.

ULTRAVIOLET SPECTROPHOTOMETRY.   Use the procedure on page 67 to determine the absorbance.

Ceric Ammonium Nitrate
Ammonium Hexanitratocerate(IV)

$(NH_4)_2Ce(NO_3)_6$ $\qquad\qquad\qquad$ Formula Wt 548.22

CAS Number 16774–21–3

REQUIREMENTS

Assay . . . . $\geq$ 98.5%
$\qquad\qquad\qquad\qquad\qquad\qquad\qquad\qquad$ $(NH_4)_2Ce(NO_3)_6$

MAXIMUM ALLOWABLE

Insoluble in dilute sulfuric acid. 0.05%
Chloride (Cl) 0.01%
Phosphate (PO$_4$) 0.02%
Iron (Fe) 0.005%

TESTS

ASSAY.   (By titration of oxidative capacity of cerium-IV). Weigh accurately 2.4 to 2.5 g and dissolve in 50 mL of water. From a pipet add 50 mL of 0.1 N ferrous sulfate solution, and swirl until the precipitate that forms is redissolved. Add 10 mL of phosphoric acid and 0.10 mL of diphenylaminesulfonic acid, sodium salt indicator solution. Titrate at once with 0.1 N standard potassium dichromate solution to a change from faint green to violet. Record this titration volume as A mL. Pipet 25 mL of 0.1 N ferrous sulfate solution into 50 mL of water, and treat it in the same way (with phosphoric acid and indicator but without sample). Titrate to a change from green to gray-blue and record this titration volume as B mL.

$$\%(NH_4)_2Ce(NO_3)_6 = \frac{(2B - A) \times 0.05482 \times 100}{\text{grams of sample}}$$

Reagents

Ferrous Sulfate Solution, 0.1 N.   Dissolve 27.8 g of clear ferrous sulfate heptahydrate crystals in about 500 mL of water containing 50 mL of sulfuric acid, and dilute with water to 1 L.

Diphenylaminesulfonic Acid, Sodium Salt Indicator Solution.   Dissolve 0.10 g of the salt in 100 mL of water.

Standard Potassium Dichromate Solution, 0.1 N.   Dissolve 4.903 g of dry potassium dichromate in water, dilute with water to 1 L in a volumetric flask, and mix.

INSOLUBLE IN DILUTE SULFURIC ACID.   To 5.0 g add 10 mL of sulfuric acid, stir, then cautiously add 90 mL of water to dissolve. Heat to boiling and digest in a covered beaker on the steam bath for 1 hour. Filter through a tared filtering crucible, wash thoroughly, and dry at 105°C.

CHLORIDE.   (Page 25). Use 0.10 g dissolved in 10 mL of water. The comparison is best made by the general method for chloride in colored solutions.

PHOSPHATE.   (Page 30, Method 2). Dissolve 0.25 g in 30 mL of dilute sulfuric acid (1 + 9), add hydrogen peroxide until the solution just turns colorless, then boil to destroy excess peroxide. Cool and dilute with water to 50 mL. Dilute 10 mL with water to 60 mL, and adjust the pH between 2 and 3 (pH paper) with ammonium hydroxide.

> NOTE. This neutralization must be made with care to avoid formation of a permanent precipitate that would vitiate the test. If this happens, discard the solution and start with another 10 mL.

Add 0.5 g of ammonium molybdate, and adjust the pH to 1.8 (using a pH meter) with dilute hydrochloric acid (1 + 9). Heat the solution to boiling and cool to room temperature. Dilute with water to 90 mL, add 10 mL of hydrochloric acid, and continue as described. Carry along a standard containing 0.01 mg of phosphate ion (PO_4) in 10 mL of dilute sulfuric acid (3 + 47) treated exactly as the 10 mL of sample.

IRON. (Page 28, Method 2). Dissolve 1.0 g in 30 mL of dilute sulfuric acid (1 + 9) and add, dropwise, 3% hydrogen peroxide solution until the yellow color disappears. For the standard add 0.05 mg of iron (Fe) to 30 mL of dilute sulfuric acid (1 + 9) and the same volume of 3% hydrogen peroxide solution used for the sample. To each solution add ammonium hydroxide until the pH is between 1 and 3, cool to room temperature, and adjust the pH to 3.5 (pH meter). Dilute each solution with water to 50 mL, and use 10 mL of each.

Ceric Ammonium Sulfate Dihydrate
Ammonium Tetrasulfatocerate Dihydrate

$(NH_4)_4Ce(SO_4)_4 \cdot 2H_2O$ Formula Wt 632.58

CAS Number 10378−47−9

REQUIREMENTS

Assay . . . . ≥ 94%
$(NH_4)_4Ce(SO_4)_4 \cdot 2H_2O$

MAXIMUM ALLOWABLE

Insoluble in dilute sulfuric acid. 0.05%
Phosphate (PO_4) 0.03%
Iron (Fe) 0.01%

TESTS

ASSAY. Weigh accurately 2.4–2.5 g and place in a beaker. Cover with 10 mL of water and add cautiously 5 mL of sulfuric acid. Stir thoroughly and add water to complete dissolution. Dilute to 75–100 mL with water, add 0.05 mL of ferroin indicator, and titrate with 0.1 N ferrous ammonium sulfate to the red end point. One milliliter of 0.1 N ferrous ammonium sulfate corresponds to 0.06326 g of $(NH_4)_4Ce(SO_4)_4 \cdot 2H_2O$

INSOLUBLE IN DILUTE SULFURIC ACID. To 2.5 g add 5 mL of sulfuric acid, stir, and then cautiously add 90 mL of water to dissolve. Heat to boiling and digest on the steam bath for 1 h. Filter through a tared filtering crucible and wash thoroughly, first with 2% sulfuric acid, then with water. Dry at 105°C.

PHOSPHATE. (Page 30, Method 2). Dissolve 0.75 g by adding 1 mL of sulfuric acid and stirring in 30 mL of water. Add hydrogen peroxide dropwise until just colorless, boil to destroy the excess peroxide, cool, and dilute to 50 mL. Dilute a 2-mL aliquot to 60 mL and adjust the pH to between 2 and 3 with ammonium hydroxide. (Note: This neutralization must avoid the formation of a precipitate, which would vitiate the test.) Add 0.5 g of ammonium molybdate and adjust the pH to 1.8 (using a pH meter) with dilute hydrochloric acid. Heat to boiling and cool to room temperature. Dilute with water to 90 mL, add 10 mL of hydrochloric acid, and continue as described. Carry along a standard containing 0.01 mg of phosphate ion (PO_4) in 2.0 mL of sulfuric acid (1 + 29) treated as the 2.0-mL aliquot of the sample.

IRON. (Page 28, Method 2). Dissolve 1 g by adding 2 mL of sulfuric acid and stirring in 30 mL of water. Add dropwise 3% hydrogen peroxide until the color disappears. For the standard add 0.1 mg of iron to 30 mL of water containing 2 mL of sulfuric acid and the same volume of hydrogen peroxide used in the test. Dilute the standard and sample each to 100 mL, and use 10-mL aliquots.

Chloramine-T Trihydrate

N-Chloro-4-toluenesulfonamide, Sodium Salt, Trihydrate

$H_3CC_6H_4SO_2NClNa \cdot 3H_2O$ Formula Wt 281.69

CAS Number 127–65–1

REQUIREMENTS

Assay 98.0–103.0%
$C_7H_7ClNNaO_2S \cdot 3H_2O$
pH of a 5% solution . . . . 8.0–10.0 at 25°C
Suitability for determination of bromide . . . . Passes test

MAXIMUM ALLOWABLE

Clarity of aqueous solution . . . . Passes test
Insoluble in alcohol 1.5%

TESTS

ASSAY. (By iodometric titration). Weigh accurately 0.5 g, transfer to a glass-stoppered Erlenmeyer flask, and dissolve in 100 mL of water. Add 5 mL of acetic acid and 2 g of potassium iodide, and allow to stand in the dark for 10 min. Titrate the liberated iodine with standard 0.1 N sodium thiosulfate, using starch indicator solution. One milliliter of 0.1 N sodium thiosulfate corresponds to 0.01408 g of $C_7H_7ClNNaO_2S \cdot 3H_2O$.

pH OF A 5% SOLUTION. (Page 43). The pH should be 8.0–10.0 at 25°C. Reserve the solution for the test for clarity of aqueous solution.

CLARITY OF AQUEOUS SOLUTION. The solution prepared for the test for pH of a 5% solution should be free from turbidity and insolubles, and not more than slightly yellow in color.

INSOLUBLE IN ALCOHOL. Dissolve 5.0 g in 100 mL of reagent alcohol, and stir for 30 min. Filter the solution through a tared medium-porosity sintered-glass filter, and wash with 5 mL of reagent alcohol. Dry the filter at 105°C to constant weight.

SUITABILITY FOR DETERMINATION OF BROMIDE. Transfer 0.01 mg and 0.02 mg of bromide ion (Br) (2.0 mL and 4.0 mL of diluted bromide standard solution) to two color-comparison tubes, and dilute each to 50 mL with water. For the blank, use 50 mL of water. To each add 2.0 mL of acetate buffer solution, 2.0 mL of phenol red solution, and 0.5 mL of chloramine-T solution, mixing thoroughly immediately after each addition. Twenty minutes after adding the chloramine-T solution, add, with mixing, 0.5 mL of sodium thiosulfate solution. The color of the solution containing 0.01 mg of bromide should be distinctly more reddish purple than the blank, and the purple color should increase with the higher bromide concentration.

Chloramine-T Solution. Dissolve 0.50 g of the sample in water, and dilute to 100 mL with water.

Diluted Bromide Standard Solution, 0.005 mg of Br per mL. Dilute 5 mL of bromide standard (0.1 mg of Br per mL) to 100 mL with water.

Acetate Buffer Solution. Dissolve 9.0 g of sodium chloride and 6.8 g of sodium acetate trihydrate in about 50 mL of water. Add 3.0 mL of glacial acetic acid, and dilute to 100 mL. The pH should be 4.6–4.7.

Phenol Red Solution. Dissolve 21 mg of the sodium salt of phenol red in water, and dilute to 100 mL.

Sodium Thiosulfate Solution (2 M). Dissolve 49.6 g of sodium thiosulfate pentahydrate in water, and dilute to 100 mL.

Chloroacetic Acid

Monochloroacetic acid
Chloroethanoic acid

$ClCH_2COOH$ Formula Wt 94.50

CAS Number 79–11–8

REQUIREMENTS

Assay . . . . $\geq$ 99.0%
$ClCH_2COOH$

MAXIMUM ALLOWABLE

Insoluble matter 0.01%
Residue after ignition 0.02%
Chloride (Cl) 0.01%
Sulfate (SO_4) 0.02%
Heavy metals (as Pb) 0.001%
Iron (Fe) 0.002%
Substances darkened by sulfuric acid . . . . Passes test

TESTS

ASSAY. (By acid–base titrimetry). Weigh accurately 3 g, transfer to a conical flask, and dissolve in about 50 mL of water. Add 0.15 mL of phenolphthalein indicator solution, and titrate with 1 N sodium hydroxide. One milliliter of 1 N sodium hydroxide corresponds to 0.0945 g of $ClCH_2COOH$.

INSOLUBLE MATTER. (Page 14). Use 10 g in 100 mL of water.

RESIDUE AFTER IGNITION. (Page 15). Use 5.0 g. Retain the residue for the test for iron.

CHLORIDE. (Page 25). Use 0.1 g.

SULFATE. (Page 32, Procedure A). Dissolve 5.0 g in 30 mL of water, add 25 mg of anhydrous sodium carbonate, evaporate to dryness, and heat over a low flame until the chloroacetic acid is volatilized. Dissolve the residue in 50 mL of water and neutralize with hydrochloric acid. Filter if necessary, wash, and dilute with water to 100 mL. To 5 mL (0.25-g sample) add 1 mL of dilute hydrochloric acid (1 + 19). Dilute to 10 mL and add 1 mL of barium chloride reagent solution. Any turbidity produced in the sample solution after 10 min should not exceed that produced by 0.05 mg of sulfate ion (SO_4) in an equal volume of solution containing the quantities of reagents used in the test.

HEAVY METALS. (Page 26, Method 1). Dissolve 5.0 g in 30 mL of water, neutralize with (1 + 1) ammonium hydroxide, and dilute with water to 50 mL. For the test solution, use 30 mL. For the standard–control solution, add 0.02 mg of lead ion (Pb) to 10 mL.

IRON. (Page 28, Method 1). To the residue after ignition add 3 mL of hydrochloric acid, cover with a watch glass, and digest on the steam bath for 15 min. Remove the cover and evaporate to dryness. Dissolve the residue with 1 mL of hydrochloric acid and dilute with water to 100 mL. Use 10 mL (0.5-g sample) as the sample solution.

SUBSTANCES DARKENED BY SULFURIC ACID. Heat 1.0 g with 10 mL of $95.0 \pm 0.5\%$ sulfuric acid at 50°C for 5 min. The solution should not have more than a faintly brown color.

Chlorobenzene
Monochlorobenzene

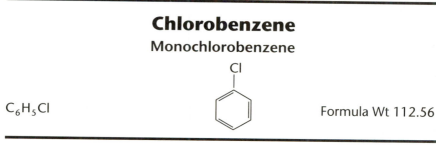

C_6H_5Cl

Formula Wt 112.56

CAS Number 108 –90 –7

REQUIREMENTS

Assay . . . . $\geq$ 99.5% C_6H_5Cl

MAXIMUM ALLOWABLE

Color (APHA). 30
Residue after evaporation 0.02%
Titrable acid . . . . 0.004 meq / g

TESTS

ASSAY.   Analyze the sample by gas chromatography, using the general parameters cited on page 56. The following specific conditions are also required.

Column: Type I, methyl silicone

Measure the area under all peaks and calculate the chlorobenzene content in area percent.

COLOR (APHA).   (Page 17).

RESIDUE AFTER EVAPORATION.   (Page 14). Evaporate 20 g (18 mL) in a tared dish on the steam bath, and dry the residue at 105°C for 30 min.

TITRABLE ACID.   Neutralize 200 mL of methanol to a methyl red end point with 0.1 N sodium hydroxide. Add 25 g (23 mL) of sample, and titrate with 0.1 N sodium hydroxide to the same methyl red end point. Not more than 1.0 mL should be required.

Chloroform
Trichloromethane

CHCl$_3$ Formula Wt 119.38

CAS Number 67–66–3

Suitable for use in high-performance liquid chromatography, extraction–concentration analysis, ultraviolet spectrophotometry, or general use. Product labeling shall designate the one or more of these uses for which suitability is represented on the basis of meeting the relevant requirements and tests. The ultraviolet spectrophotometry and liquid chromatography suitability requirements include all of the requirements for general use. The extraction–concentration suitability requirements include only the general use requirement for color.

NOTE. Chloroform should be supplied and stored in amber glass containers and protected from direct sunlight. This solvent usually contains additives to retard decomposition. Typical additives include alcohol or mixed amylenes. The substances darkened by sulfuric acid test is not required for chloroform containing mixed amylenes.

REQUIREMENTS

General Use

Assay . . . . $\geq$ 99.8% CHCl$_3$

MAXIMUM ALLOWABLE

Color (APHA). 10
Residue after evaporation 0.001%
Acetone and aldehyde . . . . Passes test [limit
                                                                 about 0.005%
                                                                 as (CH$_3$)$_2$CO]

Acid and chloride. . . . . Passes test
Free chlorine (Cl) . . . . Passes test
Lead (Pb) . . . . 0.05 ppm
Substances darkened by sulfuric acid . . . . Passes test
Suitability for use in dithizone tests . . . . Passes test

Specific Use

ULTRAVIOLET SPECTROPHOTOMETRY

Wavelength (nm)	Absorbance (AU)
290 to 400	0.01
270	0.05
260	0.15
255	0.25
245	1.00

LIQUID CHROMATOGRAPHY SUITABILITY

Absorbance	Passes test

EXTRACTION–CONCENTRATION SUITABILITY

Absorbance	Passes test
GC–FID	Passes test
GC–ECD	Passes test

TESTS

ASSAY. Analyze the sample by gas chromatography, using the general parameters cited on page 56. The following specific conditions are also required.

Column: Type I, methyl silicone

Measure the area under all peaks and calculate the chloroform and active additive content in area percent. The assay is the sum of the chloroform content plus all known active additives.

GLC INTERFERENCES. Determine the GLC interferences by using the procedures and equipment described on page 65. Omit step 1. In the sample preparation, when concentrating the residue, evaporate to about 5 mL, add 50 mL of toluene, 2,2,4-trimethylpentane, or other suitable solvent and concentrate to 5.0 mL.

COLOR (APHA).   (Page 17).

RESIDUE AFTER EVAPORATION. (Page 14). Evaporate 100 g (67 mL) to dryness in a tared dish on the steam bath and dry the residue at 105°C for 30 min.

ACETONE AND ALDEHYDE.   Shake 15 mL with 20 mL of ammonia-free water for 5 min in a separatory funnel. Allow the layers to separate and transfer 10 mL of the aqueous layer to a 125-mL glass-stoppered flask containing 40 mL of ammonia-free water. Adjust and maintain the temperature of this solution at 25 $\pm$ 1°C. Add 5 mL of Nessler reagent solution and allow to stand for 5 min. No turbidity or precipitate should develop.

ACID AND CHLORIDE.   Shake 25 g (17 mL) with 25 mL of water for 5 min, allow the liquids to separate, and draw off the aqueous phase. Add a small piece of blue litmus paper to 10 mL of the aqueous phase. The blue litmus paper should not change color. Add 0.25 mL of silver nitrate reagent solution to another 10 mL of the aqueous phase. No turbidity should be produced in this solution.

FREE CHLORINE.   Shake 10 mL for 2 min with 10 mL of water to which 0.10 mL of 10% potassium iodide reagent solution has been added, and allow to separate. The lower layer should not show a violet tint.

LEAD.   For the solution of the sample transfer 40 g (27 mL) of the sample to a separatory funnel, add 20 mL of dilute nitric acid (1 + 99), and shake vigorously for 1 min. Allow the phases to separate, draw off and discard the chloroform phase, and use the aqueous phase for the solution of the sample. For the standard prepare a solution containing 0.002 mg of lead ion (Pb) in 20 mL of dilute nitric acid (1 + 99) in another separatory funnel. To each solution add 4 mL of ammonium cyanide, lead-free, reagent solution (CAUTION. Do this in a well-ventilated hood.) and 5 mL of dithizone standard solution in chloroform, and shake vigorously for one-half min. Allow the phases to separate, draw off the dithizone–chloroform phase into clean dry comparison tubes, and compare the colors using a white background. The purplish hue in the solution of the sample due to any red lead dithizonate present should not exceed that in the standard.

> NOTE. When making this test, all glassware must be carefully cleaned and thoroughly rinsed with warm dilute nitric acid (1 + 1) to remove any adsorbed lead, and finally rinsed with distilled water. All glassware used in the preparation and storage of reagents and in performing the test must be made of lead-free glass.

SUBSTANCES DARKENED BY SULFURIC ACID.   To 40 mL in a separatory funnel add 5 mL of sulfuric acid, shake the mixture vigorously for 5 min, and allow the liquids to separate completely. The chloroform layer should be colorless. The acid layer should have no more color than 5 mL of a color standard of the following composition: 0.4 mL of cobaltous chloride solution (5.95 g of $CoCl_2 \cdot 6H_2O$ and 2.5 mL of hydrochloric acid in 100 mL), 1.6 mL of ferric chloride solution (4.50 g of $FeCl_3 \cdot 6H_2O$ and 2.5 mL of hydrochloric acid in 100 mL), 0.4 mL of cupric sulfate solution (6.24 g of $CuSO_4 \cdot 5H_2O$ and 2.5 mL of hydrochloric acid in 100 mL), and 17.6 mL of water.

SUITABILITY FOR USE IN DITHIZONE TEST

Solution A.   Dissolve 4 mg of dithizone in 200 mL of chloroform.

Alkaline Extractions.   To 50 mL of a 5% solution of sodium hydroxide in a 200-mL glass-stoppered flask add 0.025 mg of cadmium ion (Cd) and 25 mL of solution A, shake, and allow to stand for 10 min. The color after standing 10 min should be identical in hue and intensity with that in a similar, freshly prepared solution.

Acid Extractions.   Dilute 25 mL of solution A with the chloroform to 100 mL, and transfer 25 mL of this dilution to each of two 200-mL glass-stoppered flasks. Add 25 mL of 0.1 N hydrochloric acid to one flask and 25 mL of water to the other flask, and shake each intermittently for 10 min. The green color in the chloroform layers should be identical. Add 0.5 mg of mercury ion (Hg) to the flask to which 0.1 N hydrochloric acid was added, shake, and allow to stand for 10 min. The orange color should be identical in hue and intensity with that in a similar, freshly prepared solution that is allowed to stand for only 1 min instead of 10 min. Since mercury dithizonate is slightly light-sensitive, both flasks should be shaken for about 15 s before final comparison is made.

> NOTE. Dithizone is extremely sensitive to various metal ions. Therefore, the distilled water, 0.1 N hydrochloric acid, and the 5% solution of sodium hydroxide must be free of metal ions that react with dithizone. In case of doubt, the solutions must be freed of such metals before use by shaking with portions of chloroform–dithizone solution until the color of the dithizone remains unchanged.

ULTRAVIOLET SPECTROPHOTOMETRY.   Use the procedure on page 67 to determine the absorbance.

LIQUID CHROMATOGRAPHY SUITABILITY.   Use the procedure on page 67 to determine the absorbance.

EXTRACTION–CONCENTRATION SUITABILITY. Analyze the sample, using the general procedure cited on page 65. In the sample preparation step, exchange the sample with ACS-grade isooctane suitable for extraction–concentration. Use the procedure on page 67 to determine the absorbance.

Chloroplatinic Acid Hexahydrate

Hexachloroplatinic Acid Hexahydrate
Platinic Chloride Hexahydrate

$H_2PtCl_6 \cdot 6H_2O$ Formula Wt 517.90

CAS Number 16941–12–1

REQUIREMENTS

Assay . . . . $\geq$ 37.50% Pt

MAXIMUM ALLOWABLE

Solubility in alcohol . . . . Passes test
Alkali and other salts (as sulfates). 0.05%
Suitability for potassium determinations . . . . Passes test

TESTS

SOLUBILITY IN ALCOHOL. Dissolve 1 g in 10 mL of alcohol and allow to stand, with occasional stirring, for 15 min. The solution should contain no more than traces of insoluble matter.

ASSAY. (By electrolytic determination of platinum content). Using precautions to prevent absorption of moisture, weigh accurately 1 g, and transfer to a tared platinum dish of about 100-mL capacity. Dissolve in 80 mL of water and add 2 mL of sulfuric acid. Cover the dish with a split watch glass and electrolyze the solution for 4 hours at a current of 0.5 A and a temperature 55 to 60°C and $1\frac{1}{2}$ hours at 1 A, using a rotating platinum anode and the platinum dish as the cathode. Wash the cover several times during the electrolysis. When the electrolysis is complete, transfer the solution to a second tared platinum dish, and reserve for the alkali and other salts test. Rinse the platinum dish and deposited platinum with water, dry, and ignite at 600°C for 5 min. The weight of the deposit should not be less than 37.50% of the sample weight. All weighings must be determined to ± 0.00002 g.

ALKALI AND OTHER SALTS.   Evaporate the solution retained from the assay in a tared dish and ignite at 800 ± 25°C for 30 min.

SUITABILITY FOR POTASSIUM DETERMINATIONS.   Dissolve 7.456 g of potassium chloride, previously dried at 105°C, in water and dilute with water to 1 L in a volumetric flask. To a 10-mL aliquot of this solution add 0.100 to 0.110 g of sodium chloride and dilute with water to 100 mL. Heat on the steam bath and add between 0.90 and 1.00 g of the chloroplatinic acid crystals dissolved in about 1 mL of water. Evaporate the solution on the steam bath to a moist residue. (Do not evaporate to dryness, as this will render the sodium salt insoluble.) Cool and add 10 mL of absolute ethyl alcohol, crush the crystals with a glass stirring rod flattened on the end, and allow the mixture to stand for 30 min. Filter by decantation through a previously dried and weighed fritted-glass crucible. Add 10 mL of absolute ethyl alcohol to the residue, grind the residue in the beaker, and filter again by decantation. When the filtrate runs clear, transfer the precipitate to the crucible, and wash with three 10-mL portions of alcohol. Discard the filtrate. Dry the filtering crucible in an oven at 105°C for 1 h, cool, and weigh. The weight of the residue must not be less than 0.2410 g nor more than 0.2450 g.

Chromium Potassium Sulfate Dodecahydrate
Chromium(III) Potassium Sulfate Dodecahydrate

$CrK(SO_4)_2 \cdot 12H_2O$ Formula Wt 499.40

CAS Number 10279–63–7

REQUIREMENTS

Assay . . . . 98.0–102.0% as
$CrK(SO_4)_2 \cdot 12H_2O$

MAXIMUM ALLOWABLE

Insoluble matter 0.01%
Chloride (Cl) 0.002%
Aluminum (Al) 0.02%
Ammonium (NH_4) 0.01%
Heavy metals (as Pb) 0.01%
Iron (Fe) 0.01%

TESTS

ASSAY. (By iodometric titration for oxidative capacity of Cr-VI). Weigh accurately 0.65 g and transfer to a 500-mL iodine flask with 50 mL of water. Add 5 mL of 30% hydrogen peroxide and 15 mL of 10% sodium hydroxide solution. Dilute with 100 mL of water and boil until the solution color is yellow. Add dropwise 5 mL of 5% nickel sulfate hexahydrate solution. When the vigorous evolution of oxygen has ceased, boil for 3 min and cool. Add 1 g of potassium iodide and 5 mL each of phosphoric and hydrochloric acids. Mix well and titrate the liberated iodine with 0.1 N sodium thiosulfate, adding 3 mL of starch indicator solution near the end of the titration. One milliliter of 0.1 N sodium thiosulfate corresponds to 0.01665 g of $CrK(SO_4)_2 \cdot 12H_2O$.

INSOLUBLE MATTER. (Page 14). Use 10 g dissolved in 100 mL of water.

CHLORIDE. Dissolve 1.0 g in 10 mL of water, heat the solution to boiling, and add 2 mL of ammonium hydroxide to the hot, constantly stirred solution. Boil gently to expel excess ammonia, filter through a small chloride-free filter, wash with hot water until the volume of filtrate and washings is about 45 mL, dilute with water to 50 mL, and mix. To 25 mL of this solution add 1.5 mL of nitric acid and 1 mL of silver nitrate reagent solution. Any turbidity should not exceed that produced by 0.01 mg of chloride ion (Cl) in an equal volume of solution containing the quantities of reagents used in the test.

ALUMINUM. Dissolve 4.0 g in 100 mL of water. Add 20 mL of 10% sodium hydroxide reagent solution and 10 mL of hydrogen peroxide. When the reaction ceases, heat to boiling and boil for 15 min. Filter and wash with a small quantity of hot water. Reserve the precipitate for the determination of iron. Adjust the pH of the filtrate to 5.0 (using a pH meter) with hydrochloric acid and dilute with water to 200 mL. For the solution of the sample dilute 25 mL to 100 mL. For the control add 0.05 mg of aluminum ion (Al) to another 12.5 mL of the solution and dilute with water to 100 mL. To each solution add

0.2 mL of a 5% solution of cupferron. Any turbidity in the solution of the sample should not exceed that in the control.

AMMONIUM.   (Page 22). Dilute the distillate from a 1.0-g sample with water to 50 mL, and use 5 mL of the dilution. For the standard use 0.01 mg of ammonium ion (NH_4).

HEAVY METALS.   (Page 26, Method 1). Dissolve 1.0 g and 30 mg of mercuric chloride in 40 mL of water. Add 15 mL of hydrogen sulfide water, stir, and allow to stand for 30 min. Filter and wash thoroughly with hydrogen sulfide water containing 1% potassium sulfate. Ignite the precipitate at low temperature in a small porcelain dish in a well-ventilated hood to char the paper, then at $525 \pm 25°C$ for 30 min. To the residue add 1 mL of hydrochloric acid, 1 mL of nitric acid, and about 30 mg of sodium carbonate, and evaporate to dryness on the steam bath. Dissolve the residue in about 20 mL of water, and dilute with water to 100 mL. Use 20 mL to prepare the sample solution.

IRON.   (Page 28, Method 1). Wash with hot water the residue remaining on the filter in the test for aluminum until the washings are colorless. Dissolve the residue from the filter with 5 mL of dilute hydrochloric acid (1 + 1), and wash the filter paper with hot water. Cool, dilute with water to 80 mL, mix, and use 2.0 mL of this solution.

Chromium Trioxide
Chromium(VI) Oxide

CrO_3 Formula Wt 99.99

CAS Number 1333–82–0

CAUTION. Chromium trioxide should not be brought into intimate contact with organic substances or other reducing agents, as serious explosions are likely to result.

REQUIREMENTS

Assay . . . . $\geq$ 98.0% CrO_3

MAXIMUM ALLOWABLE

Insoluble matter 0.01%
Chloride (Cl) 0.005%
Nitrate (NO$_3$) 0.05%
Sulfate (SO$_4$) 0.005%
Sodium (Na) 0.2%
Iron, aluminum, barium 0.03%

TESTS

ASSAY. (By iodometric titration for oxidative capacity). Weigh accurately 5 g, transfer to a 1000-mL volumetric flask, dissolve in water, dilute with water to volume, and mix thoroughly. Place a 25-mL aliquot of this solution in a glass-stoppered conical flask and dilute with 100 mL of water. Add 5 mL of dilute sulfuric acid (1 + 1) and 3 g of potassium iodide and allow to stand in the dark for 15 min. Dilute with 100 mL of water and titrate the liberated iodine with 0.1 N sodium thiosulfate, adding 3 mL of starch indicator solution near the end of the titration. Correct for a complete blank. One milliliter of 0.1 N sodium thiosulfate corresponds to 0.003333 g of CrO_3.

INSOLUBLE MATTER. (Page 14). Use 10 g dissolved in 100 mL of water.

CHLORIDE. Dissolve 1.0 g in water, filter if necessary through a chloride-free filter, and dilute with water to 50 mL. To 10 mL of this solution add 1.5 mL of ammonium hydroxide and 1 mL of silver nitrate reagent solution, mix, and add 2 mL of nitric acid. Any turbidity should not exceed that produced in a standard containing 1 mL of ammonium hydroxide, 1 mL of silver nitrate reagent solution, 2 mL of nitric acid, and 0.01 mg of added chloride ion (Cl). The comparison is best made by the general method for chloride in colored solutions, page 25.

NITRATE

Sample Stock Solution. Dissolve 1.0 g in about 20 mL of water. Add 10% sodium hydroxide reagent solution until the color of the solution just turns light yellow, then add 0.2 mL of glacial acetic acid to change the color to orange (pH about 6.0–6.5). Slowly add a solution of 5 g of lead acetate in about 10 mL of water, dilute to 50 mL with water, mix, and allow to stand for 15 min. Transfer to a 50-mL centrifuge tube, centrifuge, and decant through a filter. *This solution must be clear and colorless.*

NOTE. Two filter papers may be required to obtain a clear, colorless filtrate. Use a Nessler tube to check.

Sample Solution A. Dissolve 0.1 g of mercuric acetate in 2.0 mL of sample stock solution plus 1.0 mL of water in a dry test tube. Add 0.05 g of urea and shake to dissolve. Put the tube in an ice bath and immediately, but slowly, add 7 mL of cold chromotropic acid reagent while swirling. Keep the tube in the ice bath an additional 2–3 min, remove, and let stand for 30 min, swirling occasionally.

Control Solution B. Dissolve 0.1 g of mercuric acetate in 2.0 mL of sample stock solution in a dry test tube. Add 1.0 mL of a solution containing 0.02 mg of nitrate ion (NO_3) per mL, then add 0.05 g of urea and shake to dissolve. Put the tube in an ice bath and immediately, but slowly, add 7 mL of cold chromotropic acid reagent while swirling. Keep the tube in the ice bath an additional 2–3 min, remove, and let stand for 30 min, swirling occasionally.

Chromotropic Acid Reagent. Dissolve 0.02 g of recrystallized 4,5-dihydroxy-2,7-naphthalenedisulfonic acid, disodium salt, in concentrated sulfuric acid and dilute to 200 mL with the acid. Store in an amber bottle.

Recrystallization of Reagent. Add solid sodium sulfate to a saturated aqueous solution of the foregoing salt. Filter the resultant crystals, using suction with a Buchner funnel, wash with ethyl alcohol, and air dry. Transfer the crystals to a beaker, add sufficient water to redissolve, and then add just enough sodium sulfate to reprecipitate the salt. Filter as before, wash with ethyl alcohol, and dry at a temperature no higher than 80°C. Store in an amber bottle.

NOTE. If an orange or red precipitate forms in the solution of the sample before the addition of the reagent, discard the solution and start again. Immediate addition of the reagent should prevent formation of such a precipitate.

Blank Solution C. Dissolve 0.1 g of mercuric acetate in 3.0 mL of water in a dry test tube. Add 0.05 g of urea, shake to dissolve, and put the tube in an ice bath. Add 7 mL of cold chromotropic acid reagent while swirling, remove from the bath, and let stand for 30 min.

Transfer sample solution A, control solution B, and blank solution C to dry 15-mL centrifuge tubes and centrifuge until the supernatant liquid is clear. Set a spectrophotometer at 405 nm and, using 1-cm cells, adjust the instrument to read 0 absorbance with blank solution C in the light path, then determine the absorbance of sample solution A. Adjust the instrument to read 0 absorbance with sample solution A in the light path and determine the absorbance of control solution B. The absorbance of sample solution A should not exceed that of control solution B.

SULFATE.   Dissolve 10 g in 350 mL of water and add 5 g of sodium carbonate. Heat to boiling and add 35 mL of a solution containing 1 g of barium chloride and 2 mL of hydrochloric acid in 100 mL of solution. Digest in a covered beaker on a hot plate at low setting for 2 h and allow to stand overnight. If a precipitate is formed, filter, wash thoroughly, and ignite. Fuse the ignited precipitate with 1 g of sodium carbonate. Extract the fused mass with water and filter out the insoluble residue. Add 5 mL of hydrochloric acid to the filtrate, dilute with water to about 200 mL, heat to boiling, and add 10 mL of alcohol. Digest on a hot plate at low setting until the reduction of chromate is complete, as indicated by the change to a clear green or colorless solution. Neutralize the solution with ammonium hydroxide and add 2 mL of hydrochloric acid. Heat to boiling and add 10 mL of barium chloride reagent solution. Digest in a covered beaker on a hot plate at low setting for 2 h and allow to stand overnight. Filter, wash thoroughly, and ignite. Correct for the weight obtained in a complete blank test.

SODIUM.   Determine the sodium by the flame atomic absorption spectrophotometric method described on page 35.

> **Sample Stock Solution.**   Dissolve 1.0 g of sample in a 100-mL volumetric flask, and dilute to the mark with water (0.01 g/mL).

Analyze the solutions by means of a suitable atomic absorption spectrophotometer, using the conditions outlined in the following table. Calculate the metal content of the sample by the method of standard additions.

Element	Wavelength (nm)	Sample Wt (g)	Standard Added (mg)	Flame Type*	Background Correction
Na	589.0	0.01	0.01; 0.02	A/A	No

*A/A is air/acetylene.

IRON, ALUMINUM, BARIUM. Dissolve 5.0 g in 100 mL of water, and add 10 mL of ammonium hydroxide and 1 mL of 30% hydrogen peroxide. Heat the solution to boiling and boil gently for 5 min. Cool, filter, and wash the precipitate with water. Dissolve the precipitate in warm dilute hydrochloric acid (1 + 1), collecting the filtrate in the original beaker. Dilute the filtrate with water to about 30 mL, adjust to pH 7 with ammonium hydroxide, and add 0.5 mL of 30% hydrogen peroxide. Heat the solution to boiling and boil gently for 5 min, cool, filter, and ignite.

Chromotropic Acid Disodium Salt
4,5-Dihydroxy-2,7-naphthalenedisulfonic Acid, Disodium Salt

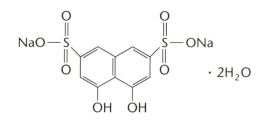

$(HO)_2C_{10}H_4(SO_3Na)_2 \cdot 2H_2O$ Formula Wt 400.29

CAS Number 5808–22–0

NOTE. Two forms of this reagent are available: the disodium salt dihydrate and the free acid, $4,5\text{-}(OH)_2C_{10}H_4\text{-}2,7\text{-}(SO_3H)_2$ (formula wt 320.29). This specification applies to both forms, but the sample weight of the acid should be changed to 80% of that for the salt.

REQUIREMENTS

Appearance	Off-white to buff-colored powder
Clarity of solution	Passes test
Sensitivity to nitrate	Passes test
Sensitivity to formaldehyde	Passes test

TESTS

CLARITY OF SOLUTION.   Use 5 g of the salt or 4 g of the acid dissolved in 100 mL of hot water. Add a few drops of sodium hydroxide solution if necessary. The solution should be free of turbidity.

Sample Solution A.   Dissolve 0.20 g of the salt (0.16 g of the acid) in 100 mL of sulfuric acid.

SENSITIVITY TO NITRATE.   Into three 50-mL volumetric flasks pipet, respectively, 1.0, 2.0, and 3.0 mL of a standard solution containing 0.01 mg/mL of nitrate ion (NO_3). For the blank add 1.0 mL of water to another 50-mL volumetric flask. To each flask add approximately 5 mL of sulfuric acid, swirl, and cool in an ice bath. Add 2.0 mL of sample solution A to each flask, swirl again, and cool. Dilute to 50 mL with sulfuric acid and shake well. Set a spectrophotometer at 410 nm and, using 5-cm cells, determine the absorbance of the standard solutions against the blank. A plot of absorbance versus concentration of nitrate should be linear, with a minimum absorbance of 0.20 for the solution containing 0.03 mg of nitrate.

SENSITIVITY TO FORMALDEHYDE.   Into three 50-mL volumetric flasks pipet, respectively, 1.0, 2.0, and 3.0 mL of standard formaldehyde solution. For the blank add 1.0 mL of water to another 50-mL volumetric flask. To each flask add 5.0 mL of sample solution A, stopper, and shake. Dilute to the mark with sulfuric acid and let stand in a hot-water bath (70–80°C) for 15 min. Cool in an ice bath to room temperature, fill to volume with sulfuric acid, stopper, and shake. Set a spectrophotometer at 580 nm and, using 1-cm cells, determine the absorbance of the standard solutions against the blank. A plot of absorbance versus concentration of formaldehyde should be linear, with a minimum absorbance of 0.30 for the solution containing 0.03 mg of formaldehyde.

Citric Acid, Anhydrous
2-Hydroxy-1,2,3-propanetricarboxylic Acid

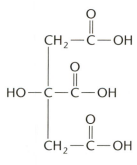

HOCOCH$_2$C(OH)(COOH)CH$_2$COOH   Formula Wt 192.13

CAS Number 77–92–9

REQUIREMENTS

Assay . . . . ≥ 99.5%
$H_3C_6H_5O_7$

MAXIMUM ALLOWABLE

Insoluble matter 0.005%
Residue after ignition 0.02%
Chloride (Cl) 0.001%
Oxalate (C$_2$O$_4$) . . . . Passes test (limit
about 0.05%)
Phosphate (PO$_4$) 0.001%
Sulfate (SO$_4$) 0.002%
Iron (Fe) . . . . 3 ppm
Lead (Pb) . . . . 2 ppm
Substances carbonizable by hot sulfuric acid
(tartrates, etc.) . . . . Passes test

TESTS

ASSAY. (By acid–base titrimetry). Weigh accurately 2.56 to 2.88 g, dissolve in about 40 mL of water, and add 0.15 mL of phenolphthalein indicator solution. Titrate with 1 N sodium hydroxide to a pink color that persists for 5 min. One milliliter of 1 N sodium hydroxide corresponds to 0.06404 g of $H_3C_6H_5O_7$.

INSOLUBLE MATTER. (Page 14). Use 20 g dissolved in 150 mL of water.

RESIDUE AFTER IGNITION. (Page 15). Ignite 5.0 g. Use 2 mL of sulfuric acid to moisten the char.

CHLORIDE. (Page 25). Use 1.0 g.

OXALATE. Dissolve 5 g in 25 mL of water and add 2 mL of 10% calcium acetate solution. No turbidity or precipitate should appear after standing 4 h.

PHOSPHATE. (Page 30, Method 1). Mix 2.0 g with 0.5 g of magnesium nitrate in a platinum dish and ignite. Dissolve the residue in 5 mL of water, add 5 mL of nitric acid, and evaporate to dryness. Dissolve the residue in 25 mL of approximately 0.5 N sulfuric acid, and continue as described.

SULFATE. To 3.0 g add 0.2–0.5 mg of ammonium vanadate and 10 mL of nitric acid. Digest in a covered beaker on a hot plate at low setting until the reaction ceases, remove the cover, and evaporate to dryness. Add 10 mL of nitric acid and repeat the digestion and evaporation. Add 5 mL of dilute hydrochloric acid (1 + 1) and evaporate to dryness. Dissolve the residue in 4 mL of water plus 1 mL of dilute hydrochloric acid (1 + 19), filter if necessary through a small filter, wash with two 2-mL portions of water, and dilute with water to 10 mL. Add 1 mL of barium chloride reagent solution. Any turbidity should not exceed that produced by 0.06 mg of sulfate ion (SO_4) treated in the same manner as the sample. Compare 10 min after adding the barium chloride to the sample and standard solutions.

IRON. (Page 28, Method 1). Use 3.3 g.

LEAD. Dissolve 1.5 g in 10 mL of water, adjust the pH to between 9 and 10 (using a pH meter) with ammonium hydroxide, and dilute with water to 15 mL. Prepare a control containing 0.003 mg of lead ion (Pb) and the amount of ammonium hydroxide required to neutralize the sample and evaporate it to dryness on the steam bath. Moisten the residue from the control with 0.1 mL of nitric acid, dissolve in 5 mL of water, and add 5 mL of ammonium citrate (lead-free) reagent solution. Adjust the pH to between 9 and 10 (using a pH meter) with ammonium hydroxide and dilute with water to 15 mL. Transfer the sample solution and the control solution to separatory funnels, add to each 5 mL of dithizone standard solution in chloroform, and shake vigorously for 2 min. Allow the layers to separate, draw off the dithizone–chloroform layers into clean, dry comparison tubes, and dilute to 10 mL with chloroform. Any red color, producing a purplish hue in the chloroform from the sample solution, should not exceed that in the chloroform from the control.

SUBSTANCES CARBONIZABLE BY HOT SULFURIC ACID.   Carefully powder about 1 g of the sample, taking precautions to prevent contamination during the powdering. Mix 0.30 g with 10 mL of sulfuric acid in a test tube previously rinsed with the acid. Heat the mixture at 110°C (in a brine bath) for 30 min, keeping the test tube covered during the heating. Any color should not exceed that produced in a color standard of the following composition: 1 part of cobaltous chloride solution (5.95 g of $CoCl_2 \cdot 6H_2O$ and 2.5 mL of hydrochloric acid in 100 mL), 2.4 parts of ferric chloride solution (4.5 g of $FeCl_3 \cdot 6H_2O$ and 2.5 mL of hydrochloric acid in 100 mL), 0.4 part of cupric sulfate solution (6.24 g of $CuSO_4 \cdot 5H_2O$ and 2.5 mL of hydrochloric acid in 100 mL), and 6.2 parts of water.

Citric Acid Monohydrate

2-Hydroxy-1,2,3-propanetricarboxylic Acid Monohydrate

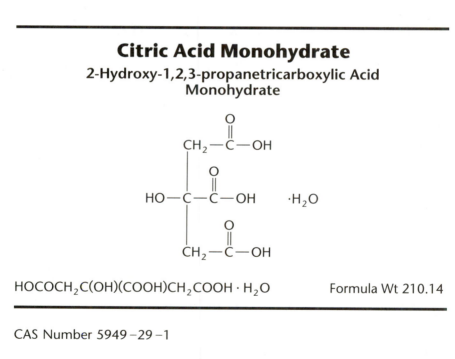

$HOCOCH_2C(OH)(COOH)CH_2COOH \cdot H_2O$ Formula Wt 210.14

CAS Number 5949–29–1

REQUIREMENTS

Assay 99.0–102.0%
$C_6H_8O_7 \cdot H_2O$

MAXIMUM ALLOWABLE

Insoluble matter 0.005%
Residue after ignition 0.02%
Chloride (Cl) 0.001%
Oxalate (C_2O_4) . . . . Passes test (limit
about 0.05%)
Phosphate (PO_4) 0.001%
Sulfate (SO_4) 0.002%
Iron (Fe) . . . . 3 ppm
Lead (Pb) . . . . 2 ppm
Substances carbonizable by hot sulfuric acid
(tartrates, etc.) . . . . Passes test

TESTS

ASSAY. (By acid–base titrimetry). Weigh accurately 2.9 g, dissolve in about 40 mL of water, and add 0.15 mL of phenolphthalein indicator solution. Titrate with 1 N sodium hydroxide to a pink color that persists for 5 min. One milliliter of 1 N sodium hydroxide corresponds to 0.07005 g of $C_6H_8O_7 \cdot H_2O$.

INSOLUBLE MATTER. (Page 14). Use 20 g dissolved in 150 mL of water.

RESIDUE AFTER IGNITION. (Page 15). Ignite 5.0 g. Use 2 mL of sulfuric acid to moisten the char.

CHLORIDE. (Page 25). Use 1.0 g.

OXALATE. Dissolve 5 g in 25 mL of water and add 2 mL of 10% calcium acetate solution. No turbidity or precipitate should appear after standing 4 h.

PHOSPHATE. (Page 30, Method 1). Mix 2.0 g with 0.5 g of magnesium nitrate in a platinum dish and ignite. Dissolve the residue in 5 mL of water, add 5 mL of nitric acid, and evaporate to dryness. Dissolve the residue in 25 mL of approximately 0.5 N sulfuric acid and continue as described.

SULFATE. To 3.0 g add 0.2–0.5 mg of ammonium vanadate and 10 mL of nitric acid. Digest in a covered beaker on the steam bath until the reaction ceases, remove the cover, and evaporate to dryness. Add 10 mL of nitric acid and repeat the digestion and evaporation. Add 5 mL of dilute hydrochloric acid (1 + 1) and evaporate to dryness.

Dissolve the residue in 4 mL of water plus 1 mL of dilute hydrochloric acid (1 + 19), filter if necessary through a small filter, wash with two 2-mL portions of water, and dilute with water to 10 mL. Add 1 mL of barium chloride reagent solution. Any turbidity should not exceed that produced by 0.06 mg of sulfate ion (SO_4) treated in the same manner as the sample. Compare 10 min after adding the barium chloride to the sample and standard solutions.

IRON.   (Page 28, Method 1). Use 3.3 g.

LEAD.   Dissolve 1.5 g in 10 mL of water, adjust the pH to 9–10 (using a pH meter) with ammonium hydroxide, and dilute with water to 15 mL. Prepare a control containing 0.003 mg of lead ion (Pb) and the amount of ammonium hydroxide required to neutralize the sample and evaporate it to dryness on the steam bath. Moisten the residue from the control with 0.1 mL of nitric acid, dissolve in 5 mL of water, and add 5 mL of ammonium citrate (lead-free) reagent solution. Adjust the pH to between 9 and 10 (using a pH meter) with ammonium hydroxide and dilute with water to 15 mL. Transfer the sample solution and the control solution to separatory funnels, add to each 5 mL of dithizone standard solution in chloroform, and shake vigorously for 2 min. Allow the layers to separate, draw off the dithizone–chloroform layers into clean, dry comparison tubes, and dilute to 10 mL with chloroform. Any red color, producing a purplish hue in the chloroform from the sample solution, should not exceed that in the chloroform from the control.

SUBSTANCES CARBONIZABLE BY HOT SULFURIC ACID.   Carefully powder about 1 g of the sample, taking precautions to prevent contamination during the powdering. Mix 0.30 g with 10 mL of sulfuric acid in a test tube previously rinsed with the acid. Heat the mixture at 110°C (in a brine bath) for 30 min, keeping the test tube covered during the heating. Any color should not exceed that produced in a color standard of the following composition: 1 part of cobaltous chloride solution (5.95 g of $CoCl_2 \cdot 6H_2O$ and 2.5 mL of hydrochloric acid in 100 mL), 2.4 parts of ferric chloride solution (4.5 g of $FeCl_3 \cdot 6H_2O$ and 2.5 mL of hydrochloric acid in 100 mL), 0.4 part of cupric sulfate solution (6.24 g of $CuSO_4 \cdot 5H_2O$ and 2.5 mL of hydrochloric acid in 100 mL), and 6.2 parts of water.

Cobalt Acetate Tetrahydrate
Cobalt(II) Acetate Tetrahydrate
Cobaltous Acetate Tetrahydrate

$Co(CH_3COO)_2 \cdot 4H_2O$ Formula Wt 249.08

CAS Number 6147–53–1

REQUIREMENTS

Assay 98.0–102.0%
$Co(CH_3COO)_2 \cdot 4H_2O$

MAXIMUM ALLOWABLE

Insoluble matter 0.01%
Chloride (Cl) 0.002%
Nitrate (NO_3) 0.01%
Sulfate (SO_4) 0.005%
Calcium (Ca) 0.005%
Copper (Cu) 0.002%
Iron (Fe) 0.001%
Lead (Pb) 0.001%
Magnesium (Mg) 0.005%
Nickel (Ni) 0.1%
Potassium (K) 0.01%
Sodium (Na) 0.05%
Zinc (Zn) 0.01%

TESTS

ASSAY. (By complexometric titration of cobalt). Weigh accurately about 1 g and transfer to a 400-mL beaker with the aid of 200 mL of water. Add 0.2 g of ascorbic acid and about 100 mg of murexide indicator mixture. Titrate with 0.1 M EDTA to a color change from yellow to violet that persists for at least 30 s. One milliliter of 0.1 M EDTA corresponds to 0.02491 g of $Co(CH_3COO)_2 \cdot 4H_2O$.

INSOLUBLE MATTER. (Page 14). Use 10 g dissolved in 100 mL of water containing 1 mL of glacial acetic acid. Reserve the filtrate without washings for the test for sulfate.

CHLORIDE.   (Page 25). Use 0.5 g. The comparison is best made by the general method for chloride in colored solutions on page 25.

NITRATE.   (Page 29). Dissolve 1.0 g in 10 mL of water. Add this solution in small portions and with constant stirring to 10 mL of 10% sodium hydroxide reagent solution. Digest in a covered beaker on the steam bath for 15 min. Cool, dilute with water to 20 mL, and filter.

> *Sample Solution A.*   To 4 mL of the preceding filtrate add 6 mL of water. Dilute to 50 mL with brucine sulfate reagent solution.

> *Control Solution B.*   To 4 mL of the preceding filtrate add 4 mL of water and 2 mL of the standard nitrate solution containing 0.01 mg of nitrate ion per mL. Dilute to 50 mL with brucine sulfate reagent solution.

Continue with the procedure described on page 29, starting with the preparation of blank solution C.

SULFATE.   (Page 32, Procedure A, Method 1). Use 1/10 of the filtrate from the test for insoluble matter. Allow 30 min for the turbidity to form.

CALCIUM, COPPER, IRON, LEAD, MAGNESIUM, NICKEL, POTASSIUM, SODIUM, AND ZINC.   Determine the calcium, copper, iron, lead, magnesium, nickel, potassium, sodium, and zinc by the flame atomic absorption spectrophotometric method described on page 35.

> *Sample Stock Solution.*   Dissolve 25.0 g in water in a 250-mL volumetric flask, add 2.5 mL of nitric acid, and dilute with water to volume (1 mL = 0.10 g).

Analyze the solutions by means of a suitable atomic absorption spectrophotometer, using the conditions outlined in the following table. Calculate the metal content of the sample by the method of standard additions.

Element	Wavelength (nm)	Sample Wt (g)	Standard Added (mg)	Flame Type*	Background Correction
Ca	422.7	0.8	0.02; 0.04	N/A	No
Cu	324.8	2.0	0.02; 0.04	A/A	Yes
Fe	248.3	2.5	0.025; 0.05	A/A	Yes
Pb	217.0	2.0	0.02; 0.04	A/A	Yes
Mg	285.2	0.2	0.005; 0.01	A/A	Yes
Ni	232.0	0.1	0.05; 0.10	A/A	Yes
K	766.5	0.2	0.01; 0.02	A/A	No
Na	589.0	0.04	0.01; 0.02	A/A	No
Zn	213.9	0.2	0.01; 0.02	A/A	Yes

*A/A is air/acetylene; N/A is nitrous oxide/acetylene.

Cobalt Chloride Hexahydrate
Cobalt(II) Chloride Hexahydrate
Cobaltous Chloride Hexahydrate

$CoCl_2 \cdot 6H_2O$ Formula Wt 237.93

CAS Number 7791–13–1

REQUIREMENTS

Assay 98.0–102.0%
$CoCl_2 \cdot 6H_2O$

MAXIMUM ALLOWABLE

Insoluble matter 0.01%
Nitrate (NO_3) 0.01%
Sulfate (SO_4) 0.01%
Ammonium (NH_4) 0.005%
Calcium (Ca) 0.005%
Copper (Cu) 0.002%
Iron (Fe) 0.005%
Magnesium (Mg) 0.005%
Nickel (Ni) 0.1%
Potassium (K) 0.01%
Sodium (Na) 0.05%
Zinc (Zn) 0.03%

TESTS

ASSAY. (By complexometric titration of cobalt). Weigh, to the nearest 0.1 mg, 1.0 g of sample and dissolve in 200 mL of water. Add about 0.2 g of ascorbic acid and 5 g of ammonium acetate followed by 10 mL of ammonium hydroxide. Add about 0.1 g of murexide indicator mixture, and titrate with 0.1 M EDTA until the end point is signaled by a color change to violet. The color change must persist for 30 s. One milliliter of 0.1 M EDTA corresponds to 0.02379 g of $CoCl_2 \cdot 6H_2O$.

INSOLUBLE MATTER. (Page 14). Use 10 g dissolved in 100 mL of water.

NITRATE. Dissolve 1.0 g in 10 mL of water. Add this solution in small portions and with constant stirring to 10 mL of 10% sodium hydroxide reagent solution. Digest in a covered beaker on the steam bath for 15 min. Cool, dilute with water to 20 mL, and filter.

> *Sample Solution A.* To 4 mL of the preceding filtrate add 6 mL of water. Dilute to 50 mL with brucine sulfate reagent solution.

> *Control Solution B.* To 4 mL of the preceding filtrate add 4 mL of water and 2 mL of the standard nitrate solution containing 0.01 mg of nitrate ion per mL. Dilute to 50 mL with brucine sulfate reagent solution.

Continue with the procedure described on page 29, starting with preparation of blank solution C.

SULFATE. (Page 32, Procedure A, Method 1). Use 1/20 of the filtrate from the test for insoluble matter. Allow 30 min for the turbidity to form.

AMMONIUM. (Page 22). Dilute the distillate from a 1.0-g sample with water to 50 mL, and use 10 mL of the dilution. For the standard use 0.01 mg of ammonium ion (NH_4).

CALCIUM, COPPER, IRON, MAGNESIUM, NICKEL, POTASSIUM, SODIUM, AND ZINC. Determine the calcium, copper, iron, magnesium, nickel, potassium, sodium, and zinc by the flame atomic absorption spectrophotometric method described on page 35.

Sample Stock Solution. Dissolve 20.0 g of sample in water, add 2 mL of nitric acid, and dilute to the mark with water in a 200-mL volumetric flask (0.1 g/mL).

Analyze the solutions by means of a suitable atomic absorption spectrophotometer, using the conditions outlined in the following table. Calculate the metal content of the sample by the method of standard additions.

Element	Wavelength (nm)	Sample Wt (g)	Standard Added (mg)	Flame Type*	Background Correction
Ca	422.7	0.8	0.02; 0.04	N/A	No
Cu	324.8	2.0	0.04; 0.08	A/A	Yes
Fe	248.3	2.0	0.05; 0.10	A/A	Yes
Mg	285.2	0.2	0.005; 0.01	A/A	Yes
K	766.5	0.2	0.01; 0.02	A/A	No
Na	589.0	0.04	0.01; 0.02	A/A	No
Ni	232.0	0.1	0.05; 0.10	A/A	Yes
Zn	213.9	0.1	0.01; 0.02	A/A	Yes

*A/A is air/acetylene; N/A is nitrous oxide/acetylene.

Cobalt Nitrate Hexahydrate
Cobalt(II) Nitrate Hexahydrate
Cobaltous Nitrate Hexahydrate

$Co(NO_3)_2 \cdot 6H_2O$ Formula Wt 291.03

CAS Number 10026–22–9

REQUIREMENTS

Assay 98.0–102.0%
$Co(NO_3)_2 \cdot 6H_2O$

MAXIMUM ALLOWABLE

Insoluble matter 0.01%
Chloride (Cl) 0.002%
Sulfate (SO$_4$) 0.005%
Ammonium (NH$_4$) 0.2%
Calcium (Ca) 0.005%
Copper (Cu) 0.002%
Iron (Fe) 0.001%
Lead (Pb) 0.002%
Magnesium (Mg) 0.005%
Nickel (Ni) 0.15%
Potassium (K) 0.01%
Sodium (Na) 0.05%
Zinc (Zn) 0.01%

TESTS

ASSAY.   (By complexometric titration of cobalt). Weigh, to the near-est 0.1 mg, 1.0 g of sample and dissolve in 200 mL of water. Add about 0.2 g of ascorbic acid and 5 g of ammonium acetate followed by 10 mL of ammonium hydroxide. Add about 0.1 g of murexide indica-tor mixture, and titrate with 0.1 M EDTA until the end point is signaled by a color change to violet. The color change must persist for 30 s. One milliliter of 0.1 M EDTA corresponds to 0.02910 g of $Co(NO_3)_2 \cdot 6H_2O$.

INSOLUBLE MATTER.   (Page 14). Use 10 g dissolved in 100 mL of water.

CHLORIDE.   Dissolve 0.50 g in 15 mL of water plus 1 mL of nitric acid, filter if necessary through a small chloride-free filter, and add 1 mL of silver nitrate reagent solution. Any turbidity should not exceed that produced by 0.01 mg of chloride ion (Cl) in an equal volume of solution containing the quantities of reagents used in the test. The comparison is best made by the general method for chloride in colored solutions, page 25.

SULFATE.   (Page 32, Procedure A, Method 2). Use 4 mL of dilute hydrochloric acid (1 + 1). Do two evaporations.

AMMONIUM.   (Page 22). Dilute the distillate from a 1.0-g sample with water to 100 mL, and use 5 mL of the dilution. For the standard use 0.1 mg of ammonium ion (NH$_4$).

**CALCIUM, COPPER, IRON, LEAD, MAGNESIUM, NICKEL, POTAS-
SIUM, SODIUM, AND ZINC.** Determine the calcium, copper, iron,
lead, magnesium, nickel, potassium, sodium, and zinc by the flame
atomic absorption spectrophotometric method described on page 35.

> **Sample Stock Solution.** Dissolve 20.0 g of sample in water,
> add 2 mL of nitric acid, and dilute to the mark with water in a
> 200-mL volumetric flask (0.1 g/mL).

Analyze the solutions by means of a suitable atomic absorption
spectrophotometer, using the conditions outlined in the following
table. Calculate the metal content of the sample by the method of
standard additions.

Element	Wavelength (nm)	Sample Wt (g)	Standard Added (mg)	Flame Type*	Background Correction
Ca	422.7	0.8	0.02; 0.04	N/A	No
Cu	324.8	2.0	0.04; 0.08	A/A	Yes
Fe	248.3	2.0	0.02; 0.04	A/A	Yes
Pb	217.0	2.0	0.04; 0.08	A/A	Yes
Mg	285.2	0.2	0.005; 0.01	A/A	Yes
Ni	232.0	0.1	0.075; 0.15	A/A	Yes
K	766.5	0.2	0.01; 0.02	A/A	No
Na	589.0	0.04	0.01; 0.02	A/A	No
Zn	213.9	0.1	0.01; 0.02	A/A	Yes

*A/A is air/acetylene; N/A is nitrous oxide/acetylene.

Copper

Cu   Atomic Wt 63.546

CAS Number 7440–50–8

REQUIREMENTS

Assay . . . . ≥ 99.90% Cu

MAXIMUM ALLOWABLE

Insoluble in dilute nitric acid 0.02%
Antimony and tin (as Sn). 0.01%
Arsenic (As). . . . . 5 ppm
Iron (Fe) 0.005%
Lead (Pb) 0.005%
Manganese (Mn) 0.001%
Phosphorus (P) 0.001%
Silver (Ag). 0.002%

TESTS

ASSAY. (By electrolytic determination of copper content). Accurately weigh 5.0050–5.0070 g of the sample and transfer this weighed sample to a tall-form lipless beaker provided with a close-fitting cover glass. Add 42 mL of a solution that contains 10 mL of sulfuric acid plus 7 mL of nitric acid and 25 mL of water, cover, and allow to stand until the active reactions have subsided. Heat at 80–90°C until the copper is completely dissolved and brown nitrous fumes are completely expelled. Wash down the cover glass and the sides of the beaker and dilute the solution sufficiently with water so that it will cover a cathode cylinder. Insert a tared cathode and an anode and electrolyze at a current density of about 0.6 A per dm^2 for about 16 h. If electrolysis of the copper is not completed, as indicated by copper plating on a new surface of the cathode stem when the level of the solution is raised, continue the electrolysis until all of the copper is deposited.

When all of the copper is deposited, slowly lower the beaker, without interrupting the current, and wash the electrodes continuously with a stream of water from a wash bottle while removing the beaker. Reserve this solution for preparing sample solution A. Rinse the electrodes by raising and lowering a beaker of water around them. Remove the beaker of rinse water, discontinue the current, and remove the electrodes. Wash the cathode in acetone and dry it in an oven at 110°C for 3–5 min. Cool, weigh, and calculate the percentage of copper from the weight of the deposit. This value may have to be corrected for a small amount of undeposited copper as determined in preparing sample solution A.

Sample Solution A. Place the anode in a 150-mL beaker and add 10 mL of dilute nitric acid (1 + 1) containing 0.20 mL of alcohol. Heat for 5 min on the steam bath, rotating the anode so that any deposit is dissolved in the dilute nitric acid. Remove the anode and wash it with a stream of water. Add this solution and washings to the electrolyte from which the copper was deposited.

Evaporate the electrolyte to fumes of sulfur trioxide and continue until the volume is reduced to approximately 5 mL. Cool, add 50 mL of water, cool, and transfer the solution to a 100-mL volumetric flask. Dilute with water to 100 mL and mix (1 mL = 0.05 g). To 20 mL of sample solution A (1-g sample) add an excess of ammonium hydroxide. If the presence of copper is indicated by a blue color, determine the amount colorimetrically, calculate the amount in sample solution A, and add to the amount determined electrolytically. Retain the remainder of sample solution A for succeeding tests.

INSOLUBLE IN DILUTE NITRIC ACID.   Dissolve 30 g in 200 mL of dilute nitric acid (1 + 1) and dilute with water to 225 mL. Filter through a tared filtering crucible, and reserve the filtrate separate from the washings (sample solution B). Wash thoroughly with hot water and dry at 105°C.

ANTIMONY AND TIN.   Dissolve 4.0 g in 35 mL of dilute nitric acid (1 + 1) in a 150-mL beaker and evaporate to between 6 and 10 mL. Dilute to 30 mL with water and digest on the steam bath for 15 min. Any insoluble residue should not be greater than that obtained when a control solution containing 0.4 mg of tin ion (Sn) dissolved in 35 mL of dilute nitric acid (1 + 1) is treated exactly like the 4.0 g of the sample.

ARSENIC*.   Dissolve 0.60 g in 20 mL of dilute nitric acid (1 + 1) in a 250-mL beaker. Add 6 mL of sulfuric acid, evaporate in a hood to dense fumes of sulfur trioxide, and continue the fuming for 15 min. Cool, cautiously add 15 mL of water to dissolve the salts, and fume again in the same way. Repeat the addition of water and the fuming. Cautiously transfer the solution to a generator flask, using a total of 30 mL of water. Add 20 mL of hydrochloric acid containing 3 g of $SnCl_2 \cdot 2H_2O$. Cool in an ice bath, add 15 g of zinc shot, connect the scrubber–absorber assembly, and allow the hydrogen evolution and the color development to proceed for 45 min, keeping the flask in the ice bath. Any red color in the silver diethyldithiocarbamate solution of the sample should not exceed that in a standard containing 0.003 mg of arsenic (As) similarly treated.

IRON, LEAD, MANGANESE, AND SILVER.   Determine the iron, lead, manganese, and silver by the flame atomic absorption spectrophotometric method described on page 35.

---

*See page 23 for description of the apparatus.

Sample Stock Solution.   Dissolve 10.0 g of sample in 80 mL of (1 + 1) nitric acid. Boil a few minutes to remove the oxides of nitrogen, cool, and dilute with water to the mark in a 100-mL volumetric flask (0.1 g/mL).

Analyze the solutions by means of a suitable atomic absorption spectrophotometer, using the conditions outlined in the following table. Calculate the metal content of the sample by the method of standard additions.

Element	Wavelength (nm)	Sample Wt (g)	Standard Added (mg)	Flame Type*	Background Correction
Fe	248.3	1.0	0.05; 0.10	A/A	Yes
Pb	217.0	1.0	0.05; 0.10	A/A	Yes
Mn	279.5	1.0	0.01; 0.02	A/A	Yes
Ag	328.1	1.0	0.02; 0.04	A/A	Yes

*A/A is air/acetylene.

PHOSPHORUS.   Dilute 10 mL of sample solution A (0.5-g sample) to 25 mL, add 1 mL of ammonium molybdate reagent solution and 1 mL of *p*-methylaminophenol sulfate reagent solution, and allow to stand at room temperature for 2 h. Any blue color should not exceed that produced by 1.5 mL of phosphate standard solution (0.005 mg of P) in 25 mL of approximately 0.5 N sulfuric acid when treated with the reagents used with the 25-mL dilution of the 10 mL of sample solution A.

Crystal Violet
Hexamethylpararosaniline Chloride

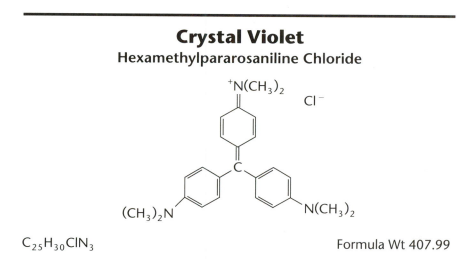

$C_{25}H_{30}ClN_3$ Formula Wt 407.99

CAS Number 548–62–9

REQUIREMENTS

Sensitivity as indicator . . . . Passes test
Loss on drying . . . . ≤ 2.5%
Assay . . . . ≥ 90.0%
Absorbance characteristics . . . . Passes test

TESTS

SENSITIVITY

Solution A. Dissolve 100 mg of sample in 10 mL of glacial acetic acid.

Solution B. Dilute 0.30 mL of 70% perchloric acid with dioxane to 25 mL.

To 50 mL of glacial acetic acid add 0.10 mL of solution A. The purplish blue color of the solution is changed to bluish green by the addition of not more than 0.05 mL of solution B.

LOSS ON DRYING. Weigh accurately 1 g and dry for 4 h at 110°C.

ASSAY. (By spectrophotometry). Weigh accurately 50.0 mg of sample previously dried for 4 h at 110°C, dissolve in about 100 mL of water in a 250-mL volumetric flask, and dilute with water to volume. Dilute 10 mL of this solution with water to 1 L, and determine the absorbance in a 1-cm cell at the wavelength of maximum absorption at about 590 nm, against water as the blank. Reserve the remainder of the solution for the determination of absorbance characteristics.

% Crystal Violet = Absorbance at 590 nm × 209

ABSORBANCE CHARACTERISTICS. Using the solution retained from the assay, determine the absorbance in a 1-cm cell at $A - 15$ nm and at $A + 15$ nm, against water as the blank. The ratio of the absorbance at $A - 15$ nm to the absorbance at $A + 15$ nm should be between 0.98 and 1.9 (A is the wavelength of maximum absorption obtained in the assay).

Cupferron

N-Hydroxy-N-Dinitrosobenzenamine, Ammonium Salt
Ammonium Salt of Nitrosophenylhydroxylamine

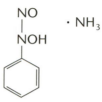

$C_6H_5N(NCO)ONH_4$ Formula Wt 155.16

CAS Number 135−20−6

> NOTE. Packages of this reagent usually contain a lump or a small cloth-wrapped package of ammonium carbonate as a preservative.

REQUIREMENTS

Appearance. . . . . Passes test
Suitability for precipitation of iron, etc. . . . . Passes test

MAXIMUM ALLOWABLE

Solubility in water . . . . Passes test
Residue after ignition 0.05%

TESTS

APPEARANCE. The material should consist of crystalline flakes, white to light buff in color. It should be free flowing and contain no tarry matter.

SOLUBILITY IN WATER. Dissolve 5.0 g in 100 mL of water. The solution should be practically clear and have no more than a pale yellow color.

RESIDUE AFTER IGNITION. Decompose 5.0 g in a tared platinum dish by means of an infrared lamp placed above the sample. Cool, moisten the residue with 1 mL of sulfuric acid, and heat gently to remove the sulfuric acid. Finally, ignite at 800 ± 25°C for 15 min.

SUITABILITY FOR PRECIPITATION OF IRON, ETC.   Accurately weigh about 3 g, dissolve in water, and dilute with water to 100 mL. Prepare a ferric chloride solution containing approximately 0.0008 g of iron (Fe) per mL. [Dissolve about 4 g of ferric chloride hexahydrate in 1 L of dilute hydrochloric acid (1 + 99).] To 40 mL of the iron solution add 20 mL of hydrochloric acid and 75 mL of water. Cool in an ice bath for about 15 min and add, from a buret, with constant stirring, 14.0 mL of the cupferron solution. Allow to stand for 15 min at room temperature and filter. (The addition of a small amount of filter paper pulp to the solution prior to filtration aids in obtaining a clear filtrate.) To the filtrate add 2 mL of the cupferron solution. No further precipitate of the brown iron compound should form. (A colorless precipitate of the acid form of the reagent may appear.)

Cupric Acetate Monohydrate
Copper(II) Acetate Monohydrate

$(CH_3COO)_2Cu \cdot H_2O$ Formula Wt 199.65

CAS Number 6046−93−1

REQUIREMENTS

Assay . . . . 98.0−102.0%
$(CH_3COO)_2Cu \cdot H_2O$

MAXIMUM ALLOWABLE

Insoluble matter 0.01%
Chloride (Cl) 0.003%
Sulfate (SO_4) 0.01%
Substances not precipitated by
   hydrogen sulfide (as sulfates) 0.2%
Iron (Fe) 0.002%
Ammonium sulfide metals other than iron. 0.01%

TESTS

ASSAY.   (By iodometric titration of oxidative capacity of Cu^{II}). Weigh accurately 0.80 g and dissolve in 50 mL of water. Add 4 mL of glacial acetic acid, 1 mL of 10% sulfuric acid, and 3 g of potassium iodide. Swirl. Titrate the liberated iodine with 0.1 N sodium thiosulfate, adding about 2 g of potassium thiocyanate and 3 mL of starch indicator solution near the end of the titration. Correct for a complete blank. One milliliter of 0.1 N sodium thiosulfate corresponds to 0.01996 g of $(CH_3COO)_2Cu \cdot H_2O$.

INSOLUBLE MATTER.   (Page 14). Use 10 g dissolved in 150 mL of water containing 1 mL of glacial acetic acid. Omit heating. Reserve the filtrate without washings for the sulfate test.

CHLORIDE.   Dissolve 0.50 g in 15 mL of water, filter if necessary through a small chloride-free filter, and add 1 mL of nitric acid and 1 mL of silver nitrate reagent solution. Any turbidity should not exceed that produced by 0.015 mg of chloride ion (Cl) in an equal volume of solution containing the quantities of reagents used in the test. The comparison is best made by the general method for chloride in colored solutions, page 25.

SULFATE.   (Page 32, Procedure A, Method 1). Use 1/20 of the filtrate from the test for insoluble matter. Allow 30 min for the turbidity to form.

SUBSTANCES NOT PRECIPITATED BY HYDROGEN SULFIDE.   Dissolve 5.0 g in about 200 mL of water. Add 2 mL of sulfuric acid, heat to about 70°C, and pass hydrogen sulfide through the solution until the copper is completely precipitated. Dilute with water to 250 mL, mix, allow to settle, and filter without washing. Evaporate 200 mL of the filtrate to dryness in a tared dish and ignite the residue at 800 ± 25°C for 15 min. Reserve the residue for preparing sample solution A.

> *Sample Solution A.*   To the residue obtained in the preceding test add 3 mL of dilute hydrochloric acid (1 + 1) and 0.10 mL of nitric acid, cover with a watch glass, and digest on the steam bath for 15–20 min. Remove the watch glass and evaporate to dryness. Dissolve in 1 mL of hydrochloric acid and dilute with water to 40 mL (1 mL = 0.1 g).

IRON.   Determine the iron by the flame atomic absorption spectrophotometric method described on page 35.

> *Sample Stock Solution.*   Dissolve 10.0 g of sample in water and add 5 mL of nitric acid. Transfer to a 100-mL volumetric flask and dilute to the mark with water (0.1 g/mL).

Analyze the solutions by means of a suitable atomic absorption spectrophotometer, using the conditions outlined in the following table. Calculate the metal content of the sample by the method of standard additions.

Element	Wavelength (nm)	Sample Wt (g)	Standard Added (mg)	Flame Type*	Background Correction
Fe	248.3	5.0	0.05; 0.10	A/A	Yes

*A/A is air/acetylene.

AMMONIUM SULFIDE METALS OTHER THAN IRON. To 10 mL of sample solution A (1-g sample) add a slight excess of ammonium hydroxide, boil for 1 min, filter, and wash with a small quantity of hot water. Dilute the filtrate and washings with water to 25 mL. Adjust the pH of 5 mL of this solution to 7 (using a pH meter) with dilute hydrochloric acid (1 + 19), and dilute with water to 20 mL. Add 0.10 mL of ammonium hydroxide and 1 mL of hydrogen sulfide water. Any darkening of the solution should not be greater than that produced by 0.02 mg of nickel ion (Ni) in an equal volume of solution containing the quantities of reagents used in the test.

Cupric Chloride Dihydrate
Copper(II) Chloride Dihydrate

$CuCl_2 \cdot 2H_2O$ Formula Wt 170.48

CAS Number 10125–13–0

REQUIREMENTS

Assay . . . . $\geq$ 99.0%
$CuCl_2 \cdot 2H_2O$

MAXIMUM ALLOWABLE

Insoluble matter 0.01%
Nitrate (NO_3) 0.015%
Sulfate (SO_4) 0.005%
Substances not precipitated by 0.10%
    hydrogen sulfide (as sulfates)
Iron (Fe) 0.005%
Ammonium sulfide metals other than iron. 0.01%

TESTS

ASSAY. (By iodometric titration of oxidative capacity of Cu^{II}). Weigh accurately 0.7 g, transfer to a glass-stoppered conical flask, and dissolve in 100 mL of water. Add 2 mL of glacial acetic acid and 3 g of potassium iodide. Stopper and swirl. Titrate the liberated iodine with 0.1 N sodium thiosulfate, adding about 2 g of potassium thiocyanate and 3 mL of starch indicator solution near the end of the titration. One milliliter of 0.1 N sodium thiosulfate corresponds to 0.01705 g of $CuCl_2 \cdot 2H_2O$.

INSOLUBLE MATTER. (Page 14). Use 10 g in 100 mL of dilute hydrochloric acid (1 + 50). Omit heating. Retain the filtrate for the test for sulfate.

NITRATE.

> *Sample Stock Solution.* Dissolve 1.0 g in 10 mL of water, add the solution with stirring to 50 mL of 10% sodium hydroxide reagent solution, and digest on the steam bath for 15 min. Cool, and filter through a retentive filter. Neutralize the filtrate with sulfuric acid, and dilute with water to 50 mL.

For sample solution A, dilute 10 mL of sample stock solution with brucine sulfate reagent solution to 50 mL. For control solution B, add 0.03 mg of nitrate ion (NO_3), and dilute with brucine sulfate reagent solution to 50 mL. Continue with the procedure described on page 29, starting with the preparation of blank solution C.

SULFATE. Heat the filtrate from the test for insoluble matter to boiling, add 10 mL of barium chloride reagent solution, digest in a covered beaker on a hot plate at low setting for 2 hours, and allow to stand overnight. If a precipitate is formed, filter, wash thoroughly, and ignite it. The weight of the precipitate should not be more than 0.0012 g greater than the weight obtained in a complete blank test.

SUBSTANCES NOT PRECIPITATED BY HYDROGEN SULFIDE. Dissolve 4.0 g in 100 mL of water, add 1 mL of sulfuric acid, heat to about 70°C, and pass hydrogen sulfide through the solution until the copper is completely precipitated. Dilute with water to 200 mL, allow the precipitate to settle, and filter without washing. Evaporate 150 mL of the filtrate to dryness in a tared dish, and ignite at 800 ± 25°C for 15 min. Retain the residue.

> *Sample Solution A.* To the residue obtained in the test for substances not precipitated by hydrogen sulfide, add 3 mL of

dilute hydrochloric acid (1 + 1) and 0.15 mL of nitric acid, cover with a watch glass, and digest on the steam bath for 15–20 min. Remove the watch glass and evaporate to dryness. Dissolve the residue in 1 mL of hydrochloric acid, and dilute with water to 30 mL (1 mL = 0.1 g).

IRON. Determine the iron by the flame atomic absorption spectrophotometric method described on page 35.

> **Sample Stock Solution.** Dissolve 10.0 g of sample in water and add 5 mL of nitric acid. Transfer to a 100-mL volumetric flask and dilute to the mark with water (0.1 g/mL).

Analyze the solutions by means of a suitable atomic absorption spectrophotometer, using the conditions outlined in the following table. Calculate the metal content of the sample by the method of standard additions.

Element	Wavelength (nm)	Sample Wt (g)	Standard Added (mg)	Flame Type*	Background Correction
Fe	248.3	1.0	0.05; 0.10	A/A	Yes

*A/A is air/acetylene.

AMMONIUM SULFIDE METALS OTHER THAN IRON. To 10 mL of sample solution A (1-g sample), add a slight excess of ammonium hydroxide, boil for 1 min, filter, and wash with a small quantity of hot water. Dilute the filtrate and washings with water to 25 mL. Adjust the pH of 5 mL of this solution to 7 (using a pH meter) with dilute hydrochloric acid (1 + 19), and dilute with water to 20 mL. Add 0.10 mL of ammonium hydroxide and 1 mL of hydrogen sulfide water. Any color should not exceed that produced by 0.02 mg of nickel ion (Ni) in an equal volume of solution containing the quantities of reagents used in the test.

Cupric Nitrate Hydrate
Copper(II) Nitrate Hydrate

$Cu(NO_3)_2 \cdot 2.5H_2O$ Formula Wt 232.59

$Cu(NO_3)_2 \cdot 3H_2O$ Formula Wt 241.60

CAS Number 3251 –23 –8

> *NOTE.* This salt is available containing either 2.5 or 3 molecules of water.

REQUIREMENTS

Assay (as the labeled hydrate) . . . . 98.0 –102.0%

MAXIMUM ALLOWABLE

Insoluble matter 0.01%
Chloride (Cl) 0.002%
Sulfate (SO_4) 0.01%
Substances not precipitated by hydrogen sulfide
   (as sulfates) 0.05%
Lead (Pb) 0.001%
Iron (Fe) 0.005%
Ammonium sulfide metals other than iron. . . . . Passes test (limit
                                                         about 0.01% as Ni)

TESTS

ASSAY.   (By iodometric titration of oxidative capacity of Cu^{II}). Weigh accurately 0.9–1.0 g, dissolve in 5 mL of hydrochloric acid in a 100-mL beaker, and evaporate to dryness on a steam bath. Add 2 mL of sulfuric acid and heat on a hot plate to dense fumes of sulfur trioxide. Cool, add 5 mL of water and 5 mL of hydrochloric acid, and again evaporate to dryness on the steam bath. Dissolve the residue in water, transfer to a conical flask, and dilute with water to about 75 mL. Add 2 mL of glacial acetic acid and 3 g of potassium iodide. Titrate the liberated iodine with 0.1 N sodium thiosulfate, adding about 2 g of potassium thiocyanate and 3 mL of starch indicator solution near the end of the titration. Correct for a complete blank. One milliliter of 0.1 N sodium thiosulfate corresponds to 0.02326 g of $Cu(NO_3)_2 \cdot 2.5H_2O$ or 0.02416 g of $Cu(NO_3)_2 \cdot 3H_2O$.

INSOLUBLE MATTER.   (Page 14). Use 10 g dissolved in 100 mL of dilute nitric acid (1 + 200). Omit heating.

CHLORIDE. Dissolve 0.50 g in 15 mL of dilute nitric acid (1 + 15), filter if necessary through a small chloride-free filter, and add 1 mL of silver nitrate reagent solution. Any turbidity should not exceed that produced by 0.01 mg of chloride ion (Cl) in an equal volume of solution containing the quantities of reagents used in the test. The comparison is best made by the general method for chloride in colored solution, page 25.

SULFATE. (Page 32, Procedure A, Method 2). Use 1.5 mL of hot dilute hydrochloric acid (1 + 1). Do two evaporations. Allow 30 min for the turbidity to form.

SUBSTANCES NOT PRECIPITATED BY HYDROGEN SULFIDE. Dissolve 4.0 g in about 190 mL of dilute sulfuric acid (1 + 95), heat to about 70°C, and pass in hydrogen sulfide until all the copper is precipitated. Dilute with water to 200 mL and filter. Evaporate 150 mL of the filtrate to dryness in a tared dish and ignite. Reserve the residue for the preparation of sample solution A.

IRON AND LEAD. Determine the iron and lead by the flame atomic absorption spectrophotometric method described on page 35.

> ***Sample Stock Solution.*** Dissolve 10.0 g of sample in water and add 5 mL of nitric acid. Transfer to a 100-mL volumetric flask and dilute to the mark with water (0.1 g/mL).

Analyze the solutions by means of a suitable atomic absorption spectrophotometer, using the conditions outlined in the following table. Calculate the metal content of the sample by the method of standard additions.

Element	Wavelength (nm)	Sample Wt (g)	Standard Added (mg)	Flame Type*	Background Correction
Fe	248.3	1.0	0.05; 0.10	A/A	Yes
Pb	217.0	2.0	0.02; 0.04	A/A	Yes

*A/A is air/acetylene.

> ***Sample Solution A.*** To the residue obtained in the test for substances not precipitated by hydrogen sulfide, add 3 mL of dilute hydrochloric acid (1 + 1) and 0.15 mL of nitric acid, cover with a watch glass, and digest on the steam bath for 15–20 min. Remove the watch glass and evaporate to dryness. Dissolve the residue in 1 mL of hydrochloric acid and dilute with water to 30 mL (1 mL = 0.1 g).

AMMONIUM SULFIDE METALS OTHER THAN IRON. To 10 mL of sample solution A (1-g sample) add a slight excess of ammonium hydroxide, boil for 1 min, filter, and wash with a small quantity of hot water. Dilute the filtrate and washings with water to 25 mL. Adjust the pH of 5 mL of this solution to 7 (using a pH meter) with dilute hydrochloric acid (1 + 19), and dilute with water to 20 mL. Add 0.10 mL of ammonium hydroxide and 1 mL of hydrogen sulfide water. Any color should not exceed that produced by 0.02 mg of nickel ion (Ni) in an equal volume of solution containing the quantities of reagents used in the test.

Cupric Oxide, Powdered
Copper(II) Oxide

CuO   Formula Wt 79.55

CAS Number 1317–38–0

REQUIREMENTS

Assay . . . . $\geq$ 99.0% CuO

MAXIMUM ALLOWABLE

Insoluble in dilute hydrochloric acid	0.02%
Carbon compounds (as C)	0.01%
Chloride (Cl)	0.005%
Nitrogen compounds (as N)	0.002%
Sulfur compounds (as SO_4)	0.02%
Free alkali	Passes test
Substances not precipitated by hydrogen sulfide (as sulfates)	0.2%
Ammonium hydroxide precipitate	0.1%

TESTS

ASSAY. (By complexometric titration of copper). Weigh, to the nearest 0.1 mg, 0.2 g of sample. Dissolve in a mixture of 20 mL of water and 5 mL of nitric acid, and boil for a few min to dissolve the sample completely. Cool, dilute with 50 mL of water in a 400-mL beaker, and add 50 mL of alcohol and 50.0 mL of 0.1 M EDTA. Then adjust the pH to about 5 (using a pH meter) with pH 5 buffer, add 0.2 mL of PAN indicator, and titrate with 0.1 M copper sulfate while maintaining the

pH at 5 by the addition of more buffer during the titration. The color
change is from green to deep blue.

$$\% \, \text{CuO} = \frac{[(\text{mL} \times \text{M EDTA}) - (\text{mL} \times \text{M CuSO}_4)] \times 7.954}{\text{sample wt (g)}}$$

INSOLUBLE IN DILUTE HYDROCHLORIC ACID. Dissolve 5.0 g by
warming on the steam bath with 30 mL of dilute hydrochloric acid
(2 + 1). Dilute with about 100 mL of water, heat to boiling, and digest
in a covered beaker on the steam bath for 1 h. Filter through a tared
filtering crucible, wash thoroughly, and dry at 105°C.

CARBON COMPOUNDS. Ignite 1.2 g in a stream of carbon dioxide-
free air or oxygen, and pass the effluent gases into 20 mL of dilute
ammonium hydroxide (2.5% NH_3) plus 2 mL of water. Prepare a
standard of 20 mL of the dilute ammonium hydroxide and 1.06 mg of
sodium carbonate (Na_2CO_3) (0.12 mg of C). Add 2 mL of barium
chloride reagent solution to each and compare promptly. Any turbidity in the solution of the sample should not exceed that in the
standard.

CHLORIDE. Shake 1.0 g with 25 mL of dilute nitric acid (1 + 3) for
10 min, filter through a chloride-free filter and dilute with water to
50 mL (sample solution A). To 10 mL of sample solution A add 5 mL
of water and 1 mL of silver nitrate reagent solution. Any turbidity
should not exceed that produced by 0.01 mg of chloride ion (Cl) in an
equal volume of solution containing 1 mL of nitric acid and 1 mL of
silver nitrate reagent solution. The comparison is best made by the
general method for chloride in colored solutions, page 25.

NITROGEN COMPOUNDS. (Page 30). Use 3.0 g dissolved in 30 mL
of freshly boiled 10% sodium hydroxide. For the standard use 0.06 mg
of nitrogen (N).

SULFUR COMPOUNDS. (Page 32, Procedure A, Method 2). Use 1.5
mL of dilute aqua regia (1 volume nitric acid + 4 volumes hydrochloric acid + 6 volumes water). Allow 30 min for the turbidity to form.

FREE ALKALI. Boil gently 3 g with 30 mL of water in a flask for 10
min. Allow to cool, add water to restore the original volume, mix well,
and allow to settle. Decant 20 mL of the liquid and add 0.10 mL of
phenolphthalein indicator solution. No red color should be produced.

SUBSTANCES NOT PRECIPITATED BY HYDROGEN SULFIDE. Dissolve 4.0 g by heating with 30 mL of dilute hydrochloric acid (2 + 1) and dilute with water to 150 mL. Heat to 70°C, pass in hydrogen sulfide to precipitate the copper completely, and filter, without washing. Evaporate 75 mL of the filtrate to near dryness in a tared dish. Cool, add 0.25 mL of sulfuric acid, and heat gently to volatilize the excess acids. Finally, ignite at 800 ± 25°C for 15 min. Reserve the residue.

AMMONIUM HYDROXIDE PRECIPITATE. To the residue obtained in the preceding test add 5 mL of dilute hydrochloric acid (1 + 1), cover with a watch glass, and digest on the steam bath for 15–20 min. Dilute with 10 mL of water, filter through a small filter, wash with two 2-mL portions of water, dilute with water to 20 mL, and add a slight excess of ammonium hydroxide. Heat to coagulate the precipitate, filter, wash, and ignite.

Cupric Oxide, Wire
Copper(II) Oxide, Wire

CuO   Formula Wt 79.55

CAS Number 1317–38–0

REQUIREMENTS

MAXIMUM ALLOWABLE

Carbon compounds (as C). 0.002%
Nitrogen compounds (as N) 0.002%
Sulfur compounds (as SO_4) 0.012%

TESTS

CARBON COMPOUNDS. Ignite 6.0 g in a stream of carbon dioxide-free air or oxygen and pass the effluent gases into 20 mL of dilute ammonium hydroxide (2.5% NH_3) plus 2 mL of water. Prepare a standard of 20 mL of the dilute ammonium hydroxide and 1.06 mg of sodium carbonate (Na_2CO_3) (0.12 mg of C). Add 2 mL of barium chloride reagent solution to each and compare promptly. Any turbidity in the solution representing the sample should not exceed that in the standard.

NITROGEN COMPOUNDS. (Page 30). Use 3.0 g dissolved in 30 mL of freshly boiled 10% sodium hydroxide. For the standard use 0.06 mg of nitrogen (N).

SULFUR COMPOUNDS. (Page 32, Procedure A, Method 2). Use 2 mL of dilute aqua regia (1 volume nitric acid + 4 volumes hydrochloric acid + 6 volumes water). Allow 30 min for the turbidity to form.

Cupric Sulfate Pentahydrate
Copper(II) Sulfate Pentahydrate

$CuSO_4 \cdot 5H_2O$ Formula Wt 249.68

CAS Number 7758-99-8

REQUIREMENTS

Assay 98.0–102.0%
$CuSO_4 \cdot 5H_2O$

MAXIMUM ALLOWABLE

Insoluble matter 0.005%
Chloride (Cl) 0.001%
Nitrogen compounds (as N) 0.001%
Substances not precipitated by hydrogen sulfide
   (as sulfates) 0.1%
Iron (Fe) 0.003%
Ammonium sulfide metals
   other than iron 0.005%

TESTS

ASSAY. (By iodometric titration of oxidative capacity of Cu^{II}). Weigh accurately 0.9–1.0 g, dissolve in water, transfer to a conical flask, and dilute with water to about 50 mL. Add 4 mL of glacial acetic acid,

1 mL of 10% sulfuric acid, and 3 g of potassium iodide. Titrate the liberated iodine with 0.1 N sodium thiosulfate, adding about 2 g of potassium thiocyanate and 3 mL of starch indicator solution near the end of the titration. Correct for a blank. One milliliter of 0.1 N sodium thiosulfate corresponds to 0.02497 g of $CuSO_4 \cdot 5H_2O$.

INSOLUBLE MATTER.   (Page 14). Use 20 g dissolved in 200 mL of dilute sulfuric acid (1 + 40). Omit heating.

CHLORIDE.   Dissolve 1.0 g in 15 mL of water, filter if necessary through a small chloride-free filter, and add 1 mL of nitric acid and 1 mL of silver nitrate reagent solution. Any turbidity should not exceed that produced by 0.01 mg of chloride ion (Cl) in an equal volume of solution containing the quantities of reagents used in the test. The comparison is best made by the general method for chloride in colored solutions, page 25.

NITROGEN COMPOUNDS.   (Page 30). Use 2.0 g. For the standard use 0.02 mg of nitrogen (N).

SUBSTANCES NOT PRECIPITATED BY HYDROGEN SULFIDE.   Dissolve 5.0 g in 200 mL of dilute sulfuric acid (1 + 99), heat to about 70°C, and pass hydrogen sulfide through the solution until the copper is completely precipitated. Dilute with water to 250 mL, mix, allow to settle, and filter without washing. Evaporate 200 mL of the filtrate to dryness in a tared dish and ignite at 800 ± 25°C for 15 min. Reserve the residue for preparing sample solution A.

> *Sample Solution A.*   To the residue from the test for substances not precipitated by hydrogen sulfide add 3 mL of dilute hydrochloric acid (1 + 1) and 0.10 mL of nitric acid, cover with a watch glass, and digest on the steam bath for 15–20 min. Remove the watch glass and evaporate to dryness. Dissolve the residue in 1 mL of hydrochloric acid and dilute with water to 60 mL (1 mL = 0.067 g).

IRON.   Determine the iron by the flame atomic absorption spectrophotometric method described on page 35.

> *Sample Stock Solution.*   Dissolve 10.0 g of sample in water and add 5 mL of nitric acid. Transfer to a 100-mL volumetric flask and dilute to the mark with water (0.1 g/mL).

Analyze the solutions by means of a suitable atomic absorption spectrophotometer, using the conditions outlined in the following

table. Calculate the metal content of the sample by the method of standard additions.

Element	Wavelength (nm)	Sample Wt (g)	Standard Added (mg)	Flame Type*	Background Correction
Fe	248.3	2.0	0.06; 0.12	A/A	Yes

*A/A is air/acetylene.

AMMONIUM SULFIDE METALS OTHER THAN IRON. To 12 mL (0.8-g sample) of sample solution A add a slight excess of ammonium hydroxide, boil for 1 min, filter, and wash with a small quantity of hot water. Dilute the filtrate and washings with water to 30 mL. Adjust the pH of 15 mL of this solution to 7 (using a pH meter) with dilute hydrochloric acid (1 + 9), and dilute with water to 20 mL. Add 0.10 mL of ammonium hydroxide and 1 mL of hydrogen sulfide water. Any color should not exceed that produced by 0.02 mg of nickel ion (Ni) in an equal volume of solution containing the quantities of reagents used in the test.

Cuprous Chloride
Copper(I) Chloride

CuCl   Formula Wt 99.00

CAS Number 7758–89–6

REQUIREMENTS

Assay . . . . ≥ 90.0% CuCl

MAXIMUM ALLOWABLE

Insoluble in acid. 0.02%
Sulfate (SO_4) 0.1%
Arsenic (As). 0.001%
Substances not precipitated by hydrogen sulfide
   (as sulfates) 0.2%
Iron (Fe) 0.005%

TESTS

ASSAY.   (By iodometric titration of reductive capacity of Cu^I). Weigh accurately 0.4 g and dissolve in the cold in 30 mL of ferric ammonium sulfate solution (10 g of ferric ammonium sulfate dissolved in 100 mL of 20% hydrochloric acid reagent solution). Add 5 mL of phosphoric acid, dilute with 200 mL of water, and titrate with 0.1 N potassium permanganate. The solution should be prepared and titrated in a flask in which the air is replaced with carbon dioxide. A stream of carbon dioxide should be passed through the flask during the titration. One milliliter of 0.1 N permanganate corresponds to 0.009900 g of CuCl.

INSOLUBLE IN ACID.   Heat 5.0 g with 50 mL of dilute hydrochloric acid (1 + 1) and add nitric acid in small portions until the sample is dissolved. Dilute with 50 mL of water. If insoluble matter is present, filter through a tared filtering crucible, wash thoroughly, and dry at 105°C.

> *Sample Solution A.*   Dilute the combined filtrate and washings from the test for insoluble in acid with water to 200 mL (1 mL = 0.025 g).

SULFATE.   (Page 32, Procedure A, Method 1). Use 2 mL of sample solution A. Allow 30 min for the turbidity to form.

ARSENIC.   (Page 23). Add 10 g to 40 mL of hydrochloric acid in a 250-mL distilling flask and distill, collecting the distillate in a 250-mL conical flask containing 30 mL of water, which is kept cool in an ice bath. When a volume of about 25 mL has distilled, add 20 mL more of hydrochloric acid to the distilling flask, and distill again until the volume in the distilling flask is about 15 mL. Dilute the distillate with water to 500 mL and mix. Dilute 15 mL of this solution with water to 35 mL. For the standard use 0.003 mg of arsenic (As).

SUBSTANCES NOT PRECIPITATED BY HYDROGEN SULFIDE.   Dilute 80 mL of sample solution A (2-g sample) with 20 mL of water, heat to about 70°C, and pass in hydrogen sulfide to precipitate the copper. Filter without washing. To 50 mL of the filtrate (1-g sample) add 0.25 mL of sulfuric acid, evaporate to dryness in a weighed dish, and ignite at 800 ± 25°C for 15 min.

IRON.   Determine the iron by the flame atomic absorption spectrophotometric method described on page 35.

Sample Stock Solution.   Dissolve 5.0 g in 50 mL of water and add 2 mL of hydrochloric acid. Transfer to a 100-mL volumetric flask and dilute to the mark with water (0.05 g/mL). To a set of three 25-mL volumetric flasks, pipet 20.0 mL (1.0 g) of sample stock solution. To two of the flasks add the specified amount of iron from the following table. Dilute each flask to the mark with water.

Analyze the solutions by means of a suitable atomic absorption spectrophotometer, using the conditions outlined in the following table. Calculate the metal content of the sample by the method of standard additions.

Element	Wavelength (nm)	Sample Wt (g)	Standard Added (mg)	Flame Type*	Background Correction
Fe	248.3	1.0	0.05; 0.10	A/A	Yes

*A/A is air/acetylene.

Cyclohexane

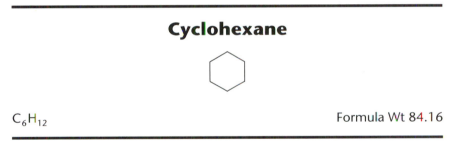

C_6H_{12}

Formula Wt 84.16

CAS Number 110−82−7

Suitable for use in ultraviolet spectrophotometry or general use. Product labeling shall designate the one or both of these uses for which suitability is represented on the basis of meeting the relevant requirements and tests. The ultraviolet spectrophotometry requirements include all of the requirements for general use.

REQUIREMENTS

General Use

Assay . . . . ≥ 99.0% C_6H_{12}
Appearance. . . . . Clear

MAXIMUM ALLOWABLE

Color (APHA). 10
Residue after evaporation 0.002%
Substances darkened by sulfuric acid . . . . Passes test
Water (H$_2$O) 0.02%

Specific Use

ULTRAVIOLET SPECTROPHOTOMETRY

Wavelength (nm)	Absorbance (AU)
300–400	0.01
260	0.02
250	0.03
240	0.08
230	0.20
220	0.50
210	1.00

TESTS

ASSAY. Analyze the sample by gas chromatography, using the general parameters cited on page 56. The following specific conditions are also required.

Column: Type I, methyl silicone

Measure the area under all peaks and calculate the cyclohexane content in area percent. Correct for water content.

COLOR (APHA). (Page 17).

RESIDUE AFTER EVAPORATION. (Page 14). Evaporate 100 g (129 mL) to dryness in a tared dish on the steam bath and dry the residue at 105°C for 30 min.

SUBSTANCES DARKENED BY SULFURIC ACID. Shake 25 mL with 15 mL of sulfuric acid for 15–20 s and allow to separate. Neither the cyclohexane nor the acid should be darkened.

WATER. (Page 55, Method 2). Use 200 µL (160 g) of the sample.

ULTRAVIOLET SPECTROPHOTOMETRY. Use the procedure on page 67 to determine the absorbance.

Cyclohexanone

C₆H₁₀O

$C_6H_{10}O$

Formula Wt 98.14

CAS Number 108−94−1

REQUIREMENTS

Assay . . . . ≥ 99.0% $C_6H_{10}O$

MAXIMUM ALLOWABLE

Color (APHA). 10
Residue after evaporation 0.05%
Water (H₂O) 0.05%

TESTS

ASSAY. Analyze the sample by gas chromatography, using the parameters cited on page 56. The following specific conditions are also required.

Column: Type I, methyl silicone

Measure the area under all peaks and calculate the cyclohexanone content in area percent. Correct for water content.

COLOR (APHA). (Page 17).

RESIDUE AFTER EVAPORATION. (Page 14). Evaporate 100 g (105 mL) to dryness in a tared dish on the steam bath and dry the residue at 105°C for 30 min.

WATER. (Page 54, Method 1). Use 30 mL (24 g) of the sample and a methanol-free system to prevent ketal formation with liberation of water.

(1,2-Cyclohexylenedinitrilo)tetraacetic Acid

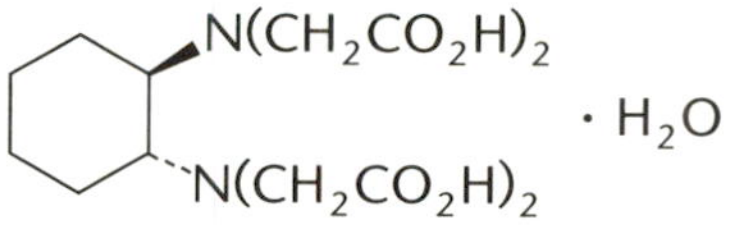

$$C_{14}H_{22}N_2O_8 \cdot H_2O \qquad\qquad \text{Formula Wt 364.35}$$

CAS Number 13291 –61 –7

REQUIREMENTS

Assay . . . . 98.0 –100.5%
$$C_{14}H_{22}N_2O_8 \cdot H_2O$$

MAXIMUM ALLOWABLE

Residue after ignition 0.2%
Heavy metals (as Pb) 0.001%
Iron (Fe) 0.005%

TESTS

ASSAY. (By complexometry). Accurately weigh 0.7–0.8 g, and dissolve it in 50 mL of 0.1 N sodium hydroxide in a 250-mL beaker. Add 25 mL of water, 10 mL of ammonium acetate buffer solution, and 0.15 mL of phenolphthalein indicator solution. Neutralize the solution to the pink color of phenolphthalein by adding dilute ammonium hydroxide (10% NH_3) dropwise. Add 1 mL of glacial acetic acid and 0.10 mL of xylenol orange indicator solution. Titrate with standard zinc chloride solution (0.1 M) to the change from yellow to off-yellow orange that remains for at least 1 min.

$$\%C_{14}H_{22}N_2O_8 \cdot H_2O = \frac{3.644 \times V}{W}$$

where

V = volume, in mL, of 0.1 M $ZnCl_2$
W = weight, in grams, of sample

Reagents

Ammonium Acetate Buffer Solution. Dissolve 200 g of ammonium acetate in water and dilute with water to 1 L.

Xylenol Orange Indicator Solution. Dissolve 0.50 g of the tetrasodium salt of xylenol orange in 100 mL of water.

Zinc Chloride Standard Solution, 0.1 M. Weigh accurately 6.537 g of pure zinc, and dissolve it in 80 mL of 10% hydrochloric acid. Warm if necessary to complete dissolution, cool, and dilute with water to 1 L.

RESIDUE AFTER IGNITION. (Page 15). Ignite 3.0 g.

Sample Solution A for the Determination of Heavy Metals and Iron. Char 3.0 g thoroughly and heat in an oven at 500°C until most of the carbon is volatilized. Cool, add 0.15 mL of nitric acid, and heat at 500°C until all the carbon is volatilized. Dissolve the residue in 2 mL of dilute hydrochloric acid (1 + 1), digest in a covered dish on the steam bath for 10 min, remove the cover, and evaporate to dryness. Dissolve in 1 mL of 1 N acetic acid and 20 mL of hot water, digest for 5 min, cool, and dilute with water to 30 mL (1 mL = 0.1 g).

HEAVY METALS. (Page 26, Method 1). Dilute 20 mL of sample solution A (2-g sample) with water to 25 mL.

IRON. (Page 28, Method 1). Use 2.0 mL of sample solution A (0.2-g sample).

1,2-Dichloroethane
Ethylene Dichloride

CH_2ClCH_2Cl Formula Wt 98.96

CAS Number 107−06−2

Suitable for use in ultraviolet spectrophotometry or general use. Product labeling shall designate one or both of these uses for which suitability is represented on the basis of meeting the relevant requirements and tests. The ultraviolet spectrophotometry requirements include all of the requirements for general use.

REQUIREMENTS

General Use

Assay . . . . $\geq$ 99.0%
CH_2ClCH_2Cl
Appearance. . . . . Clear

MAXIMUM ALLOWABLE

Color (APHA). 10
Residue after evaporation 0.002%
Titrable acid . . . . 0.0003 meq / g
Water (H_2O) 0.03%

Specific Use

ULTRAVIOLET SPECTROPHOTOMETRY

Wavelength (nm)	Absorbance (AU)
255–400	0.01
250	0.02
245	0.05
240	0.10
235	0.20
230	0.50
228	1.00

TESTS

ASSAY. Analyze the sample by gas chromatography, using the general parameters cited on page 56. The following specific conditions are also required.

Column: Type I, methyl silicone

Measure the area under all peaks and calculate the 1,2-dichloroethane content in area percent. Correct for water content.

COLOR (APHA). (Page 17).

RESIDUE AFTER EVAPORATION. (Page 14). Evaporate 100 g (82 mL) to dryness in a tared dish on the steam bath and dry the residue at 105°C for 30 min.

TITRABLE ACID. To 25 mL of alcohol in a 100-mL glass-stoppered flask, add 0.10 mL of phenolphthalein indicator solution and 0.01 N sodium hydroxide until a faint pink color persists after shaking for 30 s. Add 31 g (25 mL) of sample, mix well, and titrate with 0.01 N sodium hydroxide until the pink color is restored. Not more than 0.93 mL of 0.01 N sodium hydroxide should be required.

> *NOTE.* Special care should be taken during the addition of the sample and titration to avoid contamination from carbon dioxide.

WATER.   (Page 55, Method 2). Use 100 μL (125 mg) of the sample.

ULTRAVIOLET SPECTROPHOTOMETRY.   Use the procedure on page 67 to determine the absorbance.

2′,7′-Dichlorofluorescein
2′,7′-Dichloro-3′,6′-dihydroxyspiro[isobenzofuran-1(3 *H*),9′-[9 *H*]xanthen]-3-one
2′,7′-Dichloro-3,6-fluorandiol

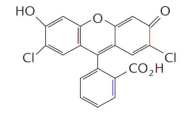

$C_{20}H_{10}Cl_2O_5$ Formula Wt 401.20

CAS Number 76 – 54 – 0

REQUIREMENTS

Appearance.	Orange to red-brown powder
Clarity of alcohol solution	Passes test
Suitability as adsorption indicator	Passes test

TESTS

CLARITY OF ALCOHOL SOLUTION.   Dissolve 100 mg in 100 mL of 70% alcohol. The solution should be clear and greenish yellow. Reserve the solution for the suitability test.

SUITABILITY AS ABSORPTION INDICATOR.   Prepare a 0.10 M solution of sodium chloride by dissolving 0.58 g of NaCl in water, and diluting with water to 100.0 mL. Dilute 10.0 mL of the 0.10 M sodium chloride with 30 mL of water, add 0.5 mL of the solution reserved from the preceding test, and titrate with 0.10 N silver nitrate, protecting the titration vessel from direct sunlight. The color change from greenish yellow to rose should require between 9.9 and 10.1 mL of 0.10 N silver nitrate.

2,6-Dichloroindophenol Sodium Salt

2,6-Dichloro-*N*-[(4-hydroxyphenyl)imino]-
2,5-cyclohexadien-1-one, Sodium Salt

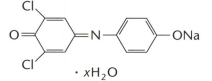

· *x*H$_2$O

O:C$_6$H$_2$Cl$_2$:NC$_6$H$_4$ONa   Formula Wt 290.08

CAS Number 620 –45 –1

REQUIREMENTS

MAXIMUM ALLOWABLE

Loss on drying at 120°C. 12.0%
Interfering dyes . . . . Passes test

TESTS

LOSS ON DRYING AT 120°C.   Weigh accurately about 1 g and dry at 120°C to constant weight.

INTERFERING DYES.

   Solution A.   Dissolve 50 mg of the sample and 42 mg of sodium bicarbonate in water and dilute with water to 200 mL. Filter through a dry filter, rejecting the first 20 mL of the filtrate.

Solution B. Dissolve 50 mg of ascorbic acid in 50 mL of a solution composed of 1.5 g of metaphosphoric acid plus 4 mL of glacial acetic acid and dilute to 50 mL with water.

To 15 mL of solution A add 2.5 mL of solution B. The mixture should become nearly colorless, with no trace of red or blue.

Dichloromethane
Methylene Chloride

CH_2Cl_2 Formula Wt 84.93

CAS Number 75–09–2

Suitable for use in high-performance liquid chromatography, extraction–concentration analysis, ultraviolet spectrophotometry, or general use. Product labeling shall designate the one or more of these uses for which suitability is represented on the basis of meeting the relevant requirements and tests. The ultraviolet spectrophotometry and liquid chromatography suitability requirements include all of the requirements for general use. The extraction–concentration suitability requirements include only the general use requirement for color.

REQUIREMENTS

General Use

Appearance	Clear
Assay	$\geq 99.5\%\ CH_2Cl_2$

MAXIMUM ALLOWABLE

Color (APHA)	10
Residue after evaporation	0.002%
Titrable acid	0.0003 meq / g
Free halogens	Passes test
Water (H_2O)	0.02%

Specific Use

ULTRAVIOLET SPECTROPHOTOMETRY

Wavelength (nm)	Absorbance (AU)
340–400	0.01
260	0.04
250	0.10
240	0.35
235	1.0

LIQUID CHROMATOGRAPHY SUITABILITY

Absorbance	Passes Test

EXTRACTION–CONCENTRATION SUITABILITY

Absorbance	Passes Test
GC–FID	Passes Test
GC–ECD	Passes Test

TESTS

ASSAY.   Analyze the sample by gas chromatography, using the parameters cited on page 56. The following specific conditions are also required.

Column: Type I, methyl silicone

Measure the area under all peaks and calculate the dichloromethane content in area percent. Correct for water content.

COLOR (APHA).   (Page 17).

RESIDUE AFTER EVAPORATION.   (Page 14). Evaporate 100 g (76 mL) in a tared dish on the steam bath and dry the residue at 105°C for 30 min.

TITRABLE ACID

> *NOTE.* Special care should be taken in the test during the addition of the sample and the titration to avoid contamination from carbon dioxide.

To 25 mL of alcohol in a 100-mL glass-stoppered flask, add 0.10 mL of phenolphthalein indicator solution and 0.01 N sodium hydroxide solution until a pink color persists for at least 30 s after vigorous shaking. Add 25 mL (33 g) of sample from a pipet and mix thoroughly with the neutralized alcohol. If no pink color remains, titrate with 0.01 N sodium hydroxide to the end point where the pink color persists for at least 30 s. Not more than 0.90 mL of 0.01 N sodium hydroxide should be required.

FREE HALOGENS.   Shake 10 mL for 2 min with 10 mL of water to which 0.10 mL of 10% potassium iodide reagent solution has been added and allow to separate. The lower layer should not show a violet tint.

WATER.   (Page 55, Method 2). Use 250 μL (330 mg) of the sample.

ULTRAVIOLET SPECTROPHOTOMETRY.   Use the procedure on page 67 to determine the absorbance.

EXTRACTION–CONCENTRATION SUITABILITY.   Analyze the sample, using the general procedure cited on page 65. In the sample preparation step, exchange the sample with ACS-grade hexane suitable for extraction–concentration.

LIQUID CHROMATOGRAPHY SUITABILITY.   Use the procedure on page 67 to determine the absorbance.

Diethanolamine
2,2′-Iminodiethanol

$(HOCH_2CH_2)_2NH$ Formula Wt 105.14

CAS Number 111 −42 −2

REQUIREMENTS

Assay . . . . $\geq$ 98.5%
$(HOCH_2CH_2)_2NH$
Apparent equivalent weight . . . . 104.0 −106.0

MAXIMUM ALLOWABLE

Color (APHA). 15
Residue after ignition 0.005%
Monoethanolamine 1.0%
Triethanolamine 1.0%
Water (H_2O) 0.15%

TESTS

ASSAY, MONOETHANOLAMINE, AND TRIETHANOLAMINE. Analyze the sample by gas chromatography, using the general parameters cited on page 56. The following specific conditions are also required.

Column: Type I, methyl silicone

Measure the area under all peaks and calculate the diethanolamine, monoethanolamine, and triethanolamine content in weight percent. Correct for water content and subtract the ethanol blank.

APPARENT EQUIVALENT WEIGHT. To 50 mL of water in a 250-mL conical flask, add 0.3 mL of mixed indicator and neutralize by adding 0.1 N hydrochloric acid just to disappearance of the green color. From a suitable weighing pipet introduce 1.8–2.0 g of sample, weighing to within 0.5 mg. Mix and titrate with standard 0.5 N hydrochloric acid to disappearance of the green color.

$$\text{Apparent equivalent weight} = \frac{\text{grams of sample} \times 1000}{\text{mL of HCl} \times N}$$

Mixed Indicator Solution. Prepare separate solutions of bromcresol green and methyl red in methanol (0.1 g/100 mL). When needed, mix freshly each day 5 parts of the bromcresol green solution with 1 part of the methyl red solution.

COLOR (APHA).   (Page 17).

RESIDUE AFTER IGNITION.   (Page 15). Evaporate 50 g (45 mL) to dryness in a tared dish. Add 0.10 mL of sulfuric acid, again take to dryness, and finally ignite at $800 \pm 25°C$.

WATER.   (Page 55, Method 2). Use 0.1 g of the sample.

4-(Dimethylamino)benzaldehyde

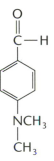

(CH₃)₂NC₆H₄CHO   Formula Wt 149.19

CAS Number 100–10–7

REQUIREMENTS

Melting point . . . . 73–75°C

MAXIMUM ALLOWABLE

Solubility in alcohol . . . . Passes test
Color (APHA) of alcohol solution 60
Solubility in hydrochloric acid . . . . Passes test
Color of hydrochloric acid solution. . . . . Passes test
Residue after ignition 0.1%

TESTS

SOLUBILITY IN ALCOHOL.   Dissolve 1 g in 25 mL of alcohol. The
solution should be clear.

COLOR (APHA) OF ALCOHOL SOLUTION.   (Page 17). Dissolve 4 g
in alcohol and dilute with alcohol to 100 mL.

SOLUBILITY IN HYDROCHLORIC ACID.   Dissolve 1 g in 20 mL of
dilute hydrochloric acid (1 + 10). Dissolution should be complete; the
solution, clear. Reserve the solution for the test for color of hydrochloric acid solution.

COLOR OF HYDROCHLORIC ACID SOLUTION. Transfer the solution prepared in the test for solubility in hydrochloric acid to a clean, dry color-comparison tube. The solution should have no more color than a color standard of the following composition: 0.3 mL of cobaltous chloride solution (5.95 g of $CoCl_2 \cdot 6H_2O$ and 2.5 mL of hydrochloric acid in 100 mL), 3.0 mL of ferric chloride solution (4.50 g of $FeCl_3 \cdot 6H_2O$ and 2.5 mL of hydrochloric acid in 100 mL), 0.2 mL of cupric sulfate solution (6.24 g of $CuSO_4 \cdot 5H_2O$ and 2.5 mL of hydrochloric acid in 100 mL), and 16.5 mL of water.

RESIDUE AFTER IGNITION.   (Page 15). Ignite 1.0 g.

N, N-Dimethylformamide

$HCON(CH_3)_2$ Formula Wt 73.09

CAS Number 68 –12 –2

Suitable for use in ultraviolet spectrophotometry or general use. Product labeling shall designate the one or both of these uses for which suitability is represented on the basis of meeting the relevant requirements and tests. The ultraviolet spectrophotometry requirements include all of the requirements for general use.

REQUIREMENTS

General Use

Assay . . . . ≥ 99.8%
$HCON(CH_3)_2$
Appearance. . . . . Clear

MAXIMUM ALLOWABLE

Color (APHA). 15
Residue after evaporation 0.005%
Titrable base . . . . 0.003 meq / g
Titrable acid . . . . 0.0005 meq / g
Water (H_2O) 0.15%

Specific Use

ULTRAVIOLET SPECTROPHOTOMETRY

Wavelength (nm)	Absorbance (AU)
340–400	0.01
310	0.05
295	0.10
275	0.30
270	1.00

TESTS

ASSAY.   Analyze the sample by gas chromatography, using the general parameters cited on page 56. The following specific conditions are also required.

  Column: Type I, methyl silicone

Measure the area under all peaks and calculate the *N*,*N*-dimethylformamide content in area percent. Correct for water content.

COLOR (APHA).   (Page 17).

RESIDUE AFTER EVAPORATION.   (Page 14). Evaporate 20 g (21 mL) to dryness in a tared dish on the steam bath and dry the residue at 105°C for 30 min.

> *NOTE.* In the tests for titrable base–acid, take care to minimize exposure of the sample to the atmosphere because of rapid absorption of carbon dioxide. Flush the flask with nitrogen during the titrations.

TITRABLE BASE.   Mix 10 g (10.6 mL) with 25 mL of carbon dioxide-free water in a glass-stoppered flask, and add 0.05 mL of methyl red indicator solution. If the solution becomes yellow, titrate with 0.01 N hydrochloric acid until a red color appears. Not more than 3.0 mL of the hydrochloric acid solution should be required.

TITRABLE ACID.   Quickly add 19 g (20 mL) of the sample from a graduated cylinder to a 125-mL conical flask. Add 0.10 mL of thymol blue indicator solution. The color of the solution should be yellow. Titrate with 0.01 N sodium methoxide solution to a blue end point. Not more than 1.0 mL should be consumed in the titration.

Reagents

Thymol Blue Indicator Solution. Dissolve 0.3 g of thymol blue in 100 mL of dry methanol.

Sodium Methoxide in Methanol, 0.01 N. Dissolve 0.540 g of sodium methylate in dry methanol and dilute with the methanol to 1 L in a volumetric flask. Standardize by titrating 40 mL of 0.01 N sulfuric acid, using thymol blue indicator.

WATER.   (Page 54, Method 1). Use 25 mL (24 g) of the sample.

ULTRAVIOLET SPECTROPHOTOMETRY.   Use the procedure on page 67 to determine the absorbance.

Dimethylglyoxime
2,3-Butanedione Dioxime

$CH_3C{:}NOHC{:}NOHCH_3$ Formula Wt 116.12

CAS Number 95 –45 –4

REQUIREMENTS

Melting point . . . . About 240°C
Suitability for nickel determination . . . . Passes test

MAXIMUM ALLOWABLE

Insoluble in alcohol 0.05%
Residue after ignition 0.05%

TESTS

INSOLUBLE IN ALCOHOL.   Gently boil 2.0 g with 100 mL of alcohol under a reflux condenser until no more dissolves. Filter through a tared filtering crucible, wash with 50 mL of alcohol in small portions, and dry at 105°C.

RESIDUE AFTER IGNITION.   (Page 15). Ignite 2.0 g.

SUITABILITY FOR NICKEL DETERMINATION. Dissolve 0.665 g of nickel sulfate hexahydrate ($NiSO_4 \cdot 6H_2O$) in water and dilute with water to 50 mL. Dilute 20 mL of this solution with water to 100 mL, heat to boiling, and add 0.25 g of the dimethylglyoxime in 25 mL of alcohol. Add dilute ammonium hydroxide (1 + 4) dropwise until the solution is alkaline to litmus, cool, and filter. To the filtrate add 1 mL of the nickel sulfate solution and heat to boiling. A substantial precipitate of red nickel dimethylglyoxime should appear.

Dimethyl Sulfoxide
Methyl Sulfoxide
Sulfinylbismethane

$(CH_3)_2SO$ Formula Wt 78.13

CAS Number 67–68–5

Suitable for use in ultraviolet spectrophotometry or general use. Product labeling shall designate the one or both of these uses for which suitability is represented on the basis of meeting the relevant requirements and tests. The ultraviolet spectrophotometry requirements include all of the requirements for general use.

REQUIREMENTS

General Use

Appearance. . . . . Clear, colorless
                                                        liquid
Assay . . . . $\geq$ 99.9% $(CH_3)_2SO$

MAXIMUM ALLOWABLE

Residue after evaporation 0.01%
Titrable acid . . . . 0.001 meq / g
Substances darkened by potassium hydroxide . . . . . Passes test
Water (H_2O) 0.1%

Specific Use

ULTRAVIOLET SPECTROPHOTOMETRY

Wavelength (nm)	Absorbance (AU)
350–400	0.01
330	0.02
310	0.06
290	0.18
270	0.40

TESTS

ASSAY. Analyze the sample by gas chromatography, using the general parameters cited on page 56. The following specific conditions are also required.

   Column: Type III, polyethylene glycol

Measure the area under all peaks and calculate the dimethylsulfoxide content in area percent. Correct for water content.

RESIDUE AFTER EVAPORATION. (Page 14). Evaporate 100 g (91 mL) to dryness in a tared dish on a hot plate.

TITRABLE ACID. Dissolve 50.0 g in 100 mL of water and add 0.1 mL of phenolphthalein indicator solution. If the solution remains colorless, titrate with 0.01 N sodium hydroxide until a pink color appears. Not more than 5.0 mL of 0.01 N sodium hydroxide should be consumed.

SUBSTANCES DARKENED BY POTASSIUM HYDROXIDE. Dissolve 2.0 g of potassium hydroxide in 10.0 mL of methanol in a small beaker. (*NOTE*: Use only solid potassium hydroxide that is white with no discoloration.) Cover with a watch glass and warm on the steam bath to dissolve. Cool and allow any potassium carbonate to settle. Decant 5 mL of clear solution and add it to 25 mL of sample in a glass-stoppered 50-mL flask. Insert the stopper and heat on a steam bath for 20 min. Cool to room temperature. Measure the absorbance of the solution and of the sample at 350 nm in a 1-cm cell with water as the blank. The absorbance of the solution of the sample should not be more than 0.023 AU greater than the absorbance of the sample.

WATER. (Page 55, Method 2). Use 50 μL (55 mg) of the sample.

ULTRAVIOLET SPECTROPHOTOMETRY. Use the procedure on page 67 to determine the absorbance.

Dioxane
1,4-Dioxane

$C_5H_8O_2$ Formula Wt 88.11

CAS Number 123–91–1

CAUTION.   Dioxane tends to form explosive peroxides, especially when anhydrous. It should not be allowed to evaporate to dryness unless the absence of peroxides has been shown.

NOTE.   Dioxane usually contains a stabilizer. If a stabilizer is present, its identity and quantity must be stated on the label.

REQUIREMENTS

Assay . . . . $\geq$ 99.0% $C_5H_8O_2$
Freezing point . . . . Not below 11.0°C

MAXIMUM ALLOWABLE

Color (APHA). 20
Peroxide (as H_2O_2) 0.005%
Residue after evaporation 0.005%
Titrable acid . . . . 0.0016 meq / g
Carbonyl (as HCHO) 0.01%
Water (H_2O) 0.05%

TESTS

ASSAY.   Analyze the sample by gas chromatography, using the general parameters cited on page 56. The following specific conditions are also required.

Column: Type I, methyl silicone

Measure the area under all peaks and calculate the dioxane content in area percent. Correct for water content.

FREEZING POINT

Apparatus. The sample container is a test tube (25 × 100 mm) supported by a cork in a water-tight glass cylinder (50 × 110 mm). The cylinder is mounted in a water bath that provides at least a 37-mm layer of water surrounding the sides and bottom of the cylinder. An accurate thermometer with 0.1°C subdivisions is centered in the test tube and a thermometer with 1°C subdivisions is mounted in the water bath. The stirrer, about 30 cm long, is a wire with a loop at the bottom.

Pour sufficient sample into the test tube to make a 50-mm column. Assemble the apparatus with the bulb of the 0.1° thermometer immersed halfway between the top and bottom of the sample in the test tube. Fill the water bath with a mixture of ice and water to within 12 mm of the top, and adjust the temperature to 0°C by the addition of more ice, if necessary. Stir the sample continuously during the test by moving the wire loop up and down throughout the entire depth of the sample at a regular rate of 20 cycles per min. Record the reading of the 0.1° thermometer every 30 s. Anticipate that the temperature will fall gradually at first, then become constant for 1–2 min at the freezing point, and finally fall gradually again. The freezing point is the average of 4 consecutive readings that lie within a range of 0.2°C.

COLOR (APHA).   (Page 17).

PEROXIDE.   Dilute 10 mL of the dioxane with water to 50 mL and use 5.0 mL of the solution (1-g sample). For the standard dilute a solution containing 0.05 mg of hydrogen peroxide standard with water to 5.0 mL. Add 5.0 mL of titanium tetrachloride reagent solution to each and mix. After standing for 5 ± 1 min, any yellow color in the solution of the sample should not exceed that in the standard. The color intensities may be determined with a spectrophotometer in 1-cm cells at a wavelength of 410 nm.

> CAUTION. If peroxide is present, do not make the test for residue after evaporation.

RESIDUE AFTER EVAPORATION.   (Page 14). Evaporate 20 g (19 mL) to dryness in a tared platinum dish on the steam bath, and dry the residue at 105°C for 30 min.

TITRABLE ACID.   Mix 31 g (30 mL) with 30 mL of carbon dioxide-free water, and titrate with 0.01 N sodium hydroxide, using phenolphthalein as the indicator. Not more than 5.0 mL of the titrant is required.

CARBONYL. To 1.0 mL of sample in a polarograph cell add 2.0 mL of water, 5.0 mL of pH 6.5 buffer solution, 1.0 mL of 0.20% Triton X-100 (polyethylene glycol ether of isooctylphenol) solution, and 1.0 mL of 2.0% hydrazine sulfate solution. Prepare a similar mixture in which 1.0 mL of the water is replaced by 1.0 mL of a water solution containing 0.10 mg of formaldehyde. Deaerate each solution with nitrogen for 10 min and record the polarograms from -0.6 to -1.6 V versus SCE at a sensitivity of 20 μA full scale. The total diffusion current for the sample should not be greater than one-half the diffusion current for sample plus formaldehyde.

The approximate half-wave potential for the formaldehyde hydrazone is -1.06 V.

> **Buffer Solution.** Dissolve 10.1 g of anhydrous dibasic sodium phosphate (Na_2HPO_4) and 3.05 g of citric acid monohydrate in water and dilute to 500 mL.

WATER. (Page 55, Method 2). Use 100 μL (103 mg) of the sample.

Diphenylamine
N-Phenylbenzeneamine

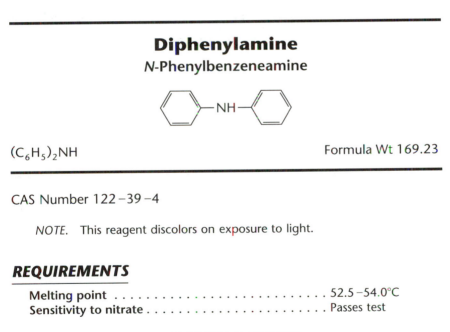

$(C_6H_5)_2NH$ Formula Wt 169.23

CAS Number 122–39–4

NOTE. This reagent discolors on exposure to light.

REQUIREMENTS

Melting point . . . . 52.5–54.0°C
Sensitivity to nitrate . . . . Passes test

MAXIMUM ALLOWABLE

Solubility in alcohol . . . . Passes test
Residue after ignition 0.03%
Nitrate (NO_3) . . . . Passes test

TESTS

SOLUBILITY IN ALCOHOL.   Dissolve 1.0 g in 50 mL of alcohol. The solution should be clear and colorless.

RESIDUE AFTER IGNITION.   (Page 15). Ignite 4.0 g. Moisten the char with 2 mL of sulfuric acid.

NITRATE.   To 20 mL of water add 60 mL of nitrate-free sulfuric acid, adjust the temperature to about 60°C, and add 0.5 mL of hydrochloric acid and 0.0100 g of the sample. No blue color should be produced in 5 min. Retain the solution for the test for sensitivity to nitrate.

SENSITIVITY TO NITRATE.   To 8 mL of the solution retained from the test for nitrate add 0.01 mg of nitrate ion (NO_3) and allow to stand at 60°C for 5 min. A blue color should be produced.

Diphenylaminesulfonic Acid Sodium Salt
Sodium Diphenylaminesulfonate

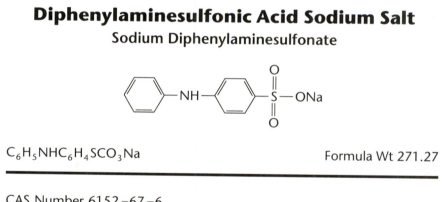

$C_6H_5NHC_6H_4SCO_3Na$ Formula Wt 271.27

CAS Number 6152–67–6

REQUIREMENTS

Sensitivity as indicator . . . . Passes test

TESTS

SENSITIVITY AS INDICATOR.   Dissolve 0.15 g in 100 mL of water. To 25 mL of water in a 50-mL test tube, add 10 mL of 4 N sulfuric acid, 5 mL of phosphoric acid, 0.05 mL of 0.01 N ferrous ammonium sulfate, and 0.05 mL of the sample solution. The addition of 0.10 mL of 0.01 N potassium dichromate should produce a violet color. The color should be completely discharged by the addition of 0.10 mL of 0.01 N ferrous ammonium sulfate.

Diphenylcarbazone Compound with s-Diphenylcarbazide

(Phenylazo)formic Acid 2-Phenylhydrazide Compound with 1,5-Diphenylcarbohydrazide

"s-Diphenylcarbazone"

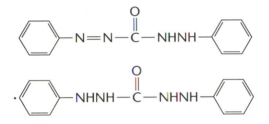

CAS Number 538−62−5 with 140−22−7

REQUIREMENTS

Sensitivity . . . . Passes test

MAXIMUM ALLOWABLE

Residue after ignition 0.1%
Solubility in acetone . . . . Passes test

TESTS

RESIDUE AFTER IGNITION.   (Page 15). Ignite 2.0 g. Moisten the char with 2 mL of sulfuric acid.

SOLUBILITY IN ACETONE.   Dissolve 0.2 g in 25 mL of acetone. The sample should dissolve completely, and the solution should be clear and red. Reserve this solution for the sensitivity test.

SENSITIVITY. Dissolve 0.135 g of mercuric chloride in 500 mL of water. To 20 mL of this solution add 0.25 mL of the solution reserved from the test for solubility. A violet color should develop.

1,5-Diphenylcarbohydrazide

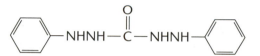

$C_6H_5NHNHCONHNHC_6H_5$ Formula Wt 242.28

CAS Number 140–22–7

REQUIREMENTS

Melting point . . . . 173–176°C
Sensitivity to chromate . . . . Passes test

MAXIMUM ALLOWABLE

Solubility in aqueous acetone . . . . Passes test
Residue after ignition 0.05%

TESTS

SOLUBILITY IN AQUEOUS ACETONE. Mix 250 mL of acetone and 50 mL of water, and adjust the pH to 6.0 ± 0.2 (using a pH meter) with 0.02 M acetic acid or 0.02 M sodium hydroxide. Dissolve 10 g in the mixture without heating. Dissolution should be complete and the solution clear, free of any red color, and not more than faintly yellow.

RESIDUE AFTER IGNITION. (Page 15). Ignite 2.0 g. Moisten the char with 1 mL of sulfuric acid.

SENSITIVITY TO CHROMATE. Dissolve 0.25 g in a mixture of 50 mL of acetone and 50 mL of water. To each of two 50-mL volumetric flasks transfer 20 mL of water and 10 mL of 1 N sulfuric acid. To one flask add 5.8 mL of standard chromate solution (1 mL = 0.01 mg of CrO_4) and to both add 2.0 mL of the sample solution, mix, and allow to stand for 1 min. To both flasks add 5 mL of 4 M monobasic sodium phosphate, dilute with water to the mark, and allow to stand for 5 min. Determine the absorbance of the solution containing chromate in a 1-cm cell at 540 nm, against the other solution in a similar matched cell set at zero absorbance as the blank. The absorbance should be not less than 0.42.

Dithizone
(Phenylazo)thioformic Acid 2-Phenylhydrazide
Diphenylthiocarbazone

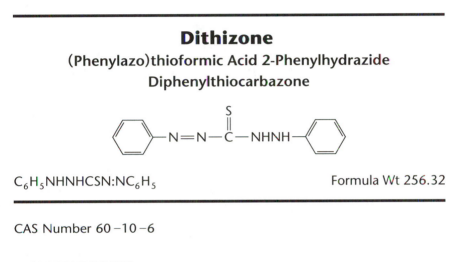

$C_6H_5NHNHCSN:NC_6H_5$ Formula Wt 256.32

CAS Number 60−10−6

REQUIREMENTS

Assay . . . . ≥ 85.0%
                                                   $C_6H_5NHNHCSN:NC_6H_5$
Ratio of absorbances . . . . ≥ 1.55

MAXIMUM ALLOWABLE

Residue after ignition 0.3%
Heavy metals (as Pb) 0.002%

TESTS

ASSAY. (By spectrophotometry). Weigh accurately 10.0 mg, transfer to a 100-mL volumetric flask, dissolve in about 75 mL of carbon tetrachloride, and dilute to volume with carbon tetrachloride. Dilute 5.0 mL of this solution to 100.0 mL with carbon tetrachloride, and determine the absorbance of this solution in 1.00-cm cells at 620 nm, using carbon tetrachloride as the blank. Reserve the remainder of the solution for determining the ratio of absorbances. Determine the molar absorptivity by dividing the absorbance by 1.95×10^{-5} (the molar concentration of the dithizone) and calculate the assay value as follows:

$$\frac{\text{Molar absorptivity at 620 nm}}{34,600} \times 100 = \% \text{ Dithizone}$$

RATIO OF ABSORBANCES.   Determine the molar absorptivity at 450 nm of the solution prepared in the test for assay, using the method described in the test. The ratio of the value at 620 nm to that at 450 nm should not be less than 1.55.

RESIDUE AFTER IGNITION.   (Page 15). Ignite 1.0 g. Moisten the char with 1 mL of nitric acid and 1 mL of sulfuric acid.

HEAVY METALS.   (Page 27, Method 2). Use 1.0 g.

Eriochrome Black T
3-Hydroxy-4-[(1-hydroxy-2-naphthalenyl)azo]-7-nitro-1-naphthalenesulfonic Acid, Monosodium Salt
C.I. Mordant Black 11

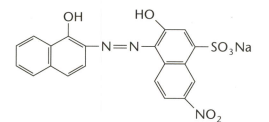

$C_{20}H_{12}N_3NaO_7S$ Formula Wt 461.38

CAS Number 1787–61–7

REQUIREMENTS

Clarity of solution . . . . Passes test
Suitability as complexometric indicator. . . . . Passes test

TESTS

CLARITY OF SOLUTION.   Dissolve 0.1 g in 100 mL of water, warming if necessary. Not more than a faint trace of turbidity or insoluble matter should remain. Reserve the solution for the test for suitability.

SUITABILITY AS COMPLEXOMETRIC INDICATOR. Add 10 mL of ammoniacal buffer (pH 10) to 100 mL of water. Add 0.5 mL of the solution reserved from the test for clarity of solution and 10.0 mL of magnesium standard solution (0.01 mg of Mg per mL). The color of the solution should be red-violet. Add 0.10 mL of 0.1 M EDTA solution. The color of the solution should change to blue.

Ethyl Acetate

$$CH_3 - \overset{\overset{\displaystyle O}{\|}}{C} - OCH_2CH_3$$

$CH_3COOCH_2CH_3$ Formula Wt 88.11

CAS Number 141−78−6

Suitable for use in ultraviolet spectrophotometry or general use. Product labeling shall designate the one or both of these uses for which suitability is represented on the basis of meeting the relevant requirements and tests. The ultraviolet spectrophotometry requirements include all of the requirements for general use.

REQUIREMENTS

General Use

Assay . . . . ≥ 99.5%
$CH_3COOCH_2CH_3$

MAXIMUM ALLOWABLE

Color (APHA). 10
Residue after evaporation 0.003%
Water (H_2O) 0.2%
Titrable acid . . . . 0.0009 meq / g
Substances darkened by sulfuric acid . . . . Passes test

Specific Use

ULTRAVIOLET SPECTROPHOTOMETRY

Wavelength (nm)	Absorbance (AU)
330–400	0.01
275	0.05
263	0.10
257	0.50
255	1.00

TESTS

ASSAY. Analyze the sample by gas chromatography, using the general parameters cited on page 56. The following specific conditions are also required.

    Column: Type I, methyl silicone

Measure the area under all peaks and calculate the ethyl acetate content in area percent. Correct for water content.

COLOR (APHA). (Page 17).

RESIDUE AFTER EVAPORATION. (Page 14). Evaporate 40 g (45 mL) to dryness in a tared dish on the steam bath and dry the residue at 105°C for 30 min.

WATER. (Page 54, Method 1). Use 20 mL (18 g) of the sample.

TITRABLE ACID. To 10 mL of alcohol add 0.10 mL of phenolphthalein indicator solution, and neutralize with 0.01 N sodium hydroxide. Add 9 g (10 mL) of the sample, mix gently, and titrate with 0.01 N sodium hydroxide until the pink color persists for 15 s. Not more than 0.80 mL should be required.

SUBSTANCES DARKENED BY SULFURIC ACID. Superimpose 5 mL of ethyl acetate upon 5 mL of sulfuric acid. No dark coloration should be produced at the zone of contact.

ULTRAVIOLET SPECTROPHOTOMETRY. Use the procedure on page 67 to determine the absorbance.

Ethyl Alcohol
Ethanol

CH_3CH_2OH Formula Wt 46.07

CAS Number 64 –17 –5

Suitable for use in ultraviolet spectrophotometry or general use. Product labeling shall designate the one or both of these uses for which suitability is represented on the basis of meeting the relevant requirements and tests. The ultraviolet spectrophotometry requirements include all of the requirements for general use.

REQUIREMENTS

General Use

Assay . . . . ≥ 95.0%
by volume

MAXIMUM ALLOWABLE

Solubility in water . . . . Passes test
Acetone, isopropyl alcohol . . . . Passes test (limit
about 0.001%
acetone, 0.003%
isopropyl alcohol)

Methanol (CH_3OH) 0.1%
Substances darkened by sulfuric acid . . . . Passes test
Substances reducing permanganate . . . . Passes test
Color (APHA) 10
Residue after evaporation 0.001%
Titrable acid . . . . 0.0005 meq / g
Titrable base . . . . 0.0002 meq / g

Specific Use

ULTRAVIOLET SPECTROPHOTOMETRY

Wavelength (nm)	Absorbance (AU)
270 –400	0.01
240	0.05
230	0.15
220	0.25
210	0.40

TESTS

ASSAY. Analyze the sample by gas chromatography using the general parameters cited on page 56. The following specific conditions are also required.

Column: Type I, methyl silicone

Measure the area under all peaks and calculate the ethyl alcohol content in area percent. Correct for water content.

COLOR (APHA). (Page 17).

SOLUBILITY IN WATER. Mix 15 mL with 45 mL of water and allow to stand for 1 h. The mixture should be as clear as an equal volume of water.

RESIDUE AFTER EVAPORATION. (Page 14). Evaporate 100 g (124 mL) to dryness in a tared dish on the steam bath and dry the residue at 105°C for 30 min.

ACETONE, ISOPROPYL ALCOHOL. Dilute 1.0 mL with 1 mL of water. Add 1 mL of a saturated solution of disodium hydrogen phosphate and 3 mL of a saturated solution of potassium permanganate. Warm the mixture to between 45 and 50°C and allow to stand until the permanganate color is discharged. Add 3 mL of 10% sodium hydroxide reagent solution and filter, without washing, through glass. Prepare a control containing 1 mL of the saturated solution of disodium hydrogen phosphate, 3 mL of 10% sodium hydroxide reagent solution, and 0.008 mg of acetone in 9 mL. To each solution add 1 mL of 1% furfural reagent solution, and allow to stand for 10 min. To 1.0 mL of each solution add 3 mL of hydrochloric acid. Any pink color produced in the solution of the sample should not exceed that in the control.

TITRABLE ACID. Add 10 mL to 25 mL of water in a glass-stoppered flask and add 0.15 mL of phenolphthalein indicator solution. Add 0.01 N sodium hydroxide until a slight pink color persists after shaking for 30 s. Add 25 mL of the sample, mix well, and titrate with 0.01 N sodium hydroxide until the pink color is restored. Not more than 1.0 mL of the hydroxide solution should be required.

TITRABLE BASE. Dilute 25 mL with 25 mL of water and add 0.15 mL of methyl red indicator solution. Not more than 0.40 mL of 0.01 N sulfuric acid should be required to produce a pink color.

METHANOL.   Dilute 5 mL with water to 100 mL. To 1 mL of the solution add 0.2 mL of dilute phosphoric acid (1 + 9) and 0.25 mL of a 5% potassium permanganate solution. Allow to stand for 15 min, add 0.3 mL of 10% sodium bisulfite solution, and shake until colorless. Add slowly 5 mL of ice-cold 80% sulfuric acid (3 volumes of acid plus 1 volume of water), keeping the mixture cool during the addition. Add 0.1 mL of a 1% aqueous solution of chromotropic acid, mix, and digest on the steam bath for 20 min. Any violet color should not exceed that produced by 0.04 mg of methanol in 1 mL of water treated in the same way as the 1 mL of the diluted sample.

SUBSTANCES DARKENED BY SULFURIC ACID.   Cool 10 mL of sulfuric acid, contained in a small conical flask, to 10°C. Add dropwise, with constant agitation, 10 mL of the sample, while keeping the temperature of the mixture below 20°C. The resulting solution should have no more color than either of the two liquids before mixing.

SUBSTANCES REDUCING PERMANGANATE.   Cool 20 mL to 15°C, add 0.10 mL of 0.1 N potassium permanganate, and allow to stand at 15°C for 5 min. The pink color should not be entirely discharged.

ULTRAVIOLET SPECTROPHOTOMETRY.   Use the procedure on page 67 to determine the absorbance.

Ethyl Alcohol, Absolute
Ethanol, Absolute

CH_3CH_2OH Formula Wt 46.07

CAS Number 64 –17 –5

REQUIREMENTS

Assay . . . . ≥ 99.5%
CH_3CH_2OH
by volume (about
99.2% by weight)

MAXIMUM ALLOWABLE

Water (H_2O) 0.2%
Color (APHA). 10
Solubility in water . . . . Passes test
Residue after evaporation 0.001%
Acetone, isopropyl alcohol . . . . Passes test (limit
  about 0.001%
  acetone, 0.003%
  isopropyl alcohol)
Titrable acid . . . . 0.0005 meq / g
Titrable base . . . . 0.0002 meq / g
Methanol (CH_3OH) 0.1%
Substances darkened by sulfuric acid . . . . Passes test
Substances reducing permanganate . . . . Passes test

TESTS

WATER.   (Page 55, Method 2). Use 20 mL (15.8 g).

ASSAY and other tests are the same as for ethyl alcohol, page 319.

Ethyl Ether
Diethyl Ether

$(CH_3CH_2)_2O$ Formula Wt 74.12

CAS Number 60 –29 –7

> *CAUTION.* Ethyl ether tends to form explosive peroxides, especially when anhydrous. It should not be allowed to evaporate to dryness unless the absence of peroxides has been shown. The presence of water or appropriate reducing agents lessens peroxide formation.

> *NOTE.* Ethyl ether normally contains about 2% of alcohol and 0.5% of water as stabilizers.

REQUIREMENTS

Assay . . . . ≥ 98.0%
  $(CH_3CH_2)_2O$

MAXIMUM ALLOWABLE

Color (APHA). 10
Peroxide (as H_2O_2) . . . . 1 ppm
Residue after evaporation 0.001%
Titrable acid . . . . 0.0002 meq / g
Carbonyl (as HCHO) 0.001%
Substances darkened by sulfuric acid . . . . Passes test

TESTS

ASSAY.   Assay the sample by gas chromatography, using the general parameters cited on page 56. The following specific conditions are also required.

Column: Type I, methyl silicone

Measure the area under all peaks and calculate the ethyl ether content in area percent. Correct for water content. The assay value is the combination of ethyl ether and added stabilizers.

COLOR (APHA).   (Page 17).

PEROXIDE.   To 35 g (50 mL) of the ethyl ether in a separatory funnel, add 5.0 mL of titanium tetrachloride reagent solution. Shake vigorously, allow the layers to separate, and drain the lower layer into a 25-mL glass-stoppered graduated cylinder. For the standard transfer 5.0 mL of titanium tetrachloride reagent solution to a similar graduated cylinder, and add a volume of solution containing 0.035 mg of hydrogen peroxide standard. Dilute both solutions with water to 10.0 mL and mix. Any yellow color in the solution of the sample should not exceed that in the standard. The color intensities may be determined with a spectrophotometer in 1-cm cells at a wavelength of 410 nm.

> *NOTE.* Fresh ethyl ether should meet this test, but after storage for several months peroxide may be formed.

> *CAUTION.* If peroxide is present, do not make the test for residue after evaporation.

RESIDUE AFTER EVAPORATION.   (Page 14). Evaporate 100 g (141 mL) to dryness in a tared dish and dry the residue at 105°C for 30 min.

TITRABLE ACID

> *NOTE.* Great care should be taken in the test during the addition of the sample and the titration to avoid contamination from carbon dioxide.

To 10 mL of water in a glass-stoppered flask, add 0.10 mL of bromthymol blue indicator solution and 0.01 N sodium hydroxide until a blue color persists after vigorous shaking. Add 25 mL of sample from a pipet and shake briskly to mix the two layers. If no blue color remains, titrate with 0.01 N sodium hydroxide until the blue color is restored and persists for several minutes. Not more than 0.30 mL of 0.01 N sodium hydroxide should be required.

CARBONYL.   To 5 g (7.0 mL) of sample in a 30-mL separatory funnel add 5.0 mL of pH 6.5 buffer solution, 1.0 mL of 2.0% hydrazine sulfate solution, 1.0 mL of 0.20% Triton X-100 (polyethylene glycol ether of isooctylphenol) solution, and 3.0 mL of water. Prepare a similar mixture in which part of the water is replaced by an equal volume of aqueous solution containing 0.05 mg of formaldehyde. Shake each funnel well to mix the contents, allow the phases to separate, and draw off each lower layer into a polarographic cell. Deaerate with nitrogen for 10 min. Record the polarogram from -0.6 to -1.6 V versus SCE at a sensitivity of 5 μA full scale. The total diffusion current for the sample should not be greater than one-half the diffusion current for sample plus aldehyde.

Approximate half-wave potentials for the hydrazones of known carbonyl compounds are: formaldehyde, -1.06 V; acetaldehyde, -1.21 V; and acetone, -1.4 V.

> ***Buffer Solution.***   Dissolve 10.1 g of anhydrous dibasic sodium phosphate (Na_2HPO_4) and 3.05 g of citric acid monohydrate in water and dilute with water to 500 mL.

SUBSTANCES DARKENED BY SULFURIC ACID.   Cool 10 mL of sulfuric acid to about 10°C and add dropwise with gentle stirring 10 mL of the ethyl ether. Any color should not exceed that produced in a color standard of the following composition: 0.4 mL of cobaltous chloride solution (5.95 g of $CoCl_2 \cdot 6H_2O$ and 2.5 mL of hydrochloric acid in 100 mL), 0.6 mL of ferric chloride solution (4.50 g of $FeCl_3 \cdot 6H_2O$ and 2.5 mL of hydrochloric acid in 100 mL), 0.6 mL of cupric sulfate solution (6.24 g of $CuSO_4 \cdot 5H_2O$ and 2.5 mL of hydrochloric acid in 100 mL), and 18.4 mL of water.

Ethyl Ether, Anhydrous
Diethyl Ether, Anhydrous

$(CH_3CH_2)_2O$ Formula Wt 74.12

CAS Number 60 – 29 – 7

> *CAUTION.* Ethyl ether tends to form explosive peroxides, especially when anhydrous. It should not be allowed to evaporate to dryness or near dryness unless the absence of peroxides has been shown. The formation of peroxides is more rapid in ethyl ether kept in containers that have been opened and partly emptied. Some ethyl ether may contain a stabilizer. If it does, the amount and type should be marked on the label.

REQUIREMENTS

Assay . . . . ≥ 99.0%
$(CH_3CH_2)_2O$

MAXIMUM ALLOWABLE

Color (APHA). 10
Peroxide (as H_2O_2) . . . . 1 ppm
Residue after evaporation 0.001%
Titrable acid . . . . 0.0002 meq / g
Carbonyl (as HCHO) 0.001%
Substances darkened by sulfuric acid . . . . Passes test
Alcohol (CH_3CH_2OH) . . . . Passes test (limit about 0.05%)
Water (H_2O) 0.03%

TESTS

ASSAY.   Analyze the sample by gas chromatography, using the general parameters cited on page 56. The following specific conditions are also required.

Column: Type I, methyl silicone

Measure the area under all peaks and calculate the ethyl ether content in area percent. Correct for water content.

COLOR (APHA). (Page 17).

PEROXIDE. To 35 g (50 mL) of the ethyl ether in a separatory funnel, add 5.0 mL of titanium tetrachloride reagent solution. Shake vigorously, allow the layers to separate, and drain the lower layer into a 25-mL glass-stoppered graduated cylinder. For the standard transfer 5.0 mL of titanium tetrachloride reagent solution to a similar graduated cylinder, and add a volume of solution containing 0.035 mg of hydrogen peroxide standard. Dilute both solutions with water to 10.0 mL and mix. Any yellow color in the solution of the sample should not exceed that in the standard. The color intensities may be determined with a spectrophotometer in 1-cm cells at a wavelength of 410 nm.

> NOTE. Fresh ethyl ether should meet this test, but after storage for several months peroxide may be formed.

> CAUTION. If peroxide is present, do not make the test for residue after evaporation.

RESIDUE AFTER EVAPORATION. (Page 14). Evaporate 100 g (141 mL) to dryness in a tared dish and dry the residue at 105°C for 30 min.

TITRABLE ACID

> NOTE. Great care should be taken in the test during the addition of the sample and the titration to avoid contamination from carbon dioxide.

To 10 mL of water in a glass-stoppered flask, add 0.10 mL of bromthymol blue indicator solution and 0.01 N sodium hydroxide until a blue color persists after vigorous shaking. Add 25 mL of sample from a pipet and shake briskly to mix the two layers. If no blue color remains, titrate with 0.01 N sodium hydroxide until the blue color is restored and persists for several minutes. Not more than 0.30 mL of 0.01 N sodium hydroxide should be required.

CARBONYL. To 5 g (7.0 mL) of sample in a 30-mL separatory funnel add 5.0 mL of pH 6.5 buffer solution, 1.0 mL of 2.0% hydrazine sulfate solution, 1.0 mL of 0.20% Triton X-100 (polyethylene glycol ether of isooctylphenol) solution, and 3.0 mL of water. Prepare a similar mixture in which part of the water is replaced by an equal volume of aqueous solution containing 0.05 mg of formaldehyde.

Shake each funnel well to mix the contents, allow the phases to separate, and draw off each lower layer into a polarographic cell. Deaerate with nitrogen for 10 min. Record the polarogram from -0.6 to -1.6 V versus SCE at a sensitivity of 5 μA full scale. The total diffusion current for the sample should not be greater than one-half the diffusion current for sample plus aldehyde.

Approximate half-wave potentials for the hydrazones of known carbonyl compounds are: formaldehyde, -1.06 V; acetaldehyde, -1.21 V; and acetone, -1.4 V.

> **Buffer Solution.** Dissolve 10.1 g of anhydrous dibasic sodium phosphate (Na_2HPO_4) and 3.05 g of citric acid monohydrate in water and dilute with water to 500 mL.

SUBSTANCES DARKENED BY SULFURIC ACID.   Cool 10 mL of sulfuric acid to about 10°C and add dropwise with gentle stirring 10 mL of the ethyl ether. Any color should not exceed that produced in a color standard of the following composition: 0.4 mL of cobaltous chloride solution (5.95 g of $CoCl_2 \cdot 6H_2O$ and 2.5 mL of hydrochloric acid in 100 mL), 0.6 mL of ferric chloride solution (4.50 g of $FeCl_3 \cdot 6H_2O$ and 2.5 mL of hydrochloric acid in 100 mL), 0.6 mL of cupric sulfate solution (6.24 g of $CuSO_4 \cdot 5H_2O$ and 2.5 mL of hydrochloric acid in 100 mL), and 18.4 mL of water.

ALCOHOL.   Transfer 100 mL to a separatory funnel and shake with five successive portions, 20 mL, 10 mL, 10 mL, 5 mL, and 5 mL, respectively, of distilled water at about 25°C. Shake each portion for 2 min and separate the water layer carefully. Finally, pour the combined water extract from one flask to another six times to assure minimum contamination with ether. Transfer 1 mL of the water extract with a pipet to a comparison tube and add 4 mL of water. For a standard take 5 mL of a solution of 0.2 mL of absolute alcohol in a liter of water. Add 10 mL of nitrochromic acid reagent to each solution, mix, and allow to stand for 1 h. At the end of this time the color of the sample solution should show no more change from yellow to green or blue than is shown by the standard. The test and the standard must be kept at the same temperature.

> **Nitrochromic Acid Reagent.**   Mix 1 volume of 5% potassium chromate solution with 133 volumes of water and 66 volumes of colorless nitric acid. This reagent should not be used if more than 1 month old.

WATER.   (Page 55, Method 2). Use 0.2 mL (0.14 g) of the sample.

(Ethylenedinitrilo)tetraacetic Acid
N,N,N',N'-Ethylenediaminetetraacetic Acid
N,N'-1,2-Ethanediylbis[N-carboxymethyl]glycine
EDTA, Free Acid

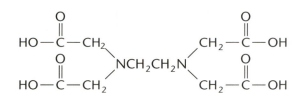

$C_{10}H_{16}N_2O_8$ Formula Wt 292.25

CAS Number 60–00–4

REQUIREMENTS

Assay 99.4–100.6%
$C_{10}H_{16}N_2O_8$

MAXIMUM ALLOWABLE

Insoluble in dilute ammonium hydroxide 0.005%
Residue after ignition 0.2%
Nitrilotriacetic acid [(HOCOCH$_4$)$_3$N]. 0.1%
Calcium (Ca) 0.001%
Heavy metals (as Pb) 0.001%
Iron (Fe) 0.005%
Magnesium (Mg) . . . . 5 ppm

TESTS

ASSAY. (By complexometric titration). Weigh accurately 4 g and transfer to a 250-mL volumetric flask. Dissolve in 25 mL of 1 M sodium hydroxide, dilute to volume with water, and mix thoroughly. Fill a 50-mL buret with this sample solution. Weigh accurately 0.2 g of calcium carbonate, chelometric standard, dried in an oven at 300°C for 3 h and cooled in a charged desiccator for 2 h, and transfer to a 400-mL beaker. Cover the beaker with a watch glass. Add 2 mL of dilute hydrochloric acid from a pipet placed between the lip of the beaker and the watch glass. Swirl the beaker to aid dissolution. With water, wash down the inner wall of the beaker, the outer surface of the pipet, and the watch glass. Dilute to about 100 mL with water. While stirring, add from the buret about 30 mL of the sample solution. Add 15 mL of 1 M sodium hydroxide and 300 mg of hydroxy naphthol blue indicator mixture. Continue the titration immediately with the sample solution to a blue color. Let W_c be the weight, in grams, of calcium carbonate; let W_s be the weight, in grams, of the sample corresponding to the aliquot of the sample solution taken; and let V be the volume, in milliliters, of the sample solution required in the titration.

$$\% \ C_{10}H_{16}N_2O_8 = (W_c \times 73{,}000)/(V \times W_s)$$

INSOLUBLE IN DILUTE AMMONIUM HYDROXIDE.   To 20 g add 190 mL of water and 0.10 mL of methyl red indicator solution. Slowly add ammonium hydroxide, keeping the solution acidic, until all the sample is dissolved. Warm if necessary. Heat to boiling and digest in a covered beaker on the steam bath for 1 h. Filter through a tared filtering crucible, wash thoroughly, and dry at 105°C.

RESIDUE AFTER IGNITION.   (Page 15). Ignite 3.0 g. Moisten the char with 1 mL of sulfuric acid.

NITRILOTRIACETIC ACID.   For a stock solution transfer 10.0 g of the sample to a 100-mL volumetric flask. Dissolve in 87 mL of potassium hydroxide solution (100 g of KOH per liter), dilute with water to volume, and mix.

> *Sample Solution A.*   Dilute 10 mL of the stock solution with water to 100 mL in a volumetric flask.

> *Control Solution B.*   Add 1.0 mL (10 mg) of the nitrilotriacetic acid standard to 10 mL of the stock solution and dilute with water to 100 mL in a volumetric flask.

Transfer 20 mL of sample solution A to a 150-mL beaker and 20 mL of control solution B to a second 150-mL beaker. Add to each beaker 1 mL of potassium hydroxide solution (100 g of KOH per L) and 2 mL of ammonium nitrate solution (100 g of NH_4NO_3 per L). Add approximately 0.05 g of Eriochrome Black T indicator and titrate with an aqueous cadmium nitrate solution [30 g of $Cd(NO_3)_2 \cdot 4H_2O$ per L] to a red end point.

Sample Solution C.   Transfer 20 mL of sample solution A to a 100-mL volumetric flask.

Control Solution D.   Transfer 20 mL of control solution B to a 100-mL volumetric flask.

Add to each flask a volume of the cadmium nitrate solution equal to the volume determined by the titration plus 0.05 mL in excess. Add 1.5 mL of potassium hydroxide solution (100 g of KOH per L), 10 mL of ammonium nitrate solution (100 g of NH_4NO_3 per L), and 0.5 mL of a 0.1% solution of methyl red in alcohol. Dilute each solution with water to 100 mL and mix.

Transfer a portion of control solution D to an H-type polarographic cell equipped with a saturated calomel electrode and deaerate with nitrogen for 10 min. Record the polarogram from -0.6 to -1.2 V versus SCE at a sensitivity of 0.006, $\mu A/mm$. Repeat with sample solution C. The diffusion current in sample solution C should not exceed 0.1 times the difference in diffusion current between control solution D and sample solution C.

CALCIUM AND MAGNESIUM.   Determine the calcium and magnesium by the flame atomic absorption spectrophotometric method described on page 35.

Sample Stock Solution.   Dissolve 5 g of sample in 50 mL of water and 5 mL of ammonium hydroxide in a 150-mL beaker, and heat if necessary. Transfer to a 100-mL volumetric flask and dilute to the mark with water (0.05 g/mL).

Analyze the solutions by means of a suitable atomic absorption spectrophotometer, using the conditions outlined in the following table. Calculate the metal content of the sample by the method of standard additions.

Element	Wavelength (nm)	Sample Wt (g)	Standard Added (mg)	Flame Type*	Background Correction
Ca	422.7	1.0	0.01; 0.02	N/A	No
Mg	285.2	1.0	0.005; 0.01	A/A	Yes

*N/A is nitrous oxide/acetylene; A/A is air/acetylene.

Sample Solution E for the Determination of Heavy Metals and Iron.   Char 3.0 g thoroughly and heat in an oven at 500°C until most of the carbon is volatilized. Cool, add 0.15 mL of nitric acid, and heat at 500°C until all of the carbon is volatilized.

Dissolve the residue in 2 mL of dilute hydrochloric acid (1 + 1), digest in a covered dish on the steam bath for 10 min, remove the cover, and evaporate to dryness. Dissolve in 1 mL of 1 N acetic acid and 20 mL of hot water, digest for 5 min, cool, and dilute with water to 30 mL (1 mL = 0.1 g).

HEAVY METALS. (Page 26, Method 1). Dilute 20 mL of sample solution E (2-g sample) with water to 25 mL.

IRON. (Page 28, Method 1). Use 2.0 mL of sample solution E (0.2-g sample).

(Ethylenedinitrilo)tetraacetic Acid Disodium Salt Dihydrate

N,N,N',N'-Ethylenediaminetetraacetic Acid, Disodium Salt, Dihydrate

N,N'-1,2-Ethanediylbis[N-carboxymethyl]glycine, Disodium Salt Dihydrate

EDTA (Dihydrate)

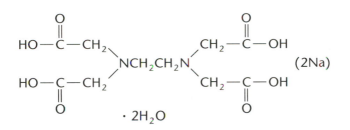

$C_{10}H_{14}N_2O_8Na_2 \cdot 2H_2O$ Formula Wt 372.24

CAS Number 6381−92−6

REQUIREMENTS

Assay . . . . 99.0−101.0%
$C_{10}H_{14}N_2O_8Na_2 \cdot 2H_2O$
pH of a 5% solution at 25°C . . . . 4.0−6.0

MAXIMUM ALLOWABLE

Insoluble matter 0.005%
Nitrilotriacetic acid [(HOCOCH$_2$)$_2$N]. 0.1%
Heavy metals (as Pb) 0.005%
Iron (Fe) 0.01%

TESTS

ASSAY. (By complexometric titration). Weigh accurately 5 g and transfer to a 250-mL volumetric flask. Dissolve in water, dilute to volume with water, and mix thoroughly. Fill a 50-mL buret with this sample solution. Weigh accurately 0.2 g of calcium carbonate, chelometric standard, dried in an oven at 300°C for 3 hours and cooled in a charged desiccator for 2 h, and transfer to a 400-mL beaker. Cover the beaker with a watch glass. Add 2 mL of dilute hydrochloric acid from a pipet placed between the lip of the beaker and the watch glass. Swirl the beaker to aid dissolution. With water, wash down the inner wall of the beaker, the outer surface of the pipet, and the watch glass. Dilute to about 100 mL with water. While stirring, add from the buret about 30 mL of the sample solution. Add 15 mL of 1 M sodium hydroxide and 300 mg of hydroxy naphthol blue indicator mixture. Continue the titration immediately with the sample solution to a blue color. Let W_c be the weight, in grams, of calcium carbonate; let W_s be the weight, in grams, of the sample corresponding to the aliquot of the sample solution taken; and let V be the volume, in milliliters, of the sample solution required in the titration.

$$\% \; C_{10}H_{14}N_2O_8Na_2 \cdot 2H_2O = (W_c \times 92{,}980)/(V \times W_s)$$

PH OF A 5% SOLUTION. (Page 43). The pH should be 4.0–6.0 at 25°C.

INSOLUBLE MATTER. (Page 14). Use 20 g dissolved in 200 mL of hot water.

NITRILOTRIACETIC ACID. For a stock solution transfer 10 g of the sample to a 100-mL volumetric flask. Dissolve in 40 mL of potassium hydroxide solution (100 g of KOH per L), dilute with water to volume, and mix.

>   *Sample Solution A.* Dilute 10 mL of the stock solution with water to 100 mL in a volumetric flask.

Control Solution B.   Add 1.0 mL (10 mg) of nitrilotriacetic acid standard to 10 mL of the stock solution and dilute with water to 100 mL in a volumetric flask.

Transfer 20 mL of sample solution A to a 150-mL beaker and 20 mL of control solution D to a second 150-mL beaker. Add to each beaker 1 mL of potassium hydroxide solution (100 g of KOH per L) and 2 mL of ammonium nitrate solution (100 g of NH_4NO_3 per L). Add approximately 0.05 g of Eriochrome Black T indicator and titrate with an aqueous cadmium nitrate solution [30 g of $Cd(NO_3)_2 \cdot 4H_2O$ per liter] to a red end point.

Sample Solution C.   Transfer 20 mL of sample solution A to a 100-mL volumetric flask.

Control Solution D.   Transfer 20 mL of control solution B to a 100-mL volumetric flask.

Add to each flask a volume of the cadmium nitrate solution equal to the volume determined by the titration plus 0.05 mL in excess. Add 1.5 mL of potassium hydroxide solution (100 g of KOH per L), 10 mL of ammonium nitrate solution (100 g of NH_4NO_3 per L), and 0.5 mL of a 0.1% solution of methyl red in alcohol. Dilute each solution with water to 100 mL and mix.

Transfer a portion of control solution D to an H-type polarographic cell equipped with a saturated calomel electrode and deaerate with nitrogen for 10 min. Record the polarogram from -0.6 to -1.2 V versus SCE at a sensitivity of 0.006 $\mu A/mm$. Repeat with sample solution C. The diffusion current in sample solution C should not exceed 0.1 times the difference in diffusion current between control solution D and sample solution C.

Sample Solution E for the Determination of Heavy Metals and Iron.   To 1.0 g add 1 mL of sulfuric acid, heat cautiously until the sample is charred, and ignite in an oven at 500°C until most of the carbon is volatilized. Cool, add 1 mL of nitric acid, heat until the acid is evaporated, and ignite again at 500°C until all of the carbon is volatilized. Cool, add 4 mL of dilute hydrochloric acid (1 + 1), digest on the steam bath for 10 min, and evaporate to dryness. Add 10 mL of 10% ammonium acetate solution and digest on the steam bath for 30 min. Filter, wash thoroughly, and dilute with water to 50 mL (1 mL = 0.02 g).

HEAVY METALS.   (Page 26, Method 1). Dilute 20 mL of sample solution E (0.4-g sample) with water to 25 mL.

IRON. (Page 28, Method 1). Use 5.0 mL of sample solution A (0.1-g sample).

Ferric Ammonium Sulfate Dodecahydrate
Ammonium Iron(III) Sulfate Dodecahydrate
Ammonium Ferric Sulfate Dodecahydrate

$NH_4Fe(SO_4)_2 \cdot 12H_2O$ Formula Wt 482.20

CAS Number 10045–89–3

REQUIREMENTS

Assay 98.5–102.0%
$NH_4Fe(SO_4)_2 \cdot 12(H_2O)$
Appearance . . . . Pale violet crystals

MAXIMUM ALLOWABLE

Insoluble matter 0.01%
Chloride (Cl) 0.001%
Nitrate (NO_3) 0.01%
Copper (Cu) 0.003%
Ferrous iron (Fe^{++}) . . . . Passes test (limit
about 0.001%)

Substances not precipitated by
ammonium hydroxide 0.05%
Zinc (Zn) 0.003%

TESTS

ASSAY. (By iodometric titration of oxidative capacity of Fe^{III}). Weigh accurately 1.8 g and transfer to an iodine flask with 50 mL of water. Add 3 mL of hydrochloric acid and 3 g of potassium iodide. Stopper, swirl, and allow to stand in the dark for 30 min. Dilute with 100 mL of water and titrate the liberated iodine with 0.1 N sodium thiosulfate, adding 3 mL of starch indicator solution near the end of the titration. Correct for a blank. One milliliter of 0.1 N sodium thiosulfate corresponds to 0.04822 g of $NH_4Fe(SO_4)_2 \cdot 12H_2O$.

INSOLUBLE MATTER.   (Page 14). Use 10 g dissolved in 100 mL of dilute hydrochloric acid (1 + 99).

CHLORIDE.   Dissolve 4.0 g in 40 mL of dilute nitric acid (1 + 9), filter if necessary through a chloride-free filter, and dilute with water to 60 mL. For the sample use 30 mL of the solution. For the control add 0.01 mg of chloride ion (Cl) and 1 mL of nitric acid to 15 mL of the solution and dilute with water to 30 mL. To each solution add 1 mL of silver nitrate reagent solution. Any turbidity in the solution of the sample should not exceed that in the control.

NITRATE.   Dissolve 2.0 g in 1 mL of water by swirling in a test tube. Add 10 mL of dilute ammonium hydroxide (1 + 1), mix well, and filter through a dry filter paper. Do not wash the precipitate.

> *Sample Solution A.*   Add 1.0 mL of the filtrate to 2 mL of water, dilute to 50 mL with brucine sulfate reagent solution, and mix.

> *Control Solution B.*   Add 1.0 mL of the filtrate to 2 mL of the standard nitrate solution containing 0.01 mg of nitrate ion (NO_3) per mL, dilute to 50 mL with brucine sulfate reagent solution, and mix.

Continue with the procedure described on page 29, starting with the preparation of blank solution C.

COPPER AND ZINC.   Determine the copper and zinc by the flame atomic absorption spectrophotometric method described on page 35.

> *Sample Stock Solution.*   Dissolve 10.0 g of sample in 90 mL of nitric acid (1 + 19) in a 100-mL volumetric flask, and dilute to the mark with water (0.1 g/mL).

Analyze the solutions by means of a suitable atomic absorption spectrophotometer, using the conditions outlined in the following table. Calculate the metal content of the sample by the method of standard additions.

Element	Wavelength (nm)	Sample Wt (g)	Standard Added (mg)	Flame Type*	Background Correction
Cu	324.7	1.0	0.03; 0.06	A/A	Yes
Zn	213.9	0.2	0.005; 0.01	A/A	Yes

*A/A is air/acetylene.

FERROUS IRON.   Dissolve 1.0 g in a mixture of 20 mL of water and 1 mL of hydrochloric acid and add 0.05 mL of a freshly prepared 5% solution of potassium ferricyanide. No blue or green color should be produced in 1 min.

SUBSTANCES NOT PRECIPITATED BY AMMONIUM HYDROXIDE. Dissolve 2.0 g in 50 mL of water. Heat to boiling and pour slowly with constant stirring into 20 mL of dilute ammonium hydroxide (1 + 4). Filter while hot and wash with hot water until the filtrate measures 100 mL. Evaporate the filtrate to dryness in a tared dish and ignite at 800 ± 25°C for 15 min.

Ferric Chloride Hexahydrate
Iron(III) Chloride Hexahydrate

$FeCl_3 \cdot 6H_2O$ Formula Wt 270.30

CAS Number 10025–77–1

REQUIREMENTS

Assay	97.0–102.0%
	$FeCl_3 \cdot 6H_2O$

MAXIMUM ALLOWABLE

Insoluble matter	0.01%
Nitrate (NO_3)	0.01%
Phosphorus compounds (as PO_4)	0.01%
Sulfate (SO_4)	0.01%
Arsenic (As)	0.002%
Copper (Cu)	0.003%
Ferrous iron (Fe^{++})	Passes test (limit about 0.002%)
Substances not precipitated by ammonium hydroxide (as sulfates)	0.1%
Zinc (Zn)	0.003%

TESTS

ASSAY. (By iodometric titration of oxidative capacity of Fe^{III}). Weigh accurately 1.0 g and transfer to an iodine flask with 50 mL of water. Add 3 mL of hydrochloric acid and 3 g of potassium iodide. Stopper, swirl, and allow to stand in the dark for 30 min. Dilute with 100 mL of water and titrate the liberated iodine with 0.1 N sodium thiosulfate, adding 3 mL of starch indicator solution near the end of the titration. Correct for a blank. One milliliter of 0.1 N sodium thiosulfate corresponds to 0.02703 g of $FeCl_3 \cdot 6H_2O$.

INSOLUBLE MATTER. (Page 14). Use 10 g dissolved in 50 mL of dilute hydrochloric acid (1 + 49).

> **Sample Solution A for the Determination of Nitrate, Sulfate, and Substances Not Precipitated by Ammonium Hydroxide.** Dissolve 10 g in 75 mL of water, heat to boiling, and pour slowly with constant stirring into 200 mL of dilute ammonium hydroxide (1 + 3). Filter through a folded filter while still hot, wash with hot water until the volume of the filtrate plus washings measures 300 mL, and thoroughly mix the solution (1 mL = 0.033 g).

NITRATE.

> **Sample Solution A-1.** Add 42 mL of brucine sulfate reagent solution and 2 mL of water to 6 mL of sample solution A (0.2-g sample).

> **Control Solution B.** Add 42 mL of brucine sulfate reagent solution and 2 mL of the standard nitrate solution containing 0.01 mg of nitrate ion (NO_3) per mL to 6 mL of sample solution A.

> **Blank Solution C.** Add 42 mL of brucine sulfate reagent solution to 8 mL of water.

Continue with the procedure described on page 29.

PHOSPHORUS COMPOUNDS. Dissolve 5.0 g in 40 mL of dilute nitric acid (1 + 1) and evaporate on the steam bath to a sirupy residue. Dissolve the residue in about 80 mL of water, dilute with water to 100 mL, and dilute 20 mL of this solution with water to 100 mL. For the test dilute 10 mL (0.1-g sample) with water to 70 mL. Prepare a standard containing 0.01 mg of phosphate ion (PO_4) in 70 mL of water. To each solution add 5 mL of ammonium molybdate solution (5 g in 50 mL) and adjust the pH to 1.8 (using a pH meter) by adding dilute hydrochloric acid (1 + 1) or dilute ammonium hydroxide (1 + 1). Cautiously heat the solutions to boiling but do not boil, and cool to room temperature. If a precipitate forms, it will dissolve when the solution is acidified in the next step. To each solution add 10 mL of hydrochloric acid and dilute each with water to 100 mL. Transfer the solutions to separatory funnels, add 35 mL of ether to each, shake vigorously, and allow to separate. Draw off and discard the aqueous phases. Wash the ether phases twice with 10-mL portions of dilute hydrochloric acid (1 + 9), discarding the washings each time. To the washed ether phases add 10 mL of dilute hydrochloric acid (1 + 9) to which has just been added 0.2 mL of a freshly prepared 2% solution of stannous chloride in hydrochloric acid. Shake the solutions and allow the phases to separate. Any blue color in the ether phase from the solution of the sample should not exceed that in the ether phase from the standard.

SULFATE. (Page 32, Procedure A, Method 1). Evaporate 15 mL of sample solution A to 10 mL. Allow 30 min for the turbidity to form.

ARSENIC. (Page 23). Dissolve 1.5 g in water and dilute with water to 100 mL. To 10 mL of the solution add 0.4 g of hydrazine hydrochloride, and dilute with water to 35 mL. For the standard use 0.003 mg of arsenic (As).

COPPER AND ZINC. Determine the copper and zinc by the flame atomic absorption spectrophotometric method described on page 35.

> *Sample Stock Solution.* Dissolve 10.0 g of sample with water in a 100-mL volumetric flask, and dilute to the mark with water (0.1 g/mL).

Analyze the solutions by means of a suitable atomic absorption spectrophotometer, using the conditions outlined in the following table. Calculate the metal content of the sample by the method of standard additions.

Element	Wavelength (nm)	Sample Wt (g)	Standard Added (mg)	Flame Type*	Background Correction
Cu	324.7	1.0	0.05; 0.10	A/A	Yes
Zn	213.9	0.2	0.005; 0.01	A/A	Yes

*A/A is air/acetylene.

FERROUS IRON. Dissolve 0.5 g in 20 mL of dilute hydrochloric acid (1 + 19) and add 0.05 mL of a freshly prepared 5% solution of potassium ferricyanide. No blue or green color should be produced in 1 min.

SUBSTANCES NOT PRECIPITATED BY AMMONIUM HYDROXIDE.
To 30 mL of sample solution A (1-g sample) add 0.50 mL of sulfuric acid, evaporate in a tared dish or crucible, ignite cautiously until ammonium salts are volatilized, and finally ignite at $800 \pm 25°C$ for 15 min.

Ferric Nitrate Nonahydrate
Iron(III) Nitrate Nonahydrate

$Fe(NO_3)_3 \cdot 9H_2O$ Formula Wt 404.00

CAS Number 10421−48−4

REQUIREMENTS

Assay . . . . 98.0−101.0%
$Fe(NO_3)_3 \cdot 9H_2O$

MAXIMUM ALLOWABLE

Insoluble matter 0.005%
Chloride (Cl) . . . . 5 ppm
Sulfate (SO_4) 0.01%
Substances not precipitated by
  ammonium hydroxide (as sulfates) 0.1%

TESTS

ASSAY. (By iodometric titration of oxidative capacity of Fe^{III}). Weigh accurately 1.6 g and transfer to an iodine flask with 50 mL of water. With stirring, cautiously add 5 mL of sulfuric acid and evaporate to dense fumes of sulfur trioxide. Cool, add 5 mL of water, and again evaporate to dense fumes of sulfur trioxide. Cool, add 100 mL of water, and again cool. Add 3 mL of hydrochloric acid and 3 g of potassium iodide. Stopper, swirl, and allow to stand in the dark for 30 min. Titrate the liberated iodine with 0.1 N sodium thiosulfate, adding 3 mL of starch indicator solution near the end of the titration. Correct for a blank. One milliliter of 0.1 N sodium thiosulfate corresponds to 0.04040 g of $Fe(NO_3)_3 \cdot 9H_2O$.

INSOLUBLE MATTER.   (Page 14). Use 20 g dissolved in 200 mL of dilute nitric acid (1 + 99).

CHLORIDE.   Dissolve 2.0 g in 20 mL of water, filter if necessary through a chloride-free filter, and add 1 mL of nitric acid plus 1 mL of phosphoric acid and 1 mL of silver nitrate reagent solution. Any turbidity should not exceed that produced by 0.01 mg of chloride ion (Cl) in an equal volume of solution containing the quantities of reagents used in the test.

> ***Sample Solution A.***   Dissolve 5.0 g in 70 mL of water, heat to boiling, and pour slowly with constant stirring into 100 mL of dilute ammonium hydroxide (1 + 6). Filter through a folded filter while still hot and wash with hot water until the filtrate (sample solution A) measures 250 mL. Mix thoroughly (1 mL = 0.02 g).

SULFATE.   To 25 mL (0.5-g sample) of sample solution A add 5 mL of hydrochloric acid and 10 mL of nitric acid. For the standard add 0.05 mg of sulfate ion (SO_4) and 1.5 mL of ammonium hydroxide to 25 mL of water, and add 5 mL of hydrochloric acid and 10 mL of nitric acid. Digest each solution in a covered beaker on a hot plate at low setting until reaction ceases, uncover, and evaporate to dryness. Dissolve the residues in 4 mL of water plus 1 mL of dilute hydrochloric acid (1 + 19), filter if necessary through a small filter, wash with two 2-mL portions of water, and dilute with water to 10 mL. To each solution add 1 mL of barium chloride reagent solution. Any turbidity in the solution of the sample should not exceed that in the standard. Compare 10 min after adding the barium chloride to the sample and standard solutions.

SUBSTANCES NOT PRECIPITATED BY AMMONIUM HYDROXIDE.
Evaporate 50 mL of sample solution A to dryness in a tared dish and add 0.50 mL of sulfuric acid. Heat carefully to volatilize the ammonium salts and excess sulfuric acid, and finally ignite at 800 ± 25°C for 15 min.

Ferrous Ammonium Sulfate Hexahydrate
Ammonium Ferrous Sulfate Hexahydrate
Ammonium Iron(II) Sulfate Hexahydrate

$Fe(NH_4)_2(SO_4)_2 \cdot 6H_2O$ Formula Wt 392.14

CAS Number 7783–85–9

> *NOTE.* Due to inherent oxidation, ferric ion (Fe^{3+}) content may increase on storage.

REQUIREMENTS

Assay 98.5–101.5%
$Fe(NH_4)_2(SO_4)_2 \cdot 6H_2O$

MAXIMUM ALLOWABLE

Insoluble matter 0.01%
Phosphate (PO_4) 0.003%
Copper (Cu) 0.003%
Ferric ion (Fe^{3+}) 0.01%
Manganese (Mn) 0.01%
Substances not precipitated by
   ammonium hydroxide 0.05%
Zinc (Zn). 0.003%

TESTS

ASSAY.   (By titration of reductive capacity of Fe^{II}). Weigh accurately 1.6 g and dissolve in a mixture of 100 mL of water and 3 mL of sulfuric acid that has previously been freed from oxygen by bubbling with an inert gas. Titrate with 0.1 N potassium permanganate to a permanent faint pink end point. One milliliter of 0.1 N permanganate corresponds to 0.03921 g of $Fe(NH_4)_2(SO_4)_2 \cdot 6H_2O$.

INSOLUBLE MATTER.   (Page 14). Use 10 g dissolved in 100 mL of freshly boiled dilute sulfuric acid (1 + 99).

PHOSPHATE.   Dissolve 1.0 g in 10 mL of water and add 2 mL of nitric acid. Boil to oxidize the iron and expel the excess gases. Dilute with water to 90 mL. Dilute 30 mL with water to 80 mL, add 5 mL of ammonium molybdate solution (5 g in 50 mL), and adjust the pH to

1.8 (using a pH meter) with dilute ammonium hydroxide (1 + 1). Prepare a standard containing 0.01 mg of phosphate ion (PO_4), the residue from evaporation of the amount of ammonium hydroxide used in adjusting the pH of the sample, and 5 mL of the ammonium molybdate solution in 85 mL. Adjust the pH to 1.8 (using a pH meter) with dilute hydrochloric acid (1 + 9). Cautiously heat the solutions to boiling, but do not boil, and cool to room temperature. If a precipitate forms, it will dissolve when acidified in the next operation. Add 10 mL of hydrochloric acid and dilute with water to 100 mL. Transfer the solutions to separatory funnels, add 35 mL of ether to each, shake vigorously, and allow to separate. Draw off and discard the aqueous phase. Wash the ether phase twice with 10-mL portions of dilute hydrochloric acid (1 + 9), discarding the washings each time. Add 10 mL of dilute hydrochloric acid (1 + 9) to which has just been added 0.2 mL of a freshly prepared 2% solution of stannous chloride in hydrochloric acid, shake, and allow to separate. Any blue color in the ether extract from the sample should not exceed that in the ether extract from the standard.

COPPER, MANGANESE, AND ZINC. Determine the copper, manganese, and zinc by the flame atomic absorption spectrophotometric method described on page 35.

> *Sample Stock Solution.* Dissolve 10.0 g of sample with water in a 100-mL volumetric flask and dilute to the mark with water (0.1 g/mL).

Analyze the solutions by means of a suitable atomic absorption spectrophotometer, using the conditions outlined in the following table. Calculate the metal content of the sample by the method of standard additions.

Element	Wavelength (nm)	Sample Wt (g)	Standard Added (mg)	Flame Type*	Background Correction
Cu	324.7	1.0	0.03; 0.06	A/A	Yes
Mn	279.5	0.2	0.02; 0.04	A/A	Yes
Zn	213.9	0.2	0.005; 0.01	A/A	Yes

*A/A is air/acetylene.

FERRIC ION. Place 0.20 g of the sample and 0.5 g of sodium bicarbonate in a dry 125-mL glass-stoppered conical flask. For the control place 0.10 g of the sample and 0.5 g of sodium bicarbonate in a similar flask. To each flask add 94 mL of a freshly boiled and cooled solution of dilute sulfuric acid (1 + 25) and loosely stopper the flasks until effervescence stops. To the control add 0.01 mg of ferric ion (Fe^{3+}). Add 6 mL of ammonium thiocyanate reagent solution to each flask. Any red color in the solution of the sample should not exceed that in the control.

SUBSTANCES NOT PRECIPITATED BY AMMONIUM HYDROXIDE.

Dissolve 2.0 g in 25 mL of dilute nitric acid (1 + 4) and boil gently to oxidize the iron. Pour the solution slowly with constant stirring into 50 mL of dilute ammonium hydroxide (1 + 4). Filter, wash moderately with hot water, and evaporate the solution to dryness in a tared dish. Heat gently to volatilize the excess ammonium salts and finally ignite at $800 \pm 25°C$ for 15 min.

Ferrous Sulfate Heptahydrate

Iron(II) Sulfate Heptahydrate

$FeSO_4 \cdot 7H_2O$
Formula Wt 278.01

CAS NUMBER 7782−63−0

NOTE. Due to inherent oxidation, ferric ion (Fe^{3+}) content may increase on storage.

REQUIREMENTS

Assay . . . . $\geq$ 99.0%
$FeSO_4 \cdot 7H_2O$

MAXIMUM ALLOWABLE

Insoluble matter 0.01%
Chloride (Cl) 0.001%
Phosphate (PO_4) 0.001%
Copper (Cu) 0.005%
Ferric ion (Fe^{3+}) 0.1%
Manganese (Mn) 0.05%
Substances not precipitated by
    ammonium hydroxide 0.05%
Zinc (Zn) 0.005%

TESTS

ASSAY. (By titration of reductive capacity of Fe^{II}). Weigh accurately 1 g, and dissolve in a mixture of 100 mL of water and 3 mL of sulfuric acid that has previously been freed from oxygen by bubbling with an inert gas. Titrate with 0.1 N potassium permanganate to a permanent faint pink end point. One milliliter of 0.1 N potassium permanganate corresponds to 0.02780 g of $FeSO_4 \cdot 7H_2O$.

INSOLUBLE MATTER. (Page 14). Use 10 g dissolved in 100 mL of freshly boiled dilute sulfuric acid (1 + 99).

CHLORIDE. Dissolve 1.0 g in 10 mL of water and add slowly 2 mL of nitric acid. After the evolution of nitrogen oxides has ceased, dilute with water to 15 mL, filter if necessary through a small chloride-free filter, and add 1 mL of silver nitrate reagent solution. Any turbidity should not exceed that produced by 0.01 mg of chloride ion (Cl) in an equal volume of solution containing the quantities of reagents used in the test. The comparison is best made by the general method for chloride in colored solutions, page 25.

PHOSPHATE. (Page 30, Method 2). Dissolve 1.0 g in 10 mL of dilute nitric acid (1 + 5) and boil to expel the excess gases. Dilute the solution with water to 80 mL, add 5 mL of ammonium molybdate solution (5 g in 50 mL), and adjust the pH to 1.8 (using a pH meter) with dilute ammonium hydroxide (1 + 1). Cautiously heat the solution to boiling, but do not boil, and cool to room temperature. If a precipitate forms, it will dissolve in the next operation. Add 10 mL of hydrochloric acid, dilute with water to 100 mL, and continue as described. Carry along a standard containing 0.01 mg of phosphate ion (PO_4), the residue from evaporation of the amount of ammonium hydroxide used in adjusting the pH of the sample, and 5 mL of the ammonium molybdate solution in 85 mL.

COPPER, MANGANESE, AND ZINC. Determine the copper, manganese, and zinc by the flame atomic absorption spectrophotometric method described on page 35.

> *Sample Stock Solution.* Dissolve 10.0 g of sample with water in a 100-mL volumetric flask and dilute to the mark with water (0.1 g/mL).

Analyze the solutions by means of a suitable atomic absorption spectrophotometer, using the conditions outlined in the following

table. Calculate the metal content of the sample by the method of standard additions.

Element	Wavelength (nm)	Sample Wt (g)	Standard Added (mg)	Flame Type*	Background Correction
Cu	324.7	1.0	0.05; 0.10	A/A	Yes
Mn	279.5	0.1	0.025; 0.05	A/A	Yes
Zn	213.9	0.1	0.005; 0.01	A/A	Yes

*A/A is air/acetylene.

FERRIC ION.   Place 0.20 g of the sample and 0.5 g of sodium bicarbonate in a dry 125-mL glass-stoppered conical flask. For the control place 0.10 g of the sample and 0.5 g of sodium bicarbonate in a similar flask. To each flask add 94 mL of a freshly boiled and cooled solution of dilute sulfuric acid (1 + 25) and loosely stopper the flasks until effervescence stops. To the control add 0.1 mg of ferric ion (Fe^{3+}). Add 6 mL of ammonium thiocyanate reagent solution to each flask. Any red color in the solution of the sample should not exceed that in the control.

SUBSTANCES NOT PRECIPITATED BY AMMONIUM HYDROXIDE.
Dissolve 2.0 g in 25 mL of dilute nitric acid (1 + 4) and boil gently to oxidize the iron. Pour the solution slowly with constant stirring into 50 mL of dilute ammonium hydroxide (1 + 4). Filter, wash moderately with hot water, and evaporate the solution to dryness in a tared dish. Heat gently to volatilize the excess ammonium salts and finally ignite at 800 ± 25°C for 15 min.

Formaldehyde Solution
(with Stabilizer)

HCHO   Formula Wt 30.03

CAS Number 50–00–0

NOTE. This reagent contains 10–15% of methanol as a stabilizer.

REQUIREMENTS

Assay . . . . 36.5–38.0% HCHO

MAXIMUM ALLOWABLE

Color (APHA). 10
Residue after ignition 0.005%
Titrable acid . . . . 0.006 meq / g
Chloride (Cl) . . . . 5 ppm
Sulfate (SO$_4$) 0.002%
Heavy metals (as Pb) . . . . 5 ppm
Iron (Fe) . . . . 5 ppm

TESTS

> *NOTE.* The assay procedure specified herein is intended to determine both formaldehyde and its paraformaldehyde polymerization product that may form upon standing. A different procedure (*see*, for example, *Reagent Chemicals*, 5th edition) should be used if only the formaldehyde content is desired.

ASSAY. (By bisulfite addition and acid–base titrimetry). Transfer a glass-stoppered weighing vial containing 2 g of sample, accurately weighed, to 100 mL of sodium sulfite solution in a 500-mL conical flask. Avoid getting any of the sample on the wall of the flask. Add 0.15 mL of thymolphthalein indicator, and titrate to a colorless end point with 1 N sulfuric acid. Correct for a reagent blank and any significant quantity of acid (as HCOOH) in the sample. One milliliter of 1 N sulfuric acid corresponds to 0.03003 g of HCHO.

> *Sodium Sulfite Solution.* Dissolve 125 g of sodium sulfite, Na_2SO_3, in water and dilute with water to 1 L. Keep in a tightly stoppered bottle. For best results freshly prepared solution should be used.

COLOR (APHA). (Page 17).

RESIDUE AFTER IGNITION. (Page 15). Evaporate 20 g (20 mL) to dryness in a tared dish on the steam bath, add 0.05 mL of sulfuric acid to the residue, and ignite gently to volatilize the excess sulfuric acid. Finally, ignite at 800 ± 25°C for 15 min.

TITRABLE ACID. To 10 mL in a flask containing 20 mL of water, add 0.15 mL of bromthymol blue indicator solution, and titrate with 0.1 N sodium hydroxide. Not more than 0.60 mL of 0.1 N sodium hydroxide should be required.

CHLORIDE. (Page 25). Dilute 2.0 mL with 20 mL of water.

SULFATE. (Page 32, Procedure A, Method 1). Use 2.5 g (2.5 mL).

HEAVY METALS. (Page 26, Method 1). To 4.0 mL add about 10 mg of sodium carbonate, evaporate to dryness, and heat gently to volatilize any organic matter. Dissolve the residue in about 20 mL of water and dilute with water to 25 mL.

IRON. (Page 28, Method 2). Dilute 2.0 mL with water to 15 mL. Adjust the pH to 5 with ammonium hydroxide after addition of the 1,10-phenanthroline reagent solution.

Formamide

HCONH$_2$ Formula Wt 45.04

CAS Number 75–12–7

REQUIREMENTS

Assay . . . . $\geq$ 99.5% HCONH$_2$
Freezing point . . . . 2.0–3.0°C

MAXIMUM ALLOWABLE

Color (APHA). 10

TESTS

ASSAY. Analyze the sample by gas chromatography, using the general parameters cited on page 56. The following specific conditions are also required.

   Column: Type I, methyl silicone

Measure the area under all peaks and calculate the formamide content in area percent.

COLOR (APHA). (Page 17).

FREEZING POINT. Place 15 mL in a test tube (20 × 150 mm) in which is centered an accurate thermometer. The sample tube is centered by corks in an outer tube about 38 × 200 mm. Cool the whole apparatus, without stirring, in a bath of shaved or crushed ice with enough water to wet the outer tube. When the temperature is about 0°C, stir to start the freezing, and read the thermometer every 30 s. The temperature that remains constant for 1–2 min is the freezing point.

Formic Acid, 96%

HCOOH   Formula Wt 46.03

CAS Number 64 –18 –6

> *CAUTION.* Slow decomposition of this reagent may produce pressure in the bottle. Loosen cap occasionally to vent the gas.

REQUIREMENTS

Assay . . . . $\geq$ 96.0% HCOOH

MAXIMUM ALLOWABLE

Color (APHA). 15
Dilution test . . . . Passes test
Residue after evaporation 0.003%
Acetic acid (CH$_3$COOH) 0.4%
Ammonium (NH$_4$) 0.005%
Chloride (Cl) 0.001%
Sulfate (SO$_4$) 0.003%
Sulfite (SO$_3$) . . . . Passes test
Heavy metals (as Pb) 0.001%
Iron (Fe) 0.001%

TESTS

ASSAY.   (By acid–base titrimetry). Tare a small glass-stoppered conical flask containing 15 mL of water. Quickly introduce 1.0–1.5 mL of the sample and weigh. Dilute with water to about 50 mL and titrate with 1 N sodium hydroxide, using 0.15 mL of phenolphthalein indicator. One milliliter of 1 N sodium hydroxide corresponds to 0.04603 g of HCOOH.

COLOR (APHA).   (Page 17).

DILUTION TEST.   Dilute 1 volume of the acid with 3 volumes of water. No turbidity should be observed within 1 h.

RESIDUE AFTER EVAPORATION. (Page 14). Evaporate 50 g (42 mL) to dryness in a tared dish on the steam bath and dry the residue at 105°C for 30 min.

ACETIC ACID. Analyze the sample by gas chromatography as described on page 56. The following specific conditions are satisfactory.

> **Column:** 1.8 m × 6.4 mm, stainless steel, packed with 80/100-mesh Porapak Q
>
> **Column Temperature:** 150°C
>
> **Injection Port Temperature:** 150°C
>
> **Detector Temperature:** 175°C
>
> **Detector Current:** 200 mA
>
> **Carrier Gas:** Helium at 40 mL/min
>
> **Sample Size:** 5 μL
>
> **Detector:** Thermal conductivity
>
> **Approximate Retention Times (min):** Water, 1.2; formic acid, 3.5; acetic acid, 8.1

Measure the area under all peaks and calculate the acetic acid content in area percent.

AMMONIUM. (Page 48). Dilute 10.0 g (8.2 mL) of sample to 50 mL with water; use 5.0 mL in each test, neutralizing with 3.2 mL of ammonia-free 6 N sodium hydroxide. For the standard use 0.05 mg of ammonium (NH_4).

CHLORIDE. (Page 25). Use 1.0 g (0.83 mL).

SULFATE. (Page 32, Procedure A, Method 3). Use 1.7 g (1.4 mL).

SULFITE. Dilute 25 mL with 25 mL of water; add 0.10 mL of 0.1 N iodine solution. The mixture should retain a distinct yellow color.

HEAVY METALS. (Page 26, Method 1). To 2 g (1.6 mL) in a beaker add about 10 mg of sodium carbonate and evaporate to dryness. Dissolve the residue in about 20 mL of water and dilute with water to 25 mL.

IRON. (Page 28, Method 1). To 6 g (5.0 mL) in a beaker, add about 10 mg of sodium carbonate, and evaporate to dryness. Dissolve the residue in 6 mL of hydrochloric acid, dilute with water to 60 mL, and use 10 mL of this solution without further acidification.

Formic Acid, 88%

HCOOH   Formula Wt 46.03

CAS Number 64 –18 –6

REQUIREMENTS

Assay . . . . ≥ 88.0% HCOOH

MAXIMUM ALLOWABLE

Color (APHA) 15
Dilution test . . . . Passes test
Residue after evaporation 0.002%
Acetic acid (CH_3COOH) 0.4%
Ammonium (NH_4) 0.005%
Chloride (Cl) 0.001%
Sulfate (SO_4) 0.002%
Sulfite (SO_3) . . . . Passes test
Heavy metals (as Pb) . . . . 5 ppm
Iron (Fe) . . . . 5 ppm

TESTS

Assay and other tests are the same as for Formic Acid, 96%, page 348.

2-Furancarboxyaldehyde
2-Furaldehyde
Furfural

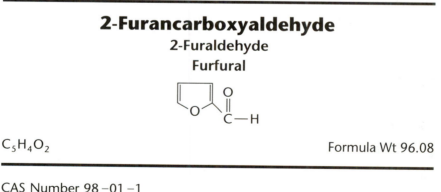

$C_5H_4O_2$ Formula Wt 96.08

CAS Number 98 –01 –1

REQUIREMENTS

Assay . . . . ≥ 98.0%
                                                            C_4H_3OCHO

MAXIMUM ALLOWABLE

Titrable acid . . . . 0.02 meq / g
Residue after evaporation 0.5%

TESTS

ASSAY.   Analyze the sample by gas chromatography, using the general parameters cited on page 56. The following specific conditions are also required.

Column: Type I, methyl silicone

Measure the area under all peaks and calculate the 2-furancarboxyaldehyde content in area percent.

TITRABLE ACID.   To 300 mL of water in an Erlenmeyer flask, add 0.15 mL of phenolphthalein indicator solution, and titrate with 0.1 N sodium hydroxide to a faint pink color that is stable for about 1 min. Add 10.0 g (9.0 mL) of sample to the flask, mix well, and titrate with 0.1 N sodium hydroxide to the original faint pink color. Not more than 2.0 mL of 0.1 N sodium hydroxide should be consumed.

RESIDUE AFTER EVAPORATION.   (Page 14). Evaporate 25 g (22 mL) to dryness in a tared dish on the steam bath, and dry the residue at 105°C for 30 min.

Gallic Acid

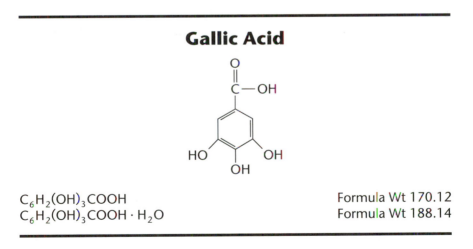

$C_6H_2(OH)_3COOH$ Formula Wt 170.12
$C_6H_2(OH)_3COOH \cdot H_2O$ Formula Wt 188.14

CAS Number 149–91–7 (anhydrous); 5995–86–8 (monohydrate)

NOTE. This reagent is available in both the monohydrate and the anhydrous form. The label should identify the degree of hydration.

REQUIREMENTS

Assay . . . . $\geq$ 98.0%
$$C_6H_2(OH)_3COOH \text{ or}$$
$$C_6H_2(OH)_3COOH \cdot H_2O$$

MAXIMUM ALLOWABLE

Insoluble matter 0.01%
Residue after ignition 0.05%
Sulfate (SO$_4$) 0.02%

TESTS

ASSAY. (By acid–base titrimetry). Weigh accurately about 0.7 g, transfer to a beaker, add 100 mL of water, and stir to dissolve. When dissolution is complete, titrate potentiometrically with 0.1 N sodium hydroxide. Correct for a blank determination. One milliliter of 0.1 N sodium hydroxide corresponds to 0.017012 g of $C_6H_2(OH)_3COOH$ or 0.018814 g of $C_6H_2(OH)_3COOH \cdot H_2O$.

INSOLUBLE MATTER. (Page 14). Use 10 g dissolved in 300 mL of water.

RESIDUE AFTER IGNITION. (Page 15). Ignite 4 g.

SULFATE. (Page 32, Procedure A, Method 1). Dissolve 2.5 g in 50 mL of hot water, cool in ice water while stirring, and filter. Dilute the filtrate with water to 50 mL and use 5 mL.

D-**Glucose, Anhydrous**

Dextrose, Anhydrous

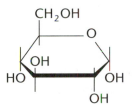

$CH_2OH(CHOH)_4CHO$ Formula Wt 180.16

CAS Number 50–99–7

REQUIREMENTS

Specific rotation $[\alpha]_D^{25°}$ . . . . +52.5° to +53.0°

MAXIMUM ALLOWABLE

Insoluble matter 0.005%
Loss on drying at 105°C. 0.2%
Residue after ignition 0.02%
Titrable acid . . . . 0.002 meq / g
Chloride (Cl) 0.01%
Sulfate and sulfite (as SO_4) 0.005%
Starch. . . . . Passes test
Arsenic (As). . . . . 0.4 ppm
Heavy metals (as Pb) . . . . 5 ppm
Iron (Fe) . . . . 5 ppm

TESTS

SPECIFIC ROTATION.   Weigh accurately about 10 g and dissolve in
90 mL of water in a 100-mL volumetric flask. Add 0.2 mL of ammonium hydroxide and dilute with water to volume at 25°C. Observe the
optical rotation in a polarimeter at 25°C using sodium light, and
calculate the specific rotation. It should not be less than +52.5° nor
more than +53.0°.

INSOLUBLE MATTER.   (Page 14). Use 20 g dissolved in 150 mL of
water.

LOSS ON DRYING.   Weigh accurately about 1 g and dry at 105°C for 6 h.

RESIDUE AFTER IGNITION.   (Page 15). Ignite 5.0 g until charred. Moisten the char with 2 mL of sulfuric acid.

TITRABLE ACID.   To 100 mL of carbon dioxide-free water add 0.15 mL of phenolphthalein indicator solution and 0.02 N sodium hydroxide until a pink color is produced. Dissolve 10.0 g of the sample in this solution and titrate with 0.01 N sodium hydroxide to the same end point. Not more than 2.5 mL should be required.

CHLORIDE.   (Page 25). Dissolve 0.50 g in 50 mL of water, and use 10 mL of this solution.

SULFATE AND SULFITE.   (Page 32, Procedure A, Method 1). After sample dissolution, add 1 mL of bromine water and boil.

STARCH.   Dissolve 1.0 g in 10 mL of water and add 0.05 mL of 0.1 N iodine solution. No blue color should appear.

ARSENIC.   (Page 23). Use 7.5 g. For the standard use 0.003 mg of arsenic (As).

HEAVY METALS.   (Page 26, Method 1). Dissolve 6.0 g in about 20 mL of water, and dilute with water to 30 mL. Use 25 mL to prepare the sample solution, and use the remaining 5.0 mL to prepare the control solution.

IRON.   (Page 28, Method 1). Use 2.0 g.

Glycerol
1,2,3-Propanetriol

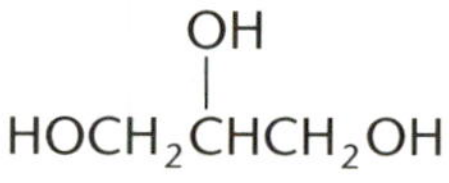

$$\underset{\displaystyle HOCH_2CHCH_2OH}{\overset{\displaystyle OH}{|}}$$

$CH_2OHCHOHCH_2OH$ Formula Wt 92.09

CAS Number 56–81–5

REQUIREMENTS

Assay (by volume) . . . . ≥ 99.5%

$$C_3H_5(OH)_3$$

MAXIMUM ALLOWABLE

Color (APHA). 10
Residue after ignition 0.005%
Neutrality . . . . Passes test
Chlorinated compounds (as Cl) 0.003%
Sulfate (SO₄) 0.001%
Acrolein, glucose, and ammonium compounds . . . . Passes test
Fatty acid esters (as butyric acid) 0.05%
Silver-reducing substances . . . . Passes test
Substances darkened by sulfuric acid . . . . Passes test
Heavy metals (as Pb) . . . . 2 ppm
Water (H₂O) 0.5%

TESTS

ASSAY.   Analyze the sample by gas chromatography, using the general parameters cited on page 56. The following specific conditions are also required.

   Column: Type II, mixed cyano, phenyl methyl silicone

Measure the area under all peaks and calulate the glycerol content in area percent. Correct for water content.

COLOR (APHA).   (Page 17).

RESIDUE AFTER IGNITION.   (Page 15). Heat 20 g in a tared open dish, ignite the vapors, and when the glycerol has been entirely consumed, ignite at 800 ± 25°C for 15 min.

NEUTRALITY.   A 10% aqueous solution of glycerol should not affect the color of either red or blue litmus paper in 1 min.

CHLORINATED COMPOUNDS.   Transfer 5.0 g of the glycerol to a dry 100-mL round-bottom flask fitted with a ground joint. Add 15 mL of morpholine, connect a matching standard taper reflux condenser, and reflux the mixture gently for 3 h. Rinse the condenser with 10 mL of water, collecting the washing in the flask. Cautiously acidify the solution with nitric acid, add 1 mL of silver nitrate reagent solution, and dilute with water to 50 mL. Any turbidity should not exceed that produced by 0.15 mg of chloride ion (Cl) in an equal volume of solution containing the quantities of reagents used in the test, omitting the refluxing.

SULFATE.   (Page 32, Procedure A, Method 1). Use 5.0 g.

ACROLEIN, GLUCOSE, AND AMMONIUM COMPOUNDS.   Heat a mixture of 5 mL of the glycerol and 5 mL of 10% potassium hydroxide solution at 60°C for 5 min. Neither a yellow color nor an odor of ammonia should develop.

FATTY ACID ESTERS.   To 40 g in a 250-mL conical flask add 50 mL of hot, freshly boiled water. Add 10.0 mL of 0.1 N sodium hydroxide, cover with a loosely fitting pear-shaped bulb, and digest on the steam bath for 45 min. Cool and titrate the excess alkali with 0.1 N hydrochloric acid, using 0.15 mL of bromthymol blue indicator solution and titrating to a bluish green end point. Run a blank with 50 mL of the same water and 10.0 mL of 0.1 N sodium hydroxide solution, heating for the same length of time and titrating to the same end point. The difference between the volumes of acid used in the titration of the blank and of the sample should be less than 2.3 mL.

SILVER-REDUCING SUBSTANCES.   Dilute 6 g (5 mL) with 2.5 mL of water, add 2.5 mL of ammonium hydroxide (10% NH_3), and heat to 60°C. Add 0.5 mL of silver nitrate reagent solution and allow to stand at 60°C for 5 min, protected from light. Any color should not exceed that produced in a standard containing 0.02 mg of lead ion (Pb) in 2.5 mL of water to which are added 2.5 mL of ammonium hydroxide (10% NH_3) and 5 mL of hydrogen sulfide water.

SUBSTANCES DARKENED BY SULFURIC ACID.   Vigorously shake 5 mL of glycerol with 5 mL of 94.5–95.5% sulfuric acid in a glass-stoppered 25-mL cylinder for 1 min and allow the liquid to stand for 1 h. The liquid should not be darker than a standard made up of 0.4 mL of cobaltous chloride color solution, 3.0 mL of ferric chloride color solution, and 6.6 mL of water.

Reagents

Ferric Chloride Color Solution.   Dissolve ferric chloride in a mixture of 25 volumes of hydrochloric acid and 975 volumes of water and adjust the strength so that the iron content corresponds to 45.0 mg of $FeCl_3 \cdot 6H_2O$ per mL.

Cobaltous Chloride Color Solution. Dissolve cobaltous chloride in a mixture of 25 volumes of hydrochloric acid and 975 volumes of water and adjust the strength so that the cobalt content corresponds to 59.5 mg of $CoCl_2 \cdot 6H_2O$ per mL.

HEAVY METALS. (Page 26, Method 1). Dilute 13 g (10 mL) with water to 36 mL. Use 32 mL to prepare the sample solution, and use the remaining 4.0 mL to prepare the control solution.

WATER. (Page 54, Method 1). Use 4.0 mL (5.0 g) of sample.

Glycine

Aminoacetic Acid

H_2NCH_2COOH Formula Wt 75.07

CAS Number 56–40–6

REQUIREMENTS

Assay . . . . $\geq$ 98.5%
H_2NCH_2COOH

MAXIMUM ALLOWABLE

Residue after ignition 0.1%
Heavy metals (as Pb) 0.002%
Chloride (Cl) 0.005%
Sulfate (SO_4) 0.005%
Ammonium (NH_4) 0.005%
Substances darkened by sulfuric acid . . . . Passes test
Hydrolyzable substances . . . . Passes test

TESTS

ASSAY. (Total alkalinity by nonaqueous titration). Weigh accurately 0.25 g and dissolve in 100 mL of glacial acetic acid. Add 0.5 mL of crystal violet indicator solution and titrate with 0.1 N perchloric acid in glacial acetic acid to a green end point. Correct for a blank. One milliliter of 0.1 N perchloric acid corresponds to 0.007507 g of H_2NCH_2COOH.

RESIDUE AFTER IGNITION.   (Page 15). Use 2.0 g.

HEAVY METALS.   (Page 26, Method 1). Use 1.0 g.

CHLORIDE.   (Page 25). Use 0.2 g.

SULFATE.   (Page 32, Procedure A, Method 1). Use 1.0 g.

AMMONIUM.   (Page 22). Use 0.2 g. For the standard use 0.01 mg of ammonium ion (NH_4) treated exactly as the sample.

SUBSTANCES DARKENED BY SULFURIC ACID.   Dissolve 0.5 g in 5 mL of sulfuric acid. Allow the solution to stand for 15 min. The color of the sample solution should not exceed that of an equal volume of the acid.

HYDROLYZABLE SUBSTANCES.   Boil 10 mL of a 1 in 10 solution of glycine for 1 min and set aside for 2 h. The solution, when compared to an equal volume of an untreated sample solution, appears as clear and as mobile.

Gold Chloride
Tetrachloroauric(III) Acid Trihydrate

$HAuCl_4 \cdot 3H_2O$ Formula Wt 393.83

CAS Number 16961−25−4

REQUIREMENTS

Assay (Au). . . . . ≥ 49.0%

MAXIMUM ALLOWABLE

Insoluble in ether. 0.1%
Alkalies and other metals (as sulfates). 0.2%

TESTS

ASSAY.   (Gold content by gravimetry). Evaporate the solution from the test for insoluble in ether to dryness or a sirupy residue. Dissolve the residue in 100 mL of dilute hydrochloric acid (1 + 19), disregarding any undissolved material. Add 10 mL of sulfurous acid and digest on the steam bath until the precipitate is well coagulated. Filter, wash with dilute hydrochloric acid (1 + 99), and ignite in a tared crucible. Retain the filtrate and washings. The weight of the residue should not be less than 49.0% of the weight of the sample.

INSOLUBLE IN ETHER. Weigh accurately about 1 g. Add 10 mL of ether, allow to stand for 10 min with occasional stirring, filter through a tared filtering crucible of fine porosity, wash with small portions of ether, and dry at 105°C. Retain the filtrate and washings.

ALKALIES AND OTHER METALS (AS SULFATES). Evaporate the filtrate and washings reserved from the assay, in a tared dish, to dryness. Ignite the residue cautiously, and finally ignite at $800 \pm 25°C$ for 15 min.

Hexamethylenetetramine

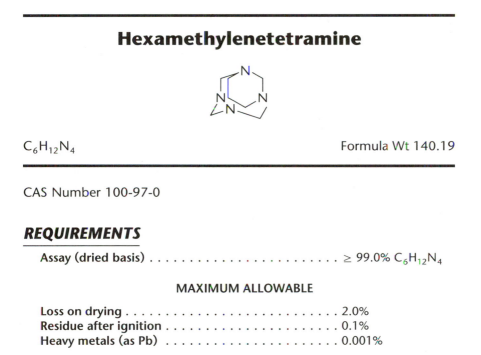

$C_6H_{12}N_4$ Formula Wt 140.19

CAS Number 100-97-0

REQUIREMENTS

Assay (dried basis) . . . . $\geq 99.0\%$ $C_6H_{12}N_4$

MAXIMUM ALLOWABLE

Loss on drying 2.0%
Residue after ignition 0.1%
Heavy metals (as Pb) 0.001%

TESTS

ASSAY. Transfer 1 g of previously dried sample, accurately weighed, to a 100-mL beaker. Add 40 mL of 1 N sulfuric acid and boil gently for at least 1 h to remove formaldehyde. Cool, add 20 mL of water, and titrate the excess acid with 1 N sodium hydroxide, using 0.15 mL of methyl red indicator solution. One milliliter of 1 N sulfuric acid corresponds to 0.03505 g of $C_6H_{12}N_4$.

LOSS ON DRYING. Accurately weigh about 1 g of sample and dry over phosphorus pentoxide for 4 h. Save the dried sample for the assay determination.

RESIDUE AFTER IGNITION. (Page 15). Ignite 1–2 g, accurately weighed.

HEAVY METALS. (Page 26, Method 1). Dissolve 2 g in 10 mL of water, add 2 mL of 3 N hydrochloric acid, and dilute with water to 25 mL. Proceed as directed, except use glacial acetic acid to adjust the pH.

Hexanes

Suitable for use in high-performance liquid chromatography, extraction–concentration analysis, ultraviolet spectrophotometry, or general use. Product labeling shall designate the one or more of these uses for which suitability is represented on the basis of meeting the relevant requirements and tests. The ultraviolet spectrophotometry and liquid chromatography suitability requirements include all of the requirements for general use. The extraction–concentration suitability requirements include only the general use requirement for color.

NOTE. This reagent is generally a mixture of several isomers of hexane (C_6H_{14}), predominantly *n*-hexane, 2-methylpentane, and 3-methylpentane, plus methylcyclopentane (C_6H_{12}).

REQUIREMENTS

General Use

Assay . . . . ≥ 98.5% hexanes
(sum of 5 isomers, total hexanes,
plus methylcyclopentane)

MAXIMUM ALLOWABLE

Color (APHA). 10
Residue after evaporation 0.001%
Water-soluble titrable acid . . . . 0.0003 meq / g
Sulfur compounds (as S) 0.005%
Thiophene . . . . Passes test

Specific Use

ULTRAVIOLET SPECTROPHOTOMETRY

Wavelength (nm)	Absorbance (AU)
280–400	0.01
250	0.02
240	0.04
230	0.10
220	0.20
210	1.0

LIQUID CHROMATOGRAPHY SUITABILITY

Absorbance. . . . . Passes test

EXTRACTION–CONCENTRATION SUITABILITY

Absorbance. . . . . Passes test
GC–FID. . . . . Passes test
GC–ECD . . . . Passes test

TESTS

ASSAY.   Analyze the sample by gas chromatography, using the general parameters cited on page 56. The following specific conditions are also required.

Column: Type I, methyl silicone

Measure the area under all peaks and calculate the hexanes and methylcyclopentane in area percent. The retention times are (min): 2-methylpentane, 7.5; 3-methylpentane, 8.5; *n*-hexane, 9.2; methylcyclopentane, 10.2.

COLOR (APHA).   (Page 17).

RESIDUE AFTER EVAPORATION.   (Page 14). Evaporate 100 g (145 mL) to dryness in a tared dish on the steam bath and dry the residue at 105°C for 30 min.

WATER-SOLUBLE TITRABLE ACID.   To 30 g (44 mL) in a separatory funnel, add 50 mL of water and shake vigorously for 2 min. Allow the layers to separate, draw off the aqueous layer, and add 0.15 mL of phenolphthalein indicator solution to the aqueous layer. Not more than 1.0 mL of 0.01 N sodium hydroxide should be required to produce a pink color.

SULFUR COMPOUNDS.   To 30 mL of 0.5 N potassium hydroxide in methanol in a conical flask, add 5.5 g (8 mL) of the sample, and boil the mixture gently for 30 min under a reflux condenser, avoiding the use of a rubber stopper or connection. Detach the condenser, dilute with 50 mL of water, and heat on a hot plate at low setting until the hexanes and methanol are evaporated. Add 50 mL of bromine water, heat for 15 min longer, and transfer to a beaker. Neutralize with dilute hydrochloric acid (1 + 3), add an excess of 1 mL of the acid, evaporate to about 50 mL, and filter if necessary. Heat the filtrate to boiling, add 5 mL of barium chloride reagent solution, digest in a covered beaker on the hot plate at low setting for 2 h, and allow to stand overnight. If a precipitate is formed, filter, wash thoroughly, and ignite. Correct for the weight obtained in a complete blank test.

THIOPHENE.   To 25 mL in a glass-stoppered flask add 15 mL of sulfuric acid to which has been added a few milligrams of isatin, shake the mixture for 15–20 s, and allow to stand for 1 h. The acid layer should not be colored blue or green.

ULTRAVIOLET SPECTROPHOTOMETRY.   Use the procedure on page 67 to determine the absorbance.

EXTRACTION–CONCENTRATION SUITABILITY.   Analyze the sample by using the general procedure cited on page 65. Use the procedure on page 67 to determine the absorbance.

Hydrazine Sulfate

$(NH_2)_2 \cdot H_2SO_4$ Formula Wt 130.12

CAS Number 10034–93–2

REQUIREMENTS

Assay . . . . ≥ 99.0%
$(NH_2)_2H_2SO_4$

MAXIMUM ALLOWABLE

Insoluble matter 0.005%
Residue after ignition 0.05%
Chloride (Cl) 0.005%
Heavy metals (as Pb) 0.002%
Iron (Fe) 0.001%

TESTS

ASSAY. (By titration of reducing power). Weigh accurately about 1 g, dissolve in water, and dilute with water to 500 mL in a volumetric flask. To 50.0 mL add 1 g of sodium bicarbonate, dissolve the reagent, and add 50.0 mL of 0.1 N iodine solution. Titrate the excess iodine with 0.1 N sodium thiosulfate, adding 3 mL of starch indicator near the end point. One milliliter of 0.1 N iodine corresponds to 0.003253 g of $(NH_2)_2 \cdot H_2SO_4$.

INSOLUBLE MATTER. (Page 14). Use 20 g dissolved in 300 mL of water.

RESIDUE AFTER IGNITION. (Page 15). Ignite 2.0 g. Reserve the residue for the test for iron.

CHLORIDE. (Page 25). Dissolve 1.0 g in water, dilute with water to 100 mL, and use 20 mL of this solution.

HEAVY METALS. (Page 26, Method 1). Dissolve 2.0 g in 40 mL of warm water. Use 30 mL to prepare the sample solution, and use the remaining 10 mL to prepare the control solution.

IRON. (Page 28, Method 1). To the residue after ignition add 3 mL of dilute hydrochloric acid (1 + 1) and 0.10 mL of nitric acid, cover with a watch glass, and digest on the steam bath for 15–20 min. Remove the watch glass and evaporate to dryness. Dissolve the residue in a mixture of 4 mL of hydrochloric acid and 10 mL of water and dilute with water to 100 mL. Use 50 mL of this solution without further acidification.

Hydriodic Acid, 47%
(with Stabilizer)

HI   Formula Wt 127.91

CAS Number 10034–85–2

> *NOTE.* To avoid danger of explosions, this acid should be distilled only in an inert atmosphere. The reagent may have a slight yellow color. It contains hypophosphorus acid as a stabilizer.

REQUIREMENTS

Assay . . . . $\geq$ 47.0% HI

MAXIMUM ALLOWABLE

Chloride and bromide (as Cl) 0.05%
Sulfate (SO_4) 0.005%
Arsenic (As) . . . . 5 ppm
Stabilizer (H_3PO_2) 1.5%
Heavy metals (as Pb) 0.001%
Iron (Fe) 0.001%

TESTS

ASSAY. (By argentometric titration). Weigh about 50 mL of water in a 250-mL glass-stoppered flask; add about 0.7 mL of the acid and weigh again. Add 50.0 mL of 0.1 N silver nitrate and shake the mixture well. Then add 5 mL of nitric acid and heat on the steam bath until the precipitate has acquired a bright yellow color. Cool, and titrate the residual silver nitrate with 0.1 N thiocyanate solution, using 2 mL of ferric ammonium sulfate indicator solution. One milliliter of 0.1 N silver nitrate corresponds to 0.01279 g of HI.

CHLORIDE AND BROMIDE.   Dilute 1 g (0.67 mL) with water to 100 mL in a 250-mL conical flask. Add 1 mL of hydrogen peroxide and 1 mL of phosphoric acid. Heat to boiling and boil gently until all the iodine is expelled and the solution is colorless. Cool, wash down the sides of the flask, and add 0.5 mL of hydrogen peroxide. If an iodine color develops, boil until the solution is colorless and for 10 min longer. If no color develops, boil for 10 min. Dilute with water to 100 mL, take a 2-mL portion of the solution, and dilute with water to 23 mL. Prepare a standard containing 0.01 mg of chloride ion (Cl) in 23 mL of water. Add 1 mL of nitric acid and 1 mL of silver nitrate reagent solution to each. Any turbidity produced in the solution of the sample should not exceed that in the standard.

SULFATE. (Page 32, Procedure B). Transfer 1.0 mL to a 40-mL reaction tube, add 15 mL of reducing mixture, and continue with the procedure.

ARSENIC. (Page 23). Dilute 6 g (4.0 mL) with water to 100 mL. To 10 mL of this solution in a 250-mL beaker add 1 mL of nitric acid and 4.5 mL of sulfuric acid, evaporate in a hood to dense fumes of sulfur trioxide, and continue the fuming for 15 min. Cool, cautiously wash down the beaker with about 10 mL of water, evaporate again to dense fumes of sulfur trioxide, and continue the fuming for 15 min. Cool, add 0.25 g of hydrazine sulfate, cautiously wash down the beaker with about 10 mL of water, and repeat the fuming. Cool, and cautiously wash the solution into a generator flask with sufficient water to make a volume of 35 mL. For the standard use 0.003 mg of arsenic (As).

> **Sample Solution A.** To 5 g (3.4 mL) add 10 mL of water, 7 mL of nitric acid, and 9 mL of hydrochloric acid. Evaporate to dryness on the steam bath, dissolve in water, and dilute with water to 50 mL (1 mL = 0.1 g).

STABILIZER. Dilute 1.0 mL of sample solution A (0.1-g sample) with water to 1 L. Dilute 4.8 mL of the solution with water to 85 mL. Add 5 mL of a 10% solution of ammonium molybdate and adjust the pH to 1.8 (using a pH meter) with dilute hydrochloric acid (1 + 9). Heat to boiling and cool to room temperature. Add 10 mL of hydrochloric acid and transfer to a separatory funnel. Add 35 mL of ether and shake vigorously. Allow the layers to separate and draw off and discard the aqueous layer. Wash the ether layer twice with 10-mL portions of dilute hydrochloric acid (1 + 9), drawing off and discarding the aqueous portion each time. Add 10 mL of dilute hydrochloric acid (1 + 9) to which has just been added 0.2 mL of a 2% stannous chloride solution in hydrochloric acid and shake. Any blue color in the ether should not exceed that produced by 0.01 mg of phosphate ion (PO_4) treated exactly like the 4.8-mL portion of the sample.

HEAVY METALS. (Page 26, Method 1). Dilute 20 mL of sample solution A (2-g sample) with water to 25 mL.

IRON. (Page 28, Method 1). Use 10 mL of sample solution A (1-g sample).

Hydriodic Acid, 55%
(Suitable for Use in Methoxyl Determinations)

HI   Formula Wt 127.91

CAS Number 10034−85−2

> *NOTE.* This reagent is uninhibited and may turn yellow on standing.

REQUIREMENTS

Assay . . . . 55.0−58.0% HI

MAXIMUM ALLOWABLE

Free iodine (I_2) 0.75%
Residue after ignition 0.01%
Chloride and bromide (as Cl). 0.05%
Phosphate (PO_4) 0.001%
Sulfate (SO_4) 0.005%
Heavy metals (as Pb) 0.001%
Iron (Fe) 0.001%

TESTS

ASSAY. (By argentometric titration). Weigh, to the nearest 0.1 mg, about 50 mL of water in a 250-mL glass-stoppered flask; add about 0.7 mL of the sample and reweigh. Add exactly 50.0 mL of 0.1 N silver nitrate and shake the mixture well. Then add 5 mL of nitric acid and heat on the steam bath until the precipitate has acquired a bright yellow color. Cool, and titrate the residual silver nitrate with 0.1 N ammonium thiocyanate solution, using 2 mL of ferric ammonium sulfate indicator solution. One milliliter of 0.1 N silver nitrate corresponds to 0.01279 g of HI.

FREE IODINE. Weigh a glass-stoppered flask containing 15 mL of water. Add about 3 mL of sample, stopper, and reweigh. Dilute with 40 mL of water and titrate with 0.1 N sodium thiosulfate to the disappearance of the yellow color. Not more than 0.60 mL of the thiosulfate should be required per gram of the acid.

RESIDUE AFTER IGNITION.   (Page 15). Evaporate 10 g (6.0 mL) to dryness in a tared platinum crucible and ignite.

CHLORIDE AND BROMIDE.   Dilute 1 g (0.6 mL) with water to 100 mL in a 250-mL conical flask. Add 1 mL of hydrogen peroxide and 1 mL of phosphoric acid. Heat to boiling and boil gently until all the iodine is expelled and the solution is colorless. Cool, wash down the sides of the flask, and add 0.5 mL of hydrogen peroxide. If an iodine color develops, boil until the solution is colorless and for 10 min longer. If no color develops, boil for 10 min. Dilute with water to 100 mL, take a 2-mL portion, and dilute with water to 23 mL. Prepare a standard containing 0.01 mg of chloride ion (Cl) in 23 mL of water. Add 1 mL of nitric acid and 1 mL of silver nitrate reagent solution to each. Any turbidity produced in the solution of the sample should not exceed that in the standard.

> ***Sample Solution A for the Determination of Phosphate, Heavy Metals, and Iron.***   Evaporate 5 g (3.0 mL) of sample with 2 mL of nitric acid to dryness on the steam bath. Reevaporate with several portions of water until all the iodine has been volatilized. Warm the residue with 2 mL of 1 N hydrochloric acid and take up with 40 mL of hot water. Cool, dilute with water to 50 mL, and mix well (1 mL = 0.1 g).

PHOSPHATE.   To 20 mL of sample solution A (2-g sample) add 2 mL of 25% sulfuric acid, 10 mL of water, 1 mL of ammonium molybdate–sulfuric acid reagent solution, and 1 mL of 4-(triethyl-amino)phenol sulfate reagent solution. Heat to 60°C for 10 min. Any blue color should not exceed that produced by 0.02 mg of phosphate ion (PO_4) in an equal volume of solution containing the quantities of reagents used in the test.

SULFATE.   (Page 32, Procedure B). Transfer 1.7 g (1.0 mL) to a 40-mL reaction tube, add 15 mL of reducing mixture, and continue with the procedure.

HEAVY METALS.   (Page 26, Method 1). Dilute 20 mL of sample solution A (2-g sample) with water to 25 mL.

IRON.   (Page 28, Method 1). Use 10 mL of sample solution A (1-g sample).

Hydrobromic Acid, 48%

HBr   Formula Wt 80.91

CAS Number 10035–10–6

REQUIREMENTS

Assay . . . . 47.0–49.0% HBr

MAXIMUM ALLOWABLE

Residue after ignition 0.002%
Chloride (Cl) 0.05%
Iodide (I) 0.003%
Phosphate (PO$_4$) 0.001%
Sulfate and sulfite (as SO$_4$) 0.003%
Arsenic (As) . . . . 0.5 ppm
Heavy metals (as Pb) . . . . 5 ppm
Iron (Fe) . . . . 1 ppm
Selenium (Se) . . . . 0.01 ppm

TESTS

ASSAY. (By acid–base titrimetry). Tare a glass-stoppered conical flask containing about 15 mL of water. Quickly add about 4 mL of the acid and weigh. Dilute with water to 50 mL and titrate with 1 N sodium hydroxide, using 0.15 mL of phenolphthalein indicator. One milliliter of 1 N sodium hydroxide corresponds to 0.08091 g of HBr.

RESIDUE AFTER IGNITION. (Page 15). To 50 g (34 mL) in a tared dish (not platinum) add 0.05 mL of sulfuric acid, evaporate as far as possible on the steam bath, then heat gently to volatilize the solution. Finally, ignite at 800 ± 25°C for 15 min.

CHLORIDE. To 1 g (0.67 mL) in a 150-mL conical flask add 50 mL of dilute nitric acid (1 + 3) and digest on the steam bath until the solution is colorless. Wash down the sides of the flask with a little water and digest for an additional 15 min. Cool, filter if necessary through a chloride-free filter, and dilute with water to 100 mL. Dilute 2.0 mL of the solution with water to 23 mL. Prepare a standard containing 0.01 mg of chloride ion (Cl) in 23 mL of water. Add 1 mL of nitric acid and 1 mL of silver nitrate reagent solution to each. Any turbidity produced in the solution of the sample should not exceed that in the standard.

IODIDE. Dilute 6 g (4.0 mL) with 20 mL of water. Add 5 mL of chloroform, 0.2 mL of 10% ferric chloride solution, and 0.1 mL of dilute sulfuric acid (1 + 15) and mix gently in a separatory funnel. Draw off the chloroform layer into a comparison tube of about 1.5-cm diameter. No violet color should be observed on looking down through the chloroform.

PHOSPHATE. (Page 30, Method 1). Add 5 mL of nitric acid to 6 g (4.0 mL) of the sample, and evaporate to dryness on the steam bath. Dissolve the residue in 75 mL of approximately 0.5 N sulfuric acid, and use 25 mL of the solution.

SULFATE AND SULFITE. (Page 32, Procedure B). Transfer 1.0 g to a reaction tube, and continue with the procedure.

ARSENIC. (Page 23). To 6 g (4.0 mL) in a 250-mL beaker add cautiously 5 mL of nitric acid and 4.5 mL of sulfuric acid, evaporate in a hood to dense fumes of sulfur trioxide, and continue the fuming for 15 min. Cool, cautiously wash down the beaker with about 10 mL of water, evaporate again to dense fumes of sulfur trioxide, and continue the fuming for 15 min. Cool, add 0.4 g of hydrazine sulfate, cautiously wash down the beaker with about 10 mL of water, and repeat the fuming. Cool, cautiously wash the solution into a generator flask with sufficient water to make a volume of 35 mL. For the standard use 0.003 mg of arsenic (As).

HEAVY METALS. (Page 26, Method 1). To 4 g (2.7 mL) in a beaker, add about 10 mg of sodium carbonate, and evaporate to dryness on the steam bath. Dissolve the residue in about 20 mL of water, and dilute with water to 25 mL.

IRON. (Page 28, Method 1). To 10 g (6.6 mL) add about 10 mg of sodium carbonate, and evaporate to dryness on the steam bath. Dissolve the residue in 2 mL of hydrochloric acid, dilute with water to 50 mL, and use the solution without further acidification.

SELENIUM. Place 150 g (100 mL) in an all-glass distilling apparatus, add 2 mL of bromine, and distill off approximately 25 mL. Dilute the distillate with an equal volume of water and decolorize with sulfurous acid. Add 0.1 g of hydroxylamine hydrochloride and warm gently on the steam bath for 30 min. Cool to room temperature, filter through a white glass-wool mat in a small Gooch crucible, without washing, and examine immediately. No pink or red color should be observed on the mat.

Hydrochloric Acid

HCl   Formula Wt 36.46

CAS Number 7647–01–0

REQUIREMENTS

Appearance . . . . Free from sus-
                                                                    pended matter or
                                                                    sediment
Assay . . . . 36.5–38.0% HCl

MAXIMUM ALLOWABLE

Color (APHA). 10
Residue after ignition . . . . 5 ppm
Bromide (Br) 0.005%
Sulfate (SO_4) . . . . 1 ppm
Sulfite (SO_3) . . . . 1 ppm
Extractable organic substances . . . . 5 ppm
Free chlorine (Cl) . . . . 1 ppm
Ammonium (NH_4) . . . . 3 ppm
Arsenic (As) . . . . 0.01 ppm
Heavy metals (as Pb) . . . . 1 ppm
Iron (Fe) . . . . 0.2 ppm

TESTS

APPEARANCE.   Mix the material in the original container, pour 10 mL into a test tube (20 × 150 mm), and compare with distilled water in a similar tube. The liquids should be equally clear and free from suspended matter.

ASSAY.   (By acid–base titrimetry). Tare a glass-stoppered flask containing about 30 mL of water. Quickly add about 3 mL of the sample, stopper, and weigh accurately. Dilute to about 50 mL, add 0.15 mL of methyl orange indicator solution, and titrate with 1 N sodium hydroxide. One milliliter of 1 N sodium hydroxide corresponds to 0.03646 g of HCl.

COLOR (APHA). (Page 17).

RESIDUE AFTER IGNITION. (Page 15). To 200 g (170 mL) in a tared platinum dish, add 0.05 mL of sulfuric acid, and evaporate as far as possible on the steam bath. Heat gently to volatilize the excess sulfuric acid, and ignite.

BROMIDE. Pipet 1 mL into 25 mL of water in a 100-mL beaker, add 2 mL of phenolsulfonphthalein indicator solution (0.020 g of the sodium salt in 100 mL of water), and titrate with 1 N sodium hydroxide. Add just sufficient titrant to change the yellow color to red. Transfer the solution to a 100-mL volumetric flask with about 25 mL of water. For the standard add identical amounts of indicator and water to 0.06 mg of bromide ion (Br) in a 100-mL volumetric flask. To each flask add 5 mL of acetate buffer solution (68 g of sodium acetate, $CH_3COONa \cdot 3H_2O$, and 30 mL of glacial acetic acid per liter) and 2.0 mL of freshly prepared chloramine-T solution (0.125 g per 100 mL). Mix, allow to stand for 20.0 min, and add 20 mL of 0.1 N sodium thiosulfate solution. Swirl, dilute with water to volume, and mix. Compare visually or spectrophotometrically at 590 nm in 1.0-cm cells. Any blue-violet color in the sample should not exceed that in the standard.

SULFATE. (Page 32, Procedure B). Transfer 20 g (17 mL) to a reaction tube, add 0.05 mL of 0.01 N potassium permanganate, and evaporate to dryness. Dissolve the residue in 1 mL of water, and continue the procedure.

SULFITE. Transfer 400 mL of freshly boiled and cooled water to each of two 500-mL conical flasks. Adjust the temperature of one flask (for the reagent blank) to 25°C and the other flask (for the sample solution) to 15°C. To each flask add 3 mL of 10% potassium iodide reagent solution, 5 mL of hydrochloric acid, and 2 mL of starch indicator solution. To the sample flask add 85 mL of hydrochloric acid sample, and adjust the temperature of the contents of the two flasks to within 1°C of each other. Using a magnetic stirring apparatus and a Teflon-coated stirring bar, titrate each with 0.01 N iodine until a faint permanent blue color is produced. Subtract the reagent blank titrant volume from the sample titrant volume to obtain the net sample titration. The net titration should not exceed 0.25 mL.

EXTRACTABLE ORGANIC SUBSTANCES. Place 120 g (100 mL) of sample in each of two 125-mL separatory funnels. To one funnel, add

7.0 mL of 2,2,4-trimethylpentane as a sample blank. To the second funnel, add 5.0 mL of 2,2,4-trimethylpentane and 2.0 mL of the internal standard solution. Stopper the funnels, shake for 2 min, allow the layers to separate, and drain the hydrochloric acid layer. Prepare a standard solution by adding 2.0 mL of the internal standard solution to 5.0 mL of the chloroform standard solution.

>   ***Internal Standard Solution.***   Add by syringe 30 mg (22.7 μL) of 1,1,1-trichloroethane to 100 mL of 2,2,4-trimethylpentane.

>   ***Chloroform Standard Solution.***   Add by syringe 12 mg (8.1 μL) of chloroform to 100 mL of 2,2,4-trimethylpentane. The 2,2,4-trimethylpentane used for both the standard and the analysis should be free from impurities that interfere with the chromatographic analysis.

Analyze the sample blank and the sample 2,2,4-trimethylpentane layers and the standard solution by gas chromatography as described on page 56. With a flame ionization detector, most organic compounds can be detected at levels of 1 ppm or less. The use of an electron-capture or other selective detector would make it possible to detect chlorinated aliphatics or other selected compounds at much lower levels. The following parameters have given satisfactory results.

Column: Type I, methyl silicone, 30 m × 0.53 mm i.d., 5.0-μm film thickness

Column Temperature: 35°C initial hold, 5 min, then programmed at 10°C per min to 230°C, final hold 21 min. Total run time, 45.5 min.

Injector Temperature: 230°C

Detector Temperature: 250°C

Carrier Gas: Helium, with flow rate 3–4 mL per min

Detector: Flame ionization, range 10^{-11} A

Sample Size: 0.5 μL. Adjust the attenuation to produce a peak height of at least 10% of full scale for the internal standard peak.

Approximate Retention Times (min.):   Dichloromethane, 3.8; chloroform, 6.8; 1,2-dichloroethane, 7.7; 1,1,1-trichloroethane, 8.0; benzene, 8.6; carbon tetrachloride, 8.8; 2,2,4-trimethylpentane, 10–13; chlorobenzene, 13.8; 1,2,4-trichlorobenzene, 20.5; lindane, 37.

Use the sample blank extraction to correct for impurities in the solvent or sample that interfere with the internal standard peak for 1,1,1-trichloroethane. The ratio of the total area of any peaks in the extraction (other than the internal standard and solvent) to the area of the internal standard peak in the sample should not exceed the ratio of the area of the chloroform peak to the area of the internal standard peak in the standard solution.

FREE CHLORINE.   To 50 mL of the sample add 50 mL of freshly boiled water and cool. Add 0.1 mL of 2% potassium iodide solution and 1 mL of carbon disulfide, and mix. The carbon disulfide should not acquire a pink color within 30 s. The potassium iodide should be free from iodate.

AMMONIUM.   (Page 48). Determine the ammonium by differential pulse polarography. Use 3.3 g (2.8 mL) of sample and 5.5 mL of ammonia-free 6 N sodium hydroxide. For the standard use 0.010 mg of ammonium (NH_4).

ARSENIC.   (Page 23). Mix 300 g (255 mL) with 10 mL of dilute sulfuric acid (1 + 1) in a generator flask. Add potassium chlorate crystals in small increments until a yellow color is produced, and evaporate on a hot plate just to fumes of sulfur trioxide, adding a few more crystals of potassium chlorate whenever the yellow color starts to fade. Cool, cautiously wash down the flask with 5–10 mL of water, and again heat just to fumes of sulfur trioxide. Cool, cautiously wash down the flask with 5–10 mL of water, and again heat just to fumes of sulfur trioxide. Cool, cautiously wash down the flask with 5–10 mL of water, and repeat the fuming. (The volume of sulfuric acid at this point should be about 4 mL.) Cool, cautiously dilute with water to 55 mL, and use this solution without further acidification. For the standard use 0.003 mg of arsenic (As).

HEAVY METALS.   (Page 26, Method 1). To 20 g (17 mL) in a 150-mL beaker, add about 10 mg of sodium carbonate, and evaporate to dryness on the steam bath. Dissolve the residue in about 20 mL of water, and dilute with water to 25 mL.

IRON.   (Page 28, Method 1). To 50 g (42 mL) in a porcelain or glass vessel add about 10 mg of sodium carbonate and evaporate to dryness on the steam bath. Dissolve in 2 mL of hydrochloric acid, dilute with water to 50 mL, and use the solution without further acidification.

Hydrofluoric Acid

HF   Formula Wt 20.01

CAS Number 7664–39–3

REQUIREMENTS

Assay . . . . 48.0–51.0% HF

MAXIMUM ALLOWABLE

Fluosilicic acid (H_2SiF_4) 0.01%
Residue after ignition . . . . 5 ppm
Chloride (Cl) . . . . 5 ppm
Phosphate (PO_4) . . . . 1 ppm
Sulfate and sulfite (as SO_4) . . . . 5 ppm
Arsenic (As). . . . . 0.05 ppm
Copper (Cu) . . . . 0.1 ppm
Heavy metals (as Pb) . . . . 0.5 ppm
Iron (Fe) . . . . 1 ppm

TESTS

ASSAY. (By acid–base titrimetry). Tare a stoppered plastic flask, containing about 25 mL of water, and deliver about 1.5 mL of sample under the water surface, using a plastic pipet. Weigh accurately, wash down the sides, and dilute the solution to 50–60 mL with water. Add 0.15 mL of phenolphthalein indicator solution, and titrate with 1 N sodium hydroxide. One milliliter of 1 N sodium hydroxide corresponds to 0.02001 g of HF.

FLUOSILICIC ACID. Weigh about 37.5 g (33 mL) into a large platinum dish. Add 2 g of potassium chloride and 3 mL of hydrochloric acid and evaporate to dryness on the steam bath in the hood. Wash down the sides of the dish with a small amount of water, add 3 mL of hydrochloric acid, and repeat the evaporation. Dissolve the residue in about 100 mL of water, cool to 0°C, and add 0.15 mL of phenolphthalein indicator solution. Neutralize any free acid with 0.1 N sodium hydroxide solution, keeping the temperature of the solution near 0°C. Heat the solution to boiling and titrate with 0.1 N sodium hydroxide solution. One milliliter of 0.1 N sodium hydroxide corresponds to 0.0036 g of H_2SiF_6.

RESIDUE AFTER IGNITION. (Page 15). To 200 g (180 mL) in a tared platinum dish, add 0.05 mL of sulfuric acid, evaporate as far as possible on the steam bath in the hood, heat gently to volatilize the excess sulfuric acid, and ignite.

CHLORIDE. Add 1.8 mL to 45 mL of water, filter if necessary through a chloride-free filter, and add 1 mL of nitric acid and 1 mL of silver nitrate reagent solution. Any turbidity should not exceed that produced by 0.01 mg of chloride ion (Cl) in an equal volume of solution containing the quantities of reagents used in the test.

PHOSPHATE. Evaporate 20 g (18 mL) to dryness on the steam bath in the hood. Dissolve in 25 mL of approximately 0.5 N sulfuric acid, add 1 mL of ammonium molybdate–sulfuric acid reagent solution and 1 mL of 4-(methylamino)phenol sulfate reagent solution, and allow to stand for 2 h at room temperature. Any blue color should not exceed that produced by 0.02 mg of phosphate ion (PO_4) in an equal volume of solution containing the quantities of reagents used in the test.

SULFATE AND SULFITE. To 20 g (18 mL) in a platinum or other suitable evaporating dish, add about 10 mg of sodium carbonate and 1 mL of 30% hydrogen peroxide. Evaporate to dryness on a hot plate at low setting in the hood, wash down the sides of the dish with a small volume of water, and add 3 mL of perchloric acid. Evaporate to about 1 mL, dilute with about 15 mL of water, and add 0.15 mL of phenolphthalein indicator solution. Neutralize with ammonium hydroxide, dilute with water to 20 mL, and add 2 mL of dilute hydrochloric acid (1 + 19) and 2 mL of barium chloride reagent solution. Any turbidity should not exceed that produced by 0.1 mg of sulfate ion (SO_4) in an equal volume of solution containing the quantities of reagents used in the test. Compare 10 min after adding the barium chloride to the sample and standard solutions.

ARSENIC. To 40 g (33 mL) of sample in a Teflon beaker add 20 mL of nitric acid, 5 mL of sulfuric acid, and 5 mL of hydrochloric acid, swirling the beaker gently after each addition. Allow to stand at room temperature for 5 min. Evaporate in a hood in a sand bath to dense fumes of sulfur trioxide. Cool, wash down the sides of the beaker with 15–20 mL of water, and evaporate again in a sand bath to dense fumes of sulfur trioxide. Cool and transfer to a 125-mL generator flask, rinsing the beaker thoroughly with water. Place the flask on a hot plate, evaporate to dense fumes of sulfur trioxide, add sulfuric acid if necessary to make a volume of about 4 mL, and cool. Add 50 mL of water, 2 mL of 16.5% potassium iodide solution, and 0.5 mL of stannous chloride solution (40% $SnCl_2 \cdot 2H_2O$ in hydrochloric acid), and mix. Proceed as described in the general method for arsenic on page 23 under Procedure, starting with the second sentence which begins "Allow the mixture to stand..." Any red color in the silver diethyldithiocarbamate solution of the sample should not exceed that in a standard containing 0.002 mg of arsenic (As).

COPPER.

> *Sample Solution A.* To 100 g (85 mL) in a platinum dish add 10 mg of sodium carbonate and evaporate to dryness in the hood. Dissolve the residue in 1 mL of 1 N hydrochloric acid and about 10 mL of water, add 3.8 mL of 1% sodium carbonate solution, and dilute with water to 45 mL. If necessary, adjust the pH between 6.5 and 6.9 (using a pH meter) with 1% sodium carbonate or 1 N hydrochloric acid, and dilute with water to 50 mL.

Transfer 30 mL of sample solution A (60-g sample) to a separatory funnel. Add 10 mL of standard dithizone solution in chloroform, shake vigorously, allow the solutions to separate, and draw off the chloroform extract. Any red color, producing a purplish hue in the chloroform, should not exceed that produced by 0.006 mg of copper ion (Cu) treated exactly as the 30 mL of sample solution A.

HEAVY METALS. (Page 26, Method 1). To 40 g (34 mL) in a platinum dish, add about 10 mg of sodium chloride, and evaporate to dryness on the steam bath in the hood. Dissolve the residue in about 20 mL of water, and dilute with water to 25 mL.

IRON. (Page 28, Method 1). To 10 g (8.7 mL) in a platinum dish, add about 10 mg of sodium carbonate, and evaporate to dryness on the steam bath in the hood. Add 0.2 mL of sulfuric acid and ignite gently. Dissolve the residue in 4 mL of dilute hydrochloric acid (1 + 1), cover, digest for 15 min on the steam bath, and dilute with water to 50 mL. Use the solution without further acidification.

Hydrogen Peroxide

H_2O_2 Formula Wt 34.01

CAS Number 7722-84-1

NOTE. This reagent should be stored in a cool place in containers with a vent in the stopper. The assay requirement applies to material stored properly for a reasonable time.

REQUIREMENTS

Assay . . . . 29.0–32.0% H_2O_2

MAXIMUM ALLOWABLE

Color (APHA). 10
Residue after evaporation 0.002%
Titrable acid . . . . 0.0006 meq / g
Chloride (Cl) . . . . 3 ppm
Nitrate (NO_3) . . . . 2 ppm
Phosphate (PO_4) . . . . 2 ppm
Sulfate (SO_4). . . . . 5 ppm
Ammonium (NH_4) . . . . 5 ppm
Heavy metals (as Pb) . . . . 1 ppm
Iron (Fe) . . . . 0.5 ppm

TESTS

ASSAY. (By titration of the reductive capacity). Weigh accurately about 1 mL in a tared 100-mL volumetric flask, dilute to volume with water, and mix thoroughly. To 20.0 mL of this solution add 20 mL of dilute sulfuric acid (1 + 15) and titrate with 0.1 N potassium permanganate. One milliliter of 0.1 N potassium permanganate corresponds to 0.001700 g of H_2O_2.

COLOR (APHA). (Page 17).

RESIDUE AFTER EVAPORATION. (Page 14). Evaporate 50 g (45 mL) to dryness in a tared porcelain or silica dish on the steam bath and dry the residue at 105°C for 30 min.

TITRABLE ACID. Dilute 10 g (9 mL) with 90 mL of carbon dioxide-free water. Add 0.15 mL of methyl red indicator solution and titrate with 0.01 N sodium hydroxide. The volume of sodium hydroxide solution consumed should not be more than 0.6 mL greater than the volume required for a blank test on 90 mL of the water used for dilution.

CHLORIDE. (Page 25). Use 3.3 g (3.0 mL).

NITRATE. Add about 10 mg of sodium carbonate to 4.5 mL (5 g) of the sample and evaporate to dryness on the steam bath. Wash down the inside of the container with about 5 mL of water and again evaporate to dryness. Add 2 mL of phenoldisulfonic acid reagent solution and heat on the steam bath for 15 min. Cool, dilute with water to 30 mL, and make alkaline with ammonium hydroxide. Any yellow color should not exceed that produced when a solution containing 0.01 mg of nitrate ion (NO_3) is evaporated to dryness and treated exactly as the sample.

PHOSPHATE. (Page 30, Method 1). Evaporate 10 g (9 mL) to dryness on the steam bath. Dissolve the residue in 25 mL of approximately 0.5 N sulfuric acid, and continue as described.

SULFATE. (Page 32, Procedure A, Method 2). Use 10 g (9 mL) of sample.

AMMONIUM. Add 0.1 mL of sulfuric acid to 2 g (1.8 mL) of the sample and evaporate on the steam bath. Dissolve the residue in 45 mL of water and add 3 mL of 10% sodium hydroxide reagent solution and 2 mL of Nessler reagent. Any color should not exceed that produced by 0.01 mg of ammonium ion (NH_4) in an equal volume of solution containing the quantities of reagents used in the test.

HEAVY METALS. (Page 26, Method 1). To 20 g (18 mL) in a 100-mL beaker add about 10 mg of sodium chloride and evaporate to dryness on the steam bath. Dissolve the residue in about 20 mL of water and dilute with water to 25 mL.

IRON. (Page 28, Method 1). To 20 g (18 mL) in a 100-mL beaker, add about 10 mg of sodium chloride, and evaporate to dryness on the steam bath. Dissolve in 2 mL of hydrochloric acid, dilute with water to 50 mL, and use the solution without further acidification.

Hydroxylamine Hydrochloride

$NH_2OH \cdot HCl$ Formula Wt 69.49

CAS Number 5470–11–1

REQUIREMENTS

Assay . . . . $\geq 96.0\%$
$NH_2OH \cdot HCl$

MAXIMUM ALLOWABLE

Clarity of alcohol solution . . . . Passes test
Residue after ignition 0.05%
Titrable free acid . . . . 0.25 meq / g
Ammonium (NH$_4$) 0.1%
Sulfur compounds (as SO$_4$) 0.005%
Heavy metals (as Pb) . . . . 5 ppm
Iron (Fe) . . . . 5 ppm

TESTS

ASSAY. (By titration of the reductive capacity). Weigh accurately 1.5 g, previously dried for 24 h over magnesium perchlorate or phosphorus pentoxide, and transfer to a 250-mL volumetric flask. Dissolve in oxygen-free water, dilute to volume with oxygen-free water, and take 20.0 mL of the sample solution. Dissolve 5 g of ferric ammonium sulfate in 30 mL of oxygen-free dilute sulfuric acid (1 + 50) and add this solution to the sample solution. Protect the solution from oxygen in the air, boil gently for 5 min, and cool. Dilute with 150 mL of oxygen-free water and titrate with 0.1 N potassium permanganate. One milliliter of 0.1 N potassium permanganate corresponds to 0.003474 g of $NH_2OH \cdot HCl$.

CLARITY OF ALCOHOL SOLUTION. Dissolve 1.0 g in 25 mL of alcohol and compare with an equal volume of alcohol. The sample solution should have no more color or turbidity than the alcohol. Retain the solution for the test for ammonium.

RESIDUE AFTER IGNITION. (Page 15). Ignite 2.0 g.

TITRABLE FREE ACID. Dissolve 10 g in 50 mL of water, add 0.15 mL of bromphenol blue, and titrate with 1 N sodium hydroxide to the neutral (green) end point. Not more than 2.5 mL of 1 N sodium hydroxide should be required.

AMMONIUM. To the solution obtained in the test for insoluble in alcohol, add 1 mL of a solution of chloroplatinic acid (2.6 g in 20 mL). The solution should remain clear for 10 min.

SULFUR COMPOUNDS. Dissolve 2.0 g in 10 mL of water containing 10 mg of sodium carbonate and add 4 mL of nitric acid. Slowly add 12 mL of 10% hydrogen peroxide solution (mix 4 mL of 30% hydrogen peroxide with 8 mL of water). Digest in a covered beaker until reaction ceases, uncover, and evaporate to dryness on a hot plate at low setting. Dissolve the residue in 10 mL of water, add 2 mL of dilute hydrochloric acid (1 + 19), filter if necessary, and dilute with water to 20 mL. To 10 mL add 1 mL of barium chloride reagent solution. Any turbidity should not exceed that produced by 0.05 mg of sulfate ion (SO_4) in an equal volume of solution containing the quantities of reagents used in the test. Compare 10 min after adding the barium chloride to the sample and standard solutions.

HEAVY METALS.   (Page 26, Method 1). Dissolve 6.0 g in about 20 mL of water, and dilute with water to 30 mL. Use 25 mL to prepare the sample solution, and use the remaining 5.0 mL to prepare the control solution.

IRON.   (Page 28, Method 1). To the residue after ignition add 3 mL of dilute hydrochloric acid (1 + 1), cover the dish, and digest on the steam bath for 15–20 min. Remove the cover, evaporate to dryness, dissolve the residue in 2 mL of hydrochloric acid, and dilute with water to 50 mL. Use the solution without further acidification.

Hydroxy Naphthol Blue
1-(2-Naphtholazo-3,6-disulfonic acid)-2-naphthol-4-sulfonic Acid, Disodium Salt

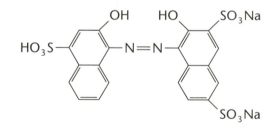

CAS Number 63451 –35 –4

NOTE. This compound is deposited on crystals of sodium chloride. The small, blue crystals are freely water-soluble. In the pH range between 12 and 13, the solution of the indicator is reddish pink in the presence of calcium ion and deep blue in the presence of excess (ethylenedinitrilo)tetraacetate.

REQUIREMENTS

Suitability for calcium determination . . . . Passes test

TEST

SUITABILITY FOR CALCIUM DETERMINATION. Dissolve 0.3 g in 100 mL of water, add 10 mL of 1 N sodium hydroxide and 1.0 mL of calcium chloride solution (1 g in 200 mL), and dilute with water to 165 mL. The solution is reddish pink in color. Add 1.0 mL of 0.05 M (ethylenedinitrilo)tetraacetic acid disodium salt. The solution becomes deep blue.

Iodic Acid
Iodic(V) Acid

HIO_3 Formula Wt 175.91

CAS Number 7782–68–5

REQUIREMENTS

Assay . . . . $\geq$ 99.5% HIO_3

MAXIMUM ALLOWABLE

Insoluble matter 0.01%
Residue after ignition 0.02%
Chloride and bromide (as Cl) 0.02%
Iodide (I) 0.01%
Nitrogen compounds (as N) 0.1%
Sulfate (SO_4) 0.015%
Heavy metals (as Pb) 0.001%
Iron (Fe) 0.002%

TESTS

ASSAY. (By indirect iodometric titration of iodate). Weigh accurately 0.5 g, transfer to a 250-mL volumetric flask, dissolve in water, dilute with water to volume, and mix thoroughly. Place a 50.0-mL aliquot of this solution in a glass-stoppered conical flask and add 2 g of potassium iodide and 5 mL of 15% hydrochloric acid. Stopper, swirl, allow to stand in the dark for 10 min, and add 100 mL of cold water. Titrate the liberated iodine with 0.1 N sodium thiosulfate, adding 3 mL of starch indicator solution near the end of the titration. Correct for a blank. One milliliter of 0.1 N sodium thiosulfate corresponds to 0.02932 g of HIO_3.

INSOLUBLE MATTER.   (Page 14). Use 10 g dissolved in 100 mL of water.

RESIDUE AFTER IGNITION.   (Page 15). Ignite 5.0 g.

CHLORIDE AND BROMIDE.   Dissolve 1.0 g in 100 mL of water in a distilling flask. Add 1 mL of hydrogen peroxide and 1 mL of phosphoric acid, heat to boiling, and boil gently until all the iodine is expelled and the solution is colorless. Cool, wash down the sides of the flask, and add 0.5 mL of hydrogen peroxide. If an iodine color develops, boil until the solution is colorless and for 10 min longer. If no color develops, boil for 10 min, filter if necessary through a chloride-free filter, and dilute with water to 100 mL. Dilute 5.0 mL of this solution with 18 mL of water, and add 1 mL of nitric acid and 1 mL of silver nitrate reagent solution. Any turbidity should not exceed that produced by 0.01 mg of chloride ion (Cl) in an equal volume of solution containing the quantities of nitric acid and silver nitrate used in the test.

IODIDE.   Dissolve 1.0 g in 20 mL of water. Add 1 mL of chloroform and 0.5 mL of 1 N sulfuric acid. No violet color should appear in the chloroform within 1 min.

NITROGEN COMPOUNDS.   Dissolve 0.50 g in water and dilute with water to 200 mL. Transfer 40 mL to a flask connected through a spray trap to a condenser, the end of which dips beneath the surface of 10 mL of 0.1 N hydrochloric acid. Add to the flask 10 mL of 10% sodium hydroxide reagent solution and 0.5 g of aluminum wire in small pieces, allow to stand for 1 h, and slowly distill about 35 mL. To the distillate add 1 mL of 10% sodium hydroxide reagent solution, dilute with water to 50 mL, and add 2 mL of Nessler reagent. Any color should not exceed that produced when a quantity of an ammonium salt containing 0.1 mg of nitrogen (N) is treated exactly like the sample.

> *Sample Solution A.*   Dissolve 10 g in 20 mL of water and add about 10 mg of sodium carbonate. Add 20 mL of hydrochloric acid and evaporate to dryness on the steam bath. Repeat the evaporation twice using 10 mL of hydrochloric acid.

Blank Solution B.   Evaporate to dryness the quantities of acid and sodium carbonate used to prepare sample solution A.

Dissolve the residues in separate 20-mL portions of water, filter if necessary, and dilute each with water to 100 mL (1 mL of sample solution A = 0.1 g).

SULFATE.   (Page 32, Procedure A, Method 1). Use 3.3 mL of sample solution A (0.33-g sample).

HEAVY METALS.   (Page 26, Method 1). Dilute 20 mL of sample solution A (2-g sample) to 25 mL. Use 20 mL of blank solution B to prepare the control solution.

IRON.   (Page 28, Method 1). Use 5.0 mL of sample solution A (0.5-g sample).

Iodine

I_2 Formula Wt 253.81

CAS Number 7553−56−2

REQUIREMENTS

Assay . . . . ≥ 99.8% I_2

MAXIMUM ALLOWABLE

Nonvolatile matter 0.01%
Chlorine and bromine (as Cl) 0.005%

TESTS

ASSAY.   (By titration of oxidizing power). Weigh accurately 0.5 g and transfer to a glass-stoppered conical flask. Add 50 mL of water containing 3 g of potassium iodide and 1 mL of sulfuric acid. Stopper, swirl, and allow to stand for 2–3 min. Titrate the iodine with 0.1 N sodium thiosulfate, adding 3 mL of starch indicator solution near the end of the titration. One milliliter of 0.1 N sodium thiosulfate corresponds to 0.01269 g of I_2.

NONVOLATILE MATTER.   Transfer 10 g to a tared porcelain crucible or dish. Heat on the steam bath until the sample is volatilized. Heat to 105°C for 1 h, cool in a desiccator, and weigh.

CHLORINE AND BROMINE.   Add 1.0 g to 100 mL of hot water containing 0.6 g of hydrazine sulfate in a 250-mL Erlenmeyer flask. Heat on a steam bath until dissolution is effected. Cool and neutralize with 1 N sodium hydroxide solution, using an external indicator. Add 2 mL of hydrogen peroxide and 1 mL of phosphoric acid. Heat to boiling and boil gently until the solution is colorless. Cool, wash down the sides of the flask, and add 0.5 mL of hydrogen peroxide. If an iodine color develops, boil until the solution is colorless and for 10 min longer. If no color develops, boil for 10 min. Filter if necessary through a chloride-free filter, dilute with water to 100 mL, and take a 20-mL portion. For the standard dilute a solution containing 0.01 mg of chloride ion (Cl) with water to 20 mL. Add 1 mL of nitric acid and 1 mL of silver nitrate reagent solution to each. Any turbidity in the solution of the sample should not exceed that in the standard.

Iodine Monochloride
(Suitable for Use in Wijs Solution)

ICl   Formula Wt 162.36

CAS Number 7790–99–0

REQUIREMENTS

Assay . . . . The I / Cl ratio of the
                                                     Wijs solution pre-
                                                     pared from the
                                                     reagent must
                                                     be 1.10 ± 0.1.
Insoluble matter . . . . ≤ 0.005%

TESTS

INSOLUBLE MATTER

Stock Solution.   Dissolve 79.3 ± 0.1 g in 250 mL of glacial acetic acid, and filter through a tared filtering crucible. Store the solution in a dry, low-actinic glass bottle.

Wash the stock solution filter thoroughly with acetic acid and dry at 105°C.

ASSAY. (By selective titration of reducing and oxidizing power). Prepare Wijs solution (1.4–1.5% ICl in acetic acid) by diluting 11.7 ± 0.1 mL of the stock solution with 217 mL of glacial acetic acid. Determine the I/Cl ratio of this solution as follows:

Iodine. To 150 mL of saturated chlorine water in a 500-mL conical flask, add some glass beads, and pipet 5 mL of the Wijs solution into the flask. Shake, heat to boiling, and boil briskly for 10 min. Cool, add 30 mL of 2% sulfuric acid and 15 mL of 15% potassium iodide solution, and mix well. Titrate immediately with 0.1 N sodium thiosulfate, adding 3 mL of starch indicator solution near the end point.

Total Halogen. Transfer 150 mL of freshly boiled water to a dry 500-mL conical flask. Add 15 mL of 15% potassium iodide solution, pipet 20 mL of Wijs solution into the flask, and mix well. Titrate immediately with 0.1 N sodium thiosulfate, adding 3 mL of starch indicator solution near the end point.

Calculate the halogen ratio as follows:

$$I/Cl = \frac{2A}{3B - 2A}$$

where

A = volume, in mL, of thiosulfate used to titrate iodine
B = volume, in mL, of thiosulfate used to titrate total halogen (as iodine)

Iron, Low in Magnesium and Manganese

Fe   Atomic Wt 55.847

CAS Number 7439–89–6

REQUIREMENTS

Magnesium (Mg) . . . . $\leq$ 5 ppm
Manganese (Mn) . . . . $\leq$ 0.002%

TESTS

MAGNESIUM.   Determine the magnesium by the flame atomic absorption spectrophotometric method described on page 35.

> ***Sample Stock Solution.***   Dissolve 10 g in 60 mL of dilute hydrochloric acid (1 + 1) and 20 mL of nitric acid in a covered 400-mL beaker. Substitute a ribbed cover glass, evaporate to dryness, and bake at moderate heat for 5 min. Add 30 mL of hydrochloric acid, heat gently until the salts are dissolved, and dilute with water to 100 mL (0.1 g/mL).

Analyze the solutions by means of a suitable atomic absorption spectrophotometer, using the conditions outlined in the following table. Calculate the metal content of the sample by the method of standard additions.

Element	Wavelength (nm)	Sample Wt (g)	Standard Added (mg)	Flame Type*	Background Correction
Mg	285.2	2.0	0.01; 0.02	A/A	Yes

*A/A is air/acetylene.

MANGANESE.   To 1.0 g in a 150-mL beaker add 25 mL of water, 10 mL of nitric acid, 5 mL of sulfuric acid, and 5 mL of phosphoric acid. When dissolution is complete, add 0.25 g of potassium periodate, boil for 5 min, and cool. Transfer the oxidized solution to a 50-mL volumetric flask, dilute to volume with water, and transfer a portion to a 1-cm absorption cell. To the remaining portion add 0.05 mL of sodium nitrite solution (freshly prepared by dissolving 0.2 g of sodium nitrite in 10 mL of water), mix, and transfer to a matching 1-cm cell. Balance the spectrophotometer at 545 nm at 100% transmittance with the reduced solution in the light path and read the transmittance of the oxidized solution. The transmittance of the sample solution should be greater than that produced by a standard containing 0.02 mg of manganese ion (Mn) that has been treated exactly like the sample solution. (For 1-cm cells the percent transmittance of the standard is approximately 96.1.)

Isobutyl Alcohol

2-Methyl-1-propanol

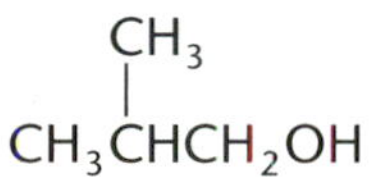

$$CH_3$$
$$|$$
$$CH_3CHCH_2OH$$

$(CH_3)_2CHCH_2OH$ Formula Wt 74.12

CAS Number 78–83–1

Suitable for use in ultraviolet spectrophotometry or general use. Product labeling shall designate the one or more of these uses for which suitability is represented on the basis of meeting the relevant requirements and tests. The ultraviolet spectrophotometry requirements include all of the requirements for general use.

REQUIREMENTS

General Use

Assay . . . . $\geq$ 99.0%
$(CH_3)_2CHCH_2OH$

MAXIMUM ALLOWABLE

Color (APHA). 10
Residue after evaporation 0.001%
Solubility in water . . . . Passes test
Titrable acid . . . . 0.0005 meq / g
Water (H_2O) 0.1%

Specific Use

ULTRAVIOLET SPECTROPHOTOMETRY

Wavelength (nm)	Absorbance (AU)
400–330.	0.01
260	0.05
250	0.07
240	0.25
230	1.00

TESTS

ASSAY. Analyze the sample by gas chromatography, using the parameters cited on page 56. The following specific conditions are also required.

> **Column:** Type I, methyl silicone

Measure the area under all peaks and calculate the isobutyl alcohol content in area percent. Correct for water content.

COLOR (APHA). (Page 17).

RESIDUE AFTER EVAPORATION. (Page 14). Evaporate 100 g (125 mL) to dryness in a tared dish on the steam bath and dry the residue at 105°C for 30 min.

SOLUBILITY IN WATER. Dilute 1 mL with 16 mL of water. The isobutyl alcohol should dissolve completely and the solution should be clear.

TITRABLE ACID. To 60 g (75 mL) in a glass-stoppered flask add 0.15 mL of phenolphthalein indicator solution and shake for 1 min. No pink color should form. Then titrate the solution with 0.01 N alcoholic potassium hydroxide to a faint pink color that persists for at least 15 s. Not more than 3.0 mL of 0.01 N potassium hydroxide should be consumed.

WATER. (Page 54, Method 1). Use 50 mL (40 g) of the sample.

ULTRAVIOLET SPECTROPHOTOMETRY. Use the procedure on page 67 to determine the absorbance.

Isopentyl Alcohol
Isoamyl alcohol
3-Methyl-1-butanol

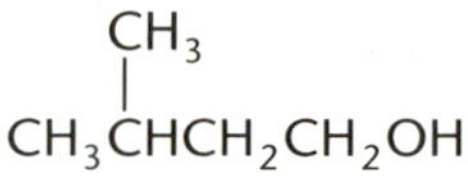

$$CH_3CHCH_2CH_2OH$$
(with CH_3 branch)

$(CH_3)_2CHCH_2CH_2OH$ Formula Wt 88.15

CAS Number 123–51–3

REQUIREMENTS

Assay . . . . $\geq$ 98.5% $C_5H_{11}OH$

MAXIMUM ALLOWABLE

Water (H_2O) 0.5%
Titrable acid . . . . 0.002 meq / g
Residue after evaporation 0.003%
Acids and esters (as amyl acetate) 0.2%
Carbonyl (as HCHO) 0.1%

TESTS

ASSAY. Analyze the sample by gas chromatography, using the general parameters cited on page 56. The following specific conditions are also required.

Column: Type I, methyl silicone

Measure the area under all peaks and calculate the isopentyl alcohol content in area percent. Correct for water content.

WATER. (Page 54, Method 1). Use 25 mL (20 g) of the sample.

TITRABLE ACID. Transfer 20 g (25 mL) of sample to a glass-stoppered flask and add 0.15 mL of phenolphthalein indicator solution. Titrate with 0.01 N sodium hydroxide to a faint pink color that persists for at least 30 s. Not more than 3.3 mL of the sodium hydroxide solution should be required.

RESIDUE AFTER EVAPORATION. (Page 14). Evaporate 40 g (50 mL) to dryness in a tared dish on a steam bath and dry the residue at 105°C for 30 min.

ACIDS AND ESTERS. Dilute 100 mL of sample with 100 mL of anhydrous isopropyl alcohol, add 25.0 mL of 0.1 N methanolic potassium hydroxide, and heat gently under a reflux condenser for 10–15 min. Cool and add 20 mL of carbon dioxide-free water through the reflux condenser. Detach the flask, add 0.15 mL of phenolphthalein indicator solution, and titrate the excess potassium hydroxide with 0.1 N hydrochloric acid to a colorless end point. The volume of 0.1 N potassium hydroxide consumed in the test should not be more than 12.3 mL greater than the volume consumed in a complete blank test.

CARBONYL COMPOUNDS. (Page 49). Analyze the sample by differential pulse polarography. Use 0.25 g (0.30 mL) of sample. For the standard use 0.25 mg of formaldehyde.

Isopropyl Alcohol
2-Propanol

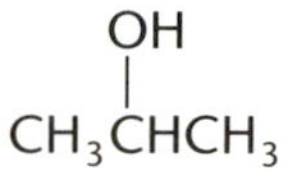

$$CH_3CHCH_3$$
with OH on the central carbon

$CH_3CHOHCH_3$ Formula Wt 60.10

CAS Number 67–63–0

Suitable for use in ultraviolet spectrophotometry or general use. Product labeling shall designate the one or both of these uses for which suitability is represented on the basis of meeting the relevant requirements and tests. The ultraviolet spectrophotometry requirements include all of the requirements for general use.

REQUIREMENTS

General Use

Assay . . . . ≥ 99.5%
$CH_3CHOHCH_3$

MAXIMUM ALLOWABLE

Color (APHA). 10
Residue after evaporation 0.001%
Solubility in water . . . . Passes test
Water (H_2O) 0.2%
Titrable acid or base. . . . . 0.0001 meq / g

Specific Use

ULTRAVIOLET SPECTROPHOTOMETRY

Wavelength (nm)	Absorbance (AU)
400–330	0.01
300	0.02
275	0.03
260	0.04
245	0.08
230	0.20
220	0.40
210	1.00

TESTS

ASSAY.   Analyze the sample by gas chromatography, using the general parameters cited on page 56. The following specific conditions are also required.

    Column: Type I, methyl silicone

Measure the area under all peaks and calculate the isopropyl alcohol content in area percent. Correct for water content.

COLOR (APHA).   (Page 17).

RESIDUE AFTER EVAPORATION.   (Page 14). Evaporate 100 g (128 mL) to dryness in a tared dish on the steam bath and dry the residue at 105°C for 30 min.

SOLUBILITY IN WATER.   Mix 10 mL with 40 mL of water and allow to stand 1 h. The solution should be as clear as an equal volume of water.

WATER.   (Page 54, Method 1). Use 10 mL (7.8 g) of the sample.

TITRABLE ACID OR BASE.   Mix 10 mL with 10 mL of carbon dioxide-free water and add 0.15 mL of bromthymol blue indicator solution. The solution should be neutral or require not more than 0.10 mL of 0.01 N hydrochloric acid or 0.10 mL of 0.01 N sodium hydroxide to render it so.

ULTRAVIOLET SPECTROPHOTOMETRY.   Use the procedure on page 67 to determine the absorbance.

Isopropyl ether
Diisopropyl ether

$$CH_3CH-O-CHCH_3$$

with methyl groups:

$$\overset{\displaystyle CH_3}{\overset{|}{CH_3CH}}-O-\overset{\displaystyle CH_3}{\overset{|}{CHCH_3}}$$

$(C_3H_7)_2O$ Formula Wt 102.18

CAS Number 108–20–3

> *WARNING.* Isopropyl ether is a volatile and flammable liquid with a tendency to form an explosive peroxide. No evaporation or distillation should be performed unless peroxide is shown to be absent. Generally, a stabilizer is present to retard peroxide formation.

REQUIREMENTS

Assay . . . . $\geq$ 99.0%
$(C_3H_7)_2O$

MAXIMUM ALLOWABLE

Color (APHA). 25
Peroxide (as $C_6H_{14}O_2$) 0.05%
Residue after evaporation 0.01%
Titrable acid . . . . 0.0007 meq / g

TESTS

ASSAY. Analyze the sample by gas chromatography, using the general parameters cited on page 56. The following specific conditions are also required.

Column: Type I, methyl silicone

Measure the area under all peaks and calculate the isopropyl ether content in area percent. Correct for water content.

COLOR (APHA). (Page 17).

PEROXIDE. Place 150 mL of 10% sulfuric acid in each of two 500-mL glass-stoppered Erlenmeyer flasks. Reserve one for a blank determination. Into the other, pipet 7.2 g (10 mL) of sample and mix well. To each flask, add 0.15–0.25 mL of 1% ammonium molybdate solution and 15 mL of 10% potassium iodide solution. Mix and allow to stand in the dark for 15 min at room temperature. Titrate each with 0.05 N sodium thiosulfate until the yellow color begins to fade. Add 3 mL of starch indicator solution, and continue the titration just to disappearance of the blue color. Correct for a blank determination.

$$\%C_6H_{14}O_2 = (V \times N \times 5.91)/W$$

where

V = net volume, in mL, of sodium thiosulfate solution consumed
N = normality of sodium thiosulfate
W = weight, in grams, of sample

RESIDUE AFTER EVAPORATION. (Page 14). See the warning on page 392. Evaporate 70 mL (50 g) of sample.

TITRABLE ACID. Titrate 36 g (50 mL) with 0.02 N alcoholic potassium hydroxide, with 0.15 mL of phenolphthalein solution as the indicator, to a pink end point that remains for at least 15 s. Not more than 1.25 mL should be required.

Lactic Acid, 85 %
2-Hydroxypropanoic Acid

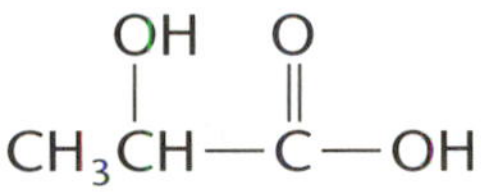

$$CH_3\overset{\overset{\displaystyle OH}{|}}{C}H-\overset{\overset{\displaystyle O}{\|}}{C}-OH$$

CAS Number 50–21–5

NOTE. The reagent generally available is a mixture of lactic acid, $CH_3CHOHCOOH$, and lactic acid lactate, $C_6H_{10}O_5$.

REQUIREMENTS

Assay 85.0–90.0%
$C_3H_6O_3$
Substances darkened by sulfuric acid . . . . Passes test

MAXIMUM ALLOWABLE

Residue after ignition 0.02%
Chloride (Cl) 0.001%
Sulfate (SO$_4$) 0.002%
Heavy metals (as Pb) . . . . 5 ppm
Iron (Fe) . . . . 5 ppm

TESTS

ASSAY. (By indirect acid–base titrimetry). Weigh accurately 3 mL in a tared, glass-stoppered flask. Add 50.0 mL of 1 N sodium hydroxide, boil gently for 20 min, and titrate the excess alkali with 1 N sulfuric acid, using 0.15 mL of phenolphthalein indicator solution. Correct for a complete blank test. One milliliter of 1 N sodium hydroxide corresponds to 0.09008 g of $C_3H_6O_3$.

RESIDUE AFTER IGNITION. (Page 15). Ignite 10 g (8 mL) and moisten the char with 1 mL of sulfuric acid. Retain the final residue for the preparation of sample solution A.

CHLORIDE. (Page 25). Dilute 12 g (10 mL) with water to 60 mL, mix, and use 5 mL of this solution. Retain the remainder of the solution for the test for sulfate.

SULFATE. (Page 32, Procedure A, Method 1). Use 12.5 mL of the solution prepared in the test for chloride.

> **Sample Solution A.** To the residue from the test for residue after ignition, add 2 mL of water and 2 mL of hydrochloric acid, and evaporate to dryness on the steam bath. Take up the residue in 1 mL of hydrochloric acid and 20 mL of water, filter if necessary, and dilute with water to 50 mL (1 mL = 0.2 g).

HEAVY METALS. (Page 27, Method 2). Test a 4.0-g sample.

IRON. (Page 28, Method 1). Use 10 mL of sample solution A.

SUBSTANCES DARKENED BY SULFURIC ACID. Cool 5 mL of sulfuric acid to 15°C and transfer to a clean, dry test tube. Cool 5 mL of the sample to 15°C and add to the test tube in such a way as to form a distinct layer over the sulfuric acid. A dark color should not develop at the interface of the two layers within 10 min.

Lactose Monohydrate

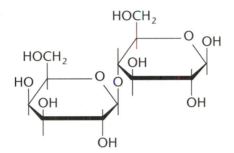

$C_{12}H_{22}O_{11} \cdot H_2O$ Formula Wt 360.32

CAS Number 64044−51−5

REQUIREMENTS

Water (H_2O) . . . . 4.0−6.0%

MAXIMUM ALLOWABLE

Insoluble matter 0.005%
Residue after ignition 0.03%
Dextrose . . . . Passes test
Sucrose . . . . Passes test
Heavy metals (as Pb) . . . . 5 ppm
Iron (Fe) . . . . 5 ppm

TESTS

WATER. (Page 54, Method 1). Use 2.0 g of the sample.

INSOLUBLE MATTER. (Page 14). Use 20 g dissolved in 150 mL of water.

RESIDUE AFTER IGNITION. (Page 15). Ignite 5.0 g. Moisten the char with 2 mL of sulfuric acid.

DEXTROSE. To 10 g, finely powdered, add 50 mL of 70% ethyl alcohol. Shake frequently for 30 min and filter through a small, dry filter paper. Evaporate 10 mL of the filtrate to dryness on the steam bath, and reserve the remainder of the filtrate for the test for sucrose. Dissolve the residue in 5 mL of water, filter if necessary, and transfer the filtrate to a test tube. Add 5 mL of cupric acetate solution, mix, immerse the test tube in a boiling-water bath for 3 min, and allow the solution to stand at room temperature for 25 min. No red precipitate should appear in the test tube.

> ***Cupric Acetate Solution.*** Dissolve 13.3 g of cupric acetate monohydrate, $(CH_3COO)_2Cu \cdot H_2O$, in a mixture of 195 mL of water and 5 mL of glacial acetic acid.

SUCROSE. Evaporate 25 mL of the filtrate reserved from the test for dextrose to dryness on the steam bath, and dissolve the residue in 9 mL of water. Add 1 mL of dilute hydrochloric acid (20 + 8) and 100 mg of resorcinol, transfer the mixture to a small test tube, and immerse the tube in a boiling-water bath for 8 min. The solution should remain colorless or acquire not more than a slight yellow coloration.

HEAVY METALS. (Page 26, Method 1). Dissolve 6.0 g in about 20 mL of water, and dilute with water to 30 mL. Use 25 mL to prepare the sample solution, and use the remaining 5.0 mL to prepare the control solution.

IRON. (Page 28, Method 1). Use 2.0 g.

Lanthanum Chloride, Hydrated

$LaCl_3 \cdot (6-7)H_2O$

CAS Number 10099-58-8

> *NOTE.* This salt is available in degrees of hydration ranging from 6 to 7 molecules of water.

REQUIREMENTS

Assay . . . . 64.5–70.0% $LaCl_3$

MAXIMUM ALLOWABLE

Insoluble matter 0.01%
Calcium (Ca) 0.001%
Magnesium (Mg) . . . . 1 ppm

TESTS

ASSAY.   (By gravimetric determination of La). Weigh 0.7–0.8 g accurately, dissolve in 25 mL of water, filter any undissolved material, and wash with water. Add 100 mL of 5% oxalic acid solution, digest on the steam bath until the supernatant solution is clear, and allow to stand overnight. Filter through a fine ashless filter paper, wash the precipitate with 10-mL portions of 1% oxalic acid solution, and ignite over an open flame. Finally, ignite in a muffle furnace at 900°C for 30 min. Cool and weigh as La_2O_3. One gram of lanthanum oxide corresponds to 1.506 g of $LaCl_3$.

INSOLUBLE MATTER.   (Page 14). Use 10 g dissolved in 100 mL of water.

CALCIUM AND MAGNESIUM.   Determine the calcium and magnesium by the flame atomic absorption spectrophotometric method described on page 35.

> *Sample Stock Solution.*   Dilute the filtrate from the insoluble matter test with water to 200.0 mL in a volumetric flask (0.05 g/mL).

Analyze the solutions by means of a suitable atomic absorption spectrophotometer, using the conditions outlined in the following table. Calculate the metal content of the sample by the method of standard additions.

Element	Wavelength (nm)	Sample Wt (g)	Standard Added (mg)	Flame Type*	Background Correction
Ca	422.7	1.0	0.01; 0.02	N/A	No
Mg	285.2	4.0	0.004; 0.008	A/A	Yes

*A/A is air/acetylene; N/A is nitrous oxide/acetylene.

Lead

Pb   Atomic Wt 207.2

CAS Number 7439–92–1

REQUIREMENTS

Assay . . . . ≥ 99.5% Pb

MAXIMUM ALLOWABLE

Antimony and tin (as Sn). 0.005%
Arsenic (As) . . . . 1 ppm
Bismuth (Bi) . . . . 5 ppm
Copper (Cu) . . . . 3 ppm
Iron (Fe) 0.001%
Nickel (Ni). 0.001%
Silver (Ag). . . . . 2 ppm

TESTS

ASSAY. (By complexometric titration of Pb). Weigh, to the nearest 0.1 mg, 0.7 g of sample, and add 5 mL of water plus 2 mL of nitric acid. Heat and stir until the lead is completely dissolved, and dilute with 200 mL of water. Add 15 mL of a saturated aqueous solution of hexamethylenetetramine and a few milligrams of xylenol orange indicator. Titrate with 0.1 M EDTA to a yellow end point. One milliliter of 0.1 M EDTA corresponds to 0.02072 g of Pb.

ANTIMONY AND TIN. Dissolve 20 g in 120 mL of dilute nitric acid (1 + 3), warming if necessary to hasten dissolution. Restore the volume to 120 mL if necessary, agitate well, and while any insoluble residue is uniformly dispersed, divide the solution into portions of 30 mL and 90 mL. To the 30-mL portion, as a control, add a solution containing 0.5 mg of tin ion (Sn). Evaporate both solutions to moist residues on the steam bath. Dissolve each residue in 100 mL of water and digest on the steam bath for 15 min. The amount of turbidity or sediment in the solution representing the sample should not exceed that in the control.

Sample Solution A.   Dissolve 110 g in 700 mL of dilute nitric acid (1 + 3), warming if necessary to hasten dissolution. Cool the solution, transfer to a 1-L volumetric flask, and cautiously add 66 mL of dilute sulfuric acid (1 + 1). Cool, dilute with water to the mark, and mix. Allow the precipitate to settle and decant 910 mL. This volume is equivalent to 100 g of the sample. Evaporate until dense fumes of sulfur trioxide appear, but avoid any considerable loss of sulfuric acid. Cool, carefully add 50 mL of water, cool again, filter, and wash slightly. Dilute the filtrate and washings with water to 100 mL (1 mL = 1 g).

ARSENIC.   (Page 23). To 3.0 mL of sample solution A (3-g sample) in a generator flask add 5 mL of sulfuric acid, and evaporate to fumes of sulfur trioxide. Cool, cautiously wash down the sides of the flask, and again evaporate to fumes of sulfur trioxide. Cool, cautiously dilute with 50 mL of water, and use this solution without further acidification. For the standard use 0.003 mg of arsenic (As).

BISMUTH.   Neutralize 5.0 mL of sample solution A (5-g sample) with ammonium hydroxide, dilute with water to 25 mL, and add 5 mL of dilute nitric acid (1 + 9). Add 10 mL of freshly prepared 10% solution of thiourea and mix. Any yellow color should not exceed that produced by 0.025 mg of bismuth ion (Bi) in an equal volume of solution containing the quantities of reagents used in the test.

COPPER, NICKEL, AND SILVER.   Determine the copper, nickel, and silver by the flame atomic absorption spectrophotometric method described on page 35.

Sample Stock Solution.   Dissolve 20.0 g of sample in 60 mL of nitric acid (1 + 3), heat to dissolve, cool, and dilute with water to the mark in a 100-mL volumetric flask (0.2 g/mL).

Analyze the solutions by means of a suitable atomic absorption spectrophotometer, using the conditions outlined in the following table. Calculate the metal content of the sample by the method of standard additions.

Element	Wavelength (nm)	Sample Wt (g)	Standard Added (mg)	Flame Type*	Background Correction
Cu	324.8	6.0	0.015; 0.03	A/A	Yes
Ni	232.0	6.0	0.06; 0.012	A/A	Yes
Ag	328.1	6.0	0.012; 0.024	A/A	Yes

*A/A is air/acetylene.

IRON. (Page 28, Method 1). Use 1.0 mL of sample solution A (1-g sample).

Lead Acetate Trihydrate
Lead(II) Acetate Trihydrate

$(CH_3COO)_2Pb \cdot 3H_2O$ Formula Wt 379.3

CAS Number 6080–56–4

REQUIREMENTS

Assay 99.0–103.0%
$(CH_3COO)_2Pb \cdot 3H_2O$

MAXIMUM ALLOWABLE

Insoluble matter 0.01%
Chloride (Cl) . . . . 5 ppm
Nitrate and nitrite (as NO_3). 0.005%
Copper (Cu) 0.002%
Substances not precipitated by
   hydrogen sulfide (as sulfates) 0.05%
Iron (Fe) 0.001%

TESTS

ASSAY. (By complexometric titration of Pb). Weigh accurately 1.5 g, transfer to a 400-mL beaker, and dissolve in 250 mL of water containing 0.15 mL of acetic acid. Add 15 mL of a saturated aqueous solution of hexamethylenetetramine. Add a few milligrams of xylenol orange indicator mixture and titrate with standard 0.1 M EDTA to a yellow color. One milliliter of 0.1 M EDTA corresponds to 0.03793 g of $(CH_3COO)_2Pb \cdot 3H_2O$.

INSOLUBLE MATTER. (Page 14). Use 10 g dissolved in 100 mL of dilute glacial acetic acid (1 + 99) free from carbon dioxide.

CHLORIDE. (Page 25). Use 2.0 g.

NITRATE AND NITRITE.

Sample Solution A.   Dissolve 0.20 g in 1 mL of water. Add 20 mL of nitrate-free sulfuric acid (2 + 1), centrifuge, and decant the liquid portion into a 200-mm test tube. Dilute to 50 mL with brucine sulfate reagent solution.

Control Solution B.   Dissolve 0.20 g in 1 mL of the standard nitrate solution containing 0.01 mg of nitrate ion (NO_3) per mL. Add 20 mL of nitrate-free sulfuric acid (2 + 1), centrifuge, and decant the liquid portion into a 200-mm test tube. Dilute to 50 mL with brucine sulfate reagent solution. Continue with the procedure described on page 29, starting with the preparation of blank solution C.

COPPER AND IRON.   Determine the copper and iron by the flame atomic spectrophotometric method described on page 35.

Sample Stock Solution.   Dissolve 10.0 g of sample in 50 mL of water and 1 mL of acetic acid. Transfer to a 100-mL volumetric flask and dilute to the mark with water (0.1 g/mL).

Analyze the solutions by means of a suitable atomic absorption spectrophotometer, using the conditions outlined in the following table. Calculate the metal content of the sample by the method of standard additions.

Element	Wavelength (nm)	Sample Wt (g)	Standard Added (mg)	Flame Type*	Background Correction
Cu	324.7	2.0	0.04; 0.08	A/A	Yes
Fe	248.3	2.0	0.02; 0.04	A/A	Yes

*A/A is air/acetylene.

SUBSTANCES NOT PRECIPITATED BY HYDROGEN SULFIDE.   Dissolve 4.0 g in 100 mL of dilute acetic acid (1 + 49). Pass hydrogen sulfide through the solution to precipitate lead completely and filter without washing. To 50 mL of the filtrate (2-g sample) add 0.25 mL of sulfuric acid, evaporate to dryness in a tared dish, heat gently to volatilize the excess acids, and finally ignite at 800 ± 25°C for 15 min.

Lead Carbonate

CAS Number 598 –63 –0 (normal); 1344 –36 –1 (basic)

> *NOTE.* This specification applies to both the normal lead carbonate and the basic lead carbonate.

REQUIREMENTS

MAXIMUM ALLOWABLE

Insoluble in dilute acetic acid. 0.02%
Chloride (Cl) 0.002%
Nitrate and nitrite (as NO_3). . . . . Passes test (limit
about 0.005%)
Substances not precipitated by
hydrogen sulfide (as sulfates) 0.2%
Cadmium (Cd). 0.002%
Iron (Fe) 0.005%
Zinc (Zn). 0.003%

TESTS

INSOLUBLE IN DILUTE ACETIC ACID. Digest 5.0 g in 50 mL of water plus 7 mL of glacial acetic acid in a covered beaker on a steam bath for 1 h. Filter through a tared filtering crucible, wash thoroughly with dilute acetic acid (2 + 98), and dry at 105°C.

CHLORIDE. (Page 25). Dissolve 1.0 g in 20 mL of 10% nitric acid reagent solution, filter if necessary through a chloride-free filter, and dilute with water to 50 mL. Use 25 mL of this solution.

NITRATE AND NITRITE

Sample Solution A. Dissolve 0.20 g in 1 mL of water and 8 mL of dilute acetic acid (1 + 7) by heating in a boiling-water bath. Add 20 mL of nitrate-free sulfuric acid (2 + 1), cool, centrifuge, and decant the liquid portion into a 200-mm test tube. Dilute to 50 mL with brucine sulfate reagent solution.

Control Solution B. Dissolve 0.20 g in 8 mL of dilute acetic acid (1 + 7) and 1 mL of the standard nitrate solution containing

0.01 mg of nitrate ion (NO_3) per mL by heating in a boiling-water bath. Add 20 mL of nitrate-free sulfuric acid (2 + 1), cool, centrifuge, and decant the liquid portion into a 200-mm test tube. Dilute to 50 mL with brucine sulfate reagent solution.

Continue with the procedure described on page 29, starting with the preparation of blank solution C.

SUBSTANCES NOT PRECIPITATED BY HYDROGEN SULFIDE. Dissolve 2.0 g in 25 mL of dilute nitric acid (3 + 7), heating if necessary. Dilute with water to 100 mL and precipitate the lead with hydrogen sulfide. Filter without washing. To 50 mL of the filtrate add 0.15 mL of sulfuric acid, evaporate to dryness in a tared dish, and ignite at 800 ± 25°C for 15 min.

CADMIUM, IRON, AND ZINC. Determine the cadmium, iron, and zinc by the flame atomic absorption spectrophotometric method described on page 35.

Sample Stock Solution. Dissolve 10.0 g of sample in 50 mL of water and sufficient nitric acid to achieve complete dissolution. Transfer the solution to a 100-mL volumetric flask and dilute to the mark with water (0.1 g/mL).

Analyze the solutions by means of a suitable atomic absorption spectrophotometer, using the conditions outlined in the following table. Calculate the metal content of the sample by the method of standard additions.

Element	Wavelength (nm)	Sample Wt (g)	Standard Added (mg)	Flame Type*	Background Correction
Cd	228.8	1.0	0.02; 0.04	A/A	Yes
Fe	248.3	1.0	0.05; 0.10	A/A	Yes
Zn	213.9	0.4	0.01; 0.02	A/A	Yes

*A/A is air/acetylene.

Lead Chromate

PbCrO$_4$ Formula Wt 323.2

CAS Number 7758–97–6

REQUIREMENTS

Assay . . . . $\geq$ 98.0% PbCrO$_4$

MAXIMUM ALLOWABLE

Soluble matter. 0.15%
Carbon compounds (as C) 0.01%

TESTS

ASSAY. (By titration of oxidizing power of chromate). Weigh accurately 0.5 g of powdered sample and dissolve by warming with 40 mL of 10% sodium hydroxide reagent solution in a glass-stoppered conical flask. Add 2 g of potassium iodide; when it is dissolved, dilute with 80 mL of water and 15 mL of hydrochloric acid. Stopper, swirl, and allow to stand in the dark for 5 min. Titrate the liberated iodine with 0.1 N sodium thiosulfate, adding 3 mL of starch indicator solution near the end of the titration. Correct for a blank. One milliliter of 0.1 N sodium thiosulfate corresponds to 0.01077 g of PbCrO$_4$.

SOLUBLE MATTER. Boil 0.50 g of the powdered sample for 5 min with 100 mL of dilute acetic acid (1 + 20), stirring well during the heating. Cool and filter. Evaporate 25 mL of the filtrate to dryness in a tared dish on the steam bath and dry at 105°C. Heat 50 mL of the filtrate with 2 g of the powdered sample for 5 min at 80–90°C, cool, dilute with water to 50 mL, and filter. Evaporate 25 mL of this filtrate in a tared dish on the steam bath and dry at 105°C. The difference between the weights of the two residues should not be more than 0.0015 g.

CARBON COMPOUNDS. This procedure uses induction heating of the sample and a conductometric measurement of evolved carbon dioxide.

Transfer 1.0 g of sample and one tin capsule to a combustion crucible. Add 1.5 g each of iron and copper accelerator. Ignite in an induction furnace in a stream of carbon dioxide-free oxygen until combustion is complete. Pass the effluent gas through an approximately 0.1% solution of reagent-grade barium hydroxide. Measure the change in conductance, in ohms, using a suitable cell and detection system. Subtract the reading from a blank and calculate the carbon content of the sample from a calibration curve.

Prepare the calibration curve as follows: transfer 0, 10, 25, 75, and 100 μL of carbon standard solution (1 μL = 4 μg of carbon) to tin capsules. Evaporate the solutions to dryness in the capsules at 80°C. Transfer the capsules to combustion crucibles, add the accelerators, ignite, and measure the conductance under the same conditions as described for running the sample. Subtract the reading of the blank from that of each of the standards and plot the calibration curve as ohms versus micrograms of carbon.

Lead Dioxide
Lead(IV) Oxide

PbO_2 Formula Wt 239.2

CAS Number 1309–60–0

REQUIREMENTS

Assay . . . . $\geq$ 97.0% PbO_2

MAXIMUM ALLOWABLE

Acid-insoluble matter 0.2%
Carbon compounds (as C) 0.04%
Chloride (Cl) 0.002%
Nitrate (NO_3) 0.02%
Sulfate (SO_4) 0.05%
Manganese (Mn) . . . . 5 ppm
Other hydrogen sulfide metals (as Cu) 0.05%
Substances not precipitated by hydrogen sulfide . . . 0.5%

TESTS

ASSAY. (By titration of oxidizing power). Dissolve 25 g of sodium chloride in 100 mL of water in a stoppered flask. Add 2 g of potassium iodide and 20 mL of hydrochloric acid. Add, accurately weighed, 0.25 g of finely powdered sample and stir until the sample is dissolved. Titrate the liberated iodine with 0.1 N sodium thiosulfate, using 3 mL of starch indicator added near the end of the titration. Run a complete blank determination using the same quantities of reagents, diluting the sample, and stirring for the same time. One milliliter of 0.1 N sodium thiosulfate corresponds to 0.01196 g of PbO_2.

ACID-INSOLUBLE MATTER. To 1.0 g add 25 mL of water and 3 mL of nitric acid. Add carefully with stirring 5 mL of 30% hydrogen peroxide or more, if needed, to dissolve all the lead dioxide. Filter through a tared filtering crucible, wash thoroughly, and dry at 105°C. Reserve the filtrate for the test for substances not precipitated by hydrogen sulfide.

CARBON COMPOUNDS. This procedure uses induction heating of the sample and a conductometric measurement of the evolved carbon dioxide.

Transfer 1.0 g of sample and one tin capsule to a combustion crucible. Add 1.5 g each of iron and copper accelerator. Ignite in an induction furnace in a stream of carbon dioxide-free oxygen until combustion is complete. Pass the effluent gas through an approximately 0.1% solution of barium hydroxide. Measure the change in conductance, in ohms, by using a suitable cell and detection system. Subtract the reading from a blank and calculate the carbon content of the sample from a calibration curve.

Prepare the calibration curve as follows: transfer 0, 10, 25, 75, and 100 μL of carbon standard solution (1 μL = 4 μg of carbon) to tin capsules. Evaporate the solutions to dryness in the capsules at 80°C. Transfer the capsules to combustion crucibles, add the accelerators, ignite, and measure the conductance under the same conditions as described for running the sample. Subtract the reading of the blank from that of each of the standards and plot the calibration curve as ohms versus micrograms of carbon.

CHLORIDE. (Page 25). Dissolve 2.0 g in 20 mL of water plus 2 mL of glacial acetic acid and 2 mL of 30% hydrogen peroxide. Filter through a chloride-free filter, dilute with water to 100 mL, and use 25 mL of this solution.

NITRATE. Dissolve 0.25 g in a test tube containing 10 mL of water, 1 mL of acetic acid, and 0.5 mL of 30% hydrogen peroxide. Pass sulfur dioxide through the solution to precipitate the lead, filter, wash with two 2-mL portions of water, and dilute the filtrate with water to 20 mL.

Sample Solution A. To 4.0 mL of the filtrate in a test tube add 1 mL of water, dilute to 50 mL with brucine sulfate reagent solution, and mix.

Control Solution B. To 4.0 mL of the filtrate in a test tube add 1 mL of the standard nitrate solution containing 0.01 mg of nitrate ion (NO_3) per mL, dilute to 50 mL with brucine sulfate reagent solution, and mix.

Continue with the procedure described on page 29, starting with the preparation of blank solution C.

SULFATE.   Decompose 2.0 g with 10 mL of hydrochloric acid and 3 mL of nitric acid and evaporate to dryness on a hot plate at low setting. Add 25 mL of water and 2 g of sodium carbonate and digest on a hot plate at low setting for several hours, with occasional stirring. Dilute with water to 100 mL and filter. Neutralize the filtrate with hydrochloric acid and add an excess of 1 mL of the acid. Heat to boiling, add 5 mL of barium chloride reagent solution, digest in a covered beaker on the hot plate for 2 h, and allow to stand overnight. Filter, using a fine ashless paper, wash thoroughly, and ignite. Correct for the weight obtained in a complete blank test.

MANGANESE.   Add 10 g to 15 mL of dilute nitric acid (1 + 1). Add carefully, with stirring, about 10 mL of 30% hydrogen peroxide—or more, if necessary—to dissolve all of the lead dioxide. Evaporate nearly to dryness, cool, add 20 mL of dilute sulfuric acid (1 + 1), and evaporate to dense fumes of sulfur trioxide. Cool, cautiously dilute to 100 mL with water, allow to settle, and filter. Dilute 20 mL of the filtrate with water to 50 mL, add 10 mL of nitric acid and 5 mL of phosphoric acid, and boil gently for 5 min. Cool slightly, add 0.25 g of potassium periodate, and again boil gently for 5 min. Any pink color should not exceed that produced by 0.01 mg of manganese ion (Mn) in 50 mL of dilute sulfuric acid (1 + 9) treated exactly like the 50 mL of filtrate.

OTHER HYDROGEN SULFIDE METALS.   Suspend 0.40 g in 20 mL of dilute acetic acid (1 + 19) and add 1 mL of 30% hydrogen peroxide. When dissolution is complete, add 0.3 mL of sulfuric acid and allow the precipitate to settle. Filter, do not wash the precipitate, and dilute the filtrate with water to 100 mL. To 10 mL of the filtrate (0.04-g sample) add about 10 mg of sodium carbonate, evaporate to near dryness on a hot plate, add 1 mL of nitric acid, and evaporate to dryness. Dissolve the residue in about 20 mL of water and dilute with water to 25 mL. For the standard dilute a solution containing 0.02 mg of copper ion (Cu) with water to 25 mL. Adjust the pH of the standard and sample solutions to between 3 and 4 (using a pH meter) with 1 N acetic acid or ammonium hydroxide (10% NH_3), dilute with water to 40 mL, and mix. Add 10 mL of freshly prepared hydrogen sulfide water to each and mix. Any color in the solution of the sample should not exceed that in the standard.

SUBSTANCES NOT PRECIPITATED BY HYDROGEN SULFIDE. Evaporate to dryness the filtrate obtained in the test for acid-insoluble matter. Moisten the residue with 0.2 mL of hydrochloric acid, dissolve in 100 mL of water, and pass hydrogen sulfide through the solution until the lead is completely precipitated. Filter, but do not wash. Evaporate 50 mL of the clear filtrate (0.5-g sample) to dryness in a tared dish. Heat gently to volatilize the excess salts and acids, and finally ignite at 800 ± 25°C for 15 min.

Lead Monoxide
Lead(II) Oxide
Litharge

PbO   Formula Wt 223.2

CAS Number 1317–36–8

REQUIREMENTS

Assay . . . . ≥ 99.0% PbO

MAXIMUM ALLOWABLE

Insoluble in dilute acetic acid. 0.02%
Chloride (Cl) 0.002%
Nitrate (NO_3) 0.01%
Copper (Cu) 0.005%
Iron (Fe) 0.002%
Silver (Ag). . . . . 5 ppm
Substances not precipitated by H_2S (as SO_4) 0.1%

TESTS

ASSAY. (By complexometric titration of Pb). Weigh accurately 0.8 g, transfer to a 400-mL beaker, and dissolve by warming with a mixture of 1 mL of acetic acid and 10 mL of water. After solution is complete, dilute to 250 mL with water and add 15 mL of saturated aqueous hexamethylenetetramine solution and 50 mg of xylenol orange indicator mixture. Titrate with 0.1 M EDTA to a color change of purple-red to lemon yellow. One milliliter of 0.1 M EDTA corresponds to 0.02232 g of PbO.

INSOLUBLE IN DILUTE ACETIC ACID. Heat 5.0 g with a mixture of 50 mL of water and 15 mL of acetic acid until solution is complete. Filter through a tared sintered-glass crucible, wash thoroughly with hot water, dry at 105°C for 1 h, cool, and weigh. Retain the filtrate for the test for substances not precipitated by H_2S.

CHLORIDE. (Page 25). Use 0.5 g.

NITRATE. (Page 29). Dissolve 0.5 g in a mixture of 10 mL of water and 1 mL of acetic acid. Add 5 mL of 25% sulfuric acid reagent solution, mix well, and allow to stand for 15 min, mixing occasionally. Filter, wash with 5 mL of water, and dilute the filtrate to 25 mL.

> *Sample Solution A.* Dilute 5.0 mL of the nitrate-test filtrate to 50 mL with brucine sulfate reagent solution.

> *Control Solution B.* To 5.0 mL of the nitrate-test filtrate add 1.0 mL of the standard nitrate solution containing 0.01 mg of nitrate ion per mL. Dilute to 50 mL with brucine sulfate reagent solution.

Continue with the procedure described on page 29, starting with the preparation of blank solution C.

COPPER, IRON, AND SILVER. Determine the copper, iron, and silver by the flame atomic absorption spectrophotometric method described on page 35.

> *Sample Stock Solution.* Heat 10.0 g with a mixture of 60 mL of water and 30 mL of acetic acid until the sample is completely dissolved. Cool, transfer to a 100-mL volumetric flask, and dilute with water to volume (1 mL = 0.10 g).

Analyze the solutions by means of a suitable atomic absorption spectrophotometer, using the conditions outlined in the following table. Calculate the metal content of the sample by the method of standard additions.

Element	Wavelength (nm)	Sample Wt (g)	Standard Added (mg)	Flame Type*	Background Correction
Cu	324.8	0.8	0.02; 0.04	A/A	Yes
Fe	248.3	2.5	0.025; 0.05	A/A	Yes
Ag	328.1	4.0	0.01; 0.02	A/A	Yes

*A/A is air/acetylene.

SUBSTANCES NOT PRECIPITATED BY H₂S. Neutralize the filtrate reserved from the test for insoluble in dilute acetic acid with ammonium hydroxide, and then acidify with 5 mL of glacial acetic acid. Dilute to 125 mL with water, and pass hydrogen sulfide through the solution until the lead is completely precipitated. Filter, and to 50 mL (2-g sample) of the filtrate add 0.25 mL of sulfuric acid and evaporate to dryness in a tared porcelain crucible, ignite, cool, and weigh. The weight of the ignited residue should not be more than 0.002 g greater than the weight obtained in a complete blank test.

Lead Nitrate
Lead(II) Nitrate

$Pb(NO_3)_2$ Formula Wt 331.2

CAS Number 10099–74–8

REQUIREMENTS

Assay . . . . $\geq$ 99.0% $Pb(NO_3)_2$

MAXIMUM ALLOWABLE

Insoluble matter 0.005%
Chloride (Cl) 0.001%
Copper (Cu) 0.002%
Iron (Fe) 0.001%
Substances not precipitated by
    hydrogen sulfide (as sulfates) 0.1%

TESTS

ASSAY. (By complexometric titration of Pb). Weigh accurately 1.3 g, transfer to a 400-mL beaker, and dissolve in 250 mL of water containing 0.15 mL of acetic acid. Add 15 mL of a saturated aqueous solution of hexamethylenetetramine. Add a few milligrams of xylenol orange indicator mixture and titrate with standard 0.1 M EDTA to a yellow color. One milliliter of 0.1 M EDTA corresponds to 0.03312 g of $Pb(NO_3)_2$.

INSOLUBLE MATTER. (Page 14). Use 20 g dissolved in 200 mL of dilute nitric acid (1 + 99).

CHLORIDE.   (Page 25). Use 1.0 g.

COPPER AND IRON.   Determine the copper and iron by the flame atomic absorption spectrophotometric method described on page 35.

> *Sample Stock Solution.*   Dissolve 10.0 g of sample with water in a 100-mL volumetric flask, add 5 mL of nitric acid, and dilute to the mark with water (0.1 g/mL).

Analyze the solutions by means of a suitable atomic absorption spectrophotometer, using the conditions outlined in the following table. Calculate the metal content of the sample by the method of standard additions.

Element	Wavelength (nm)	Sample Wt (g)	Standard Added (mg)	Flame Type*	Background Correction
Cu	324.7	1.0	0.02; 0.04	A/A	Yes
Fe	248.3	2.0	0.02; 0.04	A/A	Yes

*A/A is air/acetylene.

SUBSTANCES NOT PRECIPITATED BY HYDROGEN SULFIDE.   Dissolve 2.0 g in 100 mL of water and pass hydrogen sulfide through the solution until the lead is completely precipitated. Filter but do not wash. To 50 mL of the filtrate add 0.25 mL of sulfuric acid, evaporate to dryness in a tared dish, and ignite at $800 \pm 25°C$ for 15 min.

Lead Perchlorate Trihydrate

$Pb(ClO_4)_2 \cdot 3H_2O$ Formula Wt 460.2

CAS Number 13637−76−8

REQUIREMENTS

Assay . . . . 97.0%−102.0%
$Pb(ClO_4)_2 \cdot 3H_2O$
pH of a 5% solution . . . . 3.0−5.0 at 25°C

MAXIMUM ALLOWABLE

Insoluble matter 0.005%
Chloride (Cl) 0.01%
Iron (Fe) 0.001%

TESTS

ASSAY. (By complexometric titration of lead). While protecting from moisture, weigh accurately 1.8 g, transfer to a 400-mL beaker, and dissolve in 250 mL of water. Add 15 mL of a saturated solution of hexamethylenetetramine and a few milligrams of xylenol orange indicator mixture, and titrate with standard 0.1 M EDTA solution to a yellow color. A precipitate that dissolves readily during the titration will form. One milliliter of 0.1 M EDTA corresponds to 0.04602 g of $Pb(ClO_4)_2 \cdot 3H_2O$.

pH OF A 5% SOLUTION.   (Page 43).

INSOLUBLE MATTER.   (Page 14). Use 20 g dissolved in water.

CHLORIDE.   (Page 25). Use 1 g in 100 mL and take a 10-mL aliquot.

IRON.   (Page 28, Method 2). Use 1 g, and filter after addition of hydroxylamine hydrochloride. Wash in the funnel and use the filtrate.

Lead Subacetate
(For Sugar analysis)

CAS Number 1335–32–6

REQUIREMENTS

Basic lead (PbO) . . . . ≥ 33.0%

MAXIMUM ALLOWABLE

Loss on drying at 105°C. 1.5%
Insoluble in dilute acetic acid 0.02%
Insoluble in water 1.0%
Chloride (Cl) 0.003%
Nitrate and nitrite (as NO_3). . . . . Passes test (limit about 0.003%)
Copper (Cu) 0.002%
Substances not precipitated by
    hydrogen sulfide (as sulfates) 0.3%
Iron (Fe) 0.002%

TESTS

BASIC LEAD. (By indirect acidimetry). Weigh accurately 5 g of sample and dissolve in 100 mL of carbon dioxide-free water in a 500-mL volumetric flask. Add 50.0 mL of 1 N acetic acid and 100 mL of carbon dioxide-free 3% solution of sodium oxalate. Mix thoroughly, dilute to volume with carbon dioxide-free water, and allow the precipitate to settle. Titrate 100.0 mL of the clear supernatant liquid with 1 N sodium hydroxide, using 0.15 mL of phenolphthalein indicator. Each milliliter of 1 N acetic acid consumed is equivalent to 0.1116 g of PbO.

LOSS ON DRYING.   Weigh accurately about 0.5 g in a tared dish or crucible and dry for 2 h at 105°C. Cool and reweigh.

INSOLUBLE IN DILUTE ACETIC ACID.   Dissolve 5.0 g in 100 mL of dilute acetic acid (1 + 19), warming if necessary to effect complete dissolution. Filter through a tared filtering crucible, wash with dilute acetic acid (1 + 19) until the washings are no longer darkened by hydrogen sulfide, and dry at 105°C.

INSOLUBLE IN WATER.   Agitate 1.0 g in a small stoppered flask with 50 mL of carbon dioxide-free water, just until dissolved, and filter at once through a tared filtering crucible. Wash with carbon dioxide-free water and dry at 105°C.

CHLORIDE.   (Page 25). Dissolve 1.0 g in 30 mL of dilute nitric acid (1 + 9). Use 10 mL of this solution.

NITRATE AND NITRITE.

> *Sample Solution A.*   Suspend 0.50 g in 3 mL of water. Dilute to 50 mL with brucine sulfate reagent solution.

> *Control Solution B.*   Suspend 0.50 g in 1.5 mL of water and 1.5 mL of the standard nitrate solution containing 0.01 mg of nitrate ion (NO_3) per mL. Dilute to 50 mL with brucine sulfate reagent solution.

> *Blank Solution C.*   Use 50 mL of brucine sulfate reagent solution.

Heat sample solution A, control solution B, and blank solution C in a preheated (boiling) water bath until the lead subacetate in A and B is dissolved, then heat for 10 min more. Cool rapidly in an ice bath to room temperature, centrifuge sample solution A and control solution

B for 10 min, and decant the liquids into 200-mm test tubes. Set a spectrophotometer at 410 nm and, using 1-cm cells, adjust the instrument to read zero absorbance with blank solution C in the light path, then determine the absorbance of sample solution A. Adjust the instrument to read 0 absorbance with sample solution A in the light path and determine the absorbance of control solution B. The absorbance of sample solution A should not exceed that of control solution B.

COPPER AND IRON.   Determine the copper and iron by the flame atomic absorption spectrophotometric method described on page 35.

> *Sample Stock Solution.*   Dissolve 10 g of sample in 50 mL of dilute acetic acid (1 + 9), warming if necessary to achieve complete dissolution. Dilute the solution with water to the mark in a 100-mL volumetric flask (0.1 g/mL).

Analyze the solutions by means of a suitable atomic absorption spectrophotometer, using the conditions outlined in the following table. Calculate the metal content of the sample by the method of standard additions.

Element	Wavelength (nm)	Sample Wt (g)	Standard Added (mg)	Flame Type*	Background Correction
Cu	324.8	1.0	0.02; 0.04	A/A	Yes
Fe	248.3	2.5	0.05; 0.10	A/A	Yes

*A/A is air/acetylene.

SUBSTANCES NOT PRECIPITATED BY HYDROGEN SULFIDE.   Dissolve 2.0 g in 100 mL of dilute acetic acid (1 + 49). Pass hydrogen sulfide through the solution to precipitate the lead completely and filter without washing. To 50 mL of the filtrate (1-g sample) add 0.25 mL of sulfuric acid, evaporate to dryness in a tared dish, and ignite at $800 \pm 25°C$ for 15 min.

Lithium Carbonate

Li_2CO_3 Formula Wt 73.89

CAS Number 554–13–2

NOTE. The formula weight of this reagent is likely to deviate from the value cited, since the natural distribution of ^{6}Li and ^{7}Li isotopes is often altered in current sources of lithium compounds.

REQUIREMENTS

Assay . . . . $\geq$ 99.0% Li_2CO_3

MAXIMUM ALLOWABLE

Insoluble in dilute hydrochloric acid 0.01%
Chloride (Cl) 0.005%
Nitrate (NO_3) . . . . 5 ppm
Sulfur compounds (as SO_4) 0.2%
Ammonium (NH_4) . . . . 5 ppm
Heavy metals (as Pb) 0.002%
Iron (Fe) 0.002%
Calcium (Ca) 0.01%
Potassium (K) 0.01%
Sodium (Na) 0.1%

TESTS

ASSAY. (By indirect acid–base titrimetry for carbonate). Weigh accurately 1.3 g of sample, add 50 mL of water, and then add cautiously 50.0 mL of 1 N hydrochloric acid. Boil gently to expel carbon dioxide, cool, add 0.15 mL of methyl red indicator solution, and titrate the excess hydrochloric acid with 1 N sodium hydroxide solution. One milliliter of 1 N hydrochloric acid corresponds to 0.03694 g of Li_2CO_3.

INSOLUBLE IN DILUTE HYDROCHLORIC ACID. Suspend 10 g in 75 mL of water, then add cautiously and slowly 25 mL of hydrochloric acid. Heat to boiling and digest in a covered beaker on the steam bath for 1 h. Filter through a tared filtering crucible, wash thoroughly with hot water, and dry at 105°C.

CHLORIDE. (Page 25). Dissolve 2.0 g in 25 mL of dilute nitric acid (1 + 5), dilute with water to 50 mL, and use 5.0 mL of this solution.

NITRATE.

Sample Solution A. Add 2.0 g to 4 mL of water in a test tube. Add brucine sulfate reagent solution, drop by drop, until the sample is completely dissolved. Then dilute to 50 mL with the same reagent solution and mix.

Control Solution B.   Add 2.0 g to 3 mL of water and 1 mL of the standard nitrate solution containing 0.01 mg of nitrate ion (NO_3) per mL. Add brucine sulfate reagent solution, drop by drop, until the sample is completely dissolved. Then dilute to 50 mL with the same reagent solution and mix.

Continue with the procedure described on page 29, starting with the preparation of blank solution C.

SULFUR COMPOUNDS.   Suspend 0.10 g in 10 mL of water and cautiously neutralize with dilute hydrochloric acid (1 + 9). Add 0.10 mL of bromine water and boil gently until the bromine is completely expelled. Add 2 mL of dilute hydrochloric acid (1 + 19), filter, and dilute with water to 40 mL. To 10 mL add 1 mL of barium chloride reagent solution. Any turbidity should not exceed that produced by 0.05 mg of sulfate ion (SO_4) in an equal volume of solution containing 1 mL of dilute hydrochloric acid (1 + 19) and 1 mL of barium chloride reagent solution. Compare 10 min after adding the barium chloride to the sample and standard solutions.

AMMONIUM.   (Page 22). Suspend 2.0 g in 50 mL of water and cautiously add 5 mL of hydrochloric acid. The distillate shows not more than 0.01 mg of ammonium ion (NH_4) plus the residue from 5 mL of hydrochloric acid.

HEAVY METALS.   (Page 26, Method 1). Suspend 1.5 g in 10 mL of water and cautiously add 10 mL of hydrochloric acid. Evaporate the solution to a moist residue on the steam bath, dissolve the residue in 20 mL of water, and dilute with water to 25 mL. For the control solution suspend 0.5 g in 10 mL of water, add 0.02 mg of lead ion (Pb), and treat exactly as the 1.5 g of sample.

IRON.   (Page 28, Method 1). Suspend 0.5 g in 30 mL of water, cautiously add 3 mL of hydrochloric acid, and dilute with water to 50 mL. Use this solution without further acidification. Use only 2 mL of hydrochloric acid for the control solution.

CALCIUM, POTASSIUM, AND SODIUM.   Determine the calcium, potassium, and sodium by the flame atomic absorption spectrophotometric method described on page 35.

Sample Stock Solution.   Suspend 5 g of sample in 50 mL of water and cautiously add 25 mL of hydrochloric acid. Tranfer the solution to a 100-mL volumetric flask and dilute to the mark with water (0.05 g/mL).

Analyze the solutions by means of a suitable atomic absorption spectrophotometer, using the conditions outlined in the following table. Calculate the metal content of the sample by the method of standard additions.

Element	Wavelength (nm)	Sample Wt (g)	Standard Added (mg)	Flame Type*	Background Correction
Ca	422.7	0.1	0.02; 0.04	N/A	No
K	766.5	1.0	0.05; 0.10	A/A	No
Na	589.0	0.1	0.01; 0.02	A/A	No

*A/A is air/acetylene; N/A is nitrous oxide/acetylene.

Lithium Chloride

LiCl   Formula Wt 42.39

CAS Number 7447–41–8

> *NOTE.* The formula weight of this reagent is likely to deviate from the value cited because the natural distribution of ^6Li and ^7Li isotopes is often altered in current sources of lithium compounds.

REQUIREMENTS

Assay . . . . ≥ 99% LiCl

MAXIMUM ALLOWABLE

Insoluble matter 0.01%
Titrable base . . . . 0.008 meq / g
Loss on drying at 105°C. 1.0%
Nitrate (NO_3) 0.001%
Sulfate (SO_4) 0.01%
Barium (Ba) 0.003%
Heavy metals (as Pb) 0.002%
Iron (Fe) 0.001%
Calcium (Ca) 0.01%
Potassium (K) 0.01%
Sodium (Na) 0.20%

TESTS

ASSAY. (By indirect argentimetric titration of chloride). Weigh accurately 0.8 g, dissolve in water in a 100-mL volumetric flask, and dilute with water to volume. Transfer 20.0 mL to a 250-mL glass-stoppered flask, add 50 mL of water, and mix. Add, slowly and with constant agitation, 50.0 mL of 0.1 N silver nitrate. Mix, add 3 mL of nitric acid and 10 mL of toluene, and mix thoroughly. Add 2 mL of ferric ammonium sulfate indicator solution, and titrate the excess silver nitrate with 0.1 N thiocyanate solution. One milliliter of 0.1 N silver nitrate corresponds to 0.004239 g of LiCl.

INSOLUBLE MATTER. (Page 14). Use 10 g dissolved in 150 mL of water.

TITRABLE BASE. Transfer 10 g to a 150-mL beaker. Dissolve in 100 mL of carbon dioxide-free water, add 0.15 mL of methyl red indicator solution, and titrate with 0.01 N hydrochloric acid. Correct for a blank determination. Not more than 8.0 mL of 0.01 N hydrochloric acid should be required to produce a pink color.

LOSS ON DRYING AT 105°C. Weigh accurately 1.0 g in a tared dish or crucible and dry to constant weight at 105°C.

NITRATE. Mix 1.0 g with 5 mL of phenoldisulfonic acid reagent solution, and cautiously dilute with water to 25 mL. Cautiously add ammonium hydroxide until the maximum yellow color develops. Any yellow color should not exceed that produced in a standard, treated similarly, containing 0.01 mg of added nitrate (NO_3).

SULFATE. (Page 32, Procedure A, Method 1). Use 0.5 g.

BARIUM. Dissolve 3.3 g in 25 mL of water, filter if necessary, and add 2 mL of potassium dichromate reagent solution. Add ammonium hydroxide until the orange color dissipates and the yellow color persists. Any turbidity produced within 15 min should not exceed that in a standard containing 0.10 mg of added barium ion (Ba).

HEAVY METALS. (Page 26, Method 1). Dissolve 2.0 g in 40 mL of water. Use 30 mL to prepare the sample solution, and use the remaining 10 mL to prepare the control solution.

IRON. (Page 28, Method 1). Use 1.0 g.

CALCIUM, POTASSIUM, AND SODIUM. Determine the calcium, potassium, and sodium by the flame atomic absorption spectrophotometric method described on page 35.

> **Sample Stock Solution A.** Dissolve 10.0 g of sample with water in a 100-mL volumetric flask, and dilute to the mark with water (0.1 g/mL).

> **Sample Stock Solution B.** Pipet 10.0 mL of sample stock solution A into a 100-mL volumetric flask, and dilute to the mark with water (0.01 g/mL).

Analyze the solutions by means of a suitable atomic absorption spectrophotometer, using the conditions outlined in the following table. Calculate the metal content of the sample by the method of standard additions.

Element	Wavelength (nm)	Sample Wt (g)	Standard Added (mg)	Flame Type*	Background Correction
Ca	422.7	0.5	0.05; 0.10	N/A	No
K	766.5	0.5	0.05; 0.10	A/A	No
Na	589.0	0.005	0.01; 0.02	A/A	No

*A/A is air/acetylene; N/A is nitrous oxide/acetylene.

Lithium Hydroxide Monohydrate

$LiOH \cdot H_2O$ Formula Wt 41.96

CAS Number 1310–66–3

> *NOTE.* The formula weight of this reagent is likely to deviate from the value cited, since the natural distribution of ^6Li and ^7Li is often altered in current sources of lithium compounds.

REQUIREMENTS

Assay . . . . $\geq$ 98.0%
$LiOH \cdot H_2O$

MAXIMUM ALLOWABLE

Lithium carbonate 2.0%
Insoluble matter 0.01%
Chloride (Cl) 0.01%
Sulfate (SO_4) 0.05%
Heavy metals (as Pb) 0.002%
Iron (Fe) 0.002%

TESTS

ASSAY AND LITHIUM CARBONATE.   (By acidimetry). Weigh accurately 1.6 g and dissolve in 50 mL of carbon dioxide-free water. Add 0.15 mL of phenolphthalein indicator solution and titrate with 1 N hydrochloric acid just to a colorless end point (A mL). Add 0.15 mL of bromphenol blue indicator solution and titrate with 0.1 N hydrochloric acid to a yellow end point (B mL). Calculate the assay using ($A - 0.1$ B) mL. One milliliter of 1 N hydrochloric acid corresponds to 0.04196 g of $LiOH \cdot H_2O$. Calculate the carbonate content using B mL. One milliliter of 0.1 N hydrochloric acid corresponds to 0.007389 g of Li_2CO_3.

INSOLUBLE MATTER.   (Page 14). Use 10 g in 100 mL of water. Do not boil the sample.

CHLORIDE.   (Page 25). Use 0.1 g.

SULFATE.   (Page 32, Procedure A, Method 1). Use 0.10 g sample and add hydrochloric acid (1 + 1) until just acid. Then proceed as directed in the general procedure. For the standard add the same amount of hydrochloric acid (1 + 1) as used for the sample.

HEAVY METALS.   (Page 26, Method 1). Use 1.0 g dissolved in 10 mL of water. Neutralize with hydrochloric acid (1 + 1) and dilute to 25 mL.

IRON.   (Page 28, Method 1). Use 0.5 g in 10 mL of water, add 3 mL of hydrochloric acid and dilute to 40 mL.

Lithium Metaborate

$LiBO_2$ Formula Wt 49.75

CAS Number 13453−69−5

NOTE. The formula weight of this reagent is likely to deviate from the value cited, since the natural distribution of ^{6}Li and ^{7}Li isotopes is often altered in current sources of lithium compounds.

REQUIREMENTS

Assay . . . . 98.0–102.0% $LiBO_2$

Bulk density. . . . . $\geq$ 0.25 g / mL

MAXIMUM ALLOWABLE

Insoluble matter 0.01%
Loss on fusion at 950°C 2.0%
Phosphorus compounds (as PO_4) 0.004%
Silicon (Si). 0.01%
Aluminum (Al) 0.001%
Calcium (Ca) 0.01%
Heavy metals (as Pb) 0.001%
Iron (Fe) 0.001%
Magnesium (Mg) . . . . 5 ppm
Potassium (K) 0.005%
Sodium (Na) 0.005%

TESTS

ASSAY. (By acidimetry). Weigh accurately 1 g, dissolve in 80 mL of water in a 125-mL conical flask, and heat to boiling. Cool to room temperature, add 0.15 mL of 0.1% bromphenol blue indicator solution, and titrate with 1 N hydrochloric acid until the first appearance of yellow.

$$\% LiBO_2 = \frac{V \times N \times 497.5}{W \times (100 - L)}$$

where
V = volume, in mL, of HCl
N = normality of HCl
W = weight, in grams, of $LiBO_2$
L = % loss on fusion at 950°C

BULK DENSITY. Weigh a dry 50-mL glass-stoppered cylinder, calibrated to contain, then add approximately 50 mL of sample. Stopper and reweigh to the nearest 0.1 g. Grasp the cylinder above its base and from a height of 1 inch lower the base sharply against a No. 12 rubber stopper. Level the surface of the powder with a gentle back and forth motion of the cylinder in mid-air. Record the volume. Repeat two more times, starting with the rotation of the cylinder. Use the mean of the three volume readings to calculate the apparent density.

INSOLUBLE MATTER. (Page 14). Use 10 g dissolved in 400 mL of water.

LOSS ON FUSION AT 950°C. Heat 1 g, accurately weighed, in a platinum crucible or dish at 950°C for 15 min.

PHOSPHORUS COMPOUNDS. Dissolve 5.0 g in 40 mL of dilute nitric acid (1 + 1) and evaporate on a steam bath to a syrupy residue. Dissolve the residue in about 80 mL of water, dilute with water to 100 mL, and dilute 20 mL of this solution with water to 100 mL. For the test dilute 10 mL (0.1-g sample) with water to 70 mL. Prepare a standard containing 0.01 mg of phosphate ion (PO_4) in 70 mL of water. To each solution add 5 mL of ammonium molybdate solution (5 g in 50 mL) and adjust the pH to 1.8, using a pH meter, by adding dilute hydrochloric acid (1 + 1) or dilute ammonium hydroxide (1 + 1). Cautiously heat the solutions to near boiling, but do not boil, and cool to room temperature. If a precipitate forms, it will dissolve when the solution is acidified in the next steps. To each solution add 10 mL of hydrochloric acid and dilute each with water to 100 mL. Transfer the solutions to separatory funnels, add 35 mL of ether to each, shake vigorously, and allow to separate. Draw off and discard the aqueous phases. Wash the ether phases twice with 10-mL portions of dilute hydrochloric acid (1 + 9) and discard the washings each time. To the washed ether phases add 10 mL of dilute hydrochloric acid (1 + 9) to which has been added 0.2 mL of a freshly prepared 2% solution of stannous chloride dihydrate in hydrochloric acid. Shake the solutions and allow the phases to separate. Any blue color in the ether phase from the solution of the sample should not exceed that in the ether phase from the standard.

SILICON. Fuse 1.00 g at 950°C for 15 min in a 15-mL platinum crucible. Remove the crucible from the furnace, swirl to spread the melt around the sides of the crucible, and cool. Add a small magnetic stirring bar to the crucible, stir while adding 6.0 mL of 3 N hydrochloric acid, and stir until the melt disintegrates. Transfer the slurry with 50 mL of water to a 150-mL beaker. Adjust the pH to 2.0 ± 0.2 with 0.06 N hydrochloric acid (or dilute ammonium hydroxide), using a pH meter. The solution should be clear at the final pH adjustment. Add 2.0 mL of ammonium molybdate solution, mix and let stand for 10 min. Add 4.0 mL of tartaric acid solution, then 1.0 mL of reducing solution. Mix the solution after each addition. Finally, dilute the solution with water to 100 mL. After 30 min the blue color should not exceed that produced by 0.21 mg of silica in an equal volume of solution containing 10.0 mL of 0.06 N hydrochloric acid, 50 mL of water, and the quantities of reagents used in the test. If desired, the absorbances can be measured with a spectrophotometer at 650 nm, using water as a reference.

Reagents

Ammonium Molybdate Solution.   Dissolve 7.5 g of ammonium molybdate tetrahydrate, $(NH_4)_6Mo_7O_{24} \cdot 4H_2O$, in 75 mL of water. Add 10.0 mL of 18 N sulfuric acid, dilute with water to 100 mL, filter, and store in a plastic bottle.

Tartaric Acid Solution.   Dissolve 50 g of tartaric acid, dilute with water to 500 mL, and filter into a plastic bottle.

Reducing Solution.   Dissolve 0.7 g of sodium sulfite in 20 mL of water, then dissolve in this solution 0.15 g of 4-amino-3-hydroxy-1-naphthalenesulfonic acid. Add with constant stirring a solution of 9 g of sodium bisulfite in 80 mL of water. Filter the resulting solution into a plastic bottle. Discard after 3 days.

ALUMINUM.   Add 25 mL of methanol and 1.8 mL of hydrochloric acid to 1.0 g in a 100-mL plastic beaker. Evaporate to dryness on the steam bath, add 15 mL of methanol and 0.3 mL of hydrochloric acid, and evaporate again to dryness. Add another 15 mL of methanol and 0.3 mL of hydrochloric acid, repeat the evaporation, and dissolve the residue in 15.0 mL of 0.06 N hydrochloric acid. Transfer the solution of the sample to a 100-mL volumetric flask with 45 mL of water. For the blank and the standard, respectively, add 15.0 mL of 0.06 N hydrochloric acid to two 100-mL volumetric flasks containing 0 and 10 μg of aluminum, then add 45 mL of water. To each of the three flasks add 5.0 mL of lithium chloride solution, 1.0 mL of hydroxylamine hydrochloride reagent solution, and 4.0 mL of ascorbic acid solution. Mix and let stand for 5 min. Add 10.0 mL of buffer solution, mix, and let stand for 10 min. Add 5.0 mL of alizarin red S solution and dilute with water to 100 mL. The final pH should be 4.6 ± 0.1. After 1 h measure the absorbance of the sample and the standard at 500 nm in 5-cm cells with the blank as the reference. The absorbance of the sample should not exceed that of the standard.

Reagents

Lithium Chloride Solution.   Dissolve 61 g of lithium chloride in water and dilute with water to 200 mL. Prepare fresh solution daily.

Ascorbic Acid Solution. Dissolve 5.0 g of ascorbic acid in water and dilute with water to 100 mL. Prepare fresh solution daily.

Sodium Acetate–Acetic Acid Buffer. Dissolve 100 g of sodium acetate trihydrate, $CH_3COONa \cdot 3H_2O$, in water. Add 30 mL of glacial acetic acid and dilute with water to 500 mL. Filter if necessary.

Alizarin Red S Solution. Dissolve 0.250 g of alizarin red S in water, dilute with water to 250 mL, filter, and store in glass.

CALCIUM, IRON, MAGNESIUM, POTASSIUM, AND SODIUM. Determine the calcium, iron, magnesium, potassium, and sodium by the flame atomic absorption spectrophotometric method described on page 35.

Sample Stock Solution and Blank Solution. Weigh 10 g of sample into a 2-L beaker, add 400 mL of methanol and 5 mL of hydrochloric acid, and evaporate to dryness. Repeat the evaporation, using 200 mL of methanol and 2 mL of hydrochloric acid, then add 100 mL of methanol and 1 mL of hydrochloric acid, and perform a third evaporation. Prepare a blank in the same manner, using the quantities of reagents used for preparation of the sample. Dissolve each residue in 50 mL of water and 1 mL of hydrochloric acid, transfer to a 100-mL volumetric flask, and dilute to the mark with water (0.1 g/mL).

Analyze the solutions by means of a suitable atomic absorption spectrophotometer, using the conditions outlined in the following table. Calculate the metal content of the sample by the method of standard additions.

Element	Wavelength (nm)	Sample Wt (g)	Standard Added (mg)	Flame Type*	Background Correction
Ca	422.7	0.2	0.02; 0.04	N/A	No
Fe	248.3	1.0	0.01; 0.02	A/A	Yes
Mg	285.2	1.0	0.005; 0.01	A/A	Yes
K	766.5	1.0	0.05; 0.10	A/A	No
Na	589.0	0.1	0.005; 0.01	A/A	No

*A/A is air/acetylene; N/A is nitrous oxide/acetylene.

HEAVY METALS. Dilute 20 mL of glacial acetic acid with water to 80 mL, add 5.0 g of sample while stirring, and heat to about 80°C to dissolve. Prepare a control with 1.0 g of sample, 5.0 mL of glacial acetic acid, and 0.04 mg of lead ion (Pb) in a volume of 80 mL. The pH of the solutions should be between 3 and 4. Keep the solutions at 80°C, add 10 mL of freshly prepared hydrogen sulfide water to each, and mix. Transfer to 100-mL Nessler tubes. Any color in the solution of the sample should not exceed that in the control. (If the solutions are permitted to cool, a white crystalline precipitate forms.)

Lithium Perchlorate

LiClO$_4$ Formula Wt 106.39

CAS Number 7791−03−9

> NOTE. The formula weight of this reagent is likely to deviate from the value cited, since the natural distribution of ^{6}Li and ^{7}Li isotopes is often altered in current sources of lithium compounds.

REQUIREMENTS

Assay . . . . ≥ 95.0% LiClO$_4$
pH of a 5% solution at 25°C . . . . 6.0−7.5

MAXIMUM ALLOWABLE

Insoluble matter 0.005%
Chloride (Cl) 0.003%
Sulfate (SO$_4$) 0.001%
Heavy metals (as Pb) . . . . 5 ppm
Iron (Fe) . . . . 5 ppm

TESTS

ASSAY. (By argentimetric titration of chloride after reduction). Weigh accurately about 0.25−0.30 g of sample into a platinum dish or crucible. Add 3 g of sodium carbonate, Na$_2$CO$_3$, and heat gently at first, then to fusion until a clear melt is obtained. Leach with minimal dilute nitric acid and boil gently to expel carbon dioxide. Cool to room temperature, add 10 mL of 30% ammonium acetate solution, and titrate with 0.1 N silver nitrate to the dichlorofluorescein end point. One milliliter of 0.1 N silver nitrate corresponds to 0.01064 g of LiClO$_4$.

pH OF A 5% SOLUTION.   (Page 43). The pH should be 6.0–7.5 at 25°C.

INSOLUBLE MATTER.   (Page 14). Use 20 g dissolved in 200 mL of water.

CHLORIDE.   (Page 25). Use 1.0 g.

SULFATE.   Dissolve 40 g in 300 mL of water, add 2 mL of hydrochloric acid, filter, and heat to boiling. Add 5 mL of barium chloride reagent solution, digest in a covered beaker on a hot plate at low setting for 2 h, and allow to stand overnight. If a precipitate is formed, filter through a fine ashless paper, wash thoroughly, and ignite. Correct for the weight obtained in a complete blank test.

HEAVY METALS.   (Page 26, Method 1). Dissolve 6.0 g in about 20 mL of water and dilute with water to 30 mL. Use 25 mL to prepare the sample solution, and use the remaining 5.0 mL to prepare the control solution.

IRON.   Dissolve 1.0 g in 20 mL of water. Add 1 mL of hydroxylamine hydrochloride reagent solution, 4 mL of 1,10-phenanthroline reagent solution, and 1 mL of 10% sodium acetate solution, and mix. Any red color should not exceed that produced by 0.005 mg of iron (Fe) in an equal volume of solution containing the quantities of reagents used in the test. Compare 1 h after adding the reagents to the sample and standard solutions.

Lithium Sulfate Monohydrate

$Li_2SO_4 \cdot H_2O$ Formula Wt 127.96

CAS Number 10102–25–7

NOTE. The formula weight of this reagent is likely to deviate from the value cited, since the natural distribution of ^6Li and ^7Li isotopes is often altered in current sources of lithium compounds.

REQUIREMENTS

Assay (dried basis) . . . . ≥ 99.0% Li_2SO_4
Loss on drying at 150°C. 13.0–15.0%

MAXIMUM ALLOWABLE

Insoluble matter 0.01%
Chloride (Cl) 0.002%
Nitrate (NO_3) 0.001%
Heavy metals (as Pb) 0.001%
Iron (Fe) 0.001%
Potassium (K) 0.05%
Sodium (Na) 0.05%

TESTS

ASSAY. (By indirect acid–base titration). *NOTE:* All water used in this assay must be ammonia- and carbon dioxide-free. Weigh, to the nearest 0.1 mg, 0.3 g of previously dried sample (*see* loss on drying test) and dissolve in 50 mL of water. Pass the solution through a cation-exchange column at the rate of about 5 mL/min, and collect the eluate in a 500-mL titration flask. Then wash the resin in the column with water at a rate of about 10 mL/min, and collect in the same titration flask. Add 0.15 mL of phenolphthalein indicator solution to the flask and titrate with 0.1 N sodium hydroxide. Continue the titration, as the washing proceeds, until 50 mL of eluate requires no further titration. One milliliter of 0.1 N sodium hydroxide corresponds to 0.06398 g of $LiSO_4 \cdot H_2O$.

> **Ion-Exchange Column:** Use Dowex 50-XB 20–50 mesh or ANGL-242 16–50 mesh. Charge the resin to a column with a 2-cm bore to a height of about 20 cm. If the resin is in the H^+ form, wash free of acid and color with the ammonia- and carbon dioxide-free water. Otherwise regenerate by passing 4 N hydrochloric acid through the column, and wash free of acid.

LOSS ON DRYING AT 150°C. Weigh accurately about 1 g, and dry at 150°C to constant weight. The theoretical loss for $Li_2SO_4 \cdot H_2O$ is 14.08%. Save the dried sample for the assay.

INSOLUBLE MATTER. (Page 14). Use 10 g dissolved in 100 mL of water.

CHLORIDE. (Page 25). Use 0.50 g.

NITRATE.

> ***Sample Solution A.***   Dissolve 1.0 g in 5 mL of water by heating in a boiling-water bath. Dilute to 50 mL with brucine sulfate reagent solution.

> ***Control Solution B.***   Dissolve 1.0 g in 4 mL of water and 1 mL of the standard nitrate solution containing 0.01 mg of nitrate ion (NO_3) per mL by heating in a boiling-water bath. Dilute to 50 mL with brucine sulfate reagent solution.

Continue with the procedure described on page 29, starting with the preparation of blank solution C.

HEAVY METALS.   (Page 26, Method 1). Dissolve 4.0 g in 40 mL of water. Use 30 mL to prepare the sample solution, and use the remaining 10 mL to prepare the control solution.

IRON.   (Page 28, Method 1). Use 1.0 g.

POTASSIUM AND SODIUM.   Determine the potassium and sodium by the flame atomic absorption spectrophotometric method described on page 35.

> ***Sample Stock Solution.***   Dissolve 1.0 g of sample with water in a 100-mL volumetric flask and dilute to the mark with water (0.01 g/mL).

Analyze the solutions by means of a suitable atomic absorption spectrophotometer, using the conditions outlined in the following table. Calculate the metal content of the sample by the method of standard additions.

Element	Wavelength (nm)	Sample Wt (g)	Standard Added (mg)	Flame Type*	Background Correction
K	766.5	0.1	0.05; 0.10	A/A	No
Na	589.0	0.02	0.01; 0.02	A/A	No

*A/A is air/acetylene.

Lithium Tetraborate

(Suitable for Use in X-ray Spectroscopy)

$Li_2B_4O_7$ Formula Wt 169.12

CAS Number 12007−60−2

NOTE. The formula weight of this reagent is likely to deviate from the value cited, since the natural distribution of ^6Li and ^7Li isotopes is often altered in current sources of lithium compounds.

REQUIREMENTS

Assay . . . . 98.0−102.0%
$Li_2B_4O_7$
Bulk density. . . . . ≥ 0.25 g / mL

MAXIMUM ALLOWABLE

Insoluble matter 0.01%
Loss on fusion at 950°C 2.0%
Phosphorus compounds (as PO_4) 0.004%
Silicon (Si). 0.01%
Aluminum (Al) 0.001%
Calcium (Ca) 0.01%
Heavy metals (as Pb) 0.001%
Iron (Fe) 0.001%
Magnesium (Mg) . . . . 5 ppm
Potassium (K) 0.005%
Sodium (Na) 0.005%

TESTS

ASSAY. (By acidimetry). Weigh accurately 1 g, dissolve in 80 mL of water in a 125-mL conical flask, and heat to boiling. Cool to room temperature, add 0.15 mL of 0.1% bromphenol blue indicator solution, and titrate with 1 N hydrochloric acid until the first appearance of yellow.

$$\%Li_2B_4O_7 = \frac{V \times N \times 845.6}{W \times (100 - L)}$$

where
  V = volume, in mL, of HCl
  N = normality of HCl
  W = weight, in grams, of $Li_2B_4O_7$
  L = % loss on fusion at 950°C

BULK DENSITY. Weigh a dry 50-mL glass-stoppered cylinder, calibrated to contain, then add approximately 50 mL of sample. Stopper and reweigh to the nearest 0.1 g. Grasp the cylinder above its base and from a height of 1 inch lower the base sharply against a No. 12 rubber stopper. Level the surface of the powder with a gentle back and forth motion of the cylinder in mid-air. Record the volume. Repeat two more times, starting with the rotation of the cylinder. Use the mean of the three volume readings to calculate the apparent density.

INSOLUBLE MATTER. (Page 14). Use 10 g dissolved in 400 mL of water.

LOSS ON FUSION AT 950°C. Heat 1 g, accurately weighed, in a platinum crucible or dish at 950°C for 15 min.

PHOSPHORUS COMPOUNDS. Dissolve 5.0 g in 40 mL of dilute nitric acid (1 + 1) and evaporate on a steam bath to a syrupy residue. Dissolve the residue in about 80 mL of water, dilute with water to 100 mL, and dilute 20 mL of this solution with water to 100 mL. For the test dilute 10 mL (0.1-g sample) with water to 70 mL. Prepare a standard containing 0.01 mg of phosphate ion (PO_4) in 70 mL of water. To each solution add 5 mL of ammonium molybdate solution (5 g in 50 mL) and adjust the pH to 1.8, using a pH meter, by adding dilute hydrochloric acid (1 + 1) or dilute ammonium hydroxide (1 + 1). Cautiously heat the solutions to near boiling, but do not boil, then cool to room temperature. If a precipitate forms, it will dissolve when the solution is acidified in the next steps. To each solution add 10 mL of hydrochloric acid and dilute each with water to 100 mL. Transfer the solutions to separatory funnels, add 35 mL of ether to each, shake vigorously, and allow to separate. Draw off and discard the aqueous phases. Wash the ether phases twice with 10-mL portions of dilute hydrochloric acid (1 + 9) and discard the washings each time. To the washed ether phases add 10 mL of dilute hydrochloric acid (1 + 9) to which has been added 0.2 mL of a freshly prepared 2% solution of stannous chloride dihydrate in hydrochloric acid. Shake the solutions and allow the phases to separate. Any blue color in the ether phase from the solution of the sample should not exceed that in the ether phase from the standard.

SILICON. Fuse 1.00 g at 950°C for 15 min in a 15-mL platinum crucible. Remove the crucible from the furnace, swirl to spread the melt around the sides of the crucible, and cool. Add a small magnetic stirring bar to the crucible, stir while adding 6.0 mL of 3 N hydrochloric acid, and stir until the melt disintegrates. Transfer the slurry with 50 mL of water to a 150-mL beaker. Adjust the pH to 2.0 ± 0.2 with 0.06 N hydrochloric acid (or dilute ammonium hydroxide) using a pH meter. The solution should be clear at the final pH adjustment. Add 2.0 mL of ammonium molybdate solution, mix and let stand for 10 min. Add 4.0 mL of tartaric acid solution, then 1.0 mL of reducing solution. Mix the solution after each addition. Finally, dilute the solution with water to 100 mL. After 30 min the blue color should not exceed that produced by 0.21 mg of silica in an equal volume of solution containing 10.0 mL of 0.06 N hydrochloric acid, 50 mL of water, and the quantities of reagents used in the test. If desired, the absorbances can be measured with a spectrophotometer at 650 nm, using water as a reference.

Reagents

Ammonium Molybdate Solution. Dissolve 7.5 g of ammonium molybdate tetrahydrate, $(NH_4)_6Mo_7O_{24} \cdot 4H_2O$, in 75 mL of water. Add 10.0 mL of 18 N sulfuric acid, dilute with water to 100 mL, filter, and store in a plastic bottle.

Tartaric Acid Solution. Dissolve 50 g of tartaric acid in about 400 mL of water, dilute with water to 500 mL, and filter into a plastic bottle.

Reducing Solution. Dissolve 0.7 g of sodium sulfite in 20 mL of water, then dissolve into this solution 0.15 g of 4-amino-3-hydroxy-1-naphthalenesulfonic acid. Add with constant stirring a solution of 9 g of sodium bisulfite in 80 mL of water. Filter the resulting solution into a plastic bottle. Discard after 3 days.

ALUMINUM. Add 25 mL of methanol and 1.8 mL of hydrochloric acid to 1.0 g in a 100-mL plastic beaker. Evaporate to dryness on the steam bath, add 15 mL of methanol and 0.3 mL of hydrochloric acid, and evaporate again to dryness. Add another 15 mL of methanol and 0.3 mL of hydrochloric acid, repeat the evaporation, and dissolve the residue in 15.0 mL of 0.06 N hydrochloric acid. Transfer the solution of the sample to a 100-mL volumetric flask with 45 mL of water. For the blank and the standard, respectively, add 15.0 mL of 0.06 N hydrochloric acid to two 100-mL volumetric flasks containing 0 and 10 μg of aluminum, then add 45 mL of water. To each of the three flasks add 5.0 g of lithium chloride solution, 1.0 mL of hydroxylamine hydrochloride reagent solution, and 4.0 mL of ascorbic acid solution. Mix and let stand for 5 min. Add 10.0 mL of buffer solution, mix, and let stand for 10 min. Add 5.0 mL of alizarin red S solution and dilute with water to 100 mL. The final pH should be 4.6 ± 0.1. After 1 h measure the absorbance of the sample and the standard at 500 nm in 5-cm cells with the blank as the reference. The absorbance of the sample should not exceed that of the standard.

Reagents

Lithium Chloride Solution.   Dissolve 61 g of lithium chloride in water and dilute with water to 200 mL. Prepare fresh solution daily.

Ascorbic Acid Solution.   Dissolve 5.0 g of ascorbic acid in water and dilute with water to 100 mL. Prepare fresh solution daily.

Sodium Acetate–Acetic Acid Buffer.   Dissolve 100 g of sodium acetate trihydrate, $CH_3COONa \cdot 3H_2O$, in water. Add 30 mL of glacial acetic acid and dilute with water to 500 mL. Filter if necessary.

Alizarin Red S Solution.   Dissolve 0.250 g of alizarin red S in water, dilute with water to 250 mL, filter, and store in glass.

CALCIUM, IRON, MAGNESIUM, POTASSIUM, AND SODIUM. Determine the calcium, iron, magnesium, potassium, and sodium by the flame atomic absorption spectrophotometric method described on page 35.

Sample Stock Solution and Blank Solution.   Weigh 10 g of sample into a 2-L beaker, add 400 mL of methanol and 5 mL of hydrochloric acid, and evaporate to dryness. Repeat the evaporation using 200 mL of methanol and 2 mL of hydrochloric acid, then add 100 mL of methanol and 1 mL of hydrochloric acid, and perform a third evaporation. Prepare a blank in the same manner, using the quantities of reagent used for preparation of the sample. Dissolve each residue in 50 mL of water and 1 mL of hydrochloric acid, transfer to a 100-mL volumetric flask, and dilute to the mark with water (0.1 g/mL).

Analyze the solutions by means of a suitable atomic absorption spectrophotometer, using the conditions outlined in the following table. Calculate the metal content of the sample by the method of standard additions.

Element	Wavelength (nm)	Sample Wt (g)	Standard Added (mg)	Flame Type*	Background Correction
Ca	422.7	0.2	0.02; 0.04	N/A	No
Fe	248.3	1.0	0.01; 0.02	A/A	Yes
Mg	285.2	1.0	0.005; 0.01	A/A	Yes
K	766.5	1.0	0.05; 0.10	A/A	No
Na	589.0	0.1	0.005; 0.01	A/A	No

*A/A is air/acetylene; N/A is nitrous oxide/acetylene.

HEAVY METALS. Dilute 20 mL of glacial acetic acid with water to 80 mL, add 5.0 g of sample while stirring, and heat to about 80°C to dissolve. Prepare a control with 1.0 g of sample, 5.0 mL of glacial acetic acid, and 0.04 mg of lead ion (Pb) in a volume of 80 mL. The pH of the solutions should be between 3 and 4. Keep the solutions at 80°C, add 10 mL of freshly prepared hydrogen sulfide water to each, and mix. Transfer to 100-mL Nessler tubes. Any color in the solution of the sample should not exceed that in the control. (If the solutions are permitted to cool, a white crystalline precipitate forms.)

Litmus Paper

NOTE. When litmus paper is used in a solution, the length taken of a 0.6-cm-wide strip should not exceed 0.5 cm.

REQUIREMENTS

MAXIMUM ALLOWABLE

Ash. . . . . 0.4 mg per strip
(about 3 sq cm)
Phosphate (PO$_4$) . . . . Passes test
Rosin acids (for blue paper only) . . . . Passes test
Sensitivity . . . . Passes test

TESTS

ASH. Carefully ignite 10 strips in a tared crucible, and finally ignite at 800 ± 25°C for 15 min.

PHOSPHATE. (Page 30, Method 1). Cut five strips into small pieces, mix with 0.5 g of magnesium nitrate in a porcelain crucible, and ignite. To the ignited residue add 5 mL of nitric acid and evaporate to dryness. Dissolve in 25 mL of approximately 0.5 N sulfuric acid, and continue as described.

ROSIN ACIDS. Immerse a strip of blue paper in a solution of 0.10 g of silver nitrate in 50 mL of water. The color of the paper should not change in 30 s.

SENSITIVITY. Drop the pieces from six strips of paper into 100 mL of test solution in a beaker and stir continuously. Blue paper should change color in 45 s in 0.0005 N acid. Red paper should change color in 30 s in 0.0005 N alkali.

Magnesium Acetate Tetrahydrate

$(CH_3COO)_2Mg \cdot 4H_2O$ Formula Wt 214.46

CAS Number 16674–78–5

REQUIREMENTS

Assay 98.0–102.0%
$(CH_3CO_2)_2Mg \cdot 4H_2O$

MAXIMUM ALLOWABLE

Insoluble matter	0.005%
Chloride (Cl)	0.001%
Nitrogen compounds (as N)	0.001%
Sulfate (SO_4)	0.005%
Barium (Ba)	0.001%
Calcium (Ca)	0.01%
Heavy metals (as Pb)	5 ppm
Iron (Fe)	5 ppm
Manganese (Mn)	0.001%
Potassium (K)	0.005%
Sodium (Na)	0.005%
Strontium (Sr)	0.005%

TESTS

ASSAY. (By complexometric titration of Mg). Weigh accurately 0.85 g and transfer to a 250-mL beaker with 50 mL of water. Add 5 mL of pH 10 ammoniacal buffer solution and about 50 mg of Eriochrome Black T indicator mixture. Titrate with standard 0.1 M EDTA to a clear blue color (and disappearance of the last trace of red). One milliliter of 0.1 M EDTA corresponds to 0.02145 g of $(CH_3COO)_2Mg \cdot 4H_2O$.

INSOLUBLE MATTER. (Page 14). Use 20 g dissolved in 200 mL of dilute acetic acid (1 + 99). Retain the filtrate, without washings, for the test for sulfate.

> *Sample Solution A for the Determination of Chloride, Nitrogen Compounds, and Iron.* Dissolve 10 g in water, filter if necessary through a chloride-free filter, and dilute with water to 100 mL (1 mL = 0.1 g).

CHLORIDE. (Page 25). Use 10 mL of sample solution A (1-g sample).

NITROGEN COMPOUNDS. (Page 30). Use 10 mL of sample solution A (1-g sample). For the standard use 0.01 mg of nitrogen (N).

SULFATE. (Page 32, Procedure A, Method 1). Use 1/20 of the filtrate from the test for insoluble matter. Allow 30 min for the turbidity to form.

BARIUM. Dissolve 6.0 g in 25 mL of dilute nitric acid (1 + 1). For the control dissolve 1.0 g in 25 mL of dilute nitric acid (1 + 1) and add 0.05 mg of barium ion (Ba). Evaporate each solution on the steam bath to a syrupy consistency and add 20 mL of water. If the solutions are not clear, add 10% nitric acid until clear. Add 1 mL of 1 N acetic acid and 2 mL of potassium dichromate reagent solution to each, and adjust the pH to 7 (using a pH meter) with ammonium hydroxide (10% NH_3). Add 25 mL of methanol to each and stir vigorously. Any turbidity in the solution of the sample should not exceed that in the control.

CALCIUM, MANGANESE, POTASSIUM, SODIUM, AND STRONTIUM.

Determine the calcium, manganese, potassium, sodium, and strontium by the flame atomic absorption spectrophotometric method described on page 35.

Sample Stock Solution. Dissolve 10.0 g of sample with water in a 100-mL volumetric flask and dilute to the mark with water (0.10 g/mL).

Analyze the solutions by means of a suitable atomic absorption spectrophotometer, using the conditions outlined in the following table. Calculate the metal content of the sample by the method of standard additions.

Element	Wavelength (nm)	Sample Wt (g)	Standard Added (mg)	Flame Type*	Background Correction
Ca	422.7	0.2	0.02; 0.04	N/A	No
Mn	279.5	2.0	0.02; 0.04	A/A	Yes
K	766.5	1.0	0.05; 0.10	A/A	No
Na	589.0	0.2	0.01; 0.02	A/A	No
Sr	460.7	1.0	0.05; 0.10	N/A	No

*A/A is air/acetylene; N/A is nitrous oxide/acetylene.

HEAVY METALS. (Page 26, Method 1). Dissolve 6.0 g in 20 mL of water, and dilute with water to 30 mL. Use 25 mL to prepare the sample solution, and use the remaining 5.0 mL to prepare the control solution.

IRON. (Page 28, Method 1). Use 20 mL of sample solution A (2-g sample).

Magnesium Chloride Hexahydrate

$MgCl_2 \cdot 6H_2O$ Formula Wt 203.30

CAS Number 7791 –18 –6

REQUIREMENTS

Assay 99.0–102.0%
$MgCl_2 \cdot 6H_2O$

MAXIMUM ALLOWABLE

Insoluble matter 0.005%
Nitrate (NO$_3$) 0.001%
Phosphate (PO$_4$) . . . . 5 ppm
Sulfate (SO$_4$) 0.002%
Ammonium (NH$_4$) 0.002%
Barium (Ba) 0.005%
Calcium (Ca) 0.01%
Heavy metals (as Pb) . . . . 5 ppm
Iron (Fe) . . . . 5 ppm
Manganese (Mn) . . . . 5 ppm
Potassium (K) 0.005%
Sodium (Na) 0.005%
Strontium (Sr) 0.005%

TESTS

ASSAY. (By complexometric titration of Mg). Weigh accurately 0.8 g and transfer to a 250-mL beaker with 50 mL of water. Add 5 mL of pH 10 ammoniacal buffer solution and about 50 mg of Eriochrome Black T indicator mixture. Titrate with standard 0.1 M EDTA to a clear blue color (and disappearance of the last trace of red). One milliliter of 0.1 M EDTA corresponds to 0.02033 g of MgCl$_2 \cdot$ 6H$_2$O.

INSOLUBLE MATTER. (Page 14). Use 20 g dissolved in 200 mL of water.

NITRATE.

> *Sample Solution A.* Dissolve 1.0 g in 3 mL of water by heating in a boiling-water bath. Dilute to 50 mL with brucine sulfate reagent solution.

> *Control Solution B.* Dissolve 1.0 g in 2 mL of water and 1 mL of the standard nitrate solution containing 0.01 mg of nitrate ion (NO$_3$) per mL by heating in a boiling-water bath. Dilute to 50 mL with brucine sulfate reagent solution.

Continue with the procedure described on page 29, starting with the preparation of blank solution C.

PHOSPHATE. (Page 30, Method 1). Dissolve 4.0 g in 25 mL of approximately 0.5 N sulfuric acid, and continue as described.

SULFATE. (Page 32, Procedure A, Method 1). Use 2.5 g of sample.

AMMONIUM.   Dissolve 1.0 g in 90 mL of water and add 10 mL of freshly boiled 10% sodium hydroxide reagent solution. Allow to settle, decant 50 mL and add 2 mL of Nessler reagent. Any color should not exceed that produced by 0.01 mg of ammonium ion (NH_4) in an equal volume of solution containing 5 mL of 10% sodium hydroxide solution and 2 mL of Nessler reagent.

BARIUM.   Dissolve 1.5 g in 20 mL of water. For the control dissolve 0.5 g in 15 mL of water, add 0.05 mg of barium ion (Ba), and dilute with water to 20 mL. To each solution add 1 mL of 1 N acetic acid and 2 mL of potassium dichromate reagent solution. Adjust the pH of the control and sample solutions to 7 (using a pH meter) with ammonium hydroxide (10% NH_3), add 25 mL of methanol, and stir vigorously. Any turbidity in the solution of the sample should not exceed that in the control.

CALCIUM, MANGANESE, POTASSIUM, SODIUM, AND STRONTIUM.
Determine the calcium, manganese, potassium, sodium, and strontium by the flame atomic absorption spectrophotometric method described on page 35.

> *Sample Stock Solution.*   Dissolve 10.0 g of sample with water in a 100-mL volumetric flask and dilute to the mark with water (0.10 g/mL).

Analyze the solutions by means of a suitable atomic absorption spectrophotometer, using the conditions outlined in the following table. Calculate the metal content of the sample by the method of standard additions.

Element	Wavelength (nm)	Sample Wt (g)	Standard Added (mg)	Flame Type*	Background Correction
Ca	422.7	0.2	0.02; 0.04	N/A	No
Mn	279.5	2.0	0.01; 0.02	A/A	Yes
K	766.5	1.0	0.05; 0.10	A/A	No
Na	589.0	0.2	0.01; 0.20	A/A	No
Sr	460.7	1.0	0.05; 0.10	N/A	No

*A/A is air/acetylene; N/A is nitrous oxide/acetylene.

HEAVY METALS.   (Page 26, Method 1). Dissolve 6.0 g in 20 mL of water, and dilute with water to 30 mL. Use 25 mL to prepare the sample solution, and use the remaining 5.0 mL to prepare the control solution.

IRON.   (Page 28, Method 1). Use 2.0 g.

Magnesium Nitrate Hexahydrate

$Mg(NO_3)_2 \cdot 6H_2O$ Formula Wt 256.41

CAS Number 13446-18-9

REQUIREMENTS

Assay 98.0-102.0%
$Mg(NO_3)_2 \cdot 6H_2O$
pH of a 5% solution at 25°C 5.0-8.2

MAXIMUM ALLOWABLE

Insoluble matter 0.005%
Chloride (Cl) 0.001%
Phosphate (PO_4) . . . . 5 ppm
Sulfate (SO_4) 0.005%
Ammonium (NH_4) 0.003%
Barium (Ba) 0.005%
Calcium (Ca) 0.01%
Heavy metals (as Pb) . . . . 5 ppm
Iron (Fe) . . . . 5 ppm
Manganese (Mn) . . . . 5 ppm
Potassium (K) 0.005%
Sodium (Na) 0.005%
Strontium (Sr) 0.005%

TESTS

ASSAY. (By complexometric titration of Mg). Weigh accurately 1 g and transfer to a 250-mL beaker with 50 mL of water. Add 5 mL of pH 10 ammoniacal buffer solution and about 50 mg of Eriochrome Black T indicator mixture. Titrate with standard 0.1 M EDTA to a clear blue color (and disappearance of the last trace of red). One milliliter of 0.1 M EDTA corresponds to 0.02564 g of $Mg(NO_3)_2 \cdot 6H_2O$.

pH OF A 5% SOLUTION. (Page 43). The pH should be 5.0-8.2 at 25°C.

INSOLUBLE MATTER.   (Page 14). Use 20 g dissolved in 200 mL of water.

CHLORIDE.   (Page 25). Use 1.0 g.

PHOSPHATE.   (Page 30). Dissolve 4.0 g in 25 mL of approximately 0.5 N sulfuric acid, and continue as described.

SULFATE.   (Page 32, Procedure A, Method 2). Use 10 mL of dilute hydrochloric acid (1 + 1), do two evaporations, and allow 30 min for the turbidity to form.

AMMONIUM.   Dissolve 1.0 g in 100 mL of water, add 10 mL of freshly boiled 10% sodium hydroxide reagent solution, and dilute with water to 150 mL. Allow the precipitate to settle. Filter through sintered-glass or porous porcelain filter. To 50 mL of the filtrate, add 2 mL of Nessler reagent solution. Any color should not exceed that produced by 0.01 mg of ammonium ion (NH_4) in an equal volume of solution containing the quantities of reagents used in the test.

BARIUM.   Dissolve 1.5 g in 20 mL of water. For the control dissolve 0.5 g in 15 mL of water, add 0.05 mg of barium ion (Ba), and dilute with water to 20 mL. To each solution add 1 mL of 1 N acetic acid and 2 mL of potassium dichromate reagent solution. Adjust the pH of the control and sample solutions to 7 (using a pH meter) with ammonium hydroxide (10% NH_3), add 25 mL of methanol, and stir vigorously. Any turbidity in the solution of the sample should not exceed that in the control.

CALCIUM, MANGANESE, POTASSIUM, SODIUM, AND STRONTIUM.
Determine the calcium, manganese, potassium, sodium, and strontium by the flame atomic absorption spectrophotometric method described on page 35.

> *Sample Stock Solution.*   Dissolve 10.0 g of sample with water in a 100-mL volumetric flask and dilute to the mark with water (0.10 g/mL).

Analyze the solutions by means of a suitable atomic absorption spectrophotometer, using the conditions outlined in the following table. Calculate the metal content of the sample by the method of standard additions.

Element	Wavelength (nm)	Sample Wt (g)	Standard Added (mg)	Flame Type*	Background Correction
Ca	422.7	0.2	0.02; 0.04	N/A	No
Mn	279.5	2.0	0.01; 0.02	A/A	Yes
K	766.5	1.0	0.05; 0.10	A/A	No
Na	589.0	0.2	0.01; 0.02	A/A	No
Sr	460.7	1.0	0.05; 0.10	A/A	No

*A/A is air/acetylene; N/A is nitrous oxide/acetylene.

HEAVY METALS. (Page 26, Method 1). Dissolve 6.0 g in 20 mL of water, and dilute with water to 30 mL. Use 25 mL to prepare the sample solution, and use the remaining 5.0 mL to prepare the control solution.

IRON. (Page 28, Method 1). Use 2.0 g.

Magnesium Oxide

MgO   Formula Wt 40.30

CAS Number 1039–48–4

REQUIREMENTS

Assay (dried basis) . . . . ≥ 95.0% MgO

MAXIMUM ALLOWABLE

Insoluble in dilute hydrochloric acid 0.02%
Water-soluble substances 0.4%
Loss on ignition 2.0%
Ammonium hydroxide precipitate 0.02%
Chloride (Cl) 0.01%
Nitrate (NO_3) 0.005%
Sulfate and sulfite (as SO_4) 0.02%
Barium (Ba) 0.005%
Calcium (Ca) 0.05%
Heavy metals (as Pb) 0.003%
Iron (Fe) 0.01%
Manganese (Mn) . . . . 5 ppm
Potassium (K) 0.005%
Sodium (Na) 0.5%
Strontium (Sr) 0.005%

TESTS

ASSAY. (By complexometric titration of Mg). Weigh accurately about 0.5 g of the sample, previously ignited at $800 \pm 25°C$. Dissolve cautiously in water and a slight excess of hydrochloric acid, transfer to a 250-mL volumetric flask, dilute to the mark with water, and mix thoroughly. Pipet 50.0 mL of this solution into a 250-mL beaker, add 50 mL of water and 0.10 mL of methyl red indicator solution, and neutralize with 10% sodium hydroxide. Add 15 mL of pH 10 buffer, about 0.1 g of ascorbic acid, and a few milligrams of Eriochrome Black T indicator. Titrate with 0.1 M EDTA to a blue end point. One milliliter of 0.1 M EDTA corresponds to 0.004030 g of MgO.

INSOLUBLE IN DILUTE HYDROCHLORIC ACID. Dissolve 5.0 g in 125 mL of dilute hydrochloric acid (1 + 4), heat to boiling, and boil gently for 5 min. Digest the solution in a covered beaker on the steam bath for 1 h. Filter through a tared filtering crucible, wash thoroughly, and dry at 105°C. Retain the filtrate for the determination of ammonium hydroxide precipitate.

WATER-SOLUBLE SUBSTANCES. Suspend 2.0 g in 50 mL of water, heat to boiling, and filter while hot. Evaporate 25 mL of the clear filtrate to dryness in a tared dish. Moisten the residue with 0.10 mL of sulfuric acid, ignite gently to remove the excess acid, and finally ignite at $800 \pm 25°C$ for 15 min.

LOSS ON IGNITION. Weigh accurately about 0.25 g in a tared covered platinum crucible. Ignite at $800 \pm 25°C$ for 15 min. Cool and reweigh.

AMMONIUM HYDROXIDE PRECIPITATE. To the filtrate from the determination of insoluble in dilute hydrochloric acid add ammonium hydroxide until neutral to methyl red and add an excess of 1 mL. Boil the solution for 5 min, allow to cool, and filter. Dissolve the precipitate on the filter with 5 mL of hot dilute hydrochloric acid (1 + 1) and wash to a volume of 25 mL. Add ammonium hydroxide until neutral to methyl red, add an excess of 0.5 mL, and boil for 5 min. Filter through the same filter, wash with hot water, and ignite.

CHLORIDE. (Page 25). Use 0.10 g dissolved in 10 mL of dilute nitric acid (1 + 9).

NITRATE.

> ***Sample Solution A.***   Suspend 0.20 g in 3 mL of water. Dilute to 50 mL with brucine sulfate reagent solution.

> ***Control Solution B.***   Suspend 0.20 g in 2 mL of water and 1 mL of the standard nitrate solution containing 0.01 mg of nitrate ion (NO_3) per milliliter. Dilute to 50 mL with brucine sulfate reagent solution.

> ***Blank Solution C.***   Use 50 mL of brucine sulfate reagent solution.

Heat sample solution A, control solution B, and blank solution C in a preheated (boiling) water bath until the magnesium oxide is dissolved in A and B, then heat for 10 min more. Continue with the procedure described on page 29.

SULFATE AND SULFITE.   (Page 32, Procedure B). Cool 85–90 mL of hydrochloric acid (3 + 1) in a 250-mL beaker in a cold-water bath. Slowly add 10 g of sample while stirring, stir until the sample is dissolved, and cool to room temperature. Dilute to 100 mL with hydrochloric acid (3 + 1). Transfer 2.0 mL to a reaction tube, and proceed as described.

BARIUM.   Dissolve 1.5 g in 15 mL of nitric acid. For the control dissolve 0.5 g in 15 mL of nitric acid and add 0.05 mg of barium ion (Ba). Evaporate each solution on the steam bath to a syrupy consistency and add 20 mL of water. If the solutions are not clear, add 10% nitric acid dropwise until clear. Add 1 mL of 1 N acetic acid and 2 mL of potassium dichromate reagent solution to each, and adjust the pH to 7 (using a pH meter) with ammonium hydroxide (10% NH_3). Add 25 mL of methanol to each and stir vigorously. Any turbidity in the solution of the sample should not exceed that in the control.

CALCIUM, MANGANESE, POTASSIUM, SODIUM, AND STRONTIUM.
Determine the calcium, manganese, potassium, sodium, and strontium by the flame atomic absorption spectrophotometric method described on page 35.

> ***Sample Stock Solution.***   Dissolve 10.0 g of sample with dilute nitric acid (1 + 4), and digest in a covered beaker on a steam bath for 20 min. Cool, transfer to a 100-mL volumetric flask, and dilute to the mark with water (0.10 g/mL).

Analyze the solutions by means of a suitable atomic absorption spectrophotometer, using the conditions outlined in the following

table. Calculate the metal content of the sample by the method of standard additions.

Element	Wavelength (nm)	Sample Wt (g)	Standard Added (mg)	Flame Type*	Background Correction
Ca	422.7	0.2	0.05; 0.10	N/A	No
Mn	279.5	2.0	0.01; 0.02	A/A	Yes
K	766.5	1.0	0.05; 0.10	A/A	No
Na	589.0	0.002	0.01; 0.02	A/A	No
Sr	460.7	1.0	0.05; 0.10	N/A	No

*A/A is air/acetylene; N/A is nitrous oxide/acetylene.

HEAVY METALS. (Page 26, Method 1). Dissolve 5.0 g in 50 mL of dilute hydrochloric acid (1 + 1), heat if necessary to obtain complete dissolution, and dilute with water to 150 mL. Evaporate 30 mL to dryness on the steam bath, dissolve the residue in 20 mL of water, and dilute with water to 25 mL. For the control solution add 0.02 mg of lead ion (Pb) to 10 mL of the sample solution, evaporate to dryness on the steam bath, dissolve the residue in 20 mL of water, and dilute with water to 25 mL.

IRON. (Page 28, Method 1). Add 1.0 g to 50 mL of dilute hydrochloric acid (1 + 1), and boil gently for 5 min. Cool, dilute with water to 50 mL, and use 5 mL of this solution without further acidification.

Magnesium Perchlorate, Desiccant

$Mg(ClO_4)_2$ Formula Wt 223.21

CAS Number 10034–81–8

REQUIREMENTS

Suitability for moisture absorption . . . . Passes test

MAXIMUM ALLOWABLE

Titrable free acid . . . . 0.005 meq / g
Titrable base . . . . 0.025 meq / g
Loss on drying at 190°C. 8%

TESTS

TITRABLE FREE ACID AND TITRABLE BASE. Dissolve 2.0 g in 25 mL of water and add 0.15 mL of methyl orange indicator solution. Any red color produced should be changed to yellow by the addition of not more than 1.0 mL of 0.01 N sodium hydroxide (titrable free acid). If a yellow color is produced, not more than 5.0 mL of 0.01 N hydrochloric acid should be required to produce a red color (titrable base).

LOSS ON DRYING AT 190°C.

> *CAUTION.* Do not dry the sample in the presence of any organic material.

Weigh accurately about 1 g and dry at $190 \pm 10°C$ to constant weight. Calculate the loss in weight as percent water.

SUITABILITY FOR MOISTURE ABSORPTION. Weigh about 2 g in a tared, 50-mL weighing bottle, having a diameter of 50 mm and a height of 30 mm. Place the bottle, with cover removed, for 16 h or overnight in a closed container in which the atmosphere possesses a relative humidity of 85%, maintained by equilibrium with sulfuric acid having a specific gravity of 1.16. The increase in weight should not be less than 25%.

Magnesium Sulfate Heptahydrate

$MgSO_4 \cdot 7H_2O$ Formula Wt 246.48

CAS Number 10034–99–8

REQUIREMENTS

Assay 98.0–102.0%
$MgSO_4 \cdot 7H_2O$
pH of a 5% solution at 25°C 5.0–8.2

MAXIMUM ALLOWABLE

Insoluble matter 0.005%
Chloride (Cl) . . . . 5 ppm
Nitrate (NO$_3$) 0.002%
Ammonium (NH$_4$) 0.002%
Arsenic (As). . . . . 2 ppm
Calcium (Ca) 0.02%
Heavy metals (as Pb) . . . . 5 ppm
Iron (Fe) . . . . 5 ppm
Manganese (Mn) . . . . 5 ppm
Potassium (K) 0.005%
Sodium (Na) 0.005%
Strontium (Sr) 0.005%

TESTS

ASSAY. (By complexometric titration of Mg). Weigh accurately 1 g and transfer to a 250-mL beaker with 50 mL of water. Add 5 mL of pH 10 ammoniacal buffer solution and about 50 mg of Eriochrome Black T indicator mixture. Titrate with standard 0.1 M EDTA to a clear blue color (and disappearance of the last trace of red). One milliliter of 0.1 M EDTA corresponds to 0.02465 g of $MgSO_4 \cdot 7H_2O$.

pH OF A 5% SOLUTION. (Page 43). The pH should be 5.0–8.2 at 25°C.

INSOLUBLE MATTER. (Page 14). Use 20 g dissolved in 200 mL of water.

CHLORIDE. (Page 25). Use 2.0 g.

NITRATE.

> ***Sample Solution A.*** Dissolve 1.0 g in 5 mL of water by heating in a water bath. Dilute to 50 mL with brucine sulfate reagent solution.

> ***Control Solution B.*** Dissolve 1.0 g in 3 mL of water and 2 mL of the standard nitrate solution containing 0.01 mg of nitrate ion (NO$_3$) per mL by heating in a boiling-water bath. Dilute to 50 mL with brucine sulfate reagent solution.

Continue with the procedure described on page 29, starting with the preparation of blank solution C.

AMMONIUM.   Dissolve 2.0 g in 90 mL of water, add 10 mL of 10% sodium hydroxide reagent solution and allow to settle. Draw off 25 mL of the clear solution, dilute with water to 50 mL, and add 2 mL of Nessler reagent. Any color should not exceed that produced by 0.01 mg of ammonium ion (NH_4) in an equal volume of solution containing 2.5 mL of 10% sodium hydroxide solution and 2 mL of Nessler reagent.

ARSENIC.   (Page 23). Use 1.5 g. For the standard use 0.003 mg of arsenic (As).

CALCIUM, MANGANESE, POTASSIUM, SODIUM, AND STRONTIUM. Determine the calcium, manganese, potassium, sodium, and strontium by the flame atomic absorption spectrophotometric method described on page 35.

> *Sample Stock Solution.*   Dissolve 10.0 g of sample with water in a 100-mL volumetric flask and dilute to the mark with water (0.10 g/mL).

Analyze the solutions by means of a suitable atomic absorption spectrophotometer, using the conditions outlined in the following table. Calculate the metal content of the sample by the method of standard additions.

Element	Wavelength (nm)	Sample Wt (g)	Standard Added (mg)	Flame Type*	Background Correction
Ca	422.7	0.2	0.04; 0.08	N/A	No
Mn	279.5	2.0	0.01; 0.02	A/A	Yes
K	766.5	0.5	0.025; 0.05	A/A	No
Na	589.0	0.2	0.01; 0.02	N/A	No
Sr	460.7	1.0	0.05; 0.10	N/A	No

*A/A is air/acetylene; N/A is nitrous oxide/acetylene.

HEAVY METALS.   (Page 26, Method 1). Dissolve 6.0 g in about 20 mL of water, and dilute with water to 30 mL. Use 25 mL to prepare the sample solution, and use the remaining 5.0 mL to prepare the control solution.

IRON.   (Page 28, Method 1). Use 2.0 g.

Manganese Chloride Tetrahydrate
Manganese(II) Chloride Tetrahydrate
Manganous Chloride Tetrahydrate

$MnCl_2 \cdot 4H_2O$ Formula Wt 197.90

CAS Number 13446–34–9

REQUIREMENTS

Assay 98.0–101.0%
$MnCl_2 \cdot 4H_2O$
pH of a 5% solution at 25°C 3.5–6.0

MAXIMUM ALLOWABLE

Insoluble matter 0.005%
Sulfate (SO_4) 0.005%
Calcium (Ca) 0.005%
Heavy metals (as Pb) . . . . 5 ppm
Iron (Fe) . . . . 5 ppm
Magnesium (Mg) 0.005%
Potassium (K) 0.01%
Sodium (Na) 0.05%
Zinc (Zn) 0.005%

TESTS

ASSAY. (By complexometric titration of Mn). Weigh accurately 0.8 g, transfer to a 500-mL beaker, and dissolve in 200 mL of water. Add a few milligrams of ascorbic acid to prevent oxidation. From a buret, add about 30 mL of standard 0.1 M EDTA. Now add 10 mL of pH 10 ammoniacal buffer solution and about 50 mg of Eriochrome Black T indicator mixture. Continue the titration with EDTA to a clear blue color (and disappearance of the last trace of red). One milliliter of 0.1 M EDTA corresponds to 0.01979 g of $MnCl_2 \cdot 4H_2O$.

pH OF A 5% SOLUTION. (Page 43). The pH should be 3.5–6.0 at 25°C.

INSOLUBLE MATTER. (Page 14). Use 20 g dissolved in 150 mL of water.

SULFATE.   Dissolve 10 g in 100 mL of water, add 1 mL of hydrochloric acid, filter, and heat the filtrate to boiling. Add 10 mL of barium chloride reagent solution, digest in a covered beaker on a hot plate at low setting for 2 h, and allow to stand overnight. If a precipitate is formed, filter through a fine ashless paper, wash thoroughly, and ignite. Correct for the weight obtained in a complete blank test.

CALCIUM, MAGNESIUM, POTASSIUM, SODIUM, AND ZINC.   Determine the calcium, magnesium, potassium, sodium, and zinc by the flame atomic absorption spectrophotometric method described on page 35.

> ***Sample Stock Solution.***   Dissolve 5.0 g in water in a 100-mL volumetric flask, add 0.5 mL of hydrochloric acid, and dilute with water to the mark (1 mL = 0.05 g).

Analyze the solutions by means of a suitable atomic absorption spectrophotometer, using the conditions outlined in the following table. Calculate the metal content of the sample by the method of standard additions.

Element	Wavelength (nm)	Sample Wt (g)	Standard Added (mg)	Flame Type*	Background Correction
Ca	422.7	0.8	0.02; 0.04	N/A	No
Mg	285.2	0.2	0.005; 0.01	A/A	Yes
K	766.5	0.2	0.01; 0.02	A/A	No
Na	589.0	0.04	0.01; 0.02	A/A	No
Zn	213.9	0.2	0.01; 0.02	A/A	Yes

*A/A is air acetylene; N/A is nitrous oxide/acetylene.

HEAVY METALS.   (Page 26, Method 1). Dissolve 6.0 g in about 20 mL of water, and dilute with water to 30 mL. Use 25 mL to prepare the sample solution, and use the remaining 5.0 mL to prepare the control solution. Adjust the pH with ammonium acetate solution.

IRON.   (Page 28, Method 1). Use 2.0 g.

Manganese Sulfate Monohydrate
Manganese(II) Sulfate Monohydrate
Manganous Sulfate Monohydrate

$MnSO_4 \cdot H_2O$ Formula Wt 169.02

CAS Number 10034 – 96 – 5

REQUIREMENTS

Assay . . . . 98.0 – 101.0%
$MnSO_4 \cdot H_2O$
Loss on ignition . . . . 10.0 – 12.0%
Substances reducing permanganate . . . . Passes test

MAXIMUM ALLOWABLE

Insoluble matter 0.01%
Chloride (Cl) 0.005%
Calcium (Ca) 0.005%
Heavy metals (as Pb) 0.002%
Iron (Fe) 0.002%
Magnesium (Mg) 0.005%
Nickel (Ni) 0.02%
Potassium (K) 0.01%
Sodium (Na) 0.05%
Zinc (Zn) 0.005%

TESTS

ASSAY. (By complexometric titration of Mn^{II}). Weigh accurately 0.7 g, transfer to a 500-mL beaker, and dissolve in 200 mL of water. Add a few milligrams of ascorbic acid to prevent oxidation. From a buret, add about 30 mL of standard 0.1 M EDTA. Now add 10 mL of pH 10 ammoniacal buffer solution and about 50 mg of Eriochrome Black T indicator mixture. Continue the titration with 0.1 M EDTA to a clear blue color. One milliliter of 0.1 M EDTA corresponds to 0.01690 g of $MnSO_4 \cdot H_2O$.

LOSS ON IGNITION. Weigh accurately about 2 g. Ignite to constant weight at 400–500°C.

INSOLUBLE MATTER. (Page 14). Use 10 g dissolved in 130 mL of water.

CHLORIDE. (Page 25). Dissolve 1.0 g in 100 mL of water, and use 20 mL of this solution.

CALCIUM, MAGNESIUM, NICKEL, POTASSIUM, SODIUM, AND ZINC. Determine the calcium, magnesium, nickel, potassium, sodium, and zinc by the flame atomic absorption spectrophotometric method described on page 35.

> **Sample Stock Solution.** Dissolve 10.0 g in water in a 100-mL volumetric flask, add 0.5 mL of hydrochloric acid, and dilute with water to the mark (1 mL = 0.1 g).

Analyze the solutions by means of a suitable atomic absorption spectrophotometer, using the conditions outlined in the following table. Calculate the metal content of the sample by the method of standard additions.

Element	Wavelength (nm)	Sample Wt (g)	Standard Added (mg)	Flame Type*	Background Correction
Ca	422.7	0.8	0.02; 0.04	N/A	No
Mg	285.2	0.2	0.005; 0.01	A/A	Yes
Ni	232.0	0.5	0.05; 0.10	A/A	Yes
K	766.5	0.2	0.01; 0.02	A/A	No
Na	589.0	0.02	0.005; 0.01	A/A	No
Zn	213.9	0.2	0.005; 0.01	A/A	Yes

*A/A is air/acetylene; N/A is nitrous oxide/acetylene.

HEAVY METALS. (Page 26, Method 1). Dissolve 1.0 g in about 20 mL of water, and dilute with water to 25 mL.

IRON. (Page 28, Method 1). Dissolve 1.0 g in water, dilute with water to 50 mL, and use 25 mL of this solution without further acidification.

SUBSTANCES REDUCING PERMANGANATE. Dissolve 7.5 g in 200 mL of water containing 3 mL of sulfuric acid and 3 mL of phosphoric acid. To this solution add 0.10 mL of 0.1 N potassium permanganate in excess of the amount required to produce a pink color in 200 mL of water containing 3 mL of sulfuric acid and 3 mL of phosphoric acid, and allow to stand for 1 min. The pink color should not be entirely discharged.

Mannitol

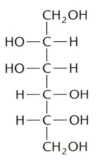

HOCH$_2$(CHOH)$_4$CH$_2$OH   Formula Wt 182.17

CAS Number 69–65–8

REQUIREMENTS

Assay . . . . $\geq$ 99.0% C$_6$H$_{14}$O$_6$
Specific rotation [α]$_D^{25°}$ . . . . +23.3° to +24.3°
Reducing sugars. . . . . Passes test

MAXIMUM ALLOWABLE

Insoluble matter 0.01%
Loss on drying at 105°C. 0.05%
Residue after ignition 0.01%
Titrable acid . . . . 0.0008 meq / g
Heavy metals (as Pb) . . . . 5 ppm

TESTS

SPECIFIC ROTATION. Weigh accurately about 10 g, transfer to a 100-mL volumetric flask, and add 12.8 g of sodium borate. Dissolve the mixture in sufficient water to make about 90 mL of solution, allow to stand with occasional shaking for 1 h, dilute with water to volume at 25°C, and mix. Observe the optical rotation in a polarimeter at 25°C using sodium light, and calculate the specific rotation.

INSOLUBLE MATTER. (Page 14). Use 20 g dissolved in 100 mL of water.

LOSS ON DRYING AT 105°C. Weigh accurately 5.0 g in a tared dish or crucible and dry at 105°C for 2 h.

RESIDUE AFTER IGNITION. (Page 15). Ignite 10 g. Moisten the char with 1 mL of sulfuric acid.

TITRABLE ACID. To 100 mL of carbon dioxide-free water add 0.15 mL of phenolphthalein indicator solution and 0.01 N sodium hydroxide until a pink color is produced. Dissolve 10 g of the sample in this solution and titrate with 0.01 N sodium hydroxide to the same end point. Not more than 0.83 mL should be required.

REDUCING SUGARS. Add 1 mL of a saturated solution of the mannitol (about 0.2 g per mL) to 5 mL of Benedict's solution in a boiling-water bath and heat for 5 min. No more than a slight precipitate should form.

> ***Benedict's Solution.*** Dissolve 173 g of sodium citrate dihydrate and 100 g of anhydrous sodium carbonate, Na_2CO_3, in 800 mL of water. Heat to aid dissolution, filter if necessary, and dilute with water to 850 mL. Dissolve 17.3 g of copper sulfate pentahydrate, $CuSO_4 \cdot 5H_2O$, in 100 mL of water. Add this solution, with constant stirring, to the alkaline citrate solution, and dilute with water to 1 L.

HEAVY METALS. (Page 26, Method 1). Dissolve 4.0 g in about 25 mL of water, and dilute with water to 30 mL.

Mercuric Acetate
Mercury(II) Acetate

$(CH_3COO)_2Hg$ Formula Wt 318.68

CAS Number 1600–27–7

REQUIREMENTS

Assay . . . . ≥ 98.0%
$(CH_3COO)_2Hg$

MAXIMUM ALLOWABLE

Insoluble matter 0.01%
Residue after reduction. 0.02%
Chloride (Cl) 0.005%
Nitrate (NO_3) 0.005%
Sulfate (SO_4) 0.005%
Foreign heavy metals (as Pb) 0.002%
Iron (Fe) 0.001%
Mercurous mercury (as Hg) 0.4%

TESTS

ASSAY. (By thiocyanate precipitation titration of mercuric ion). Weigh accurately 0.7 g of sample and dissolve in 100 mL of dilute nitric acid (1 + 19). Add 2 mL of ferric ammonium sulfate indicator solution and titrate with 0.1 N ammonium thiocyanate. One milliliter of 0.1 N ammonium thiocyanate corresponds to 0.01593 g of $(CH_3COO)_2Hg$.

INSOLUBLE MATTER. (Page 14). Use 10 g dissolved in 100 mL of dilute acetic acid (5 + 95). Also use dilute acetic acid (1 + 99) as the wash.

Sample Solution A for the Determination of Residue After Reduction, Chloride, Sulfate, Foreign Heavy Metals, and Iron. Dissolve 10.0 g in 15 mL of water, 2 mL of nitric acid, and 10 mL of formic acid (98%). Digest under total reflux until all the mercury is reduced to metal and the solution is clear. Cool, filter through a thoroughly washed filter paper, and wash with a small quantity of water. Dilute the filtrate and washings with water to 100 mL in a volumetric flask (1 mL = 0.1 g).

RESIDUE AFTER REDUCTION. Evaporate 50 mL (5-g sample) of sample solution A plus 0.10 mL of sulfuric acid to dryness in a tared dish in a well-ventilated hood. Continue heating until the excess sulfuric acid has been volatilized. Finally, ignite at 800 ± 25°C for 15 min. Correct for the weight obtained in a complete blank test.

CHLORIDE. (Page 25). Use 2.0 mL of sample solution A (0.2-g sample) for the determination of chloride.

NITRATE.

> *Sample Solution A-1.* Dissolve 0.50 g completely in 3.0 mL of water in a dry test tube. Put the tube in an ice bath and immediately, but slowly, add 7 mL of cold chromotropic acid reagent while swirling. Keep the tube in the ice bath an additional 2–3 min, remove, and let stand for 30 min, swirling occasionally.
>
> *Chromotropic Acid Reagent.* Dissolve 0.02 g of recrystallized 4,5-dihydroxy-2,7-naphthalenedisulfonic acid, disodium salt, in concentrated, nitrate-free sulfuric acid and dilute to 200 mL with the acid. Store in an amber bottle.
>
> *Recrystallization of Reagent.* Add solid sodium sulfate to a saturated aqueous solution of the foregoing salt. Filter the resultant crystals using suction with a Buchner funnel, wash with ethyl alcohol, and air dry. Transfer the crystals to a beaker, add sufficient water to redissolve, then add just enough sodium sulfate to reprecipitate the salt. Filter as before, wash with ethyl alcohol, and dry at a temperature no higher than 80°C. Store in an amber bottle.
>
> NOTE. If an orange or red precipitate forms in the solution of the sample before the addition of the reagent, discard the solution and start again. Immediate addition of the reagent should prevent formation of such a precipitate.
>
> *Control Solution B.* Dissolve 0.50 g of sample in 0.5 mL of water plus 2.5 mL of a solution containing 0.01 mg of nitrate ion (NO_3) per milliliter in a dry test tube. Put the tube in an ice bath and immediately, but slowly, add 7 mL of cold chromotropic acid reagent while stirring. Keep the tube in the ice bath an addi-

tional 2 to 3 min, remove, and let stand for 30 min, swirling occasionally.

Blank Solution C.   Add 7 mL of cold chromotropic acid reagent to 3.0 mL of water in a dry test tube immersed in an ice bath. Remove and let stand for 30 min.

Transfer sample solution A-1 and control solution B to dry 15-mL centrifuge tubes and centrifuge until the supernatant liquid is clear. Set a spectrophotometer at 405 nm and, using 1-cm cells, adjust the instrument to read zero absorbance with blank solution C in the light path, then determine the absorbance of sample solution A-1. Adjust the instrument to read zero absorbance with sample solution A-1 in the light path and determine the absorbance of control solution B. The absorbance of sample solution A-1 versus blank solution C should not exceed that of control solution B versus sample solution A-1.

SULFATE.   (Page 32, Procedure A, Method 3). Use 10 mL of sample solution A (1-g sample) for the determination of sulfate.

FOREIGN HEAVY METALS.   (Page 26, Method 1). Add 10 mg of sodium carbonate to 10 mL of sample solution A (1-g sample) for the determination of foreign heavy metals and evaporate to dryness. Dissolve the residue in about 20 mL of water, filter, and dilute with water to 25 mL.

IRON.   (Page 28, Method 1). To 10 mL of sample solution A (1-g sample) for the determination of iron, add about 10 mg of sodium carbonate, and evaporate to dryness. Dissolve the residue in 2 mL of hydrochloric acid, dilute with water to 50 mL, and use the solution without further acidification.

MERCUROUS MERCURY.   Dissolve 5.0 g in 100 mL of a 12.5% solution of potassium iodide. Add 5.0 mL of 0.1 N iodine and 2 mL of 1 N hydrochloric acid. Allow to stand, protected from the light, for 1 h with frequent agitation. Titrate the iodine with 0.1 N sodium thiosulfate, adding 3 mL of starch indicator near the end point, and correct for a complete blank. Not more than 1.0 mL of 0.1 N iodine should have been consumed.

Mercuric Bromide
Mercury(II) Bromide

$HgBr_2$ Formula Wt 360.40

CAS Number 7789–47–1

REQUIREMENTS

Appearance. . . . . White or having at most a very faint yellow tinge. If a mercurous salt is present, the product darkens on exposure to light.

MAXIMUM ALLOWABLE

Residue after reduction. 0.02%
Insoluble in methanol. 0.05%
Chloride (Cl) 0.25%

TESTS

RESIDUE AFTER REDUCTION.   Dissolve 5.0 g in 10 mL of water plus 10 mL of ammonium hydroxide. Add 40 mL of formic acid (96%) and reflux until all the mercury is reduced to metal. Cool, filter through a well-washed filter paper, and wash with a small quantity of water. Add 0.10 mL of sulfuric acid to the combined filtrate and washings. Evaporate in a tared dish in a well-ventilated hood. Continue heating until the excess sulfuric acid has been volatilized. Finally, ignite at $800 \pm 25°C$ for 15 min. Correct for the weight obtained in a complete blank test.

INSOLUBLE IN METHANOL.   Dissolve 2.0 g in 30 mL of methanol. Filter through a tared filtering crucible, wash with methanol until the washings remain unaffected by hydrogen sulfide, and dry at 105°C.

CHLORIDE. To 0.70 g add 20 mL of water, 5 mL of 10% sodium hydroxide reagent solution, and 4 mL of 30% hydrogen peroxide, and digest on the steam bath until all reaction ceases. Cool, filter through a chloride-free filter, wash with water, and dilute the filtrate and washings with water to 100 mL. To 5.0 mL in a small conical flask, add 20 mL of water and 1.5 mL of ammonium carbonate solution [20 g of ammonium carbonate and 20 mL of dilute ammonium hydroxide (10% NH_3) in sufficient water to make 100 mL]. Add, with agitation, 5 mL of silver nitrate reagent solution and allow to stand, with frequent agitation, for 10 min. Filter through a chloride-free filter, wash with water, and dilute the filtrate and washings with water to 100 mL. To 25 mL add 0.5 mL of nitric acid. Any turbidity should not exceed that in a control obtained by treating 0.20 g of sample and 1.25 mg of added chloride ion (Cl) in 20 mL of water exactly like the solution of the sample.

Mercuric Chloride
Mercury(II) Chloride

$HgCl_2$ Formula Wt 271.50

CAS Number 7487–94–7

REQUIREMENTS

Assay . . . . ≥ 99.5% $HgCl_2$
Solution in ethyl ether . . . . Passes test

MAXIMUM ALLOWABLE

Residue after reduction. 0.02%
Iron (Fe) 0.002%

TESTS

ASSAY. (By complexometric titration of Hg^{II}). Weigh accurately about 0.7 g, and dissolve completely in 250 mL of water. From a buret add 20 mL of 0.1 M EDTA, then add 10 mL of a solution containing 2 g of silver nitrate. Add 2 g of ammonium nitrate, 20 mL of a saturated aqueous solution of hexamethylenetetramine, and a few milligrams of xylenol orange indicator. Continue the titration with 0.1 M EDTA to a yellow end point. One milliliter of 0.1 M EDTA corresponds to 0.02715 g of $HgCl_2$.

RESIDUE AFTER REDUCTION.   Dissolve 5.0 g in 10 mL of water plus 10 mL of ammonium hydroxide. Add 40 mL of formic acid (96%) and reflux until all the mercury is reduced to metal. Cool, filter through a well-washed filter paper, and wash with a small quantity of water. Add 0.10 mL of sulfuric acid to the combined filtrate and washings. Evaporate in a tared dish in a well-ventilated hood. Continue heating until the excess sulfuric acid has been volatilized. Finally, ignite at 800 ± 25°C for 15 min. Correct for the weight obtained in a compete blank test.

SOLUTION IN ETHYL ETHER.   Dissolve 2.0 g in 60 mL of ethyl ether in a stoppered flask. Shake the flask to agitate the sample and ether, but do not expose the contents to atmospheric moisture. Not more than a faint trace of insoluble residue should remain.

IRON.   (Page 28, Method 1). To the residue after reduction add 3 mL of dilute hydrochloric acid (1 + 1), cover the dish, and digest on the steam bath for 15–20 min. Remove the cover and evaporate to dryness. Dissolve the residue in 10 mL of dilute hydrochloric acid (1 + 9), dilute with water to 50 mL, and use 5.0 mL of this solution.

Mercuric Iodide, Red
Mercury(II) Iodide, Red

HgI_2 Formula Wt 454.40

CAS Number 7774–29–0

REQUIREMENTS

Assay (dried basis) . . . . $\geq$ 99.0% HgI_2
Solubility in potassium
   iodide solution . . . . Passes test

MAXIMUM ALLOWABLE

Mercurous mercury (as Hg) 0.1%
Soluble mercury salts (as Hg) 0.05%

TESTS

ASSAY. (By iodometry). Dry about 1 g of sample over magnesium perchlorate overnight. Weigh accurately 0.5 g of the dried sample, place in a glass-stoppered conical flask, and add 30 mL of hydrochloric acid and 20 mL of water. Rotate the flask until the mecuric iodide is dissolved, add 5 mL of chloroform, and titrate the solution with 0.05 M potassium iodate until the iodine color is discharged from the aqueous layer. Stopper the flask, shake well for 30 s, then continue the titration, shaking vigorously after each addition of the potassium iodate until the chloroform is free of iodine color. One milliliter of 0.05 M potassium iodate corresponds to 0.02272 g of HgI_2.

SOLUBILITY IN POTASSIUM IODIDE SOLUTION. Dissolve 10 g of the sample in 100 mL of 10% potassium iodide reagent solution in a glass-stoppered flask. A complete, or practically complete, solution results. Retain the solution for the test for mercurous mercury.

MERCUROUS MERCURY. To the solution reserved from the test for solubility in potassium iodide solution add 5.0 mL of 0.1 N iodine and 3 mL of 1 N hydrochloric acid. Allow to stand in a dark place for 1 h with frequent agitation. Titrate the excess iodine with 0.1 N sodium thiosulfate, adding 3 mL of starch indicator near the end point, and correct for a complete blank. Not more than 0.50 mL of the 0.1 N iodine should be consumed.

SOLUBLE MERCURY SALTS. Shake 1.0 g of the sample with 20 mL of water for 2 min and filter. Dilute 10 mL of the filtrate with water to 40 mL and add 10 mL of hydrogen sulfide water. Any color should not exceed that produced by 0.25 mg of mercury ion (Hg) in an equal volume of solution containing the quantities of reagents used in the test.

Mercuric Nitrate Monohydrate or Dihydrate

Mercury(II) Nitrate Monohydrate or Dihydrate

$Hg(NO_3)_2 \cdot H_2O$ Formula Wt 342.62

$Hg(NO_3)_2 \cdot 2H_2O$ Formula Wt 360.63

CAS Number 22852–67–1 (dihydrate); 7783–34–8 (monohydrate)

NOTE. This reagent is available as either the mono- or dihydrate. The label should identify the degree of hydration.

REQUIREMENTS

Assay . . . . $\geq 98.0\%$
$$Hg(NO_3)_2 \cdot H_2O$$
$$\text{or } Hg(NO_3)_2 \cdot 2H_2O$$

MAXIMUM ALLOWABLE

Residue after reduction. 0.01%
Chloride (Cl) 0.002%
Sulfate (SO$_4$) 0.002%
Iron (Fe) 0.001%
Mercurous mercury (as Hg) 0.2%

TESTS

ASSAY. (By thiocyanate precipitation titration of mercuric ion). Weigh accurately about 0.6 g of sample, and transfer with 100 mL of water to a 250-mL flask. Add 5 mL of nitric acid and 2 mL of ferric ammonium sulfate indicator solution. Cool to 15°C, and titrate with 0.1 N ammonium thiocyanate to a permanent reddish brown color. One milliliter of 0.1 N ammonium thiocyanate corresponds to 0.01713 g of $Hg(NO_3)_2 \cdot H_2O$ or 0.01803 g of $Hg(NO_3)_2 \cdot 2H_2O$.

RESIDUE AFTER REDUCTION. Dissolve 10 g in 15 mL of water, 2 mL of nitric acid, and 10 mL of formic acid (96%). Digest under total reflux until all the mercury is reduced to metal and the solution is clear. Cool, filter through a well-washed filter paper, and wash with a small quantity of water. Add 0.10 mL of sulfuric acid to the combined filtrate and washings. Evaporate in a tared dish in a well-ventilated hood. Continue heating until the excess sulfuric acid has been volatilized. Finally, ignite at 800 ± 25°C for 15 min. Correct for the weight obtained in a complete blank test. Retain the residue for the determination of iron.

CHLORIDE. Dissolve 2.0 g in 50 mL of water and 2 mL of formic acid. Add, dropwise, 10% sodium hydroxide solution until a small amount of permanent precipitate is formed. Digest under a reflux condenser until the mercury is all reduced to the metal and the supernatant liquid is clear. Cool, filter through a chloride-free filter, wash thoroughly, and dilute the filtrate with water to 100 mL. To 50 mL of the dilution add 1 mL of nitric acid and 1 mL of silver nitrate reagent solution. Any turbidity should not exceed that produced by 0.02 mg of chloride ion (Cl) in an equal volume of solution containing the quantities of reagents used in the test.

SULFATE.   Dissolve 5.0 g in 50 mL of water and 5 mL of formic acid. Digest under a reflux condenser until the mercury is all reduced to the metal and the supernatant liquid is clear. Cool, filter, and wash thoroughly. To the combined filtrate and washings add 20 mg of sodium carbonate, and evaporate to dryness on a hot plate at low setting. Dissolve the residue in 10 mL of water and 1 mL of dilute hydrochloric acid (1 + 19), and filter if necessary. To the filtrate add 1 mL of barium chloride reagent solution. Any turbidity should not exceed that produced by 0.1 mg of sulfate ion (SO_4) in an equal volume of solution containing the quantities of reagents used in the test. Compare 10 min after adding the barium chloride to the sample and standard solutions.

IRON.   (Page 28, Method 1). To the residue obtained in the test for residue after reduction, add 3 mL of dilute hydrochloric acid (1 + 1), cover with a watch glass, and digest on the steam bath for 20 min. Remove the cover and evaporate to dryness. Dissolve the residue in 2 mL of dilute hydrochloric acid (1 + 1), add about 40 mL of water, filter if necessary, and dilute with water to 100 mL. Use 10 mL of this solution.

MERCUROUS MERCURY.   Dissolve 5.0 g in 100 mL of 10% potassium iodide solution, and add 5 mL of 0.1 N iodine and 3 mL of dilute hydrochloric acid (1 + 19). Allow to stand in the dark for 1 h with frequent agitation. Titrate the excess iodine with 0.1 N sodium thiosulfate, adding 3 mL of starch indicator solution near the end point. Perform a complete blank test. The volume of 0.1 N iodine consumed should not exceed 0.50 mL.

Mercuric Oxide, Red
Mercury(II) Oxide, Red

HgO   Formula Wt 216.59

CAS Number 21908–53–2

REQUIREMENTS

Assay . . . . $\geq$ 99.0% HgO

MAXIMUM ALLOWABLE

Insoluble in dilute hydrochloric acid 0.03%
Residue after reduction. 0.025%
Chloride (Cl) 0.025%
Sulfate (SO_4) 0.015%
Nitrogen compounds (as N) 0.005%
Iron (Fe) 0.005%

TESTS

ASSAY. (By complexometric titration of mercuric ion). Weigh accurately 0.8 g of sample. Dissolve in 3.5 mL of nitric acid, dilute to about 100 mL with water, and add 50 mg of xylenol orange indicator. Add a saturated solution of hexamethylenetetramine until the color changes to purple, followed by 3–4 mL in excess, then titrate with 0.1 M EDTA until the color changes to yellow. One milliliter of 0.1 M EDTA corresponds to 0.02166 g of HgO.

INSOLUBLE IN DILUTE HYDROCHLORIC ACID. Dissolve 4.0 g in 40 mL of dilute hydrochloric acid (1 + 3), heat to boiling, and digest in a covered beaker on the steam bath for 1 h. Filter through a tared filtering crucible, wash well with water, and dry at 105°C.

>*Sample Solution A for the Determination of Residue After Reduction and Sulfate.* Dissolve 10 g in 15 mL of water, 2 mL of nitric acid, and 10 mL of formic acid (96%). Digest under total reflux until all the mercury is reduced to metal and the solution is clear. Cool, filter through a well-washed filter paper, and wash with a small quantity of water. Dilute the filtrate and washings with water to 200 mL in a volumetric flask (1 mL = 0.05 g).

RESIDUE AFTER REDUCTION. Evaporate 80 mL (4-g sample) of sample solution A plus 0.10 mL of sulfuric acid to dryness in a tared dish in a well-ventilated hood. Continue heating until the excess sulfuric acid has been volatilized. Finally, ignite at 800 ± 25°C for 15 min. Correct for the weight obtained in a complete blank test. Retain the residue for the test for iron.

CHLORIDE. Dissolve 1.0 g in 50 mL of water plus 1 mL of formic acid. Add, dropwise, 10% sodium hyroxide reagent solution until a small amount of permanent precipitate is formed. Digest under a reflux condenser until all the mercury is reduced to metal and the solution is clear. Cool, filter through a chloride-free filter, and dilute with water to 100 mL. Dilute 4.0 mL of this solution with water to 20 mL and add 1 mL of nitric acid and 1 mL of silver nitrate reagent solution. Any turbidity should not exceed that produced by 0.01 mg of chloride ion (Cl) in an equal volume of solution containing the quantities of reagents used in the test.

SULFATE.   (Page 32, Procedure A, Method 3). Use 6.7 mL of sample solution A (0.33-g sample).

NITROGEN COMPOUNDS.   (Page 30). Dissolve 0.2 g in 5 mL of dilute hydrochloric acid (1 + 1). For the standard use 0.01 mg of nitrogen (N) and 5 mL of the dilute acid.

IRON.   (Page 28, Method 1). To the residue obtained in the test for residue after reduction, add 3 mL of dilute hydrochloric acid (1 + 1), cover with a watch glass, and digest on the steam bath for 20 min. Remove the cover and evaporate to dryness. Dissolve in 2 mL of dilute hydrochloric acid (1 + 1), add about 40 mL of water, filter if necessary, and dilute with water to 100 mL. Use 5 mL of this solution.

Mercuric Oxide, Yellow
Mercury(II) Oxide, Yellow

HgO   Formula Wt 216.59

CAS Number 21908–53–2

REQUIREMENTS

Assay . . . . ≥ 99.0% HgO

MAXIMUM ALLOWABLE

Insoluble in dilute hydrochloric acid 0.03%
Residue after reduction. 0.05%
Chloride (Cl) 0.025%
Sulfate (SO_4). 0.01%
Nitrogen compounds (as N) 0.005%
Iron (Fe) 0.003%

TESTS

ASSAY.   (By complexometric titration of mercuric ion). Weigh accurately 0.8 g of sample. Dissolve in 3.5 mL of nitric acid, dilute to about 100 mL with water, and add 50 mg of xylenol orange indicator. Add a saturated solution of hexamethylenetetramine until the color changes to purple, followed by 3–4 mL in excess, then titrate with 0.1 M EDTA until the color changes to yellow. One milliliter of 0.1 M EDTA corresponds to 0.02166 g of HgO.

INSOLUBLE IN DILUTE HYDROCHLORIC ACID.   Dissolve 3.0 g in 30 mL of dilute hydrochloric acid (1 + 3), heat to boiling, and digest in a covered beaker on the steam bath for 1 h. Filter through a tared filtering crucible, wash thoroughly, and dry at 105°C.

Sample Solution A for the Determination of Residue After Reduction and Sulfate.   Dissolve 10 g in 15 mL of water, 2 mL of nitric acid, and 10 mL of formic acid (96%). Digest under total reflux until all the mercury is reduced to metal and the solution is clear. Cool, filter through a well-washed filter paper, and wash with a small quantity of water. Dilute the filtrate and washings with water to 200 mL in a volumetric flask (1 mL = 0.05 g).

RESIDUE AFTER REDUCTION.   Evaporate 60 mL (3 g) of sample solution A plus 0.10 mL of sulfuric acid to dryness in a tared dish in a well-ventilated hood. Continue heating until the excess sulfuric acid has been volatilized. Finally, ignite at 800 ± 25°C for 15 min. Correct for the weight obtained in a complete blank test. Retain the residue for the test for iron.

CHLORIDE.   Dissolve 1.0 g in 50 mL of water and 1 mL of formic acid. Add, dropwise, 10% sodium hydroxide reagent solution until a small amount of permanent precipitate is formed. Digest under a reflux condenser until all the mercury is reduced to metal and the solution is clear. Cool, filter through a chloride-free filter, and dilute with water to 100 mL. Dilute 4.0 mL of this solution with water to 20 mL and add 1 mL of nitric acid and 1 mL of silver nitrate reagent solution. Any turbidity should not exceed that produced by 0.01 mg of chloride ion (Cl) in an equal volume of solution containing the quantities of reagents used in the test.

SULFATE.   (Page 32, Procedure A, Method 3). Use 10 mL of sample solution A (0.5-g sample).

NITROGEN COMPOUNDS. (Page 30). Dissolve 0.2 g in 5 mL of dilute hydrochloric acid (1 + 1). For the standard use 0.01 mg of nitrogen (N) and 5 mL of the dilute acid.

IRON. (Page 28, Method 1). To the residue obtained in the test for residue after reduction add 1 mL of hydrochloric acid, 0.15 mL of nitric acid, and about 10 mg of sodium carbonate, cover, and digest on the steam bath for 15–20 min. Uncover and evaporate to dryness. Dissolve the residue in 9 mL of hydrochloric acid and dilute with water to 90 mL. Use 10 mL of this solution and only 1 mL of additional hydrochloric acid.

Mercuric Sulfate
Mercury(II) Sulfate

$HgSO_4$ Formula Wt 296.65

CAS Number 7783−35−9

REQUIREMENTS

Assay . . . . ≥ 98.0% $HgSO_4$

MAXIMUM ALLOWABLE

Residue after reduction. 0.02%
Chloride (Cl) 0.003%
Nitrate (NO_3) . . . . Passes test (limit
                                                                  about 0.005%)
Iron (Fe) 0.005%
Mercurous mercury (as Hg). 0.15%

TESTS

ASSAY. (By thiocyanate precipitation titration of mercuric ion). Weigh accurately 0.5 g, and dissolve in 50 mL of dilute nitric acid (1 + 1). Add 1 mL of 10% ferric nitrate solution, and titrate with 0.1 N ammonium thiocyanate to a permanent reddish brown color. One milliliter of 0.1 N ammonium thiocyanate corresponds to 0.01483 g of $HgSO_4$.

RESIDUE AFTER REDUCTION.   Dissolve 5.0 g in 50 mL of water plus 10 mL of formic acid (96%), and add 25 mL of 30% ammonium hydroxide. Digest under total reflux until all the mercury is reduced to metal. Cool, filter through a thoroughly washed filter paper, and wash with a small quantity of water. Add 0.1 mL of sulfuric acid to the combined filtrate and washings, and evaporate in a tared dish in a well-ventilated hood. Continue heating until the excess sulfuric acid has been volatilized. Finally, ignite at 800 ± 25°C for 15 min. Correct for the weight obtained in a complete blank test. Retain the residue for the test for iron.

CHLORIDE.   Dissolve 1.0 g in 50 mL of water plus 1 mL of formic acid (96%). Add, dropwise, 10% sodium hydroxide reagent solution until a small amount of permanent precipitate is formed. Digest under total reflux until all the mercury is reduced to metal and the solution is clear. Cool, filter through a chloride-free filter, and dilute to 90 mL. To 30 mL of this solution add 1 mL of nitric acid and 1 mL of silver nitrate reagent solution. Any turbidity should not exceed that produced by 0.01 mg of chloride ion (Cl) in an equal volume of solution containing the quantities of reagents used in the test.

NITRATE.   Disperse 1.0 g in 9 mL of water, add 1 mL of sodium chloride solution (1 in 200), mix, and add 0.1 mL of indigo carmine reagent solution, followed by 10 mL of sulfuric acid. The blue color of the clear solution should not be discharged entirely within 5 min.

IRON.   (Page 28, Method 1). To the residue obtained in the test for residue after reduction, add 3 mL of dilute hydrochloric acid (1 + 1), cover with a watch glass, and digest on the steam bath for 20 min. Remove the watch glass and evaporate to dryness. Take up the residue in a mixture of 1 mL of dilute hydrochloric acid (1 + 1) and 30 mL of water, filter if necessary, dilute with water to 100 mL, and use 4.0 mL of the solution.

MERCUROUS MERCURY.   Transfer 5.0 g to a glass-stoppered flask. Add 100 mL of 15% potassium iodide reagent solution, 5.00 mL of 0.1 N iodine, and 3 mL of 1 N hydrochloric acid. Allow to stand in the dark, with frequent agitation, for 1 h. Titrate the excess iodine with 0.1 N sodium thiosulfate, adding 3 mL of starch indicator solution near the end point. Correct for a blank determination. Not more than 0.38 mL of the 0.1 N iodine should be consumed.

Mercurous Chloride
Mercury(I) Chloride

Hg_2Cl_2 Formula Wt 472.09

CAS Number 10112–91–1

REQUIREMENTS

Assay . . . . $\geq$ 99.5% Hg_2Cl_2

MAXIMUM ALLOWABLE

Residue after reduction. 0.02%
Mercuric chloride ($HgCl_2$). 0.01%
Sulfate (SO_4). 0.01%

TESTS

ASSAY. (By iodometric titration of mercury). Weigh accurately 0.9 g and transfer to a 250-mL glass-stoppered conical flask. Add 50.0 mL of 0.1 N iodine and 2 g of potassium iodide. Stopper and swirl until the precipitate redissolves. Titrate the excess of iodine with 0.1 N sodium thiosulfate, adding 3 mL of starch indicator solution near the end point. One milliliter of 0.1 N iodine corresponds to 0.02360 g of Hg_2Cl_2.

RESIDUE AFTER REDUCTION. Dissolve 5.0 g in 10 mL of water plus 10 mL of ammonium hydroxide. Add 40 mL of formic acid (96%) and reflux until all the mercury is reduced to metal. Cool, filter through a thoroughly washed filter paper, and wash with a small quantity of water. Add 0.10 mL of sulfuric acid to the combined filtrate and washings. Evaporate in a tared dish in a well-ventilated hood. Continue heating until the excess sulfuric acid has been volatilized. Finally, ignite at 800 $\pm$ 25°C for 15 min. Correct for the weight obtained in a complete blank test.

MERCURIC CHLORIDE. Shake 1.0 g with 10 mL of alcohol for 5 min, and filter. To 5 mL of the filtrate add 0.10 mL of hydrochloric acid and 5 mL of hydrogen sulfide water. Any darkening should not be more than that produced by 0.05 mg of mercuric chloride ($HgCl_2$) in a mixture of 5 mL of alcohol, 0.10 mL of hydrochloric acid, and 5 mL of hydrogen sulfide water.

SULFATE.   (Page 32, Procedure A, Method 3). Digest for 10 min with 10 mL of dilute hydrochloric acid (1 + 1), and filter.

Mercurous Nitrate Dihydrate
Mercury(I) Nitrate Dihydrate

$Hg_2(NO_3)_2 \cdot 2H_2O$ Formula Wt 561.22

CAS Number 14836−60−3

> *NOTE.* This salt is commercially available in a degree of hydration ranging from 0 to 2 molecules of water per molecule of $Hg_2(NO_3)_2$. The specifications refer to both anhydrous and hydrated reagents, and the assay is expressed as the dihydrate.

REQUIREMENTS

Assay . . . . ≥ 98.0%
$Hg_2(NO_3)_2 \cdot 2H_2O$

MAXIMUM ALLOWABLE

Insoluble matter 0.005%
Residue after reduction. 0.01%
Chloride (Cl) 0.005%
Sulfate (SO_4) 0.005%
Iron (Fe) 0.001%
Mercuric mercury (as Hg) 0.5%

TESTS

ASSAY.   (By iodometric titration of mercury). Weigh accurately 1 g and transfer to a 250-mL glass-stoppered conical flask. Add 50.0 mL of 0.1 N iodine and 2 g of potassium iodide. Stopper and swirl until the precipitate redissolves. Titrate the excess of iodine with 0.1 N sodium thiosulfate, adding 3 mL of starch indicator solution near the end point. One milliliter of 0.1 N iodine corresponds to 0.02806 g of $Hg_2(NO_3)_2 \cdot 2H_2O$.

INSOLUBLE MATTER.   (Page 14). Use 20 g dissolved in 200 mL of dilute nitric acid (5 + 95).

Sample Solution A for the Determination of Residue After Reduction, Chloride, and Sulfate. Dissolve 25 g in 30 mL of water, 4 mL of nitric acid, and 20 mL of formic acid (96%). Digest under total reflux until all the mercury(I) is reduced to mercury and the solution is clear. Cool, filter through a well-washed filter paper, and wash with a small quantity of water. Dilute the filtrate and washings with water to 500 mL in a volumetric flask (1 mL = 0.05 g).

RESIDUE AFTER REDUCTION.   Evaporate 400 mL (20-g sample) of sample solution A plus 0.10 mL of sulfuric acid to dryness in a tared dish in a well-ventilated hood. Continue heating until the excess sulfuric acid has been volatilized. Finally, ignite at 800 ± 25°C for 15 min. Correct for the weight obtained in a complete blank test. Retain the residue for the test for iron.

CHLORIDE.   (Page 25). Use 4.0 mL of sample solution A (0.2-g sample).

SULFATE.   (Page 32, Procedure A, Method 1). Use 20 mL of sample solution A (1-g sample).

IRON.   (Page 28, Method 1). To the residue obtained in the test for residue after reduction, add 3 mL of dilute hydrochloric acid (1 + 1), cover with a watch glass, and digest on the steam bath for 20 min. Remove the cover, and evaporate to dryness. Dissolve the residue in 2 mL of dilute hydrochloric acid (1 + 1) and 30–40 mL of water, filter if necessary, and dilute with water to 100 mL. Use 5.0 mL of this solution.

MERCURIC MERCURY.   Dissolve 1.0 g in a mixture of 5 mL of water and 0.25 mL of nitric acid in a glass-stoppered 50-mL cylinder. When dissolution is complete, add 10 mL of water and 5 mL of 1 N hydrochloric acid, and mix well. Dilute with water to 50 mL, shake well, and allow to stand for 15 min. Filter the solution and dilute 5 mL of the filtrate with water to 25 mL. For the standard mix 0.25 mL of nitric acid with about 10 mL of water, add 5 mL of 1 N hydrochloric acid, dilute with water to 50 mL, and mix. To 5 mL of the standard solution add 0.5 mg of mercury ion (Hg), prepared from mercuric chloride, and dilute with water to 25 mL. Add 10 mL of hydrogen sulfide water to both the sample and the standard solution. Any color in the solution of the sample should not exceed that in the standard.

Mercury

Hg

Atomic Wt 200.59

CAS Number 7439–97–6

REQUIREMENTS

Appearance . . . . Passes test

MAXIMUM ALLOWABLE

Nonvolatile matter . . . . 5 ppm

TESTS

APPEARANCE. The mercury should have a bright mirrorlike surface, free from film or scum. It should pour freely from a thoroughly clean, dry glass container without leaving any mercury adhering to the glass.

NONVOLATILE MATTER. Transfer 200 g to a tared boat of 25–30-mL capacity. Place the boat in the approximate center of a glass combustion tube, about 350 mm long and 35 mm in diameter, and mount in a horizontal position. The tube is fitted at one end with a removable closure and the exit end is bent downward at 90° to the horizontal and is drawn out so as to fit into a one-hole rubber stopper held in the mouth of a suction flask. The suction flask is cooled in an ice bath. After the boat and sample are placed in the tube, close the entrance end and reduce the pressure to less than 15 mm of mercury. Heat gently until all the mercury has distilled from a quiet surface (without ebullition). Remove the boat, heat it in a muffle furnace at 600°C for 15 min, and cool.

Methanol
Methyl Alcohol

CH$_3$OH

Formula Wt 32.04

CAS Number 67−56−1

Suitable for use in high-performance liquid chromatography, extraction−concentration analysis, ultraviolet spectrophotometry, or general use. Product labeling shall designate the one or more of these uses for which suitability is represented based on meeting the relevant requirements and tests. The ultraviolet spectrophotometry and liquid chromatography suitability requirements include all of the requirements for general use. The extraction−concentration suitability requirements include only the general use requirement for color.

REQUIREMENTS

General Use

Assay . . . . ≥ 99.8% CH$_3$OH
Appearance. . . . . Clear
Substances darkened by
  sulfuric acid . . . . Passes test
Substances reducing permanganate . . . . Passes test

MAXIMUM ALLOWABLE

Color (APHA). 10
Water (H$_2$O) 0.1%
Residue after evaporation 0.001%
Solubility in water . . . . Passes test
Carbonyl compounds . . . . 0.001% each
                                                     of acetone,
                                                     formaldehyde,
                                                     and acetaldehyde
Titrable acid . . . . 0.0003 meq / g
Titrable base . . . . 0.0002 meq / g

Specific Use

ULTRAVIOLET SPECTROPHOTOMETRY

Wavelength (nm)	Absorbance (AU)
280−400	0.01
260	0.04
240	0.10
230	0.20
220	0.40
210	0.80
205	1.00

LIQUID CHROMATOGRAPHY SUITABILITY

Absorbance . . . . Passes test
Gradient elution . . . . Passes test

EXTRACTION–CONCENTRATION SUITABILITY

Absorbance . . . . Passes test
GC–FID . . . . Passes test
GC–ECD . . . . Passes test

TESTS

ASSAY. Analyze the sample by gas chromatography, using the general parameters cited on page 56. The following specific conditions are also required.

Column: Type I, methyl silicone

Measure the area under all peaks and calculate the methanol content in area percent. Correct for water content.

COLOR (APHA). (Page 17).

WATER. (Page 54, Method 1). Use 25 mL (19.8 g) of the sample.

RESIDUE AFTER EVAPORATION. (Page 14). Evaporate 100 g (125 mL) in a tared dish on the steam bath and dry the residue at 105°C for 30 min.

SOLUBILITY IN WATER. Dilute 15 mL with 45 mL of water, mix, and allow to stand for 1 h. The solution should be as clear as an equal volume of water.

CARBONYL COMPOUNDS. (Page 49). Determine carbonyl compounds by differential pulse polarography. Use 5.0 g (6.4 mL) of sample. For the standard use 0.05 mg each of acetone, formaldehyde, and acetaldehyde.

TITRABLE ACID. To 25 mL of water and 10 mL of ethyl alcohol in a glass-stoppered flask, add 0.50 mL of phenolphthalein indicator solution and 0.01 N sodium hydroxide until a slight pink color persists after shaking for 30 s. Add 15 g (19 mL) of sample, mix well, and titrate with 0.01 N sodium hydroxide until the pink color is restored. Not more than 0.50 mL should be required.

> *NOTE.* Special care should be taken during the addition of the sample and titration to avoid contamination from carbon dioxide.

TITRABLE BASE.   Dilute 22.6 g (28.6 mL) with 25 mL of water and add 0.15 mL of methyl red indicator solution. Not more than 0.40 mL of 0.01 N hydrochloric acid should be required to produce a pink color.

SUBSTANCES DARKENED BY SULFURIC ACID.   Cool 10 mL of sulfuric acid, contained in a small conical flask, to 10°C and add dropwise with constant agitation 10 mL of the sample, keeping the temperature of the mixture below 20°C. The mixture should be colorless and have no more color than the acid or methanol before mixing.

SUBSTANCES REDUCING PERMANGANATE.   Cool 20 mL to 15°C, add 0.10 mL of 0.1 N potassium permanganate, and allow to stand at 15°C for 15 min. The pink color should not be entirely discharged.

ULTRAVIOLET SPECTROPHOTOMETRY.   Use the procedure on page 67 to determine the absorbance.

LIQUID CHROMATOGRAPHY SUITABILITY.   Analyze the sample, using the gradient elution procedure cited on page 64. Use the procedure on page 67 to determine the absorbance.

EXTRACTION–CONCENTRATION SUITABILITY.   Analyze the sample using the general procedure cited on page 65. Use the procedure on page 67 to determine the absorbance.

2-Methoxyethanol
Ethylene Glycol Monomethyl Ether

$CH_3OCH_2CH_2OH$ Formula Wt 76.10

CAS Number 109−86−4

REQUIREMENTS

Assay . . . . $\geq$ 99.3%
$$CH_3OCH_2CH_2OH$$

MAXIMUM ALLOWABLE

Color (APHA). 10
Titrable acid . . . . 0.002 meq / g
Water. 0.1%

TESTS

ASSAY.   Analyze the sample by gas chromatography using the general parameters cited on page 56. The following specific conditions are also required.

   Column: Type III, polyethylene glycol

Measure the area under all peaks and calculate the 2-methoxyethanol content in area percent. Correct for water content.

COLOR (APHA).   (Page 17).

TITRABLE ACID.   To 50 mL of water in a conical flask, add 0.15 mL of phenol red indicator, and adjust to a pink color with 0.01 N sodium hydroxide. Add 25 g (26 mL) of sample, and titrate with 0.01 N sodium hydroxide to the same color. Not more than 5.0 mL should be required.

WATER.   (Page 54, Method 1). Use 25 mL (24 g) of sample.

4-(Methylamino)phenol Sulfate

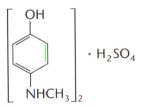

$(CH_3NHC_6H_4OH)_2 \cdot H_2SO_4$ Formula Wt 344.39

CAS Number 1936-57-8

REQUIREMENTS

   Assay 99.0-101.5%
   Residue after ignition . . . . $\leq 0.1\%$
   Suitability for determination of phosphate . . . . Passes test

TESTS

ASSAY. (By titration of reductive capacity). Weigh about 250 mg to the nearest 0.1 mg. Transfer to a 500-mL conical flask containing 100 mL of water and 10 mL of 0.1 N sulfuric acid. Dissolve, add 0.15 mL of ferroin indicator solution, and titrate with 0.1 N ceric ammonium nitrate to a light green color that persists for 15 s.

$$\%(CH_3NHC_6H_4OH)_2 \cdot H_2SO_4 = \frac{V \times N \times 8.61}{W}$$

where
  V = volume, in mL, of standard ceric ammonium nitrate
  N = normality of ceric ammonium nitrate solution
  W = weight, in grams, of sample

Reagents

Ferroin Indicator Solution.   Dissolve 0.70 g of ferrous sulfate heptahydrate, $FeSO_4 \cdot 7H_2O$, and 1.5 g of 1,10-phenanthroline in 100 mL of water.

Standard Ceric Ammonium Nitrate Solution, 0.1 N.

Preparation. Mix 59 g of ceric ammonium nitrate, $(NH_4)_2Ce(NO_3)_6$, with 31 mL of sulfuric acid in a 500-mL beaker with stirring. Cautiously add water in 20-mL portions, with stirring, allowing 2–3 min between additions. Continue the addition of water until the ceric ammonium nitrate is completely dissolved. Filter, if necessary, through a sintered glass crucible or funnel, and dilute to 1 L with water in a volumetric flask.

Standardization. Weigh about 200 mg of dry primary standard arsenic trioxide to the nearest 0.1 mg. Transfer to a 500-mL conical flask, add 15 mL of 10% sodium hydroxide reagent solution, and warm the mixture gently to hasten dissolution. When dissolution is complete, cool to room temperature and add 25 mL of dilute sulfuric acid (1 + 5). Dilute to 100 mL with water and add, as catalyst, 0.15 mL of 0.01 M osmium tetroxide (0.25 g of OsO_4 in 100 mL of 0.1 N sulfuric acid). (*WARNING:* Osmium tetroxide is poisonous; avoid contact!) Add 0.15 mL of ferroin indicator solution, and titrate with the ceric ammonium nitrate solution until the reddish orange color changes to colorless or very pale blue. (*NOTE:* As the osmium tetroxide catalyst ages,

0.15–0.65 mL may be needed to achieve the desired action. A sluggish end point indicates insufficient osmium tetroxide.)

$$\text{Normality of Ceric Ammonium Nitrate} = \frac{W \times 1000}{V \times 49.45}$$

where
W = weight, in mg, of As_2O_3
V = volume, in mL, of $(NH_4)_2Ce(NO_3)_6$ solution

RESIDUE AFTER IGNITION.   (Page 15). Ignite 5.0 g.

SUITABILITY FOR DETERMINATION OF PHOSPHATE.   Dissolve 2 g in 100 mL of water. To 10 mL of this solution add 90 mL of water and 20 g of sodium bisulfite, dissolve, and mix. Transfer 1 mL of this solution to each of two solutions containing 25 mL of 0.5 N sulfuric acid and 1 mL of ammonium molybdate–sulfuric acid reagent solution. For the control add 0.005 mg of phosphate ion (PO_4) to one of the solutions. For the sample use the remaining solution. Allow the solutions to stand at room temperature for 2 h. The control should turn perceptibly darker blue than the sample.

Methyl Orange
4-(Dimethylamino)azobenzenesulfonic Acid, Sodium Salt
4-[[4-(Dimethylamino)phenyl]azo]benzenesulfonic Acid, Sodium Salt
C.I. Acid Orange 52

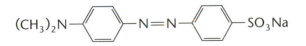

$C_{14}H_{14}N_3NaO_3S$ Formula Wt 327.34

CAS Number 547–58–0

REQUIREMENTS

Clarity of solution . . . . Passes test
Visual transition interval . . . . From pH 3.2 (pink
or red) to
pH 4.4 (yellow)

TESTS

CLARITY OF SOLUTION. Dissolve 0.1 g in 100 mL of water. Not more than a faint trace of turbidity or insoluble matter should remain. Reserve the solution for the test for visual transition interval.

VISUAL TRANSITION INTERVAL. Dissolve 1 g of potassium chloride in 100 mL of water. Adjust the pH of the solution to 3.0 (using a pH meter as described on page 43) with 0.01 N hydrochloric acid. Add 0.15 mL of the 0.1% solution reserved from the test for clarity of solution. The solution should have a definite pink color. Titrate the solution with 0.01 N sodium hydroxide to a pH of 3.2 (using a pH meter). A small amount of yellow color should appear, producing a pinkish orange color. Continue the titration to pH 4.4; the solution should have a definite yellow color. No more than 12.5 mL of 0.01 N sodium hydroxide should be consumed in the entire titration.

4-Methyl-2-pentanone
Methyl Isobutyl Ketone

$$CH_3-\overset{\overset{\displaystyle O}{\|}}{C}CH_2\overset{\overset{\displaystyle CH_3}{|}}{C}HCH_3$$

$(CH_3)_2CHCH_2COCH_3$ Formula Wt 100.16

CAS Number 108−10−1

REQUIREMENTS

Assay . . . . $\geq$ 98.5%
$(CH_3)_2CHCH_2COCH_3$
Appearance. . . . . Clear

MAXIMUM ALLOWABLE

Color (APHA). 15
Residue after evaporation 0.005%
Titrable acid . . . . 0.002 meq / g
Water. 0.1%

TESTS

ASSAY.   Analyze the sample by gas chromatography, using the general parameters cited on page 56. The following specific conditions are also required.

Column: Type I, methyl silicone

Measure the area under all peaks and calculate the 4-methyl-2-pentanone content in area percent. Correct for water content.

COLOR (APHA).   (Page 17).

RESIDUE AFTER EVAPORATION.   (Page 14). Evaporate 100 g (125 mL) to dryness in a tared dish on the steam bath and dry the residue at 125°C for 30 min.

TITRABLE ACID.   To 25 mL of alcohol in a glass-stoppered flask add 0.50 mL of phenolphthalein indicator solution. Add 0.01 N sodium hydroxide solution until a slight pink color persists after shaking for 30 s. Add 25 mL of the sample, mix, and titrate with 0.01 N sodium hydroxide until the pink color is restored. Not more than 3.33 mL of the sodium hydroxide solution should be required.

> *NOTE.* Great care should be taken in the test during the addition of the sample and the titration to avoid contamination from carbon dioxide.

WATER.   (Page 54, Method 1). Use 25 mL (20 g) of the sample and a methanol-free system to prevent ketal formation with liberation of water.

Methyl Red

2-[4-(Dimethylamino)phenylazo]benzoic Acid

C.I. Acid Red 2

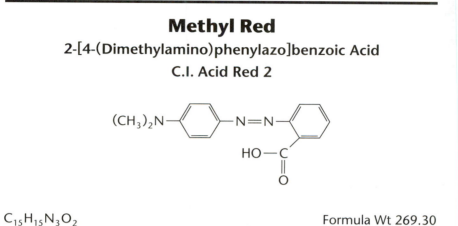

$C_{15}H_{15}N_3O_2$ Formula Wt 269.30

CAS Number 845−10−3 (sodium salt); 493−52−7 (free acid)

NOTE. Three forms of this indicator are available: the compound named above (**I**), the sodium salt of the compound (**II**), and the hydrochloride of the compound (**III**). **I** must meet the requirements for melting point (range), clarity of alcohol solution, and visual transition interval. This form is recommended for nonaqueous titrations, particularly when an aprotic solvent is used. **II** must meet the requirements for clarity of alcohol solution, clarity of aqueous solution, and visual transition interval. This form is the choice for titrations in aqueous media and is also suitable for nonaqueous titrations where the medium is an amphiprotic solvent. **III** must meet the requirements for clarity of alcohol solution and visual transition interval. This form can be used as an indicator for titrations in aqueous media and amphiprotic solvents.

REQUIREMENTS

Melting point (range). . . . . 179−182°C(**I**)
Clarity of alcohol solution . . . . Passes test
                                                        (**I, II, III**)
Clarity of aqueous solution . . . . Passes test (**II**)
Visual transition interval . . . . From pH 4.2
                                                        (pink) to 6.2
                                                        (yellow) (**I, II, III**)

TESTS

CLARITY OF ALCOHOL AND AQUEOUS SOLUTIONS.   Dissolve 0.1 g of **II** in 100 mL of water. For all three forms (**I**, **II**, and **III**), dissolve 0.1 g in 100 mL of alcohol. Not more than a faint trace of turbidity or insoluble matter should remain. Reserve the alcohol solutions for the test for visual transition interval.

VISUAL TRANSITION INTERVAL.   Dissolve 1 g of potassium chloride in 100 mL of water. Adjust the pH of the solution to 4.2 (using a pH meter as described on page 43) with 0.01 N hydrochloric acid. Add 0.15 mL of the 0.1% solution reserved from the test for clarity of solution. The color of the solution should be pink. Titrate the solution with 0.01 N sodium hydroxide to a pH of 5.5 (using the pH meter). The solution should be orange in color. Continue the titration to a pH of 6.2. The solution should be yellow in color. Not more than 1.0 mL of the 0.01 N sodium hydroxide should be consumed in the entire titration.

Molybdenum Trioxide
Molybdenum(VI) Oxide
Molybdic Acid Anhydride

MoO_3 Formula Wt 143.94

CAS Number 1313−27−5

REQUIREMENTS

Assay . . . . $\geq$ 99.5% MoO_3

MAXIMUM ALLOWABLE

Insoluble in dilute
   ammonium hydroxide 0.01%
Chloride (Cl) 0.002%
Nitrate (NO_3) . . . . Passes test (limit
                                               about 0.003%)

Arsenate, phosphate, and silicate
   (as SiO_2) 0.001%
Phosphate (PO_4) . . . . 5 ppm
Sulfate (SO_4) 0.02%
Ammonium (NH_4) 0.002%
Heavy metals (as Pb) 0.005%

TESTS

ASSAY. (Molybdenum content by gravimetry). Weigh accurately 1 g and dissolve in 10 mL of water plus 1 mL of ammonium hydroxide. Transfer to a 250-mL volumetric flask, dilute with water to volume, and mix thoroughly. To 50.0 mL of the solution in a 600-mL beaker add 250 mL of water, 20 g of ammonium chloride, 15 mL of hydrochloric acid, and 0.15 mL of methyl orange indicator solution. Heat nearly to boiling and add 18 mL of a 10% lead acetate reagent solution. To the hot and constantly stirred solution add slowly a saturated solution of ammonium acetate until the solution is alkaline, and then an excess of 15 mL of the acetate solution. Digest on the hot plate in the covered beaker at a temperature just below boiling until the precipitate has settled (from 30 min to 1 h). Filter through a Gooch crucible that has been ignited and weighed, wash seven or eight times with a solution containing 10 mL of nitric acid and 100 mL of a saturated solution of ammonium acetate in 1 L, and wash three times with hot water. Ignite to constant weight in a muffle furnace at 560–625°C. Cool and weigh as $PbMoO_4$. One gram of lead molybdate corresponds to 0.3921 g of MoO_3.

INSOLUBLE IN DILUTE AMMONIUM HYDROXIDE. Dissolve 10 g in 120 mL of dilute ammonium hydroxide (1 + 6), heat to boiling, and digest in a covered beaker on the steam bath for 2 h. Filter through a tared filtering crucible, wash thoroughly, and dry at 105°C.

CHLORIDE. Dissolve 1.0 g in 4 mL of ammonium hydroxide, heat on the steam bath until completely dissolved, and evaporate to dryness. Dissolve the residue in 20 mL of water, filter if necessary through a chloride-free filter, add 4 mL of nitric acid, and dilute with water to 30 mL. To 15 mL add 1 mL of silver nitrate reagent solution. Any turbidity should not exceed that produced by 0.01 mg of chloride ion (Cl) in an equal volume of solution containing the quantities of reagents used in the test. The comparison is best made by the general method for chloride in colored solutions, page 25.

NITRATE. Triturate 17 g with 9 mL of water and 1 mL of sodium chloride solution containing 5 mg of sodium chloride. Add 0.20 mL of indigo carmine reagent solution and 10 mL of sulfuric acid. The blue color should not be completely discharged in 5 min.

ARSENATE, PHOSPHATE, AND SILICATE. Dissolve 2.5 g in 60 mL of water and enough silica-free ammonium hydroxide to effect dissolution (about 4 mL) in a platinum dish. For the control dissolve 0.5 g in 60 mL of water and enough silica-free ammonium hydroxide to effect dissolution and add 0.02 mg of silica (SiO_2). Heat on the steam bath until the solutions are neutral as determined with an external indicator (at least 30 min). Cool and adjust the pH to between 3 and 4 with an external indicator. Transfer to beakers and dilute with water to 80 mL. Add enough bromine water to impart a distinct yellow color to the solutions. Adjust the pH of each solution to 1.8 with dilute hydrochloric acid (1 + 9) (using a pH meter). Heat just to boiling and allow to cool to room temperature. (If a precipitate forms, it will dissolve when the solution is acidified in the next operation.) Add 10 mL of hydrochloric acid and dilute with water to 100 mL. Transfer the solutions to separatory funnels and add 30 mL of 4-methyl-2-pentanone and 1 mL of butyl alcohol. Shake vigorously and allow to separate. Draw off and discard the aqueous phase. Wash the ketone phase three times with 10-mL portions of dilute hydrochloric acid (1 + 99), discarding the aqueous phase each time. To the washed ketone phase add 10 mL of dilute hydrochloric acid (1 + 99) to which has just been added 0.2 mL of a freshly prepared 2% solution of stannous chloride dihydrate in hydrochloric acid. Any blue color in the solution of the sample should not exceed that in the control.

PHOSPHATE. (Page 30). Proceed as described in Method 3.

SULFATE. Boil 1.0 g with 15 mL of dilute nitric acid (1 + 2) for 5 min. Cool thoroughly, dilute with water to 50 mL, mix well, and filter. Evaporate 10 mL of the filtrate to dryness on a hot plate at low setting. Dissolve the residue in 4 mL of water plus 1 mL of dilute hydrochloric acid (1 + 19), filter if necessary through a small filter, wash with two 2-mL portions of water, and dilute with water to 10 mL. Add 1 mL of barium chloride reagent solution. Any turbidity should not exceed that produced by 0.04 mg of sulfate ion (SO_4) in an equal volume of solution containing the quantities of reagents used in the test. Compare 10 min after adding the barium chloride to the sample and standard solutions.

AMMONIUM. Dissolve 1.0 g in 15 mL of freshly boiled 10% sodium hydroxide reagent solution and dilute with water to 100 mL. To 50 mL add 2 mL of Nessler reagent. Any color should not exceed that produced by 0.01 mg of ammonium ion (NH_4) in an equal volume of solution containing 1.5 mL of 10% sodium hydroxide reagent solution and 2 mL of Nessler reagent.

HEAVY METALS. Dissolve 1.0 g in 15 mL of 10% sodium hydroxide solution, add 2 mL of ammonium hydroxide, and dilute with water to 40 mL. For the control add 0.025 mg of lead ion (Pb) to 10 mL of the solution and dilute with water to 40 mL. For the sample dilute the remaining solution with water to 40 mL. Add 10 mL of freshly prepared hydrogen sulfide water to each and mix. Any color in the solution of the sample should not exceed that in the control.

Molybdic Acid, 85%

CAS Number 7782–91–4

NOTE. This reagent consists largely of an ammonium molybdate.

REQUIREMENTS

Assay . . . . ≥ 85.0% MoO_3

MAXIMUM ALLOWABLE

Insoluble in dilute ammonium hydroxide 0.01%
Chloride (Cl) 0.002%
Arsenate, phosphate, and silicate (as SiO_2) 0.001%
Phosphate (PO_4) . . . . 5 ppm
Sulfate (SO_4) 0.2%
Heavy Metals (as Pb) 0.003%

TESTS

ASSAY. (By gravimetric molybdenum content). Weigh accurately 1 g and dissolve in 10 mL of water plus 1 mL of ammonium hydroxide. Transfer to a 250-mL volumetric flask, dilute with water to volume, and mix thoroughly. To 50.0 mL of the solution in a 600-mL beaker add 250 mL of water, 20 g of ammonium chloride, 15 mL of hydrochloric acid, and 0.15 mL of methyl orange indicator solution. Heat nearly to boiling and add 18 mL of a 10% lead acetate reagent solution. To the hot and constantly stirred solution add slowly a saturated solution of ammonium acetate until the solution is alkaline and then an excess of 15 mL of the acetate solution. Digest on the hot plate at a temperature just below boiling until the precipitate has settled (from 30 min to 1 h). Filter through a Gooch crucible that has been ignited and weighed, wash seven or eight times with a solution containing 10 mL of nitric acid and 100 mL of a saturated solution of ammonium acetate in 1 L, and wash three times with hot water. Ignite to constant weight in a muffle furnace at 560–625°C. Cool and weigh as $PbMoO_4$. One gram of lead molybdate corresponds to 0.3921 g of MoO_3.

INSOLUBLE IN DILUTE AMMONIUM HYDROXIDE. Dissolve 10 g in 100 mL of dilute ammonium hydroxide (1 + 9), heat to boiling, and digest in a covered beaker on the steam bath for 2 h. Filter through a tared filtering crucible, wash thoroughly, and dry at 105°C.

CHLORIDE. Dissolve 1.0 g in 4 mL of ammonium hydroxide, heat on the steam bath until completely dissolved, and evaporate to dryness. Dissolve the residue in 20 mL of water, add 4 mL of nitric acid, and dilute with water to 30 mL. To 15 mL add 1 mL of silver nitrate reagent solution. Any turbidity should not exceed that produced by 0.01 mg of chloride ion (Cl) in an equal volume of solution containing the quantities of reagents used in the test. The comparison is best made by the general method for chloride in colored solutions, page 25.

ARSENATE, PHOSPHATE, AND SILICATE. Dissolve 2.5 g in 60 mL of water and enough silica-free ammonium hydroxide to effect dissolution (about 4 mL) in a platinum dish. For the control dissolve 0.5 g in 60 mL of water and enough silica-free ammonium hydroxide to effect dissolution and add 0.02 mg of silica (SiO_2). Heat on the steam bath until the solutions are neutral as determined with an external indicator (at least 30 min). Cool and adjust the pH to between 3 and 4 with an external indicator. Transfer to beakers and dilute with water to 80 mL. Add enough bromine water to impart a distinct yellow color to the solutions. Adjust the pH of each solution to 1.8 with dilute hydrochloric acid (1 + 9) (using a pH meter). Heat just to boiling and allow to cool to room temperature. (If a precipitate forms, it will dissolve when the solution is acidified in the next operation.) Add 10 mL of hydrochloric acid and dilute with water to 100 mL. Transfer the solutions to separatory funnels and add 30 mL of 4-methyl-2-pentanone and 1 mL of butyl alcohol. Shake vigorously and allow to separate. Draw off and discard the aqueous phase. Wash the ketone phase 3 times with 10-mL portions of dilute hydrochloric acid (1 + 99), discarding the aqueous phase each time. To the washed ketone phase add 10 mL of dilute hydrochloric acid (1 + 99), to which has just been added 0.2 mL of a freshly prepared 2% solution of stannous chloride in hydrochloric acid. Any blue color in the solution of the sample should not exceed that in the control.

PHOSPHATE. (Page 30). Proceed as described in Method 3.

SULFATE. Boil 1.0 g with 15 mL of dilute nitric acid (1 + 2) for 5 min. Cool thoroughly, dilute with water to 100 mL, and filter. Evaporate 2 mL of the filtrate to dryness on a hot plate at low setting, dissolve the residue in 4 mL of water plus 1 mL of dilute hydrochloric acid (1 + 19), and filter if necessary through a small filter. Wash with two 2-mL portions of water, dilute with water to 10 mL, and add 1 mL of barium chloride reagent solution. Any turbidity should not exceed that produced by 0.04 mg of sulfate ion (SO_4) in an equal volume of solution containing the quantities of reagents used in the test. Compare 10 min after adding the barium chloride to the sample and standard solutions.

HEAVY METALS.   Dissolve 1.3 g in 15 mL of 10% sodium hydroxide, add 2 mL of ammonium hydroxide, and dilute with water to 40 mL. For the control add 0.02 mg of lead ion (Pb) to 10 mL of the solution and dilute with water to 40 mL. For the sample dilute the remaining portion with water to 40 mL. Add 10 mL of freshly prepared hydrogen sulfide water to each and mix. Any color in the solution of the sample should not exceed that in the control.

Morpholine

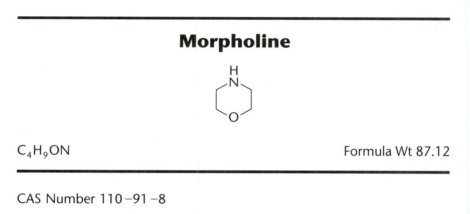

C_4H_9ON Formula Wt 87.12

CAS Number 110−91−8

REQUIREMENTS

Assay . . . . ≥ 99.0% C_4H_9NO
Color (APHA). . . . . ≤ 15
Density (g /mL) at 25°C . . . . 0.994−0.997

TESTS

ASSAY.   (By acid−base titrimetry). Add 0.30−0.40 mL of mixed indicator solution to 50 mL of water in a 250-mL flask, and neutralize by the dropwise addition of 0.1 N hydrochloric acid just to the disappearance of the green color. Weigh accurately 1.4−1.6 g of the sample into the flask and swirl to effect complete dissolution. Titrate with 0.5 N hydrochloric acid to the disappearance of the green color. One milliliter of 0.5 N hydrochloric acid corresponds to 0.04356 g of C_4H_9NO.

Mixed Indicator Solution.   Prepare separate 0.1% solutions of bromcresol green in methanol and the sodium salt of methyl red in water. Mix 5 parts by volume of the bromcresol green solution with 1 part of the methyl red solution.

COLOR (APHA).   (Page 17).

BOILING RANGE.   Distill 100 mL by the method described on page 17. The corrected initial boiling point should not lie below 126°C, and the corrected dry point should not be above 130°C. The barometric pressure correction for morpholine is +0.045°C per mm under 760 mm and −0.045°C per mm above 760 mm. The boiling point for pure morpholine at 760-mm mercury 128.2°C.

N-(1-Naphthyl)ethylenediamine Dihydrochloride

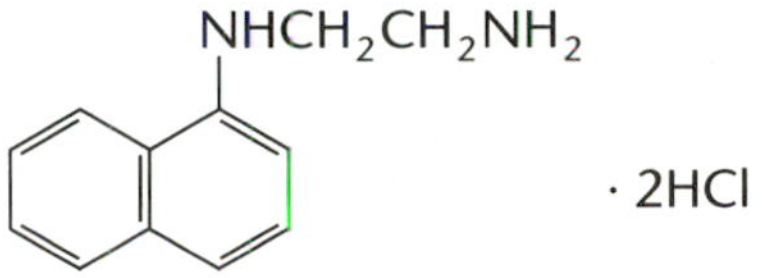

$$NHCH_2CH_2NH_2$$

· 2HCl

$C_{12}H_{14}N_2 \cdot 2HCl$ Formula Wt 259.18

CAS Number 1465–25–4

REQUIREMENTS

Assay (anhydrous basis) . . . . ≥ 98.0%
$C_{12}H_{14}N_2 \cdot 2HCl$
Sensitivity to sulfanilamide . . . . Passes test
Solubility . . . . Passes test

MAXIMUM ALLOWABLE

Water (H$_2$O) 5%

TESTS

ASSAY.   (By argentimetric titration of chloride content). Weigh accurately 0.5 g and dissolve the sample in a solution consisting of 50 mL of water and 50 mL of alcohol. Titrate potentiometrically with 0.1 N silver nitrate, using a silver electrode versus a modified calomel electrode containing a saturated solution of potassium nitrate as the electrolyte. One milliliter of 0.1 N silver nitrate corresponds to 0.01296 g of $C_{12}H_{14}N_2 \cdot 2HCl$.

$$\% \; C_{12}H_{14}N_2 \cdot 2HCl = \frac{V \times 0.01296 \times 100}{W - (W \times \%H_2O)}$$

where

V = volume, in milliliters, of 0.1 N silver nitrate solution consumed

W = weight, in grams, of sample

SENSITIVITY TO SULFANILAMIDE. In two different Nessler tubes dilute 1.0 and 2.5 mL of sulfanilamide solution with water to 100 mL. For the blank, omit the sulfanilamide from a third tube and fill with water to 100 mL. To each tube add 8 mL of trichloroacetic acid solution and 1 mL of sodium nitrite solution. After 3 min add 1 mL of ammonium sulfamate solution, then after 2 additional min add 1 mL of sample solution. The tubes containing the sulfanilamide should turn pink to red, the higher concentration producing the deeper color.

Reagents

Sulfanilamide Solution. Dissolve 0.02 g of sulfanilamide in 100 mL of water.

Trichloroacetic Acid Solution. Dissolve 15 g of trichloroacetic acid in 100 mL of water.

Sodium Nitrite Solution. Dissolve 0.1 g of sodium nitrite in 100 mL of water.

Ammonium Sulfamate Solution. Dissolve 0.5 g of ammonium sulfamate in 100 mL of water.

Sample Solution. Dissolve 0.1 g of *N*-(1-naphthyl)ethylenediamine dihydrochloride in 100 mL of water.

SOLUBILITY. A solution of 1 g in 50 mL of water is clear and dissolution is complete.

WATER. (Page 54, Method 1). Use 3.0 g of the sample.

Nickel Sulfate Hexahydrate
Nickel(II) Sulfate Hexahydrate
Nickelous Sulfate Hexahydrate

$NiSO_4 \cdot 6H_2O$ Formula Wt 262.85

CAS Number 10101 –97 –0

REQUIREMENTS

Assay . . . . 98.0 –102.0%
$NiSO_4 \cdot 6H_2O$

MAXIMUM ALLOWABLE

Insoluble matter	0.005%
Chloride (Cl)	0.001%
Nitrogen compounds (as N)	0.002%
Calcium (Ca)	0.005%
Cobalt (Co)	0.002%
Copper (Cu)	0.005%
Iron (Fe)	0.001%
Magnesium (Mg)	0.005%
Manganese (Mn)	0.002%
Potassium (K)	0.01%
Sodium (Na)	0.05%

TESTS

ASSAY. (By complexometric titration of nickel). Weigh accurately 1 g, transfer to a 400-mL beaker, and dissolve in 250 mL of water. Add 10 mL of pH 10 ammoniacal buffer solution and about 50 mg of murexide indicator mixture. Titrate immediately with standard 0.1 M EDTA to a purple color. One milliliter of 0.1 M EDTA corresponds to 0.02628 g of $NiSO_4 \cdot 6H_2O$.

INSOLUBLE MATTER. (Page 14). Use 20 g dissolved in 150 mL of water.

CHLORIDE. Dissolve 1.0 g in 15 mL of water, filter if necessary through a small chloride-free filter, and add 1 mL of nitric acid and 1 mL of silver nitrate reagent solution. Any turbidity should not exceed that produced by 0.01 mg of chloride ion (Cl) in an equal volume of solution containing the quantities of reagents used in the test. The comparison is best made by the general method for chloride in colored solutions, page 25.

NITROGEN COMPOUNDS. (Page 30). Use 1.0 g. For the standard use 0.02 mg of nitrogen (N).

CALCIUM, COBALT, COPPER, IRON, MAGNESIUM, MANGANESE, POTASSIUM, AND SODIUM. Determine the calcium, cobalt, copper, iron, magnesium, manganese, potassium, and sodium by the flame atomic absorption spectrophotometric method described on page 35.

> **Sample Stock Solution.** Dissolve 25.0 g of sample in 150 mL of water and 2.5 mL of nitric acid in a 250-mL volumetric flask, and dilute to the mark with water (0.1 g/mL).

Analyze the solutions by means of a suitable atomic absorption spectrophotometer, using the conditions outlined in the following table. Calculate the metal content of the sample by the method of standard additions.

Element	Wavelength (nm)	Sample Wt (g)	Standard Added (mg)	Flame Type*	Background Correction
Ca	422.7	0.8	0.02; 0.04	N/A	No
Co	240.7	2.0	0.02; 0.04	A/A	Yes
Cu	324.8	2.0	0.05; 0.10	A/A	Yes
Fe	248.3	4.0	0.04; 0.08	A/A	Yes
Mg	285.2	0.2	0.005; 0.01	A/A	Yes
Mn	279.5	2.0	0.02; 0.04	A/A	Yes
K	766.5	0.2	0.01; 0.02	A/A	No
Na	589.0	0.04	0.01; 0.02	A/A	No

*A/A is air/acetylene; N/A is nitrous oxide/acetylene.

Ninhydrin
1*H*-Indene-1,2,3-trione Monohydrate
1,2,3-Triketohydrindene Monohydrate

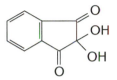

C₉H₆O₄   Formula Wt 178.14

CAS Number 485−47−2

REQUIREMENTS

Appearance. . . . . White to brownish-
                                                                    white crystals
Identification and melting point . . . . Passes test
Solubility . . . . Passes test
Sensitivity to amino acids . . . . Passes test

TESTS

IDENTIFICATION AND MELTING POINT. A sample heated above 150°C becomes purple and melts with decomposition at approximately 250 °C.

LOSS ON DRYING. Weigh accurately about 1 g in a tared dish or crucible and dry at 105°C for 4 h.

SOLUBILITY. A solution of 0.10 g of sample in 10 mL of water is clear and pale yellow in color. Dissolution is complete. Retain the solution for the sensitivity test.

SENSITIVITY TO AMINO ACIDS. Dissolve 0.10 g of beta-alanine in 10 mL of warm water and heat to boiling for 5 min. Add 0.50 mL of the ninhydrin solution reserved from the solubility test and continue heating. A rose color starts to develop within a minute and intensifies to a deep violet color on continued heating.

Nitric Acid

HNO_3 Formula Wt 63.01

CAS Number 7697–37–2

> *NOTE.* This material may darken during storage due to photochemical reactions.

REQUIREMENTS

Appearance . . . . Colorless and free
from suspended
matter or
sediment
Assay . . . . 69.0–71.0% HNO_3

MAXIMUM ALLOWABLE

Color (APHA). 10
Residue after ignition . . . . 5 ppm
Chloride (Cl) . . . . 0.5 ppm
Sulfate (SO_4) . . . . 1 ppm
Arsenic (As) . . . . 0.01 ppm
Heavy metals (as Pb) . . . . 0.2 ppm
Iron (Fe) . . . . 0.2 ppm

TESTS

APPEARANCE.   Mix the acid in the original container, transfer 10 mL to a test tube (20 mm × 150 mm) and compare with distilled water in a similar tube. The liquids should be equally clear and free from suspended matter.

ASSAY.   (By acid–base titrimetry). Tare a small glass-stoppered flask containing about 15 mL of water. Quickly pipet about 2 mL of the sample under the water surface, stopper, cool, and weigh accurately. Dilute with about 40 mL of water, add 0.15 mL of methyl orange indicator solution, and titrate with 1 N sodium hydroxide. One milliliter of 1 N sodium hydroxide corresponds to 0.06301 g of HNO_3.

COLOR (APHA).   (Page 17).

RESIDUE AFTER IGNITION.   To 200 g (140 mL) in a tared dish add 0.05 mL of sulfuric acid, evaporate as far as possible on low heat, and heat gently to volatilize the excess sulfuric acid. Finally, ignite at 800 ± 25°C for 15 min.

CHLORIDE.   Dilute 20 g (14 mL) with 10 mL of water and add 1 mL of silver nitrate reagent solution. Prepare a standard containing 0.01 mg of chloride ion (Cl) in 20 mL of water and add 1 mL of silver nitrate reagent solution. Evaporate the solutions to dryness on the steam bath. Dissolve the residues in 0.5 mL of ammonium hydroxide, dilute with water to 20 mL, and add 1.5 mL of nitric acid. Any turbidity in the solution of the sample should not exceed that in the standard.

SULFATE.   (Page 32, Procedure A, Method 3).

ARSENIC.   (Page 23). To 300 g (210 mL) in a 250-mL beaker add 5 mL of sulfuric acid, and evaporate on the steam bath to a volume of about 5 mL. Transfer the beaker to a hot plate in a hood and heat to dense fumes of sulfur trioxide. Cool, cautiously wash down the beaker with about 10 mL of water, evaporate again to dense fumes of sulfur trioxide, and continue the fuming for 15 min. Cool, wash down the beaker with about 10 mL of water, and repeat the fuming. Cool, cautiously wash the solution into a generator flask with sufficient water to make 35 mL, add 0.5 g of hydrazine sulfate, and proceed as described. For the standard use 0.003 mg of arsenic (As).

HEAVY METALS.   (Page 26, Method 1). To 100 g (70 mL) in a beaker add about 10 mg of sodium carbonate, evaporate to dryness on the steam bath, dissolve the residue in about 20 mL of water, and dilute with water to 25 mL.

IRON.   (Page 28, Method 1). Evaporate 50 g (35 mL) to dryness, dissolve the residue in 2 mL of hydrochloric acid, dilute with water to 50 mL, and use the solution without further acidification.

Nitric Acid, 90%

HNO_3 Formula Wt 63.01

CAS Number 7697–37–2

REQUIREMENTS

Assay . . . . ≥ 90.0% HNO_3
Dilution test . . . . Passes test

MAXIMUM ALLOWABLE

Residue after ignition 0.002%
Dissolved oxides . . . . Passes test (limit
                                                                   about 0.1%
                                                                   as N_2O_3)
Chloride (Cl) . . . . 0.7 ppm
Sulfate (SO_4) . . . . 5 ppm
Arsenic (As) . . . . 0.3 ppm
Heavy metals (as Pb) . . . . 5 ppm
Iron (Fe) . . . . 2 ppm

TESTS

ASSAY. (By acid–base titrimetry). Tare a small glass-stoppered flask containing about 15 mL of water. Quickly pipet about 2 mL of the sample under the water surface, stopper, cool, and weigh accurately. Dilute with about 40 mL of water, add 0.15 mL of methyl orange indicator solution, and titrate with 1 N sodium hydroxide. One milliliter of 1 N sodium hydroxide corresponds to 0.06301 g of HNO_3.

DILUTION TEST. Dilute 1 volume of acid with 3 volumes of water; mix and allow to stand for 1 h. No turbidity or precipitate should be observed.

RESIDUE AFTER IGNITION. To 50 g (33 mL) in a tared dish add 0.10 mL of sulfuric acid, evaporate as far as possible on low heat, and heat gently to volatilize the excess sulfuric acid. Finally, ignite at 800 ± 25°C for 15 min.

DISSOLVED OXIDES. Dilute 10 g (6.6 mL) to 150 mL with cold water in a casserole. Add 20 mL of dilute sulfuric acid (1 + 4) and titrate with 0.1 N potassium permanganate, rapidly at first, then slowly till the pink color lasts 3 min. Not more than 5.0 mL of the permanganate solution should be required. One milliliter of 0.1 N permanganate solution corresponds to 0.0019 g of N_2O_3.

CHLORIDE.   Dilute 15 g (10 mL) with 10 mL of water and add 1 mL of silver nitrate reagent solution. Prepare a standard of 0.01 mg of chloride ion (Cl) in 20 mL of water and add 1 mL of silver nitrate reagent solution. Evaporate the solutions to dryness on the steam bath. Dissolve the residues in 0.5 mL of ammonium hydroxide, dilute with water to 20 mL, and add 1.5 mL of nitric acid. Any turbidity in the solution of the sample should not exceed that in the standard.

SULFATE.   (Page 32, Procedure A, Method 3).

ARSENIC.   (Page 23). To 10 g (6.6 mL) in a 150-mL beaker add 5 mL of sulfuric acid and evaporate on the steam bath to a volume of about 5 mL. Transfer the beaker to a hot plate in a hood and heat to dense fumes of sulfur trioxide. Cool, cautiously wash down the sides of the beaker with about 10 mL of water, evaporate again to dense fumes of sulfur trioxide, and continue the fuming for 15 min. Cool, cautiously wash down the sides of the beaker with about 10 mL of water, and repeat the fuming. Cool, cautiously wash the solution into an arsine generator flask with 35 mL of water, add 0.5 g of hydrazine sulfate, and proceed as described. For the standard use 0.003 mg of arsenic (As).

HEAVY METALS.   (Page 26, Method 1). To 4 g (2.7 mL) add about 10 mg of sodium carbonate, evaporate to dryness on the steam bath, dissolve the residue in about 20 mL of water, and dilute with water to 25 mL.

IRON.   (Page 28, Method 1). Evaporate 5 g (3.3 mL) to dryness, dissolve the residue in 2 mL of hydrochloric acid, dilute with water to 50 mL, and use the solution without further acidification.

Nitrilotriacetic Acid

N,N-Bis(carboxymethyl)glycine

$C_6H_9O_6N$ Formula Wt 191.14

CAS Number 139–13–9

REQUIREMENTS

Assay . . . . $\geq$ 98.0% $C_6H_9O_6N$
Clarity of solution . . . . Passes test

TESTS

ASSAY (BY ALKALIMETRY) AND CLARITY OF SOLUTION. To check first for clarity of solution, weigh accurately 0.35 g, transfer to a beaker, and dissolve in 100 mL of water. The sample should dissolve completely, with heating and stirring if necessary, to produce a clear solution. Add 0.15 mL of phenolphthalein indicator solution, and titrate with 0.1 N sodium hydroxide. One milliliter of 0.1 N sodium hydroxide corresponds to 0.006372 g of $C_6H_9O_6N$.

Nitrobenzene

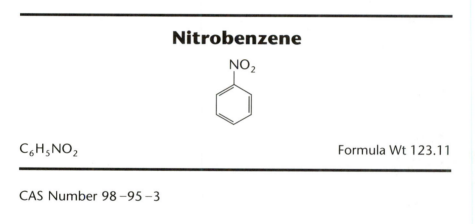

$C_6H_5NO_2$ Formula Wt 123.11

CAS Number 98–95–3

REQUIREMENTS

Assay . . . . $\geq$ 99.0% $C_6H_5NO_2$

MAXIMUM ALLOWABLE

Residue after evaporation 0.005%
Water-soluble titrable acid . . . . 0.0005 meq / g
Chloride (Cl) . . . . 5 ppm

TESTS

ASSAY. Analyze the sample by gas chromatography, using the general parameters cited on page 56. The following specific conditions are also required.

Column: Type I, methyl silicone

Measure the area under all peaks and calculate the nitrobenzene content in area percent. Correct for water content.

RESIDUE AFTER EVAPORATION.   (Page 14). Evaporate 20 g (16.7 mL) to dryness in a tared dish on the steam bath and dry the residue at 105°C for 30 min.

WATER-SOLUBLE TITRABLE ACID.   Shake 25 g (21 mL) with 60 mL of water for 1 min and allow the layers to separate. Draw off and discard the nitrobenzene. Titrate 48 mL of the water extract with 0.01 N sodium hydroxide, using 0.15 mL of bromphenol blue indicator solution. Not more than 1.0 mL of 0.01 N sodium hydroxide should be required.

CHLORIDE.   Dilute the remaining 12 mL of water from the test for acidity with water to 25 mL, filter through a chloride-free filter, and dilute 10 mL of the filtrate with water to 20 mL. Add 1 mL of nitric acid and 1 mL of silver nitrate reagent solution. Any turbidity should not exceed that produced by 0.01 mg of chloride ion (Cl) in an equal volume of solution containing the quantities of reagents used in the test.

Nitromethane

CH_3NO_2 Formula Wt 61.04

CAS Number 75–52–5

REQUIREMENTS

Appearance. . . . . Clear
Assay . . . . $\geq$ 95% CH_3NO_2

MAXIMUM ALLOWABLE

Color (APHA). 10
Water (H_2O) 0.05%

TESTS

ASSAY.   Analyze the sample by gas chromatography, using the general parameters cited on page 56. The following specific conditions are also required.

   Column: Type III, polyethylene glycol

Measure the area under all peaks and calculate the nitromethane content in area percent. Correct for water content.

COLOR (APHA).   (Page 17).

WATER.   (Page 55, Method 2). Use 100 μL (100 ng) of the sample.

1-Octanol
n-Octyl Alcohol
(Suitable for use in determining TSCA partition coefficients)

$$CH_3(CH_2)_6CH_2OH$$

$C_8H_{18}O$ Formula Wt 130.23

CAS Number 111 –87 –5

REQUIREMENTS

General Use

Assay . . . . $\geq$ 99% $C_8H_{18}O$
Appearance. . . . . Clear

MAXIMUM ALLOWABLE

Color (APHA). 10
Residue after evaporation 0.004%
Titrable acid . . . . 0.0002 meq / g

Specific Use

ULTRAVIOLET SPECTROPHOTOMETRY

Wavelength (nm)	Absorbance (AU)
280–400	0.01
260	0.05
240	0.10
230	0.5
220	1.0

TESTS

ASSAY.   Analyze the sample by gas chromatography, using the general parameters cited on page 56. The following specific conditions are also required.

Column: Type I, methyl silicone

Measure the area under all peaks and calculate the 1-octanol content in area percent. Correct for water content.

COLOR (APHA).   (Page 17).

RESIDUE AFTER EVAPORATION.   (Page 14). Evaporate 100 g (120.5 mL) to dryness in a tared platinum dish on the steam bath in a well-ventilated hood, and dry the residue at 105°C for 30 min.

TITRABLE ACID.   To 25 mL of ethyl alcohol add 0.15 mL of phenolphthalein indicator, neutralize the solution with either 0.01 N sodium hydroxide or 0.01 N hydrochloric acid. Add 21 g (25 mL) of sample and titrate with 0.01 N sodium hydroxide to a pink end point. Not more than 0.4 mL should be required.

ULTRAVIOLET SPECTROPHOTOMETRY.   Use the procedure on page 67 to determine the absorbance.

Oxalic Acid Dihydrate

Ethanedioic Acid Dihydrate

$$HO-\underset{\underset{O}{\|}}{C}-\underset{\underset{O}{\|}}{C}-OH \cdot 2H_2O$$

$C_2H_2O_4 \cdot 2H_2O$ Formula Wt 126.07

CAS Number 6153 – 56 – 6

REQUIREMENTS

Assay . . . . 99.5 –102.5%
$H_2C_2O_4 \cdot 2H_2O$

Substances darkened by
hot sulfuric acid . . . . Passes test

MAXIMUM ALLOWABLE

Insoluble matter 0.005%
Residue after ignition 0.01%
Chloride (Cl) 0.002%
Sulfate (SO$_4$) 0.005%
Calcium (Ca) 0.001%
Nitrogen compounds (as N) 0.001%
Heavy metals (as Pb) . . . . 5 ppm
Iron (Fe) . . . . 2 ppm

TESTS

ASSAY. (By titration of reductive capacity). Weigh accurately 2.5 g, dissolve in water in a 500-mL volumetric flask, dilute to volume with water, and mix. To a 50-mL aliquot of the solution add 50 mL of water and 5 mL of sulfuric acid. Titrate slowly with 0.1 N potassium permanganate until about 25 mL has been added. Heat the solution to about 70°C, and complete the titration to a permanent faint pink color. One milliliter of 0.1 N potassium permanganate consumed corresponds to 0.006303 g of $H_2C_2O_4 \cdot 2H_2O$.

INSOLUBLE MATTER. (Page 14). Use 20 g dissolved in 250 mL of water.

RESIDUE AFTER IGNITION. (Page 15). Dry 10 g in a tared crucible or dish for 1 h at 105°C, slowly ignite, and continue as described. Retain the residue for the test for iron.

CHLORIDE. (Page 25). Dissolve 0.5 g in 20 mL of nitric acid.

SULFATE. Digest 1.0 g of sample plus about 10 mg of sodium carbonate with 3 mL of 30% hydrogen peroxide and 0.5 mL of nitric acid in a covered beaker on a hot plate at low setting until reaction ceases. Remove the cover and evaporate to dryness. Dissolve the residue in 6 mL of dilute hydrochloric acid (1 + 1) and again evaporate to dryness. Dissolve the residue in 4 mL of water plus 1 mL of dilute hydrochloric acid (1 + 19), filter if necessary through a small filter, wash the precipitate with two 2-mL portions of water, and dilute with water to 10 mL. Add 1 mL of barium chloride reagent solution. Any turbidity should not exceed that produced when a solution containing 0.05 mg of sulfate ion (SO_4) is treated exactly like the 1.0 g of sample. Compare 10 min after adding the barium chloride to the sample and standard solutions.

CALCIUM. Determine the calcium by the flame atomic absorption spectrophotometric method described on page 35.

> *Sample Stock Solution.* Suspend 10.0 g of sample in 50 mL of nitric acid (1 + 24) and add 25 mL of 30% hydrogen peroxide. Cover and digest on the steam bath until reaction ceases. Add 5 mL more of the hydrogen peroxide and when reaction ceases, uncover and evaporate to dryness. Dissolve the residue in 80 mL of nitric acid (1 + 99), transfer to a 100-mL volumetric flask, and dilute to the mark with nitric acid (1 + 99) (0.1 g/mL).

Analyze the solutions by means of a suitable atomic absorption spectrophotometer, using the conditions outlined in the following table. Calculate the metal content of the sample by the method of standard additions.

Element	Wavelength (nm)	Sample Wt (g)	Standard Added (mg)	Flame Type*	Background Correction
Ca	422.7	2.0	0.02; 0.04	N/A	No

*N/A is nitrous oxide/acetylene.

NITROGEN COMPOUNDS. (Page 30). Use 1.0 g. For the standard use 0.01 mg of nitrogen (N).

HEAVY METALS. (Page 26, Method 1). Place 4.0 g in a covered beaker, and add about 10 mg of sodium carbonate, 5 mL of 30% hydrogen peroxide, and 0.5 mL of nitric acid. Digest on the steam bath until reaction ceases, remove the cover, and evaporate to dryness. Digest the residue with 0.5 mL of hydrochloric acid and 0.1 mL of nitric acid in a covered beaker for 15 min, remove the cover, and again evaporate to dryness. Dissolve the residue in about 20 mL of water, and dilute with water to 25 mL. For the standard solution treat 0.02 mg of lead ion (Pb) exactly as the 4.0 g of sample.

IRON. (Page 28, Method 1). To the residue after ignition add 3 mL of dilute hydrochloric acid (1 + 1), cover with a watch glass, and digest on the steam bath for 15 min. Remove the cover, and evaporate to dryness. Dissolve the residue in 2 mL of dilute hydrochloric acid (1 + 1), add about 40 mL of water, filter if necessary, and dilute with water to 50 mL. Use 25.0 mL (5.0 g) of the solution.

SUBSTANCES DARKENED BY HOT SULFURIC ACID. Dissolve 2.0 g in 20 mL of sulfuric acid and heat at 150°C for 30 min. Any color should not exceed that produced in a color standard of the following composition: 0.2 mL of cobalt chloride solution (5.95 g of $CoCl_2 \cdot 6H_2O$ and 2.5 mL of hydrochloric acid in 100 mL), 1.0 mL of ferric chloride solution (4.50 g of $FeCl_3 \cdot 6H_2O$ and 2.5 mL of hydrochloric acid in 100 mL), and 0.2 mL of cupric sulfate solution (6.24 g of $CuSO_4 \cdot 5H_2O$ and 2.5 mL of hydrochloric acid in 100 mL), in 20 mL of solution.

Pararosaniline Hydrochloride
4-[(4-Aminophenyl)(-4-imino-2,5-cyclohexadien-1-ylidene)methyl]benzenamine Monohydrochloride

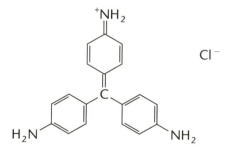

$C_{19}H_{17}N_3 \cdot HCl$

Formula Wt 323.82

CAS Number 569–61–9

REQUIREMENTS

Assay . . . . $\geq$ 99.0%

$C_{19}H_{18}N_3Cl$

Suitability for SO_2 determination

Absorbance of reagent blank . . . . $\leq$ 0.170

Calibration curve slope (μg/mL) 0.706–0.786

TESTS

ASSAY. (By spectrophotometry at 540 nm).

Sample Stock Solution. Weigh 0.200 g of sample, and dissolve completely by shaking with 100 mL of 1 N hydrochloric acid in a 100-mL graduated cylinder.

Dilute 1.0 mL of sample stock solution with water to the mark in a 100-mL volumetric flask. Pipet 5.0 mL into a 50-mL volumetric flask, add 5 mL of 1 M sodium acetate–acetic acid buffer, and dilute with water to volume. Allow to stand for 1 h at room temperature, and determine the absorbance at 540 nm with a spectrophotometer, using 1-cm cells at a slit width of 0.04 mm.

$$\% \ C_{19}H_{18}N_3Cl = (A \times 21.30)/W$$

where

A = absorbance

W = weight, in grams, of sample in 1.0 mL of sample stock solution.

Buffer Solution. Dissolve 13.61 g of sodium acetate trihydrate, $CH_3COONa \cdot 3H_2O$, in water in a 100-mL volumetric flask. Add 5.7 mL of glacial acetic acid, and dilute to volume with water.

SUITABILITY FOR SO_2 DETERMINATION.

NOTE. All absorbance measurements using this procedure are temperature-dependent (0.015 absorbance units/°C), and should be performed at 22°C.

Absorbance of Reagent Blank. Pipet 10 mL of 0.04 M potassium tetrachloromercurate and 1 mL of 0.6% sulfamic acid into a 25-mL volumetric flask. Allow to stand for 10 min, then add from a pipet 2.0 mL of 0.2% formaldehyde solution and 5.0 mL of pararosaniline reagent. Dilute to volume with water, allow to stand for 30 min, and determine the absorbance at 548 nm in 1-cm cells, using water as the reference.

Calibration Curve Slope. Pipet 1.0 mL of standard sulfite solution into a 25-mL volumetric flask. Add 9 mL of 0.04 M potassium tetrachloromercurate and 1 mL of 0.6% sulfamic acid, allow to stand for 10 min, then add from a pipet 2.0 mL of 0.2% formaldehyde solution and 5.0 mL of pararosaniline reagent. Dilute to volume with water, allow to stand for 30 min and determine the absorbance at 548 nm in 1-cm cells, using water as the reference.

$$\text{Absorbance (g/mL)} = (A_s - A_b)/W_{\text{SO}_2}$$

where

A_s = absorbance of sample
A_b = absorbance of reagent blank
W_{SO_2} = weight, in micrograms, of SO_2 in 1 mL

Reagents

All six reagents cited should be freshly prepared.

Potassium Tetrachloromercurate, K_2HgCl_4, 0.04 M. Dissolve 10.86 g of mercuric chloride ($HgCl_2$) in water in a 1-L volumetric flask. (CAUTION: If this highly poisonous reagent is spilled on the skin, flush off with water immediately!) Add 5.96 g of potassium chloride and 0.066 g of (ethylenedinitrilo)tetraacetic acid [$(HOCOCH_2)_2NCH_2CH_2N(CH_2COOH)$], dissolve, and dilute with water to volume. The pH of this reagent should not be less than 5.2.

Stock Sulfite Solution. Dissolve 0.4 g of sodium sulfite (Na_2SO_3) or 0.3 g of sodium metabisulfite ($Na_2S_2O_5$) in 500 mL of deaerated water (purged with high-purity nitrogen gas or by boiling). This solution contains 320–400 μg/mL as SO_2. The actual concentration is determined by adding excess iodine solution and back-titrating with sodium thiosulfate solution that has been standardized against potassium iodate or potassium dichromate. Sulfite solutions are unstable.

For blank solution A in the back-titration, transfer 25 mL of water to a 500-mL iodine flask, and add from a pipet 50.0 mL of 0.01 N iodine. For sample solution B in a second 500-mL iodine flask, add successively from a pipet 25.0 mL of stock sulfite solution and 50.0 mL of 0.01 N iodine. Stopper the flasks, allow to react for 5 min, and titrate each with 0.01 N sodium thiosulfate to a pale yellow color. Add 5 mL of starch indicator solution, and continue the titration to the disappearance of the blue color. Calculate the concentration of SO_2 as follows:

$$\text{SO}_2 \text{ concentration, } \mu g/mL = (A - B) \times N \times 32{,}030/V$$

where

A = volume, in milliliters, of thiosulfate solution consumed by blank

B = volume, in milliliters, of thiosulfate solution consumed by sample

N = normality of thiosulfate solution

V = volume, in milliliters, of sample

Standard Sulfite Solution. Immediately after its standardization, pipet a 2-mL aliquot of stock sulfite solution into a 100-mL volumetric flask, and dilute to volume with 0.04 M potassium tetrachloromercurate. This solution is stable for 30 days if stored at 5°C.

Sulfamic Acid, 0.6%. Dissolve 0.6 g of sulfamic acid in 100 mL of water. This solution can be kept for a few days if protected from air.

Formaldehyde, 0.2%. Dilute 5 mL of 36–38% formaldehyde with water to 1 L. Prepare this solution daily.

Pararosaniline Reagent. Pipet 20.0 mL of sample stock solution of pararosaniline (refer to assay) into a 250-mL volumetric flask. Add an additional 0.2 mL of stock for each percent the pararosaniline assay is below 100%. Add 25 mL of 3 M phosphoric acid, and dilute to volume with water.

1-Pentanol

n-Amyl Alcohol

$CH_3(CH_2)_3CH_2OH$ Formula Wt 88.15

CAS Number 71 –41 –0

REQUIREMENTS

Assay . . . . $\geq$ 98.0%
$CH_3(CH_2)_3CH_2OH$

MAXIMUM ALLOWABLE

Water (H_2O) 0.5%
Color (APHA). 30
Residue after evaporation 0.003%
Acids and esters. . . . . 0.075 meq / g
Carbonyl (as HCHO) 0.1%

TESTS

ASSAY. Analyze the sample by gas chromatography, using the parameters cited on page 56. The following specific conditions are also required.

Column: Type I, methyl silicone

Measure the area under all peaks and calculate the 1-pentanol content in area percent. Correct for water content.

WATER. (Page 54, Method 1). Use 25 mL (20 g) of the sample.

COLOR (APHA). (Page 17).

RESIDUE AFTER EVAPORATION. (Page 14). Use 40 g (50 mL) of sample, evaporate to dryness on the steam bath, and complete drying at 105°C.

ACIDS AND ESTERS. Dilute 8.1 g (10 mL) with 20 mL of ethyl alcohol, add 10 mL of 0.1 N sodium hydroxide, and reflux gently for 10 min. Cool, add 0.15 mL of phenolphthalein indicator solution, and titrate the excess sodium hydroxide with 0.1 N hydrochloric acid. Run a reagent blank. The net volume of 0.1 N sodium hydroxide consumed by the sample should not exceed 6.1 mL.

CARBONYL COMPOUNDS. (Page 49). Determine carbonyl compounds by differential pulse polarography. Use 0.25 g (0.30 mL) of sample. For the standard use 0.25 mg of formaldehyde.

Perchloric Acid, 70%

$HClO_4$ Formula Wt 100.46

CAS Number 7601–90–3

REQUIREMENTS

Assay . . . . 69.0–72.0% HClO$_4$

MAXIMUM ALLOWABLE

Color (APHA). 10
Residue after ignition 0.003%
Silicate and phosphate (as SiO$_2$) . . . . 5 ppm
Chloride (Cl) 0.001%
Nitrogen compounds (as N) 0.001%
Sulfate (SO$_4$). 0.001%
Heavy metals (as Pb) . . . . 1 ppm
Iron (Fe) . . . . 1 ppm

TESTS

CAUTION. Perchloric acid in contact with some organic materials can form explosive mixtures. Use safety goggles and rubber gloves when handling. Rinse any glassware that has been in contact with perchloric acid before setting it aside.

ASSAY.   (By alkalimetric titration). Weigh accurately 3 mL, dilute with water to 60 mL, add 0.15 mL of phenolphthalein indicator solution, and titrate with 1 N sodium hydroxide. One milliliter of 1 N sodium hydroxide corresponds to 0.1005 g of HClO$_4$.

COLOR (APHA).   (Page 17).

RESIDUE AFTER IGNITION.   (Page 15). To 67 g (42 mL) in a tared dish add 1 mL of nitric acid, evaporate to dryness, and ignite at $800 \pm 25°C$ for 15 min.

SILICATE AND PHOSPHATE.   Dilute 2 g (1.2 mL) with 50 mL of water in a platinum dish, add 1 g of sodium carbonate, and digest on the steam bath for 15–30 min. Cool, add 5 mL of a 10% solution of ammonium molybdate, acidify with hydrochloric acid to about pH 4 (indicator paper), and transfer to a beaker. Adjust the pH to 1.8 (using a pH meter) with dilute hydrochloric acid (1 + 9). Dilute with water to 90 mL, heat just to boiling, and cool to room temperature. Add 10 mL of hydrochloric acid and dilute with water to 100 mL. Transfer to a separatory funnel, add 50 mL of 4-methyl-2-pentanone, shake vigorously, and allow to separate. Draw off and discard the aqueous phase. Wash the ketone layer three times with 20-mL portions of dilute hydrochloric acid (1 + 99), drawing off and discarding the aqueous layer each time. Add 10 mL of dilute hydrochloric acid

(1 + 99), to which has just been added 0.2 mL of a freshly prepared 2% solution of stannous chloride dihydrate in hydrochloric acid. Shake vigorously and allow to separate. Any blue color should not exceed that produced by 0.01 mg of silica, SiO_2, and 1 g of sodium carbonate, Na_2CO_3, treated exactly like the sample.

CHLORIDE.   (Page 25). Dilute 1 g (0.6 mL) with 25 mL of water.

NITROGEN COMPOUNDS.   (Page 30). Use 5 g (3.0 mL). For the standard use 0.05 mg of nitrogen (N).

SULFATE.   Dilute 32 g (20 mL) with water to 160 mL and neutralize with ammonium hydroxide, using litmus paper as an indicator. Add 1 mL of hydrochloric acid, heat to boiling, add 5 mL of barium chloride reagent solution, digest in a covered beaker on the hot plate at low setting for 2 h, and allow to stand overnight. If a precipitate is formed, filter, wash thoroughly, and ignite. The weight of the precipitate should not be more than 0.0008 g greater than the weight obtained from a complete blank test, including the evaporated residue from a volume of ammonium hydroxide equal to that used to neutralize the sample.

> *Sample Solution A.*   To 40 g (25 mL) add 1 mL of nitric acid and about 20 mg of sodium carbonate. Evaporate to dryness on a hot plate in a well-ventilated hood, dissolve the residue in 2 mL of 1 N acetic acid, and dilute with water to 40 mL (1 mL = 1 g).

HEAVY METALS.   (Page 26, Method 1). Dilute 20 mL of sample solution A (20-g sample) with water to 25 mL.

IRON.   (Page 28, Method 1). Use 10 mL of sample solution A (10-g sample).

Perchloric Acid, 60%

$HClO_4$ Formula Wt 100.46

CAS Number 7601–90–3

REQUIREMENTS

Assay . . . . 60.0–62.0% $HClO_4$

MAXIMUM ALLOWABLE

Color (APHA)	10
Residue after ignition	0.003%
Silicate and phosphate (as SiO_2)	5 ppm
Chloride (Cl)	0.001%
Nitrogen compounds (as N)	0.001%
Sulfate (SO_4)	0.001%
Heavy metals (as Pb)	1 ppm
Iron (Fe)	1 ppm

TESTS

All tests the same as for Perchloric Acid, 70%.

> NOTE. The density of the 60% acid (1.54 g / mL) is less than that of the 70% reagent (1.67 g / mL), and proper allowance should be made for this when measured volumes are taken for the relevant tests.

Periodic Acid

Para-Periodic Acid

H_5IO_6 Formula Wt 227.96

CAS Number 10450−60−9

REQUIREMENTS

Assay	99.0−101.0% H_5IO_6

MAXIMUM ALLOWABLE

Insoluble matter	0.01%
Residue after ignition	0.01%
Other halogens (as Cl)	0.01%
Sulfate (SO_4)	0.01%
Heavy metals (as Pb)	0.005%
Iron (Fe)	0.003%

TESTS

ASSAY. (By titration of oxidizing power). Weigh accurately 0.10–0.13 g, transfer to an iodine flask, and dissolve in about 50 mL of water. Add 1 mL of sulfuric acid and 3 g of potassium iodide, swirl, and let stand in the dark for 10 min. Titrate the liberated iodine with 0.1 N sodium thiosulfate, adding starch indicator near the end of the titration. One milliliter of 0.1 N thiosulfate corresponds to 0.02849 g of H_5IO_6.

INSOLUBLE MATTER. (Page 14). Use 10 g dissolved in 100 mL of water.

RESIDUE AFTER IGNITION. (Page 15). Use 10-g sample.

OTHER HALOGENS. Dissolve 1 g in 20 mL of water with 1 mL of phosphoric acid. Add 1 mL of hydrogen peroxide and boil until the iodine is expelled. Wash down the sides with water, add 1 mL of hydrogen peroxide, and boil. If iodine color forms, repeat the treatment until no more iodine is evolved. Boil an additional 10 min and dilute to 100 mL. Dilute 10 mL to 23 mL, and add 1 mL of nitric acid and 1 mL of silver nitrate reagent solution. The turbidity should not exceed that produced by 0.01 mg of chloride ion in an equal volume containing the quantities of all reagents used in the test.

> *Sample Solution A.* Dissolve 10 g in 20 mL of water. Add 10 mg of sodium carbonate and 20 mL of hydrochloric acid, and evaporate to dryness on the steam bath. Repeat the evaporation twice, using 10 mL of hydrochloric acid. Dissolve the residue in 20 mL of water and dilute to 100 mL (1 mL = 0.1 g).

> *Blank Solution B.* Use the quantity of hydrochloric acid added to sample solution A and 10 mg of sodium carbonate. Evaporate to dryness, dissolve in 20 mL of water, and dilute to 100 mL.

SULFATE. (Page 32, Procedure A, Method 1). Use 5 mL of sample solution A (0.5-g sample).

HEAVY METALS. (Page 26, Method 1). Dilute 4 mL of sample solution A (0.4-g sample) to 25 mL. Use 4 mL of blank solution B to prepare the standard.

IRON. (Page 28, Method 2). Use 3.3 mL of sample solution A (0.33-g sample). For the standard use 3.3 mL of blank solution B.

Petroleum Ether
Ligroin

CAS Number 8032−32−4

Suitable for use in extraction−concentration analysis or general use. Product labeling shall designate the one or both of these uses for which suitability is represented on the basis of meeting the relevant requirements and tests. The extraction−concentration suitability requirements include only the general use requirement for color.

NOTE. Extremely flammable. Keep away from heat, sparks, and open flame. Store in a cool place. Keep container closed.

REQUIREMENTS

General Use

Color (APHA) . . . . ≤ 10
Boiling range . . . . 35−60°C
Residue after evaporation . . . . ≤ 0.001%
Acidity . . . . Passes test

Specific Use

EXTRACTION−CONCENTRATION SUITABILITY

GC−FID . . . . Passes test
GC−ECD . . . . Passes test

TESTS

COLOR (APHA).   (Page 17).

BOILING RANGE.   Distill 100 mL by the method described on page 17. The corrected initial boiling point should not be below 35 °C, and the corrected dry point should not be above 60°C. The barometric pressure correction for petroleum ether is +0.039 °C per mm under 760 mm and −0.039°C per mm above 760 mm.

RESIDUE AFTER EVAPORATION. (Page 14). Evaporate 100 g (150 mL) to dryness in a tared dish at 70–80°C and dry the residue at 105°C for 30 min.

ACIDITY. Thoroughly shake 10 mL with 5 mL of water for 2 min and allow to separate. The aqueous layer should not turn blue litmus paper red in 15 s.

EXTRACTION–CONCENTRATION SUITABILITY. Analyze the sample by using the general procedure cited on page 65.

1,10-Phenanthroline Monohydrate

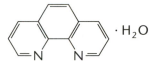 $\cdot H_2O$

$C_{12}H_8N_2 \cdot H_2O$ Formula Wt 198.22

CAS Number 5144–89–8

REQUIREMENTS

Suitability as redox indicator . . . . Passes test
Suitability for determining iron . . . . Passes test

TESTS

Sample Solution A. Dissolve 0.10 g in water and dilute with water to 100 mL.

SUITABILITY AS REDOX INDICATOR. To 2.5 mL of sample solution A add 0.1 mL of 0.1 N ferrous ammonium sulfate solution and dilute to 50 mL with 12 N sulfuric acid. A distinct red color should be produced. Add 0.1 mL of 0.1 N potassium dichromate solution. The red color should be discharged.

SUITABILITY FOR DETERMINING IRON. To 20 mL of water add 0.001 mg of iron (Fe) and 1 mL of hydroxylamine hydrochloride reagent solution. For the control add 1 mL of hydroxylamine hydrochloride reagent solution to 20 mL of water. To each solution add 2.5 mL of sample solution A and allow to stand for 1 h. The color in the solution containing the added iron should be definitely pink compared with the control. The control containing no added iron should not show any turbidity or color.

Phenol

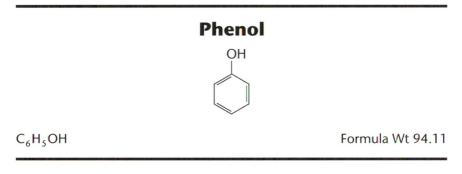

C_6H_5OH Formula Wt 94.11

CAS Number 108 –95 –2

NOTE. Phenol that conforms to this specification may contain a stabilizer. If a stabilizer is present, its presence should be stated on the label.

REQUIREMENTS

Assay . . . . ≥ 99.0%
C_6H_5OH
Freezing point . . . . Not below 40.5°C
(dry basis)
Clarity of solution . . . . Passes test

MAXIMUM ALLOWABLE

Residue after evaporation 0.05%
Water 0.5%

TESTS

ASSAY. (By iodometry after bromination of hydroxyl groups). Weigh accurately about 1.5 g and dissolve in sufficient water to make exactly 1000 mL. Transfer 25 mL into a 500-mL glass-stoppered volumetric flask, add 30 mL of 0.1 N bromine and 5 mL of hydrochloric acid, and immediately stopper the flask. Shake frequently during one-half hour and allow to stand for 15 min. Then quickly add 10 mL of 10% potassium iodide, being careful that no bromine escapes, and stopper immediately. Shake, add 1 mL of chloroform, and titrate the liberated iodine (which represents the excess of bromine) with 0.1 N sodium thiosulfate, using 3 mL of starch indicator solution. One milliliter of 0.1 N bromine corresponds to 0.001568 g of C_6H_5OH.

FREEZING POINT. Dry the sample by adding 15 g of molecular sieve zeolite* having 4A-size pores to 100 g of molten sample in a 400-mL conical flask. Stopper the flask loosely and maintain the mixture at 55–60°C for 20 min with frequent stirring. Quickly transfer a portion of the dried sample to a jacketed test tube, filling the tube to a depth of 8 cm. Insert a suitable thermometer, graduated in 0.1°C units, and stir until the temperature falls to 10°C above the expected freezing point. Immerse the tube in a water bath that is adjusted to about 4°C below the expected freezing point and clamp in a vertical position. Insert a ring-type stirrer around the thermometer that is clamped to place the bottom of the bulb about 0.5 cm above the bottom of the test tube. Stir the sample continuously until crystallization occurs and the temperature does not change more than 0.05°C in 2 min or, if supercooling occurs, the temperature rises rapidly to maximum accompanied by rapid crystallization. Record the constant temperature or the maximum temperature as the freezing point.

CLARITY OF SOLUTION. Dissolve 5.0 g in 100 mL of water. The solution should be free of turbidity and insoluble residue.

RESIDUE AFTER EVAPORATION. (Page 14). Evaporate 10 g to dryness in a tared dish on the steam bath and dry the residue at 105°C for 30 min.

WATER. (Page 54, Method 1). Use 25 g of molten sample.

---

*It may be in the form of a powder or of cylindrical granules about 3 mm in diameter.

Phenolphthalein

3,3-Bis(4-hydroxyphenyl)-1(3 H)-isobenzofuranone
3,3-Bis(p-hydroxyphenyl)phthalide

$C_{20}H_{14}O_4$ Formula Wt 318.33

CAS Number 77−09−8

REQUIREMENTS

Clarity of alcohol solution . . . . Passes test
Visual transition interval . . . . From pH 8.0
(colorless) to
pH 10 (red)

TESTS

CLARITY OF ALCOHOL SOLUTION.   Dissolve 1 g in 100 mL of ethyl alcohol. Not more than a faint trace of turbidity or insoluble matter should remain. Reserve the solution for the test for visual transition interval.

VISUAL TRANSITION INTERVAL.   Dissolve 1 g of potassium chloride in 100 mL of water. Adjust the pH of the solution to 8.0 (using a pH meter as described on page 43) with 0.01 N acid or base. Add 0.15 mL of the 1% solution reserved from the test for clarity of alcohol solution. The solution should be colorless. Titrate the solution with 0.01 N sodium hydroxide until the pH is 8.2 (using the pH meter). The solution should have a pale pink color. Continue the titration until the pH is 8.6 (using the pH meter). The solution should have a definite pink color. Not more than 0.20 mL of 0.01 N sodium hydroxide should be consumed. Each additional 0.20 mL of 0.01 N sodium hydroxide should increase the amount of red color, until at pH 10 the solution should be very red.

Phenol Red

Phenolsulfonphthalein
4,4'-(3 H-2,1-Benzoxathiol-3-ylidene)bisphenol
S, S-Dioxide

$C_{19}H_{14}O_5S$ Formula Wt 354.38

CAS Number 143–74–8 (sultone), 34487–61–1 (sodium salt)

NOTE. This standard applies both to the free acid form and to the salt form of this indicator.

REQUIREMENTS

Clarity of solution . . . . Passes test
Visual transition interval . . . . From pH 6.8
                                                   (yellow) to
                                                   pH 8.2 (red)

TESTS

CLARITY OF SOLUTION. If the indicator is the acid form, dissolve 0.1 g in 100 mL of alcohol. If the indicator is a salt form, dissolve 0.1 g in 100 mL of water. Not more than a faint trace of turbidity or insoluble matter should remain. Reserve the solution for the test for visual transition interval.

VISUAL TRANSITION INTERVAL. Dissolve 1 g of potassium chloride in 100 mL of water. Adjust the pH of the solution to 6.8 (using a pH meter as described on page 43) with 0.01 N acid or alkali. Add 0.15 mL of the 0.1% solution reserved from the test for clarity of solution. The color of the solution should be yellow, with not more than a faint trace of green color. Titrate the solution with 0.01 N sodium hydroxide to a pH of 7.0 (using a pH meter). The color of the solution should be orange (flesh). Continue the titration to pH 8.2. The color of the solution should be red. Not more than 0.20 mL of 0.01 N sodium hydroxide should be consumed in the entire titration.

Phosphomolybdic Acid
Molybdophosphoric Acid

CAS Number 51429–74–4

REQUIREMENTS

MAXIMUM ALLOWABLE

Insoluble matter 0.01%
Chloride (Cl) 0.02%
Sulfate (SO_4) 0.025%
Ammonium (NH_4) 0.01%
Calcium (Ca) 0.02%
Heavy metals (as Pb) 0.005%
Iron (Fe) 0.005%

TESTS

INSOLUBLE MATTER. (Page 14). Use 10 g dissolved in 250 mL of water.

CHLORIDE. Dissolve 1.0 g in 50 mL of water, add 1 mL of nitric acid, filter through a chloride-free filter, and divide into two equal portions. For the control add 0.5 mL of silver nitrate reagent solution to one-half of the solution, let stand for 10 min, filter, and add 0.1 mg of chloride ion (Cl). For the sample add 0.5 mL of silver nitrate reagent solution to the remaining half. Any turbidity in the solution of the sample should not exceed that in the control.

SULFATE. (Page 32, Procedure A, Method 1).

AMMONIUM. (Page 22). Dilute the distillate from a 1.0-g sample with water to 100 mL; 20 mL of the dilution shows not more than 0.02 mg of ammonium (NH_4).

CALCIUM. Determine the calcium by the flame atomic absorption spectrophotometric method described on page 35.

Sample Stock Solution. Dissolve 1.0 g of sample in 50 mL of water in a 100-mL volumetric flask and dilute to the mark with water (0.01 g/mL).

Analyze the solutions by means of a suitable atomic absorption spectrophotometer, using the conditions outlined in the following table. Calculate the metal content of the sample by the method of standard additions.

Element	Wavelength (nm)	Sample Wt (g)	Standard Added (mg)	Flame Type*	Background Correction
Ca	422.7	0.1	0.02; 0.04	N/A	No

*N/A is nitrous oxide/acetylene.

HEAVY METALS. Dissolve 3.0 g in 30 mL of water, filter, neutralize the filtrate to litmus with sodium hydroxide reagent solution, and add 5 mL of 1 N sodium hydroxide in excess. Heat the solution to boiling to destroy any blue color. For the standard boil a solution containing 5 mL of 1 N sodium hydroxide and 0.15 mg of lead ion (Pb) in 30 mL of water. Cool the solutions to about 10°C and to each add 4.5 mL of 1 N hydrochloric acid and 10 mL of 10% (W/V) sodium sulfide reagent solution. Any color in the solution of the sample should not exceed that in the standard.

IRON. Dissolve 5.0 g in 30 mL of water, make alkaline with ammonium hydroxide, and add 0.25 to 0.30 mL of bromine water. Boil for a few minutes, filter, wash the precipitate several times, and discard the filtrate and washings. Pass 25 mL of dilute hydrochloric acid (1 + 1) through the filter to dissolve any precipitate, wash thoroughly, and dilute with water to 50 mL. Dilute 20 mL of this solution with water to 45 mL and add 2 mL of ammonium thiocyanate reagent solution. Any red color should not exceed that produced by 0.1 mg of ferric ion (Fe^{3+}) in an equal volume of solution containing 10 mL of dilute hydrochloric acid (1 + 1) and 2 mL of ammonium thiocyanate reagent solution.

Phosphoric Acid
Orthophosphoric Acid
Phosphoric(V) Acid

H_3PO_4 Formula Wt 98.00

CAS Number 7664−38−2

REQUIREMENTS

Assay . . . . ≥ 85.0% H_3PO_4

MAXIMUM ALLOWABLE

Color (APHA). 10
Insoluble matter, calcium, magnesium,
   and ammonium hydroxide precipitate 0.005%
Chloride (Cl) . . . . 3 ppm
Nitrate (NO_3) . . . . 5 ppm
Sulfate (SO_4). 0.003%
Volatile acids (as CH_3COOH) 0.001%
Antimony (Sb) 0.002%
Arsenic (As). . . . . 1 ppm
Heavy metals (as Pb) 0.001%
Iron (Fe) 0.003%
Manganese (Mn) . . . . 0.5 ppm
Potassium (K) 0.005%
Sodium (Na) 0.025%
Reducing substances . . . . Passes test

TESTS

ASSAY. (By acid−base titrimetry). Weigh accurately about 2.0 g and dilute with 120 mL of water. Titrate potentiometrically with carbonate-free 1 N sodium hydroxide to the end point at a pH near 9, using the first-derivative method (page 51) One milliliter of 1 N sodium hydroxide corresponds to 0.04900 g of H_3PO_4.

COLOR (APHA). (Page 17).

INSOLUBLE MATTER, CALCIUM, MAGNESIUM, AND AMMONIUM HYDROXIDE PRECIPITATE. Dilute 20 g (12 mL) to 100 mL with water. Add ammonium hydroxide until just alkaline to litmus, hydrochloric acid until acid to litmus, and 0.5 mL of acid in excess. Add 5 mL of ammonium oxalate reagent solution and 10 mL of ammonium hydroxide, and allow to stand overnight. If a precipitate is formed, filter, wash with water containing 2.5% of ammonium hydroxide and 0.1% of ammonium oxalate, and ignite.

CHLORIDE.   (Page 25). Dilute 3.4 g (2.0 mL) with water to 25 mL and add 0.5 mL of nitric acid.

NITRATE

>   ***Sample Solution A.***   Add 4 g (2.4 mL) to 2 mL of water, dilute to 50 mL with brucine sulfate reagent solution, and mix.

>   ***Control Solution B.***   Add 4 g (2.4 mL) to 2 mL of the standard nitrate solution containing 0.01 mg of nitrate ion (NO_3) per mL, dilute to 50 mL with brucine sulfate reagent solution, and mix.

Continue with the procedure described on page 29, starting with the preparation of blank solution C.

SULFATE.   (Page 32, Procedure A, Method 1). Use 1.7 g (1.0 mL), and neutralize with ammonium hydroxide. Allow 30 min for the turbidity to form.

VOLATILE ACIDS.   Dilute 64 g (37.5 mL) with 75 mL of freshly boiled and cooled water in a distilling flask provided with a spray trap and distill off 50 mL. To the distillate add 0.15 mL of phenolphthalein indicator solution and titrate with 0.01 N sodium hydroxide. Not more than 1.0 mL of the 0.01 N sodium hydroxide should be required to produce a pink color.

ANTIMONY, POTASSIUM, AND SODIUM.   Determine the antimony, potassium, and sodium by the flame atomic spectrophotometric method described on page 35.

>   ***Sample Stock Solution.***   Dilute 6.0 mL (10 g) of sample with 20 mL of water in a 100-mL volumetric flask and dilute to the mark with water (0.1 g/mL). For antimony use a 10.0-g sample and standard additions.

Analyze the solutions by means of a suitable atomic absorption spectrophotometer, using the conditions outlined in the following

table. Calculate the metal content of the sample by the method of standard additions.

Element	Wavelength (nm)	Sample Wt (g)	Standard Added (mg)	Flame Type*	Background Correction
Sb	217.6	10.0	0.20; 0.40	A/A	Yes
K	766.5	1.0	0.04; 0.08	A/A	No
Na	589.0	0.1	0.02; 0.04	A/A	No

*A/A is air/acetylene.

ARSENIC. (Page 23). To an arsine generator flask, add 10 g (6 mL) of sample, 30 mL of water, and 10 mL of ferric ammonium sulfate solution. Then add 2% potassium permanganate reagent solution dropwise to a persistent pink color. Add 1.5 g of sodium chloride and mix well until dissolution is complete. Heat nearly to boiling, remove from the heat, and add 1 mL of 40% stannous chloride dihydrate reagent solution. Dilute to 60 mL and cool to about 25°C. For the standard, to a second arsine generator flask, add 0.01 mg of arsenic ion (As) and 20 mL of dilute sulfuric acid (1 + 4), and treat as described for the sample. Assemble the test apparatus (page 23), and continue in the manner of pages 23–24, using 10 g of No. 20-mesh granulated zinc. After 30 min, transfer the two silver diethyldithio-carbamate-absorbing solutions to 1-cm cells, and read the absorbance at 540 nm against the stock silver diethyldithiocarbamate solution. The absorbance for the sample should not exceed that for the standard.

> **Ferric Ammonium Sulfate Solution.** Mix 84 g of ferric ammonium sulfate dodecahydrate, $FeNH_4(SO_4)_2 \cdot 12H_2O$, 2 mL of sulfuric acid, and 1 g of sodium chloride, NaCl. Dissolve and dilute with water to 1 L.

HEAVY METALS. (Page 26, Method 1). Dilute 4 g (2.4 mL) with water to 32 mL. Use 24 mL to prepare the sample solution, and use the remaining 8.0 mL to prepare the control solution. Adjust the pH to 4.5.

IRON. (Page 28, Method 2). Dilute 6.7 g (4.0 mL) to 100 mL. To 5.0 mL of the solution add 8 mL of 0.5 N sodium acetate.

MANGANESE. Add 25 g (15 mL) to 100 mL of dilute sulfuric acid (1 + 9). For the control add 3 mL (5 g) of sample and 0.01 mg of manganese ion (Mn) to 100 mL of dilute sulfuric acid (1 + 9). To each solution add 20 mL of nitric acid, heat to boiling, and boil gently for 5 min. Cool slightly, add 0.25 g of potassium periodate, and again boil for 5 min. Any red color in the solution of the sample should not exceed that in the control.

REDUCING SUBSTANCES. Add 5.0 mL of 0.1 N bromine solution to each of two 500-mL "iodine determination" flasks. To each flask add 120 mL of cold, dilute sulfuric acid (1 + 59) and to one of the flasks add rapidly 17 g (10 mL) of the sample. Stopper the flasks, shake gently, seal with water, and allow to stand at room temperature for 10 min. Cool in an ice bath, add 10 mL of potassium iodide reagent solution, and reseal with water. Shake gently to dissolve all fumes and place in the ice bath for about 5 min. Wash the stoppers and inside walls of the flasks with water and titrate each with 0.01 N sodium thiosulfate, adding 3 mL of starch indicator solution near the end of the titration. The difference between the two titrations should not exceed 1.0 mL of 0.01 N sodium thiosulfate.

Phosphoric Acid, Meta-
Vitreous Sodium Acid Metaphosphate

CAS Number 37267 – 86 – 0

NOTE. This reagent contains as a stabilizer a somewhat greater proportion of sodium metaphosphate than that corresponding to the empirical formula $NaH(PO_3)_2$.

REQUIREMENTS

Assay (HPO_3). . . . . 33.5 – 36.5%
Stabilizer ($NaPO_3$) . . . . 57.0 – 63.0%

MAXIMUM ALLOWABLE

Chloride (Cl) 0.001%
Nitrate (NO_3) 0.001%
Sulfate (SO_4) 0.005%
Arsenic (As) . . . . 1 ppm
Heavy metals (as Pb) 0.005%
Iron (Fe) 0.005%
Substances reducing permanganate
  (as H_3PO_3) 0.02%

TESTS

ASSAY. (By alkalimetry). Weigh accurately 5 g, dissolve in 400 mL of water, and titrate with 1 N sodium hydroxide solution to a pH of 4.4 (using a pH meter). One milliliter of 1 N sodium hydroxide corresponds to 0.07998 g of metaphosphoric acid (HPO_3).

STABILIZER. Weigh accurately 20 g, dissolve in water, and dilute with water to 1 L. To 50.0 mL of this solution in a 200-mL volumetric flask add 50 mL of water and 50 mL of a 10% solution of lead acetate, dilute with water to 200 mL, mix thoroughly, and decant through a filter. Treat 100 mL of the clear filtrate with hydrogen sulfide sufficiently to precipitate the excess of lead, filter, and wash with about 20 mL of water. To the filtrate add 2 mL of sulfuric acid, evaporate to dryness in a tared dish, and ignite to constant weight at $800 \pm 25°C$. Calculate the weight of $NaPO_3$ by multiplying the weight of sodium sulfate by the factor 1.4356.

> *Sample Solution A for the Determination of Chloride, Arsenic, Heavy Metals, Iron, and Substances Reducing Permanganate.* Dissolve 20 g in 150 mL of water, cool, and dilute with water to 200 mL (1 mL = 0.1 g).

CHLORIDE. (Page 25). Dilute 10 mL of sample solution A (1-g sample) with water to 25 mL, and add 0.5 mL of nitric acid.

NITRATE. (Page 29). For sample solution A use 1.0 g in 4 mL of water. For control solution B use 1.0 g and 1 mL of standard nitrate solution containing 0.01 mg of nitrate ion (NO_3) per mL.

SULFATE. Dissolve 8.0 g in 180 mL of water, heat the solution to boiling, cool, filter, and heat again to boiling. Add 10 mL of barium chloride reagent solution, digest in a covered beaker on a hot plate at low setting for 2 h, and allow to stand overnight. If a precipitate is formed, filter, wash thoroughly, and ignite. The weight of the precipitate should not be more than 0.001 g greater than the weight obtained in a complete blank test, including filtration of a solution of 2 mL of hydrochloric acid in 200 mL of water.

ARSENIC. (Page 23). Dilute 30 mL of sample solution A (3-g sample) with water to 35 mL. For the standard use 0.003 mg of arsenic (As).

HEAVY METALS. To 50 mL of sample solution A (5-g sample) add 0.15 mL of phenolphthalein indicator solution and neutralize with ammonium hydroxide. Adjust the pH of this solution to 4.5 (using a pH meter) with 1 N sulfuric acid and dilute with water to 250 mL. For the sample use 40 mL of the solution. For the control add 0.02 mg of lead ion (Pb) to 20 mL of the solution and dilute with water to 40 mL. Add 10 mL of freshly prepared hydrogen sulfide water to each and mix. Any color in the solution of the sample should not exceed that in the control.

IRON.   To 2.0 mL of sample solution A (0.2-g sample) add 5 mL of water and 3 mL of hydrochloric acid, boil gently for 10 min, and cool. Add 3.5 mL of 0.5 N sodium acetate, 6 mL of hydroxylamine hydrochloride reagent solution, and 4 mL of 1,10-phenanthroline reagent solution. Adjust the pH to between 4 and 6 with ammonium hydroxide. Any red color produced within 1 h should not exceed that produced by 0.01 mg of iron (Fe) in an equal volume of solution containing the quantities of reagents used in the test.

SUBSTANCES REDUCING PERMANGANATE.   To 50 mL of sample solution A (5-g sample) add 10 mL of 10% sulfuric acid reagent solution, heat the solution to boiling, and titrate with 0.1 N potassium permanganate solution to a pink color that remains for 5 min. Not more than 0.25 mL of permanganate should be required.

Phosphorus Pentoxide
Phosphorus(V) Oxide

P_2O_5 Formula Wt 141.94

CAS Number 1314–56–3

REQUIREMENTS

Assay . . . . $\geq$ 98.0% P_2O_5

MAXIMUM ALLOWABLE

Insoluble matter 0.02%
Phosphorus trioxide (P_2O_3). . . . . Passes test (limit
                                                                  about 0.02%)
Ammonium (NH_4) 0.01%
Heavy metals (as Pb) 0.01%

TESTS

NOTE. When making a solution, the phosphorus pentoxide must be added carefully and in small portions to water that has been cooled. Quiet dissolution may be achieved by allowing the sample to remain overnight in the uncovered weighing bottle placed in a covered beaker containing the necessary water. The pentoxide will absorb enough water from the air to dissolve without sputtering.

ASSAY. (By acid–base titrimetry). Weigh accurately 1 g, carefully dissolve in 100 mL of water, and evaporate to about 25 mL. Cool, dilute to 120 mL, add 0.5 mL of thymolphthalein indicator solution, and titrate with 1 N sodium hydroxide to the first appearance of a permanent blue color. Correct for an indicator blank, if any. One milliliter of 1 N sodium hydroxide corresponds to 0.03549 g of P_2O_5.

INSOLUBLE MATTER.   Carefully dissolve 5.0 g in 40 mL of water, warming if necessary. Filter through a tared filtering crucible and set aside the filtrate for sample solution A. Wash the residue thoroughly and dry at 105°C.

Sample Solution A.   Dilute the filtrate, without washings, obtained in the preceding test with water to 100 mL (1 mL = 0.05 g).

PHOSPHORUS TRIOXIDE.   To 60 mL of sample solution A (3-g sample) add 0.20 mL of 0.1 N potassium permanganate solution. Heat to boiling and allow to digest on the steam bath for 10 min. The pink color should not be entirely discharged.

AMMONIUM.   Dilute 2.0 mL of sample solution A (0.1-g sample) with water to 40 mL, and add 10 mL of freshly boiled 10% sodium hydroxide reagent solution and 2 mL of Nessler reagent. Any color should not exceed that produced by 0.01 mg of ammonium ion (NH_4) in an equal volume of solution containing the quantities of reagents used in the test.

HEAVY METALS.   (Page 26, Method 1). Dilute 8.0 mL of sample solution A (0.4-g sample) with water to about 30 mL, boil for 5 min, cool, and dilute with water to 32 mL. Use 24 mL to prepare the sample solution, and use the remaining 8.0 mL to prepare the control solution.

Phthalic Acid

1,2-Benzenedicarboxylic Acid

$C_6H_4(COOH)_2$ Formula Wt 166.13

CAS Number 88–99–3

REQUIREMENTS

Assay . . . . ≥ 99.5% $C_6H_4(COOH)_2$

MAXIMUM ALLOWABLE

Insoluble matter 0.05%
Residue after ignition (as SO_4). 0.02%
Chloride (Cl) 0.001%
Nitrate (NO_3) 0.005%
Sulfate (SO_4) 0.005%
Heavy metals (as Pb) 0.001%
Iron (Fe) 0.001%
Water (H_2O) 0.5%

TESTS

ASSAY. (By acid–base titrimetry). Weigh accurately 2.8 g, transfer to a conical flask, cool, and add 50.0 mL of standard 1 N sodium hydroxide. Add 25 mL of water, and boil on a hot plate until dissolved. Add 0.15 mL of phenolphthalein indicator solution, and titrate the excess sodium hydroxide with standard 1 N hydrochloric acid. Perform a blank determination, and make any necessary correction. One milliliter of 1 N sodium hydroxide consumed corresponds to 0.08306 g of $C_6H_4(COOH)_2$.

INSOLUBLE MATTER. (Page 14). Use 10 g dissolved in 250 mL of 10% sodium carbonate (Na_2CO_3) solution.

RESIDUE AFTER IGNITION. (Page 15). Use 4.0 g. Retain the residue to prepare sample solution B.

> **Sample Solution A.** Dissolve 10.0 g in 100 mL of boiling water, and allow to cool to room temperature. The supernatant

liquid after decantation from the crystallized phthalic acid is sample solution A (1 mL = 0.1 g).

CHLORIDE.   (Page 25). Dilute 10 mL of sample solution A (1-g sample) to 20 mL with water. Filter if necessary through a chloride-free filter. For the standard, use 0.01 mg of chloride ion (Cl) and dilute to 20 mL with water. To each add 1 mL of nitric acid and 1 mL of silver nitrate reagent solution. Mix, allow to stand for 5 min protected from sunlight, and compare. The solutions can best be viewed visually against a black background. Any turbidity in the solution of the sample should not exceed that in the standard.

NITRATE.   (Page 29). Use 10 mL of sample solution A (1-g sample) and 0.05 mg of nitrate ion (NO_3). Follow the general procedure for nitrate by the brucine sulfate method.

SULFATE.   (Page 32, Procedure A, Method 3). Add 10 mg of sodium carbonate to 10 mL of sample solution A (1-g sample), and evaporate to dryness on a hot plate at low setting. Redissolve in a minimum volume of water and add 1 mL of dilute hydrochloric acid (1 + 19). If necessary, filter through a small filter, and wash with two 2-mL portions of water. Dilute to 10 mL and add 1 mL of barium chloride reagent solution. Any turbidity should not exceed that produced by 0.05 mg of sulfate ion (SO_4) in an equal volume of solution containing the quantities of reagents used in the test. Compare 30 min after adding the barium chloride to the sample and standard solutions.

Sample Solution B.   To the residue after ignition add 10 mL of water and dilute to 20 mL (1 mL = 0.2 g).

HEAVY METALS.   (Page 26). Use 10 mL of sample solution B (2.0-g sample).

IRON.   (Page 28, Method 2). Use 5 mL of sample solution B (1.0-g sample).

WATER.   (Page 54, Method 1). Use 5.0 g.

Phthalic Anhydride
1,3-Isobenzofurandione

C$_6$H$_4$(CO)$_2$O   Formula Wt 148.12

CAS Number 85–44–9

REQUIREMENTS

Appearance . . . . White flaky crystals
Assay 99.0–100.2%
                                                     C$_8$H$_4$O$_3$
Melting point . . . . Not more than
                                                     3°C range,
                                                     including 131°C

MAXIMUM ALLOWABLE

Residue after ignition 0.01%
Chloride (Cl) 0.002%
Sulfate (SO$_4$) 0.003%
Heavy metals (as Pb) . . . . 5 ppm
Iron (Fe) . . . . 5 ppm

TESTS

ASSAY.   (By acid–base titrimetry). Accurately weigh about 0.3 g of sample into a 150-mL beaker containing 15.0 mL of 0.5 N morpholine in methanol and approximately 80 mL of methanol. Stir at least 30 min and titrate with 0.5 N hydrochloric acid, using 0.2 mL of 0.1% naphthyl red indicator solution (or titrate potentiometrically). Similarly titrate a blank containing 15.0 mL of morpholine and 80 mL of methanol.

$$\% \text{ Phthalic anhydride} = \frac{(V_{\text{sample}} - V_{\text{blank}}) \times 148.12 \times 100}{\text{Sample wt in mg}}$$

RESIDUE AFTER IGNITION. (Page 15). Gently ignite 10 g in a tared crucible or dish. Finally, ignite at $800 \pm 25°C$ for 15 min. (Reserve this residue for the iron test.)

CHLORIDE. Mix 0.50 g of sample with 0.5 g of sodium carbonate and add 10–15 mL of water. Evaporate on the steam bath and ignite until the mass is thoroughly charred, avoiding an excessively high temperature. When the sample is completely ignited, cool, and extract the fusion with 10 mL of water and 2 mL of nitric acid. Filter through chloride-free filter paper, wash with water to a volume of 20 mL, and add 1 mL of silver nitrate reagent solution. Any turbidity should not exceed that produced by 0.01 mg of chloride ion (Cl) in an equal volume of solution containing 0.5 g of sodium carbonate, 2 mL of nitric acid, and 1 mL of silver nitrate reagent solution.

SULFATE. Mix 3.0 g of sample with 1 g of sodium carbonate in a platinum crucible, add 20 mL of water, and evaporate to dryness on a hot plate at low setting. Ignite completely, protecting the fusion from sulfur in the flame. Cool, dissolve the residue in 20 mL of water, add 2 mL of 30% hydrogen peroxide, and boil for 5 min. Add 3 mL of hydrochloric acid and evaporate to dryness on the hot plate. Dissolve the residue in 10 mL of water, filter, and wash with 10 mL of water. To the filtrate add 1 mL of 1 N hydrochloric acid and 2 mL of barium chloride reagent solution. Any turbidity should not be greater than that of a standard prepared as follows: evaporate 1 g of sodium carbonate, 2 mL of 30% hydrogen peroxide, and 3 mL of hydrochloric acid to dryness on a hot plate at low setting. Dissolve the residue and 0.09 mg of sulfate ion (SO_4) in sufficient water to make 20 mL, and add 1 mL of 1 N hydrochloric acid and 2 mL of barium chloride reagent solution. Compare 10 min after adding the barium chloride to the sample and standard solutions.

HEAVY METALS. (Page 27, Method 2). Use 4.0 g of sample.

IRON. (Page 28, Method 1). To the residue after ignition add 3 mL of hydrochloric acid, cover with a watch glass, and digest on the steam bath for 15–20 min. Remove the cover and evaporate to dryness. Dissolve the residue in 5 mL of hydrochloric acid, filter if necessary, and dilute with water to 100 mL. Use 20 mL of this solution.

Picric Acid
2,4,6-Trinitrophenol

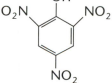

$(NO_2)_3C_6H_2OH$ Formula Wt 229.11

CAS Number 88–89–1

> *CAUTION.* Trinitrophenol explodes when heated rapidly or when subjected to percussion. For safety in transportation, trinitrophenol is usually mixed with a minimum of 10% of water.

REQUIREMENTS

Water (H_2O) . . . . ≥ 10%
Melting point (dried) . . . . Not below 121°C
nor above 123°C

MAXIMUM ALLOWABLE

Insoluble and resinous matter 0.01%
Insoluble in toluene 0.1%
Sulfate (SO_4) 0.01%

TESTS

WATER. Dissolve 1 g, accurately weighed, in 200 mL of water. Titrate with 0.1 N sodium hydroxide, using 0.15 mL of phenolphthalein indicator solution. One milliliter of 0.1 N sodium hydroxide corresponds to 0.02291 g of $(NO_2)_3C_6H_2OH$. Subtract the percent of picric acid from 100% to determine the water content.

MELTING POINT (DRIED). Carefully transfer a small amount of dried sample (prepared in the test for insoluble in toluene) to a capillary tube, and run the melting point, using an oil immersion bath.

INSOLUBLE AND RESINOUS MATTER. Dissolve a sample equivalent to 10 g on the dry basis in 500 mL of water, add 1 mL of sulfuric acid, and digest on the steam bath for 1 h. Filter through a tared filtering crucible, wash thoroughly, and dry at 105°C. No resinous material should be apparent in the crucible.

INSOLUBLE IN TOLUENE. Dry 5–10 g at 70°C and dissolve 4.0 g of the dried sample in 100 mL of warm toluene. (NOTE. Add water to the remaining dried sample before discarding it.) Filter off any insoluble material using a tared filtering crucible, wash thoroughly with warm toluene, and dry at 105°C.

SULFATE. (Page 32, Procedure A, Method 2). Use 25 mL of nitric acid. Allow 30 min for the turbidity to form.

Potassium Acetate

CH_3COOK Formula Wt 98.14

CAS Number 127–08–2

REQUIREMENTS

Assay . . . . ≥ 99.0%
                                                                    CH_3COOK
pH of 5% solution at 25°C 6.5–9.0

MAXIMUM ALLOWABLE

Insoluble matter 0.005%
Chloride (Cl) 0.003%
Phosphate (PO_4) 0.001%
Sulfate (SO_4) 0.002%
Calcium, magnesium, and
  R_2O_3 precipitate 0.01%
Heavy metals (as Pb) . . . . 5 ppm
Iron (Fe) . . . . 5 ppm
Sodium (Na) 0.03%

TESTS

ASSAY. (By nonaqueous acid–base titrimetry). Weigh, to the nearest 0.1 mg, 0.4 g of sample, and dissolve in 5 mL of acetic anhydride and 25 mL of acetic acid. Add 0.15 mL of crystal violet indicator solution, and titrate in a closed system with 0.1 N perchloric acid in glacial acetic acid to a green end point. Perform a blank determination and make any necessary correction. One milliliter of 0.1 N perchloric acid corresponds to 0.00981 g of CH_3COOK.

pH OF A 5% SOLUTION. (Page 43). The pH should be 6.5–9.0 at 25°C.

INSOLUBLE MATTER. (Page 14). Use 20 g dissolved in 200 mL of water. Add 0.10 mL of methyl red indicator solution and enough hydrochloric acid to produce a red color. Reserve the filtrate for the test for calcium, magnesium, and R_2O_3 precipitate.

CHLORIDE. (Page 25). Use 0.33 g.

PHOSPHATE. (Page 30, Method 2). Dissolve 1.0 g in 10 mL of nitric acid and evaporate to near dryness on the steam bath. Gently heat with a burner until the salt is dry. Add 10 mL of nitric acid and repeat the evaporation and drying. Dissolve in 80 mL of water, add 0.5 g of ammonium molybdate, and adjust the pH to 1.8 (using a pH meter) with dilute hydrochloric acid (1 + 9). Heat to boiling, cool, add 10 mL of hydrochloric acid, and proceed as described. Carry along a standard containing 0.01 mg of phosphate ion (PO_4) treated in the same manner as the sample after the addition of 80 mL of water.

SULFATE. (Page 32, Procedure A, Method 1).

CALCIUM, MAGNESIUM, AND R_2O_3 PRECIPITATE. To the filtrate, without washings, reserved from the test for insoluble matter, add 10 mL of ammonium oxalate reagent solution, 4 mL of ammonium phosphate reagent solution, and 30 mL of ammonium hydroxide solution, and allow to stand overnight. Filter, wash thoroughly with water containing 2.5% ammonia and about 0.1% ammonium oxalate, and ignite.

HEAVY METALS. (Page 26, Method 1). Dissolve 6.0 g in about 10 mL of water, add 15 mL of 10% hydrochloric acid, and dilute with water to 30 mL. Use 25 mL to prepare the sample solution, and use the remaining 5.0 mL to prepare the control solution.

IRON.   (Page 28, Method 1). Use 2.0 g.

SODIUM.   Determine the sodium by the flame atomic absorption spectrophotometric method described on page 35.

> ***Sample Stock Solution.***   Dissolve 1.0 g of sample with water in a 100-mL volumetric flask and dilute to the mark with water (0.01 g/mL).

Analyze the solutions by means of a suitable atomic absorption spectrophotometer, using the conditions outlined in the following table. Calculate the metal content of the sample by the method of standard additions.

Element	Wavelength (nm)	Sample Wt (g)	Standard Added (mg)	Flame Type*	Background Correction
Na	589.0	0.03	0.01; 0.02	A/A	No

*A/A is air/acetylene.

Potassium Antimony Tartrate Trihydrate
Antimony Potassium Tartrate Trihydrate

$K_2(C_4H_2O_6Sb)_2 \cdot 3H_2O$ Formula Wt 667.87

CAS Number 11071−15−1

REQUIREMENTS

Assay . . . . 99.0−103.0%
$K_2(C_4H_2O_6Sb)_2 \cdot 3H_2O$

MAXIMUM ALLOWABLE

Titrable acid or base. . . . . 0.020 meq / g
Loss on drying 2.7%
Arsenic (As). 0.015%

TESTS

ASSAY.   (By iodimetric titration of antimony). Weigh accurately 0.5 g of sample, transfer to a 250-mL beaker, and dissolve in 50 mL of water. Add 5 g of potassium sodium tartrate tetrahydrate, KOCO-CHOHCHOHCOONa $\cdot$ 4H$_2$O, 2 g of sodium borate decahydrate, Na$_2$B$_4$O$_7$ $\cdot$ 10H$_2$O, and 3 mL of starch indicator solution. Titrate immediately with 0.1 N iodine to the production of a persistent blue color. One milliliter of 0.1 N iodine corresponds to 0.01670 g of K$_2$(C$_4$H$_2$O$_6$Sb)$_2$ $\cdot$ 3H$_2$O.

TITRABLE ACID OR BASE.   Dissolve 1.0 g in 50 mL of carbon dioxide-free water, and titrate to a pH of 4.5 using 0.01 N hydrochloric acid or 0.01 N sodium hydroxide as required. Not more than 2.0 mL of either titrant should be required.

LOSS ON DRYING.   Weigh accurately 1.0 g, and dry at 105°C to constant weight.

ARSENIC.   Dissolve 0.1 g in 5 mL of hydrochloric acid, and add 10 mL of a freshly prepared solution of 20 g of stannous chloride dihydrate, SnCl$_2$ $\cdot$ 2H$_2$O, in 30 mL of hydrochloric acid. Mix, transfer to a color-comparison tube, and allow to stand for 30 min. Viewed downward over a white surface, the color of the solution is not darker than that of a standard containing 0.015 mg of arsenic (As).

Potassium Bicarbonate
Potassium Hydrogen Carbonate

KHCO$_3$ Formula Wt 100.12

CAS Number 298–14–6

REQUIREMENTS

Assay (dried basis) 99.7–100.5%
                                                  KHCO$_3$

MAXIMUM ALLOWABLE

Insoluble matter 0.01%
Chloride (Cl) 0.001%
Phosphate (PO_4) . . . . 5 ppm
Sulfur compounds (as SO_4) 0.003%
Ammonium (NH_4) . . . . 5 ppm
Calcium, magnesium, and
    R_2O_3 precipitate 0.005%
Heavy metals (as Pb) . . . . 5 ppm
Iron (Fe) . . . . 5 ppm
Sodium (Na) 0.03%

TESTS

ASSAY. (By acid–base titrimetry). Weigh accurately 3 g, previously dried over sulfuric acid for 24 h, dissolve it in 50 mL of water, add 0.15 mL of methyl orange indicator solution, and titrate with 1 N hydrochloric acid. The potassium bicarbonate content calculated from the total alkalinity, as determined by the titration, should not be less than 99.7 nor more than 100.5% of the weight taken. One milliliter of 1 N hydrochloric acid corresponds to 0.1001 g of $KHCO_3$.

INSOLUBLE MATTER. (Page 14). Dissolve 10 g in 100 mL of hot water, heat to boiling, and boil gently for 2–3 min.

CHLORIDE. (Page 25). Use 1.0 g.

PHOSPHATE. (Page 30, Method 1). Dissolve 4.0 g in 20 mL of water, add 5 mL of hydrochloric acid, and evaporate to dryness on the steam bath. Dissolve the residue in 25 mL of approximately 0.5 N sulfuric acid, and proceed as described.

SULFUR COMPOUNDS. (Page 32, Procedure A, Method 4). Allow 30 min for the turbidity to form.

AMMONIUM. Dissolve 2.0 g in 40 mL of water, and add 10 mL of 10% sodium hydroxide reagent solution and 2 mL of Nessler reagent solution. Any color should not exceed that produced by 0.01 mg of ammonium ion (NH_4) in an equal volume of solution containing the quantities of reagents used in the test.

CALCIUM, MAGNESIUM, AND R_2O_3 PRECIPITATE. Dissolve 20 g in 200 mL of hot water, neutralize with hydrochloric acid, and add 0.5 mL of the acid in excess. Filter, if necessary, add to the filtrate 5 mL of ammonium oxalate reagent solution, 2 mL of ammonium phosphate reagent solution, and 20 mL of ammonium hydroxide, and allow to stand overnight. Filter, wash thoroughly with water containing 2.5% ammonium hydroxide and about 0.1% ammonium oxalate, ignite in a tared crucible, cool, and weigh.

HEAVY METALS. (Page 26, Method 1). To 5.0 g in a 150-mL beaker add 10 mL of water, mix, and add cautiously 10 mL of hydrochloric acid. Evaporate to dryness on the steam bath, dissolve the residue in about 20 mL of water, and dilute with water to 25 mL. For the control solution add 0.02 mg of lead ion (Pb) to 1.0 g of sample, and treat exactly as the 5.0 g of sample.

IRON. (Page 28, Method 1). Dissolve 2.0 g in 20 mL of water, cautiously add 5 mL of hydrochloric acid, and dilute with water to 50 mL. Use the solution without further acidification. In the standard use only 3 mL of hydrochloric acid.

SODIUM. Determine the sodium by the flame atomic absorption spectrophotometric method described on page 35.

> **Sample Stock Solution.** Dissolve 1.0 g of sample in 10 mL of dilute hydrochloric acid (1 + 1) in a 100-mL volumetric flask, and dilute to the mark with water (0.01 g/mL).

Analyze the solutions by means of a suitable atomic absorption spectrophotometer, using the conditions outlined in the following table. Calculate the metal content of the sample by the method of standard additions.

Element	Wavelength (nm)	Sample Wt (g)	Standard Added (mg)	Flame Type*	Background Correction
Na	589.0	0.05	0.015; 0.03	A/A	No

*A/A is air/acetylene.

Potassium Bromate

$KBrO_3$ Formula Wt 167.00

CAS Number 7758–01–2

REQUIREMENTS

Assay (dried basis) . . . . ≥ 99.8% KBrO$_3$
pH of a 5% solution at 25°C 5.0–9.0

MAXIMUM ALLOWABLE

Insoluble matter 0.005%
Bromide (Br) . . . . Passes test (limit
                                            about 0.05%)
Nitrogen compounds (as N) 0.001%
Sulfate (SO$_4$) 0.005%
Heavy metals (as Pb) . . . . 5 ppm
Iron (Fe) 0.002%
Sodium (Na) 0.01%

TESTS

ASSAY. (By oxidation–reduction titration of bromate). Dry a powdered sample to constant weight at 150°C. Weigh accurately 1 g, dissolve in water, and transfer to a 250-mL volumetric flask. Dilute to volume with water and mix thoroughly. Place a 25.0-mL aliquot of this solution in a glass-stoppered conical flask and add 3 g of potassium iodide and 3 mL of hydrochloric acid. Stopper, swirl, and allow to stand for 5 min. Titrate the liberated iodine with 0.1 N sodium thiosulfate, adding 3 mL of starch indicator solution near the end of the titration. One milliliter of 0.1 N sodium thiosulfate corresponds to 0.002783 g of KBrO$_3$.

pH OF A 5% SOLUTION. (Page 43). The pH should be 5.0–9.0 at 25°C.

INSOLUBLE MATTER. (Page 14). Use 20 g dissolved in 150 mL of hot water.

BROMIDE. Dissolve 4.0 g in 80 mL of water and divide the solution into two equal portions. Add 0.15 mL of 1 N sulfuric acid to one portion. At the end of 2 min this portion should show no more yellow color than the portion to which no acid was added.

NITROGEN COMPOUNDS. (Page 30). Use 1.0 g. For the standard use 0.01 mg of nitrogen (N).

SULFATE. (Page 32, Procedure A, Method 2). Use 6 mL of dilute hydrochloric acid (1 + 1). Do two evaporations.

HEAVY METALS. (Page 26, Method 1). Dissolve 5.0 g in 20 mL of dilute hydrochloric acid (1 + 1). Evaporate the solution to dryness on the steam bath. Add 10 mL more of dilute hydrochloric acid (1 + 1), and again evaporate to dryness. Dissolve in about 20 mL of water and dilute with water to 25 mL. For the control solution add 0.02 mg of lead ion (Pb) to 1.0 g of sample, and treat exactly as the 5.0 g of sample.

IRON. (Page 28, Method 1). Dissolve 1.0 g in 20 mL of dilute hydrochloric acid (1 + 1), and evaporate on the steam bath to dryness. Add 10 mL of additional dilute hydrochloric acid (1 + 1), and again evaporate to dryness. Dissolve the residue in water, add 4 mL of hydrochloric acid, and dilute with water to 100 mL. Use 50 mL of the solution. For the control solution use 0.02 mg of iron (Fe) carried through the same procedure as the 1.0 g of sample.

SODIUM. Determine the sodium by the flame atomic absorption spectrophotometric method described on page 35.

Sample Stock Solution. Dissolve 1.0 g of sample in 25 mL of hydrochloric acid (1 + 3), and digest in a covered beaker on the steam bath until the reaction ceases. Uncover, evaporate to dryness, add 10 mL of hydrochloric acid (1 + 3), and dilute with hydrochloric acid (1 + 99) to the mark in a 100-mL volumetric flask (0.01 g/mL).

Analyze the solutions by means of a suitable atomic absorption spectrophotometer, using the conditions outlined in the following table. Calculate the metal content of the sample by the method of standard additions.

Element	Wavelength (nm)	Sample Wt (g)	Standard Added (mg)	Flame Type*	Background Correction
Na	589.0	0.1	0.01; 0.02	A/A	No

*A/A is air/acetylene.

Potassium Bromide

KBr

Formula Wt 119.00

CAS Number 7758–02–3

REQUIREMENTS

Assay . . . . $\geq$ 99.0% KBr
pH of a 5% solution at 25°C 5.0–8.8

MAXIMUM ALLOWABLE

Insoluble matter 0.005%
Bromate (BrO_3) 0.001%
Chloride (Cl) 0.2%
Iodate (IO_3) 0.001%
Iodide (I) 0.001%
Nitrogen compounds (as N) 0.005%
Sulfate (SO_4) 0.005%
Barium (Ba) 0.002%
Calcium, magnesium, and
   R_2O_3 precipitate 0.005%
Heavy metals (as Pb) . . . . 5 ppm
Iron (Fe) . . . . 5 ppm
Sodium (Na) 0.02%

TESTS

ASSAY.   (By argentimetric titration of bromide content). Weigh accurately 0.5 g of sample and dissolve in 50 mL of water. Add 50.0 mL of 0.1 N silver nitrate and 10 mL of dilute nitric acid, and titrate the excess silver nitrate with 0.1 N ammonium thiocyanate, using ferric ammonium sulfate as indicator.

$$\% \text{ KBr} = \frac{[(\text{mL} \times \text{N AgNO}_3) - (\text{mL} \times \text{N NH}_4\text{SCN})]11.90}{\text{Sample wt (g)}}$$

pH OF A 5% SOLUTION.   (Page 43). The pH should be 5.0–8.8 at 25°C.

INSOLUBLE MATTER.   (Page 14). Use 20 g dissolved in 150 mL of water. Save the filtrate separate from the washings for the test for calcium, magnesium, and R_2O_3 precipitate.

BROMATE AND IODATE.   (Page 49). Determine bromate and iodate by differential pulse polarography. Use 10.0 g of sample in 25 mL of solution. For the standard add 0.10 mg of bromate (BrO_3) and 0.10 mg of iodate (IO_3).

CHLORIDE.   Dissolve 0.50 g in 15 mL of dilute nitric acid (1 + 2) in a small flask. Add 3 mL of 30% hydrogen peroxide and digest on the steam bath until the solution is colorless. Wash down the sides of the flask with a little water, digest for an additional 15 min, cool, and dilute with water to 200 mL. Dilute 2.0 mL to 20 mL and add 1 mL of nitric acid and 1 mL of silver nitrate reagent solution. Any turbidity should not exceed that produced by 0.01 mg of chloride ion (Cl) in an equal volume of solution containing the quantities of reagents used in the test.

IODIDE.   Dissolve 2.5 g in 50 mL of water in a 100-mL volumetric flask. For the control dissolve 0.5 g in a similar flask and add 0.02 mg of iodide ion (I). Treat each solution as follows: Add 2 mL of bromine water, mix, and allow to stand for 5 min. Add 2 mL of sodium formate solution (25 g per 100 mL) and mix. Add 1 mL of 1 N sodium hydroxide or enough to bring the pH to 9 or greater, dilute to volume with water, and mix. Using differential pulse polarography, as described on page 46, starting with transfer of the solution to the polarographic cell, determine any iodate present. The wave for the sample, after correction for any iodate originally present, should not exceed that for the control (similarly corrected).

NITROGEN COMPOUNDS.   (Page 30). Use 1.0 g. For the standard use 0.05 mg of nitrogen (N).

SULFATE.   (Page 32, Procedure A, Method 1).

BARIUM.   For the sample dissolve 6.0 g in 15 mL of water. For the control dissolve 1.0 g in 15 mL of water and add 0.1 mg of barium ion (Ba). To each solution add 5 mL of acetic acid, 5 mL of 30% hydrogen peroxide, and 1 mL of hydrochloric acid. Digest in a covered beaker on the steam bath until reaction ceases, uncover, and evaporate to dryness. Dissolve the residues in 15 mL of water, filter if necessary, and dilute with water to 23 mL. Add 2 mL of 10% potassium dichromate reagent solution and add ammonium hydroxide until the orange color is just dissipated and the yellow color persists. Add 25 mL of methanol, stir vigorously, and allow to stand for 10 min. The turbidity in the solution of the sample should not exceed that in the control.

CALCIUM, MAGNESIUM, AND R_2O_3 PRECIPITATE.   To the filtrate from the test for insoluble matter add 5 mL of ammonium oxalate reagent solution, 2 mL of ammonium phosphate reagent solution, and 10 mL of ammonium hydroxide. Allow to stand overnight, filter, wash with water containing 2.5% ammonia and about 0.1% ammonium oxalate, and ignite.

HEAVY METALS. (Page 26, Method 1). Dissolve 6.0 g in about 20 mL of water, and dilute with water to 30 mL. Use 25 mL to prepare the sample solution, and use the remaining 5.0 mL to prepare the control solution.

IRON. (Page 28, Method 1). Use 2.0 g. In both sample and control solutions use 4 mL of hydrochloric acid.

SODIUM. Determine the sodium by the flame atomic absorption spectrophotometric method described on page 35.

> *Sample Stock Solution.* Dissolve 1.0 g of sample in water in a 100-mL volumetric flask and dilute to the mark with water (0.01 g/mL).

Analyze the solutions by means of a suitable atomic absorption spectrophotometer, using the conditions outlined in the following table. Calculate the metal content of the sample by the method of standard additions.

Element	Wavelength (nm)	Sample Wt (g)	Standard Added (mg)	Flame Type*	Background Correction
Na	589.0	0.05	0.01; 0.02	A/A	No

*A/A is air/acetylene.

Potassium Carbonate
Potassium Carbonate, Anhydrous

K_2CO_3 Formula Wt 138.21

CAS Number 584–08–7

REQUIREMENTS

Assay . . . . $\geq$ 99.0% K_2CO_3

MAXIMUM ALLOWABLE

Insoluble matter 0.01%
Chloride (as Cl) 0.003%
Nitrogen compounds (as N) 0.001%
Phosphate (PO_4) 0.001%
Silica (SiO_2). 0.005%
Sulfur compounds (as SO_4) 0.004%
Ammonium hydroxide precipitate 0.01%
Calcium and magnesium precipitate 0.01%
Arsenic (As). . . . . 1 ppm
Heavy metals (as Pb) . . . . 5 ppm
Iron (Fe) . . . . 5 ppm
Sodium (Na) 0.02%

TESTS

ASSAY.   (By acid–base titrimetry). Weigh, to the nearest 0.1 mg, 3 g of sample, transfer to a 125-mL glass-stoppered flask, and dissolve with 50 mL of water. Add 0.15 mL of methyl orange indicator solution and titrate with 1 N hydrochloric acid. One milliliter of 1 N hydrochloric acid corresponds to 0.0691 g of K_2CO_3.

INSOLUBLE MATTER.   (Page 14). Use 10 g dissolved in 100 mL of water.

CHLORIDE.   Dissolve 1.0 g in about 40 mL of water plus 6 mL of nitric acid. Filter through a chloride-free filter, wash, and dilute the filtrate with water to 60 mL. To 20 mL add 1 mL of silver nitrate reagent solution. Any turbidity should not exceed that produced by 0.01 mg of chloride ion (Cl) in an equal volume of solution containing the quantities of reagents used in the test.

NITROGEN COMPOUNDS.   (Page 30). Use 1.0 g. For the standard use 0.01 mg of nitrogen (N).

PHOSPHATE.   (Page 30, Method 2). Dissolve 1.0 g in 50 mL of water in a platinum dish and digest on the steam bath for 30 min. Cool, neutralize with dilute sulfuric acid (1 + 19) to a pH of about 4, and dilute with water to about 75 mL. Add 0.5 g of ammonium molybdate to the solution, and when it is dissolved, adjust the pH to 1.8 (using a pH meter) with dilute hydrochloric acid (1 + 9). Heat to boiling, cool, add 10 mL of hydrochloric acid, and dilute with water to 100 mL. Proceed as described. Carry along a standard containing 0.01 mg of phosphate ion (PO_4) and 0.05 mg of silica (SiO_2) in about 75 mL of water treated as the 75 mL of sample solution. Reserve the aqueous phase for the determination of silica.

SILICA. Add 10 mL of hydrochloric acid to the solutions reserved from the determination of phosphate and transfer to separatory funnels. Add 40 mL of butyl alcohol, shake vigorously, and allow to separate. Draw off and discard the aqueous phase. Wash the butyl alcohol three times with 20-mL portions of dilute hydrochloric acid (1 + 99), discarding the washings each time. Dilute each butyl alcohol solution to 50 mL with butyl alcohol, take 10 mL from each, and dilute each to 50 mL with butyl alcohol. Add 0.5 mL of a freshly prepared 2% solution of stannous chloride dihydrate in hydrochloric acid. The blue color in the extract from the sample should not exceed that in the control. If the butyl alcohol extracts are turbid, wash them with 10 mL of dilute hydrochloric acid (1 + 99).

SULFUR COMPOUNDS. (Page 32, Procedure A, Method 4). Allow 30 min for the turbidity to form.

AMMONIUM HYDROXIDE PRECIPITATE. Dissolve 10 g in 100 mL of water. Cautiously add 12 mL of sulfuric acid to 12 mL of water, cool, add the mixture to the solution of the sample, and evaporate to dense fumes. Cool, dissolve the residue in 130 mL of hot water, and add ammonium hydroxide until the solution is just alkaline to methyl red. Heat to boiling and filter, reserving the filtrate and washings for the test for calcium and magnesium. Continue to wash the precipitate, but reject the washings, and finally ignite. The residue includes some, but not all, of the silica in the sample.

CALCIUM AND MAGNESIUM PRECIPITATE. To the filtrate from the test for ammonium hydroxide precipitate add 0.5 mL of hydrochloric acid, 5 mL of ammonium oxalate reagent solution, 2 mL of ammonium phosphate reagent solution, and 10 mL of ammonium hydroxide. Allow to stand overnight. If any precipitate is formed, filter, wash with water containing 2.5% of ammonia and about 0.1% of ammonium oxalate, and ignite.

ARSENIC. (Page 23). Dissolve 3.0 g in a small volume of water, neutralize with sulfuric acid, and dilute with water to 35 mL. For the standard use 0.003 mg of arsenic (As).

HEAVY METALS. (Page 26, Method 1). To 5.0 g in a 150-mL beaker add 10 mL of water, mix, and cautiously add 10 mL of hydrochloric acid. Evaporate the solution to dryness on the steam bath, dissolve the residue in about 20 mL of water, and dilute with water to 25 mL.

For the control solution add 0.02 mg of lead ion (Pb) to 1.0 g of sample and treat exactly as the 5.0 g of sample.

IRON. (Page 28, Method 1). Dissolve 2.0 g in 20 mL of dilute hydrochloric acid (1 + 1), and evaporate to dryness on the steam bath. Dissolve the residue in 5 mL of dilute hydrochloric acid (1 + 1), and again evaporate to dryness. Dissolve the residue in 50 mL of dilute hydrochloric acid (1 + 24), and use the solution without further acidification. Use the residue from evaporation of 15 mL of hydrochloric acid to prepare the standard.

SODIUM. Determine the sodium by the flame atomic absorption spectrophotometric method described on page 35.

> *Sample Stock Solution.* Dissolve 1.0 g of sample with water in a 100-mL volumetric flask and dilute to the mark with water (0.01 g/mL).

Analyze the solutions by means of a suitable atomic absorption spectrophotometer, using the conditions outlined in the following table. Calculate the metal content of the sample by the method of standard additions.

Element	Wavelength (nm)	Sample Wt (g)	Standard Added (mg)	Flame Type*	Background Correction
Na	589.0	0.05	0.01; 0.02	A/A	No

*A/A is air/acetylene.

Potassium Carbonate Sesquihydrate

$K_2CO_3 \cdot 1\frac{1}{2}H_2O$ Formula Wt 165.23

CAS Number 6381−79−9

REQUIREMENTS

Assay . . . . 98.5−101.0%
$K_2CO_3 \cdot 1\frac{1}{2}H_2O$
Loss on heating at 285°C. . . . . 14.0−16.5%

MAXIMUM ALLOWABLE

Insoluble matter 0.01%
Chloride (as Cl) 0.003%
Nitrogen compounds (as N) 0.001%
Phosphate (PO_4) 0.001%
Silica (SiO_2). 0.005%
Sulfur compounds (as SO_4) 0.004%
Ammonium hydroxide precipitate 0.01%
Calcium and magnesium precipitate 0.01%
Arsenic (As). . . . . 0.5 ppm
Heavy metals (as Pb) . . . . 5 ppm
Iron (Fe) . . . . 5 ppm
Sodium (Na) 0.02%

TESTS

ASSAY. (By acidimetric titration of carbonate). Weigh accurately about 3.3 g and dissolve in 50 mL of water. Add 0.15 mL of methyl orange indicator solution, and titrate with 1 N hydrochloric acid to the change from yellow to orange. One milliliter of 1 N hydrochloric acid corresponds to 0.08262 g of $K_2CO_3 \cdot 1\frac{1}{2}H_2O$.

LOSS ON HEATING AT 285°C. Weigh accurately about 2 g and heat to constant weight at 270–300°C.

INSOLUBLE MATTER. (Page 14). Use 10 g dissolved in 100 mL of water.

CHLORIDE. Dissolve 1.0 g in about 40 mL of water plus 6 mL of nitric acid. Filter through a chloride-free filter, wash, and dilute the filtrate with water to 60 mL. To 20 mL add 1 mL of silver nitrate reagent solution. Any turbidity should not exceed that produced by 0.01 mg of chloride ion (Cl) in an equal volume of solution containing the quantities of reagents used in the test.

NITROGEN COMPOUNDS. (Page 30). Use 1.0 g. For the standard use 0.01 mg of nitrogen (N).

PHOSPHATE. (Page 30, Method 2). Dissolve 1.0 g in 50 mL of water in a platinum dish and digest on the steam bath for 30 min. Cool, neutralize with dilute sulfuric acid (1 + 19) to a pH of about 4, and dilute with water to about 75 mL. Add 0.5 g of ammonium molybdate to the solution, and when it is dissolved, adjust the pH to 1.8 (using a pH meter) with dilute hydrochloric acid (1 + 9). Heat the solution to boiling, cool to room temperature, add 10 mL of hydrochloric acid, and dilute with water to 100 mL. Proceed as described. Carry along a standard containing 0.01 mg of phosphate ion (PO_4) and 0.05 mg of silica (SiO_2) in about 75 mL of water treated as the 75 mL of sample solution. Reserve the aqueous phase for the determination of silica.

SILICA.   Add 10 mL of hydrochloric acid to the solutions reserved from the determination of phosphate and transfer to separatory funnels. Add 40 mL of butyl alcohol, shake vigorously, and allow to separate. Draw off and discard the aqueous phase. Wash the butyl alcohol three times with 20-mL portions of dilute hydrochloric acid (1 + 99), discarding the washings each time. Dilute the butyl alcohol solutions to 50 mL with butyl alcohol, take 10 mL of each, and dilute each to 50 mL with butyl alcohol. Add 0.2 mL of a freshly prepared 2% solution of stannous chloride dihydrate in hydrochloric acid. The blue color in the extract from the sample should not exceed that in the control. If the butyl alcohol extracts are turbid, wash them with 10 mL of dilute hydrochloric acid (1 + 99).

SULFUR COMPOUNDS.   (Page 32, Procedure A, Method 4). Allow 30 min for the turbidity to form.

AMMONIUM HYDROXIDE PRECIPITATE.   Dissolve 10 g in 100 mL of water. Cautiously add 12 mL of sulfuric acid to 12 mL of water, cool, add the mixture to the solution of the sample, and evaporate to strong fuming. Cool, dissolve the residue in 130 mL of hot water, and add ammonium hydroxide until the solution is just alkaline to methyl red. Heat to boiling and filter, reserving the filtrate without the washings for the test for calcium and magnesium precipitate. Wash with hot water, rejecting the washings, and ignite. The residue includes some, but not all, of the silica in the sample.

CALCIUM AND MAGNESIUM PRECIPITATE.   To the filtrate from the test for ammonium hydroxide precipitate add 0.5 mL of hydrochloric acid, 5 mL of ammonium oxalate reagent solution, 2 mL of ammonium phosphate reagent solution, and 10 mL of ammonium hydroxide. Allow to stand overnight. If any precipitate is formed, filter, wash with water containing 2.5% of ammonia and about 0.1% of ammonium oxalate, and ignite.

ARSENIC.   (Page 23). Dissolve 6.0 g in a small volume of water, neutralize with sulfuric acid, and dilute with water to 35 mL. For the standard use 0.003 mg of arsenic (As).

HEAVY METALS.   (Page 26, Method 1). To 5.0 g in a 150-mL beaker add 10 mL of water, mix, and add cautiously 10 mL of hydrochloric acid. Evaporate the solution to dryness on the steam bath, dissolve the residue in about 20 mL of water, and dilute with water to 25 mL. For the control solution add 0.02 mg of lead ion (Pb) to 1.0 g of sample and treat exactly as the 5.0 g of sample.

IRON.   (Page 28, Method 1). Dissolve 2.0 g in 20 mL of dilute hydrochloric acid (1 + 1), and evaporate to dryness on the steam bath. Dissolve the residue in 5 mL of dilute hydrochloric acid (1 + 1), and again evaporate to dryness. Dissolve the residue in 50 mL of dilute hydrochloric acid (1 + 24), and use the solution without further acidification. Use the residue from evaporation of 15 mL of hydrochloric acid to prepare the standard.

SODIUM.   Determine the sodium by the flame atomic absorption spectrophotometric method described on page 35.

> **Sample Stock Solution.**   Dissolve 1.0 g of sample with water in a 100-mL volumetric flask and dilute to the mark with water (0.01 g/mL).

Analyze the solutions by means of a suitable atomic absorption spectrophotometer, using the conditions outlined in the following table. Calculate the metal content of the sample by the method of standard additions.

Element	Wavelength (nm)	Sample Wt (g)	Standard Added (mg)	Flame Type*	Background Correction
Na	589.0	0.05	0.01; 0.02	A/A	No

*A/A is air/acetylene.

Potassium Chlorate

$KClO_3$ Formula Wt 122.55

CAS Number 3811−04−9

> *DANGER! STRONG OXIDIZER.* Causes irritation.

REQUIREMENTS

Assay . . . . $\geq$ 99.0% $KClO_3$

MAXIMUM ALLOWABLE

Insoluble matter 0.005%
Bromate (BrO_3) 0.015%
Chloride (Cl) 0.001%
Nitrogen compounds (as N) 0.001%
Sulfate (SO_4) . . . . Passes test (limit
about 0.002%)
Arsenic (As) . . . . 0.5 ppm
Calcium, magnesium, and
R_2O_3 precipitate 0.005%
Heavy metals (as Pb) . . . . 5 ppm
Iron (Fe) . . . . 3 ppm
Sodium (Na) 0.01%

TESTS

ASSAY. (By oxidation–reduction titration of chlorate). Weigh, to the nearest 0.1 mg, 0.05 g of sample, and dissolve it with 10 mL of water in a 250-mL glass-stoppered flask. Add 50.0 mL of 0.1 N acid ferrous sulfate, cover the flask to prevent exposure to the air, and boil for 10 min. Cool, add 10 mL of 10% manganous sulfate solution and 5 mL of phosphoric acid, and titrate the excess ferrous sulfate with 0.1 N potassium permanganate. Run a blank in the same manner.

$$\% \ KClO_3 = \frac{(V_b - V_s) \times N \ KMnO_4 \times 2.043}{\text{Sample wt (g)}}$$

where

V_s, V_b = mL of $KMnO_4$ in titration of blank and sample solutions, respectively

Acid Ferrous Sulfate Solution, 0.1 N. Dissolve 27.8 g of ferrous sulfate crystals in about 500 mL of water containing 50 mL of sulfuric acid, and dilute to 1 L.

INSOLUBLE MATTER. (Page 14). Use 20 g dissolved in 250 mL of water.

BROMATE. (Page 49). Determine bromate by differential pulse polarography. Use 1.0 g of sample and 0.5 g of calcium chloride dihydrate, $CaCl_2 \cdot 2H_2O$, in 25 mL of solution. For the standard add 0.15 mg of bromate (BrO_3). Also prepare a reagent blank.

CHLORIDE. Dissolve 2.0 g in 40 mL of warm water, and filter if necessary through a chloride-free filter. Add 0.25 mL of nitric acid, free from lower oxides of nitrogen, and 1 mL of silver nitrate reagent solution. Any turbidity should not exceed that produced by 0.02 mg of chloride ion (Cl) in an equal volume of solution containing the quantities of reagents used in the test.

NITROGEN COMPOUNDS. (Page 30). Use 1.0 g. For the standard use 0.01 mg of nitrogen (N).

SULFATE. Dissolve 5.0 g in 150 mL of water, filter if necessary, and add 1 mL of 10% hydrochloric acid reagent solution and 5 mL of barium chloride reagent solution. No precipitate should be produced on standing overnight.

ARSENIC. (Page 23). Dissolve 6.0 g in 30 mL of dilute hydrochloric acid (1 + 1) in a 150-mL beaker, and evaporate to dryness on the steam bath. Transfer the residue to a generator flask with the aid of 35 mL of water, and add 0.5 g of hydrazine sulfate. For the standard use 0.003 mg of arsenic (As).

CALCIUM, MAGNESIUM, AND R_2O_3 PRECIPITATE. Dissolve 10 g in 75 mL of dilute hydrochloric acid (1 + 2) and boil gently until no more chlorine is evolved. Dilute with water to about 80 mL, filter if necessary, heat to boiling, and add 5 mL of ammonium oxalate reagent solution, 2 mL of ammonium phosphate reagent solution, and 15 mL of ammonium hydroxide. Allow to stand overnight, filter, wash with water containing 2.5% of ammonia and about 0.1% of ammonium oxalate, and ignite.

HEAVY METALS. (Page 26, Method 1). Dissolve 5.0 g in 20 mL of dilute hydrochloric acid (1 + 1). Evaporate the solution to dryness on the steam bath, add 5 mL more of dilute hydrochloric acid (1 + 1), and again evaporate to dryness. Dissolve the residue in about 20 mL of water and dilute with water to 25 mL. For the control solution add 0.02 mg of lead ion (Pb) to 1.0 g of sample and treat exactly as the 5.0 g of sample.

IRON. (Page 28). Dissolve 3.3 g in 20 mL of dilute hydrochloric acid (1 + 1) and evaporate to dryness on the steam bath. Add 5 mL of dilute hydrochloric acid (1 + 1) and again evaporate to dryness. Dissolve in 50 mL of dilute hydrochloric acid (1 + 25), and use the

solution without further acidification. In preparing the control use
the residue from evaporation of 15 mL of hydrochloric acid.

SODIUM. Determine the sodium by the flame atomic absorption
spectrophotometric method described on page 35.

> *Sample Stock Solution.* Dissolve 1.0 g in 25 mL of dilute
> hydrochloric acid (1 + 3), and digest in a covered beaker on the
> steam bath until the reaction ceases. Uncover the beaker and
> evaporate to dryness. Add 10 mL of dilute hydrochloric acid
> (1 + 3) and again evaporate to dryness. Dissolve in dilute hydro-
> chloric acid (1 + 99) and dilute to 100 mL with the hydrochloric
> acid (1 + 99) (0.01 g/mL).

Analyze the solutions by means of a suitable atomic absorption
spectrophotometer, using the conditions outlined in the following
table. Calculate the metal content of the sample by the method of
standard additions.

Element	Wavelength (nm)	Sample Wt (g)	Standard Added (mg)	Flame Type*	Background Correction
Na	589.0	0.01	0.01; 0.02	A/A	No

*A/A is air/acetylene.

Potassium Chloride

KCl   Formula Wt 74.55

CAS Number 7447−40−7

REQUIREMENTS

Assay . . . . 99.0−100.5% KCl
pH of a 5% solution at 25°C . . . . 5.4−8.6

MAXIMUM ALLOWABLE

Insoluble matter	0.005%
Iodide (I)	0.002%
Bromide (Br)	Passes test (limit about 0.01%)
Chlorate and nitrate (as NO_3)	0.003%
Nitrogen compounds (as N)	0.001%
Phosphate (PO_4)	5 ppm
Sulfate (SO_4)	0.001%
Barium (Ba)	Passes test (limit about 0.001%)
Calcium, magnesium, and R_2O_3 precipitate	0.005%
Heavy metals (as Pb)	5 ppm
Iron (Fe)	3 ppm
Sodium (Na)	0.005%

TESTS

ASSAY. (By argentimetric titration of chloride content). Weigh, to the nearest 0.1 mg, about 0.25 g of sample, and dissolve with 50 mL of water in a 250-mL glass-stoppered flask. Add, while agitating, exactly 50.0 mL of 0.1 N silver nitrate reagent solution, then add 3 mL of nitric acid and 10 mL of benzyl alcohol, and shake vigorously. Add 2 mL of ferric ammonium sulfate reagent solution (as indicator), and titrate the excess silver nitrate with 0.1 N ammonium thiocyanate.

$$\% \text{ KCl} = \frac{[(50.0 \text{ mL} \times \text{N AgNO}_3) - (\text{mL} \times \text{N NH}_4\text{SCN})] \times 7.455}{\text{Sample wt (g)}}$$

pH OF A 5% SOLUTION. (Page 43). The pH should be 5.4–8.6 at 25°C.

INSOLUBLE MATTER. (Page 14). Use 20 g dissolved in 150 mL of water. Save the filtrate separate from the washings for the test for calcium, magnesium, and R_2O_3 precipitate.

IODIDE. Dissolve 11 g in 50 mL of water. Prepare a control by dissolving 1 g of the sample, 0.2 mg of iodide ion (I), and 1 mg of bromide ion (Br) in 50 mL of water. To each solution, in a separatory funnel, add 2 mL of hydrochloric acid and 5 mL of ferric chloride reagent solution. Allow to stand for 5 min. Add 10 mL of carbon tetrachloride, shake for 1 min, allow the phases to separate, and draw off the carbon tetrachloride layer. Reserve the water solution for the test for bromide. Any violet color in the carbon tetrachloride extract from the solution of the sample should not exceed that in the extract from the control.

BROMIDE. Treat both the solution of the sample and the control obtained in the test for iodide as follows: wash twice by shaking with 10-mL portions of carbon tetrachloride. Each time allow to separate, then draw off and discard the carbon tetrachloride. To each of the water solutions add 10 mL of water, 65 mL of cold dilute sulfuric acid (1 + 1), and 15 mL of a solution of chromic acid prepared by dissolving 10 g in 100 mL of dilute sulfuric acid (1 + 3). Allow to stand for 5 min. Add 10 mL of carbon tetrachloride, shake for 1 min, allow to settle, and draw off. (Half a 7-cm piece of filter paper rolled and placed in the stem of the separatory funnel will absorb any of the aqueous solution that may pass the stopcock, and thus assure a clear extract.) Any yellow-brown color in the carbon tetrachloride extract from the solution of the sample should not exceed that in the extract from the control.

CHLORATE AND NITRATE

> *Sample Solution A.* Dissolve 0.50 g in 3 mL of water by heating in a boiling-water bath. Dilute to 50 mL with brucine sulfate reagent solution.

> *Control Solution B.* Dissolve 0.50 g in 1.5 mL of water and 1.5 mL of the standard nitrate solution containing 0.01 mg of nitrate ion (NO_3) per mL by heating in a boiling-water bath. Dilute to 50 mL with brucine sulfate reagent solution.

> *Blank Solution C.* Use 50 mL of brucine sulfate reagent solution.

Heat the three solutions in a preheated (boiling) water bath for 10 min. Cool rapidly in an ice bath to room temperature. Set a spectrophotometer at 410 nm and, using 1-cm cells, adjust the instrument to read 0 absorbance with blank solution C in the light path, then determine the absorbance of sample solution A. Adjust the instrument to read 0 absorbance with sample solution A in the light path and determine the absorbance of control solution B. The absorbance of sample solution A should not exceed that of control solution B.

NITROGEN COMPOUNDS. (Page 30). Use 1.0 g. For the standard use 0.01 mg of nitrogen (N).

PHOSPHATE. (Page 30, Method 1). Dissolve 4.0 g in 25 mL of approximately 0.5 N sulfuric acid, and continue as described.

SULFATE. Dissolve 25 g in 200 mL of water, add 2 mL of hydrochloric acid, heat to boiling, add 10 mL of barium chloride reagent solution, digest in a covered beaker on a hot plate at low setting for 2 h, and allow to stand overnight. If a precipitate is formed, filter, wash thoroughly, and ignite. Correct for the weight obtained in a complete blank test.

BARIUM. Dissolve 4.0 g in 20 mL of water, filter if necessary, and divide into two portions. To one portion add 2 mL of 10% sulfuric acid reagent solution and to the other 2 mL of water. The solutions should be equally clear at the end of 2 h.

CALCIUM, MAGNESIUM, AND R$_2$O$_3$ PRECIPITATE. To the filtrate from the test for insoluble matter add 5 mL of ammonium oxalate reagent solution, 2 mL of ammonium phosphate reagent solution, and 10 mL of ammonium hydroxide. Allow to stand overnight. Filter, wash with water containing 2.5% ammonia and about 0.1% ammonium oxalate, and ignite.

HEAVY METALS. (Page 26, Method 1). Dissolve 6.0 g in about 20 mL of water and dilute with water to 30 mL. Use 25 mL to prepare the sample solution, and use the remaining 5.0 mL to prepare the control solution.

IRON. (Page 28, Method 1). Use 3.3 g.

SODIUM. Determine the sodium by the flame atomic absorption spectrophotometric method described on page 35.

> *Sample Stock Solution.* Dissolve 1.0 g of sample in water and dilute to the mark with water in a 100-mL volumetric flask (0.01 g/mL).

Analyze the solutions by means of a suitable atomic absorption spectrophotometer, using the conditions outlined in the following table. Calculate the metal content of the sample by the method of standard additions.

Element	Wavelength (nm)	Sample Wt (g)	Standard Added (mg)	Flame Type*	Background Correction
Na	589.0	0.01	0.005; 0.01	A/A	No

*A/A is air/acetylene.

Potassium Chromate

K_2CrO_4 Formula Wt 194.19

CAS Number 7789–00–6

REQUIREMENTS

Assay . . . . $\geq$ 99.0% K_2CrO_4
pH of a 5% solution at 25°C 8.6–9.8

MAXIMUM ALLOWABLE

Insoluble matter 0.005%
Chloride (Cl) 0.005%
Sulfate (SO_4) 0.03%
Calcium (Ca) 0.005%
Sodium (Na) 0.02%

TESTS

ASSAY. (By iodometric oxidation–reduction titration). Weigh, to the nearest 0.1 mg, 0.25 g of sample, and dissolve in 200 mL of carbon dioxide-free water in a 500-mL glass-stoppered iodine flask. Add 3 g of potassium iodide and 7 mL of hydrochloric acid, and allow to stand in the dark for 10 min. Titrate the liberated iodine with 0.1 N sodium thiosulfate, adding 5 mL of starch indicator near the end point.

$$\% \ K_2CrO_4 = \frac{(mL \times N \ Na_2S_2O_3) \times 6.473}{\text{Sample wt (g)}}$$

pH OF A 5% SOLUTION. (Page 43). The pH should be 8.6–9.8 at 25°C.

INSOLUBLE MATTER. (Page 14). Use 20 g dissolved in 150 mL of water.

CHLORIDE. Dissolve 0.20 g in 10 mL of water, filter if necessary through a small chloride-free filter, and add 1 mL of ammonium hydroxide, 1 mL of silver nitrate reagent solution, and 2 mL of nitric acid. Any turbidity should not exceed that produced by 0.01 mg of chloride ion (Cl) in an equal volume of solution containing the quantities of reagents used in the test. The comparison is best made by the general method for chloride in colored solutions, page 25.

SULFATE. (Page 32, Procedure B). Transfer 0.50 g to a reaction tube, and dissolve in 1 mL of water. Add 2 mL of hydrochloric acid and swirl. Evaporate to dryness on a hot plate at low setting. Repeat, dissolve the residue in 2 mL of water, and dilute with water to 10 mL. Transfer 2.0 mL to a reaction tube, and continue with the procedure.

CALCIUM AND SODIUM. Determine the calcium and sodium by the flame atomic absorption spectrophotometric method described on page 35.

> *Sample Stock Solution.* Dissolve 5.0 g in water and dilute to 100 mL in a volumetric flask with water (0.05 g/mL).

Analyze the solutions by means of a suitable atomic absorption spectrophotometer, using the conditions outlined in the following table. Calculate the metal content of the sample by the method of standard additions.

Element	Wavelength (nm)	Sample Wt (g)	Standard Added (mg)	Flame Type*	Background Correction
Ca	422.7	0.5	0.025; 0.05	N/A	No
Na	589.0	0.05	0.01; 0.02	A/A	No

*A/A is air/acetylene.

Potassium Cyanide

KCN   Formula Wt 65.12

CAS Number 151–50–8

REQUIREMENTS

Assay . . . . ≥ 96.0% KCN

MAXIMUM ALLOWABLE

Chloride (Cl) 0.5%
Phosphate (PO_4) 0.005%
Sulfate (SO_4) 0.04%
Sulfide (S) 0.003%
Thiocyanate (SCN) . . . . Passes test (limit
                                                              about 0.02%)
Iron, total (as Fe) 0.03%
Lead (Pb) . . . . 2 ppm
Sodium (Na) 0.5%

TESTS

> *CAUTION*. Because of the extremely poisonous nature of potassium cyanide and of the hydrogen cyanide gas, HCN, evolved on treatment of the salt or solution of the salt with an acid, all tests must be made in a fume hood with a strong draft. Special care must be taken to avoid inhaling any of the fumes or allowing the salt or solution of the salt to come in contact with open cuts of the skin. If safety pipets are not available for measuring aliquots, use a graduated cylinder. Under no conditions should suction by mouth be used to fill an ordinary pipet.

ASSAY. (By argentimetric titration of cyanide content). Weigh accurately 0.5 g and dissolve in 30 mL of water. Add 0.2 mL of 10% potassium iodide reagent solution and 1 mL of ammonium hydroxide and titrate with 0.1 N silver nitrate to a slight yellowish permanent turbidity. One milliliter of 0.1 N silver nitrate corresponds to 0.01302 g of KCN.

> *Sample Solution A.* Dissolve 10 g in water and dilute with water to 200 mL. Filter, if necessary, under a hood into a dry flask (1 mL = 0.05 g).

CHLORIDE. (Page 25). Dilute 1.0 mL of sample solution A with water to 50 mL. To 2.0 mL (0.002-g sample) of this solution add 2 mL of 30% hydrogen peroxide and allow to stand in a covered beaker until reaction ceases, then digest in the covered beaker on the steam bath for 20–30 min. Cool, and dilute with water to 25 mL.

PHOSPHATE. (Page 30, Method 1). To 20 mL of sample solution A add 2 mL of hydrochloric acid, and evaporate to dryness in a well-ventilated hood. Add 5 mL of dilute hydrochloric acid (1 + 1) and evaporate to dryness again. Dissolve in 50 mL of approximately 0.5 N sulfuric acid. To 20 mL of the solution add 5 mL of approximately 0.5 N sulfuric acid, and continue as described.

SULFATE. (Page 32, Procedure A, Method 2). Use 2.5 mL of sample solution A plus 1 mL of dilute hydrochloric acid (1 + 1) in a hood.

SULFIDE. To 20 mL of sample solution A (measured in a graduated cylinder) (1-g sample) add 0.15 mL of alkaline lead solution (made by adding 10% sodium hydroxide reagent solution to a 10% lead acetate reagent solution until the precipitate is redissolved). The color should not be darker than is produced by 0.03 mg of sulfide ion (S) in an equal volume of solution when treated with 0.15 mL of the alkaline lead solution.

THIOCYANATE. To 20 mL of sample solution A (measured in a graduated cylinder) (1-g sample) add 4 mL of hydrochloric acid and 0.20 mL of ferric chloride reagent solution. At the end of 5 min the solution should show no reddish tint when compared with 20 mL of water to which have been added the quantities of hydrochloric acid and ferric chloride used in the test.

IRON, TOTAL. (Page 28, Method 1). Transfer 10 mL of sample solution A (0.5-g sample) to a platinum evaporating dish, add 3 mL of hydrochloric acid, and evaporate to dryness in a well-ventilated hood. Heat the residue at 650°C for 30 min. Cool, add 2 mL of hydrochloric acid and 10 mL of water, and digest on the steam bath until dissolution is complete. Dilute with water to 30 mL, and use 2.0 mL of this solution.

LEAD. Dissolve 1.2 g in 10 mL of water in a separatory funnel. Add 5 mL of ammonium citrate reagent solution, 2 mL of hydroxylamine hydrochloride reagent solution for the dithizone test, and 0.10 mL of phenol red indicator solution, and make the solution alkaline if necessary by the addition of ammonium hydroxide. Add 5 mL of standard dithizone solution in chloroform, shake gently but well for 1 min, and allow the layers to separate. The intensity of the red color of the chloroform layer should be no greater than that of a control made with 0.002 mg of lead ion (Pb) and 0.2 g of the sample treated exactly like the solution of 1.2 g of sample in 10 mL of water.

SODIUM. Determine the sodium by the flame atomic absorption spectrophotometric method described on page 35.

Sample Stock Solution. Dissolve 0.5 g of sample in a 100-mL volumetric flask and dilute to the mark with water (0.005 g/mL).

Analyze the solution by means of a suitable atomic absorption spectrophotometer, using the conditions outlined in the following table.
Calculate the metal content of the sample by the method of standard
additions.

Element	Wavelength (nm)	Sample Wt (g)	Standard Added (mg)	Flame Type*	Background Correction
Na	589.0	0.01	0.05; 0.10	A/A	No

*A/A is air/acetylene.

Potassium Dichromate

$K_2Cr_2O_7$ Formula Wt 294.18

CAS Number 7778−50−9

REQUIREMENTS

Assay . . . . ≥ 99.0% $K_2Cr_2O_7$

MAXIMUM ALLOWABLE

Insoluble matter and ammonium hydroxide
  precipitate 0.005%
Loss on drying 0.05%
Chloride (Cl) 0.001%
Sulfate (SO_4) 0.005%
Calcium (Ca) 0.003%
Sodium (Na) 0.02%

TESTS

ASSAY.   (By iodometric oxidation−reduction titration). Weigh, to the
nearest 0.1 mg, 9.2 g of sample, transfer to a 500-mL glass-stoppered
iodine flask, and dissolve with 125 mL of water. Add 5 g of potassium
iodide and 5 mL of hydrochloric acid $(1 + 1)$ with constant swirling.
Carefully wash down the sides of the flask with water so that a layer
of water is formed on top of the solution, stopper the flask, and allow

to stand in the dark for 10 min. Add about 160 mL of water and titrate with 0.1 N sodium thiosulfate, adding 5 mL of starch indicator solution near the end point.

$$\% \ K_2Cr_2O_7 = \frac{(mL \times N \ Na_2SO_3) \times 4.903}{Sample \ wt \ (g)}$$

INSOLUBLE MATTER AND AMMONIUM HYDROXIDE PRECIPITATE. Dissolve 20 g in 200 mL of water, add 2 mL of ammonium hydroxide, heat to boiling, and digest in a covered beaker on the steam bath for 1 h. Filter, wash thoroughly, and ignite. Retain the filtrate for the determination of calcium.

LOSS ON DRYING.   Weigh accurately about 2 g and dry to constant weight at 105°C.

CHLORIDE.   Dissolve 1.0 g in 10 mL of water, filter if necessary through a small chloride-free filter, and add 1 mL of ammonium hydroxide and 1 mL of silver nitrate reagent solution. Prepare a standard containing 0.01 mg of chloride ion (Cl) in 10 mL of water, and add 1 mL of ammonium hydroxide and 1 mL of silver nitrate reagent solution. Add 2 mL of nitric acid to each. The comparison is best made by the general method for chloride in colored solutions, page 25.

SULFATE.   Dissolve 10 g in 250 mL of water, filter if necessary, and heat to boiling. Add 25 mL of a solution containing 1 g of barium chloride and 2 mL of hydrochloric acid per 100 mL of solution. Digest in a covered beaker on a hot plate at low setting for 2 h and allow to stand overnight. If a precipitate is formed, filter, wash thoroughly, and ignite. Fuse the residue with 1 g of sodium carbonate. Extract the fused mass with water and filter off the insoluble residue. Add 5 mL of hydrochloric acid to the filtrate, dilute with water to about 200 mL, heat to boiling, and add 10 mL of alcohol. Digest in a covered beaker on the hot plate at low setting until reduction of chromate is complete, as indicated by the change to a clear green or colorless solution. Neutralize the solution with ammonium hydroxide and add 2 mL of hydrochloric acid. Heat to boiling and add 10 mL of barium chloride reagent solution. Digest in a covered beaker on a hot plate at low setting for 2 h and allow to stand overnight. Filter, wash thoroughly, and ignite. Correct for the weight obtained in a complete blank test. If the original precipitate of barium sulfate weighs less than the requirement allows, the fusion with sodium carbonate is not necessary.

CALCIUM AND SODIUM.   Determine the calcium and sodium by the flame atomic absorption spectrophotometric method described on page 35.

> *Sample Stock Solution.*   Dissolve 5.0 g of sample in 20 mL of water, transfer the solution to a 100-mL volumetric flask, and dilute to the mark with water (0.05 g/mL).

Analyze the solutions by means of a suitable atomic absorption spectrophotometer, using the conditions outlined in the following table. Calculate the metal content of the sample by the method of standard additions.

Element	Wavelength (nm)	Sample Wt (g)	Standard Added (mg)	Flame Type*	Background Correction
Ca	422.7	0.5	0.015; 0.03	N/A	No
Na	589.0	0.05	0.01; 0.02	A/A	No

*A/A is air/acetylene; N/A is nitrous oxide/acetylene.

Potassium Ferricyanide
Potassium Hexacyanoferrate(III)

$K_3Fe(CN)_6$ Formula Wt 329.25

CAS Number 13746–66–2

REQUIREMENTS

Assay . . . . $\geq$ 99.0%
$K_3Fe(CN)_6$

MAXIMUM ALLOWABLE

Insoluble matter 0.005%
Chloride (Cl) 0.01%
Sulfate (SO_4) 0.01%
Ferro compounds . . . . 0.05% as
ferrocyanide radi-
cal $[Fe(CN)_6]^{4-}$

TESTS

ASSAY. (By iodometric titration). Weigh accurately 1.4 g and in a glass-stoppered conical flask dissolve in 50 mL of water. Add 3 g of potassium iodide and 1 drop of glacial acetic acid. Then add 1.5 g of zinc sulfate heptahydrate, $ZnSO_4 \cdot 7H_2O$, in 10 mL of water. Stopper, swirl, and allow to stand in the dark for 30 min. Titrate the liberated iodine with 0.1 N sodium thiosulfate, adding 3 mL of starch indicator solution near the end of the titration. Correct for a blank. One milliliter of 0.1 N sodium thiosulfate corresponds to 0.03292 g of $K_3Fe(CN)_6$.

INSOLUBLE MATTER. Dissolve 20 g in 200 mL of water at room temperature, filter promptly through a tared filtering crucible, wash thoroughly, and dry at 105°C.

CHLORIDE. Dissolve 2.0 g in 175 mL of water in a cylinder, add 2.5 g of cupric sulfate crystals dissolved in 25 mL of water, mix thoroughly, and allow the precipitate to settle. Filter the supernatant liquid, if necessary, through a chloride-free filter. To 10 mL of the clear solution add 1 mL of nitric acid, 10 mL of water, and 1 mL of silver nitrate reagent solution. Any turbidity should not exceed that produced by 0.01 mg of chloride ion (Cl) in an equal volume of solution containing the quantities of nitric acid and silver nitrate used in the test and enough cupric sulfate to match the color in the test solution.

SULFATE. Dissolve 5.0 g in 100 mL of water without heating, filter promptly, and to 10 mL (0.5 g) of the filtrate add 0.2 mL of glacial acetic acid and 5 mL of barium chloride reagent solution. Any turbidity should not exceed that produced by a standard containing 0.05 mg of sulfate ion (SO_4) in an equal volume of solution containing 0.2 mL of glacial acetic acid and 5 mL of barium chloride reagent solution. Compare 10 min after adding the barium chloride to the sample and standard solutions.

FERRO COMPOUNDS. Mix 400 mL of water with 10 mL of 25% sulfuric acid reagent solution and add 0.1 N potassium permanganate until the pink color persists for 1 min. Dissolve 4.0 g of the sample in the solution, add 0.10 mL of the 0.1 N potassium permanganate solution, and stir. The solution should retain a pink tint in comparison with a blank prepared with the same quantities of potassium ferricyanide, water, and sulfuric acid.

Potassium Ferrocyanide Trihydrate
Potassium Hexacyanoferrate(II) Trihydrate

$K_4Fe(CN)_6 \cdot 3H_2O$ Formula Wt 422.39

CAS Number 14459–95–1

REQUIREMENTS

Assay 98.5–102.0%
$K_4Fe(CN)_6 \cdot$
$3H_2O$

MAXIMUM ALLOWABLE

Insoluble matter 0.005%
Chloride (Cl) 0.01%
Sulfate (SO_4) . . . . Passes test (limit
about 0.01%)

TESTS

ASSAY. (By titration of reducing power). Weigh accurately about 1.6 g and dissolve in 250 mL of water. Add 30 mL of 25% sulfuric acid, and titrate with 0.1 N potassium permanganate to a change from yellow to orange-yellow. One milliliter of 0.1 N potassium permanganate corresponds to 0.04224 g of $K_4Fe(CN)_6 \cdot 3H_2O$.

INSOLUBLE MATTER. (Page 14). Use 20 g dissolved in 200 mL of water.

CHLORIDE. Dissolve 2.0 g in 175 mL of water, filter if necessary through a chloride-free filter, add 2.5 g of cupric sulfate crystals dissolved in 25 mL of water, mix thoroughly, and allow the precipitate to settle in a cylinder. To 10 mL of the clear supernatant liquid add 1 mL of nitric acid, 10 mL of water, and 1 mL of silver nitrate reagent solution. Any turbidity should not exceed that produced by 0.01 mg of chloride ion (Cl) in an equal volume of solution containing the quantities of nitric acid and silver nitrate used in the test and enough cupric sulfate to match the color in the test solution.

SULFATE. Dissolve 5.0 g in 100 mL of water without heating, filter promptly, and to 10 mL (0.5 g) of the filtrate add 0.2 mL of glacial acetic acid and 5 mL of barium chloride reagent solution. Any turbidity should not exceed that produced by a standard containing 0.05 mg of sulfate ion (SO_4) in an equal volume of solution containing 0.2 mL of glacial acetic acid and 5 mL of barium chloride reagent solution. Compare 10 min after adding the barium chloride to the sample and standard solutions.

Potassium Fluoride

KF   Formula Wt 58.10

CAS Number 7789–23–3

REQUIREMENTS

Assay . . . . ≥ 99.0% KF

MAXIMUM ALLOWABLE

Chloride (Cl) 0.005%
Titrable acid . . . . 0.03 meq / g
Titrable base . . . . 0.01 meq / g
Potassium fluosilicate (K_2SiF_6). 0.1%
Sulfate (SO_4). 0.005%
Heavy metals (as Pb) 0.001%
Iron (Fe) 0.001%
Sodium (Na) 0.2%

TESTS

ASSAY.   (By indirect acid–base titration).

Ion-Exchange Column: Use Amberlite IR-120 or the equivalent. Charge the resin to a polyethylene column with a 2-cm bore to a height of about 20 cm. If the resin is in the hydrogen ion form, wash free of acid and color with carbon dioxide-free water. Otherwise regenerate by passing 4 N hydrochloric acid through the column and washing free of the hydrochloric acid.

NOTE. All water used in the assay must be carbon dioxide- and ammonia-free.

Weigh accurately 0.2 g of sample and dissolve in 50 mL of water in a polyethylene vessel. Pass the solution through the cation column at a rate of about 5 mL/min and collect the eluate in a 500-mL polyethylene titration flask. Then wash the resin in the column with water at a rate of about 10 mL/min, and collect in the same titration flask. Add 0.15 mL of phenolphthalein indicator solution to the flask and titrate with 0.1 N sodium hydroxide. Continue the titration as the washing proceeds until 50 mL of eluate requires no further titration. One milliliter of 0.1 N sodium hydroxide corresponds to 0.005810 g of KF.

CHLORIDE.   (Page 25). Dissolve 1.0 g plus 0.2 g of boric acid in 80 mL of water, filter if necessary through a chloride-free filter, and dilute to 100 mL with water. Use 20 mL (0.2-g sample) of this solution.

TITRABLE ACID.   Dissolve 5.0 g in 40 mL of carbon dioxide-free water in a platinum dish, add 10 mL of a freshly-prepared saturated solution of potassium nitrate, cool the solution to 0 °C, and add 0.15 mL of phenolphthalein indicator solution. If a pink color is produced, omit the titration for titrable acid and reserve the solution for titrable base. If no pink color is produced, titrate with 0.1 N sodium hydroxide until the pink color persists for 15 s while the temperature is near 0°C. Not more than 1.50 mL of the 0.1 N sodium hydroxide should be required. Reserve the solution for the test for potassium fluosilicate.

TITRABLE BASE.   If a pink color was produced in the solution prepared for the test for titrable acid, add 0.1 N hydrochloric acid, stirring the liquid only gently, until the pink color is discharged. Not more than 0.50 mL of the 0.1 N hydrochloric acid should be required. Reserve the solution for the test for potassium fluosilicate.

POTASSIUM FLUOSILICATE.   Heat the solution reserved from the test for titrable acid or titrable base to boiling and titrate while hot with 0.01 N sodium hydroxide to a permanent pink color. Not more than 9.0 mL of 0.01 N sodium hydroxide should be required.

SULFATE.   (Page 32, Procedure A, Method 2). Use 1.0 g, and evaporate with 2 mL of hydrochloric acid in a platinum dish. Repeat four times.

HEAVY METALS.   (Page 26, Method 1). Use 2.0 g. Adjust the pH, using pH indicator paper.

IRON.   (Page 28, Method 1). To 1.0 g in a platinum dish add 10 mL of hydrochloric acid, and evaporate to dryness. Dissolve the residue in 10 mL of hydrochloric acid, and repeat the evaporation to dryness.

Warm the residue with a few drops of hydrochloric acid, and dilute with water to 40 mL. In preparing the standard, use the residue from the evaporation of 20 mL of hydrochloric acid.

SODIUM. Determine the sodium by the flame atomic absorption spectrophotometric method described on page 35.

Sample Stock Solution. Dissolve 0.1 g in 100 mL of water, using suitable plastic volumetric flasks (1 mL = 0.001 g).

Analyze the solutions by means of a suitable atomic absorption spectrophotometer, using the conditions outlined in the following table. Calculate the metal content of the sample by the method of standard additions.

Element	Wavelength (nm)	Sample Wt (g)	Standard Added (mg)	Flame Type*	Background Correction
Na	589.0	0.005	0.005; 0.01	A/A	No

*A/A is air/acetylene.

Potassium Hydrogen Phthalate, Acidimetric Standard
Potassium Acid Phthalate
Potassium Biphthalate
Phthalic Acid, Monopotassium Salt

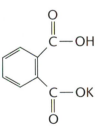

HOCOC$_6$H$_4$COOK   Formula Wt 204.22

CAS Number 877–24–7

NOTE. This reagent is satisfactory for use as a pH standard. For use as an acidimetric standard, this material should be lightly crushed and dried for 2 h at 120°C to remove any absorbed moisture.

REQUIREMENTS

Assay (dried basis) . . . . 99.95 –100.05%

$C_8H_5O_4K$

pH of a 0.05 *m* solution at 25.0 ± 0.2°C. . . . . 4.00 –4.02

MAXIMUM ALLOWABLE

Insoluble matter 0.005%
Chlorine compounds (as Cl) 0.003%
Sulfur compounds (as S) 0.002%
Heavy metals (as Pb) . . . . 5 ppm
Iron (Fe) . . . . 5 ppm
Sodium (Na) 0.005%

TESTS

ASSAY.   (By acid–base titrimetry). This is a comparative procedure in which the potassium hydrogen phthalate to be assayed is compared with NIST SRM Potassium Hydrogen Phthalate. Accurately weighed portions of anhydrous sodium carbonate, dissolved in water, are allowed to react with accurately weighed portions of the two potassium hydrogen phthalate samples of such size as to provide a small excess of the latter. The carbon dioxide is removed by boiling and the excess potassium hydrogen phthalate is determined by titration with carbonate-free sodium hydroxide using a potentiometric end point.

CAUTION. Due to the small tolerance in assay limits for this acidimetric standard, extreme care must be observed in the weighing, transferring, and titrating operations in the following procedure. Strict adherence to the specified sample weights and final titration volumes is absolutely necessary. It is recommended that the titrations be run at least in duplicate on both the sample and the SRM. Duplicate values of M, the number of milliequivalents of total alkali required per gram of potassium hydrogen phthalate, obtained on the sample and on the SRM, should agree within 2 parts in 5000 to be acceptable for averaging.

Procedure.   Place duplicate 1.300 ± 0.001-g portions of a uniform sample of anhydrous sodium carbonate, Na_2CO_3, in clean, dry weighing bottles. Dry at 285°C for 3 h, stopper the bottles, and cool in a desiccator for at least 2 h. Open the weighing bottles momentarily to equalize the pressure, weigh each bottle and its contents to the nearest 0.1 mg, and transfer the entire contents of each bottle to a clean, dry 50-mL beaker. Store the beaker in a desiccator until ready for use, weigh each bottle again, and determine by difference, to the nearest 0.1 mg, the weight of each portion of sodium carbonate.

Transfer the portions of sodium carbonate to 500-mL titration flasks* by pouring them through powder funnels. Rinse the beakers, funnels, and sides of the flasks thoroughly with several small portions of water. In the same manner add to one of these flasks 5.100 ± 0.003 g of the potassium hydrogen phthalate to be assayed and to the other flask the same weight of the NIST SRM potassium hydrogen phthalate—both of which have been previously crushed (not ground) in an agate or mullite mortar to approximately 100-mesh fineness, dried at 120°C for 2 h, and cooled in a desiccator for at least 2 h. Dilute the contents of each titration flask with water to about 90 mL, swirl to dissolve the samples, stopper, and bubble carbon dioxide-free air through the solutions at a moderately fast rate during all subsequent operations. Boil the solutions gently for 20 min to complete the removal of carbon dioxide, taking care to prevent loss of solutions by spraying. Cool the solutions to room temperature in an ice bath, and dilute to 75–80 mL with carbon dioxide-free water. Titrate each solution with 0.02 N carbonate-free sodium hydroxide using a combination electrode (glass–calomel or glass–silver, silver chloride) for the measurement of E in millivolts or pH. The end point of the titration is determined by the second-derivative method (page 51).

Calculations.   Calculate the number of milliequivalents of total alkali (sodium carbonate plus sodium hydroxide) required to neutralize exactly 1 g of the sample of potassium hydrogen phthalate and of the NIST SRM by the following equations:

$$M_u = \frac{\dfrac{B}{0.052994} + (C \times D)}{W_u}$$

$$M_s = \frac{\dfrac{B}{0.052994} + (C \times D)}{W_s \times F}$$

---

*A titration flask is a 500-mL conical flask fitted with a small glass tube sealed through the side. The tube inside the flask terminates in a constricted tip positioned as close to the bottom as possible. The tube outside the flask is bent at a right angle to permit the attachment of the carbon dioxide-free air supply.

where

M_u = number of milliequivalents of total alkali required per gram of potassium hydrogen phthalate being assayed

M_s = number of milliequivalents of total alkali required per gram of NIST standard potassium hydrogen phthalate

B = weight, in grams, of sodium carbonate

C = volume, in mL, of NaOH standard solution

D = normality of NaOH standard solution

W_u = weight, in grams, of potassium hydrogen phthalate being assayed

W_s = weight, in grams, of NIST SRM potassium hydrogen phthalate

F = assay value of NIST SRM in percent divided by 100

Calculate the assay value of the sample of potassium hydrogen phthalate as follows:

$$A = \frac{M_u \times 100}{M_s}$$

where

A = percent of potassium hydrogen phthalate in the sample assayed and M_u and M_s are defined as before.

pH OF A 0.05 *m* SOLUTION.   Dissolve 1.021 g in 100 g of carbon dioxide- and ammonia-free water (or 1.012 g in a volume of 100 mL). Standardize the pH meter and electrodes at pH 4.00 at 25.0 ± 0.2 °C with 0.05 *m* NIST SRM Potassium Hydrogen Phthalate. Determine the pH of the solution by the method described on page 43. The pH should be 4.00–4.02 at 25.0 ± 0.2°C.

INSOLUBLE MATTER.   (Page 14). Use 20 g dissolved in 200 mL of water.

CHLORINE COMPOUNDS.   Mix 1.0 g with 0.5 g of sodium carbonate, moisten the mixture, and ignite until thoroughly charred, avoiding an unduly high temperature. Treat the residue with 30 mL of water, cautiously add 3 mL of nitric acid, and filter through a chloride-free filter. Wash the residue with a few milliliters of hot water, dilute the filtrate with water to 50 mLs, and to 25 mL of the filtrate add 1 mL of silver nitrate reagent solution. Any turbidity should not exceed that produced by 0.015 mg of chloride ion (Cl) in an equal volume of solution containing the quantities of reagents used in the test.

SULFUR COMPOUNDS.   Mix 2.0 g with 1 g of sodium carbonate, add in small portions 15 mL of water, evaporate, and thoroughly ignite, protected from sulfur in the flame. Treat the residue with 20 mL of water and 2 mL of 30% hydrogen peroxide and heat on a hot plate at low setting for 15 min. Add 5 mL of hydrochloric acid and evaporate to dryness on the hot plate. Dissolve the residue in 10 mL of water, filter, wash with several portions of water, and dilute the filtrate with water to 25 mL. Add to this solution 0.5 mL of 1 N hydrochloric acid and 2 mL of barium chloride reagent solution. Any turbidity should not exceed that in a standard prepared as follows: treat 1 g of sodium carbonate with 2 mL of 30% hydrogen peroxide and 5 mL of hydrochloric acid and evaporate to dryness on a hot plate at low setting. Dissolve the residue and 0.12 mg of sulfate ion (SO_4) in sufficient water to make 25 mL and add 0.5 mL of 1 N hydrochloric acid and 2 mL of barium chloride reagent solution.

HEAVY METALS.   (Page 26, Method 1). Dissolve 6.6 g in warm water, and dilute with water to 50 mL. Use 40 mL to prepare the sample solution, and use the remaining 10 mL to prepare the control solution.

IRON.   (Page 28, Method 1). Use 2.0 g.

SODIUM.   Determine the sodium by the flame atomic absorption spectrophotometric method described on page 35.

> ***Sample Stock Solution.***   Dissolve 1.0 g of sample in water and dilute to 100 mL in a volumetric flask with water (0.01 g/mL).

Analyze the solutions by means of a suitable atomic absorption spectrophotometer, using the conditions outlined in the following table. Calculate the metal content of the sample by the method of standard additions.

Element	Wavelength (nm)	Sample Wt (g)	Standard Added (mg)	Flame Type*	Background Correction
Na	589.0	0.2	0.01; 0.02	A/A	No

*A/A is air/acetylene.

Potassium Hydrogen Sulfate, Fused

CAS Number 7790−62−7

> *NOTE.* This product is a mixture of potassium pyrosulfate, $K_2S_2O_7$, and potassium hydrogen sulfate, $KHSO_4$.

REQUIREMENTS

Acidity (as H_2SO_4) . . . . 37.5−38.6%

MAXIMUM ALLOWABLE

Water (H_2O) 2.5%
Insoluble matter and ammonium
   hydroxide precipitate 0.01%
Chloride (Cl) 0.002%
Phosphate (PO_4) 0.001%
Arsenic (As) . . . . 5 ppm
Calcium and magnesium precipitate 0.005%
Heavy metals (as Pb) 0.001%
Iron (Fe) 0.002%
Sodium (Na) 0.01%

TESTS

ACIDITY. (By acid−base titrimetry). Weigh accurately 4 g, dissolve in 50 mL of water, and titrate with 1 N sodium hydroxide, using methyl orange as indicator. One milliliter of 1 N alkali corresponds to 0.04904 g of H_2SO_4.

WATER. Weigh accurately about 5 g, dissolve in water, and dilute with water to 250 mL. Evaporate 25.0 mL of the solution to dryness in a tared platinum container and ignite to remove all sulfuric acid and leave K_2SO_4. Calculate the percentage of K_2SO_4 (B). Calculate the percentage of SO_3 (A) from the percentage of H_2SO_4 obtained in the test for acidity. One percent of H_2SO_4 is equivalent to 0.8163% of

SO_3. Calculate the percentage of H_2O (C) as the difference between 100% and the sum of the percentages of SO_3 and K_2SO_4.

$$C = 100 - (A + B)$$

INSOLUBLE MATTER AND AMMONIUM HYDROXIDE PRECIPITATE. Dissolve 20 g in 200 mL of water, add ammonium hydroxide until the solution is alkaline to methyl red, boil for 1 min, and digest in a covered beaker on the steam bath for 1 h. Filter through a tared filtering crucible, saving the filtrate separate from the washings for the test for calcium and magnesium precipitate. Wash thoroughly and dry at 105°C.

CHLORIDE.   (Page 25). Use 0.50 g.

PHOSPHATE.   Dissolve 2.0 g in 15 mL of dilute ammonium hydroxide (1 + 2), and evaporate to dryness. Dissolve the residue in 25 mL of approximately 0.5 N hydrochloric acid, add 1 mL of ammonium molybdate reagent solution and 1 mL of 4-(methylamino)phenol sulfate reagent solution, and allow to stand for 2 h at room temperature. Any blue color should not exceed that produced by 0.02 mg of phosphate ion (PO_4) in an equal volume of solution containing the quantities of reagents used in the test, including the residue from evaporation of 5 mL of ammonium hydroxide.

ARSENIC.   (Page 23). Use 0.60 g. For the standard use 0.003 mg of arsenic (As).

CALCIUM AND MAGNESIUM PRECIPITATE.   To the filtrate from the test for insoluble matter and ammonium hydroxide precipitate (without the washings) add 5 mL of ammonium oxalate reagent solution, 3 mL of ammonium phosphate reagent solution, and 10 mL of ammonium hydroxide. If any precipitate forms on standing overnight, filter, wash thoroughly with water containing 2.5% ammonia and about 0.1% ammonium oxalate, and ignite.

HEAVY METALS.   (Page 26, Method 1). Dissolve 4.0 g in about 20 mL of water, and dilute with water to 32 mL. Use 24 mL to prepare the sample solution, and use the remaining 8.0 mL to prepare the control solution.

IRON. (Page 28, Method 1). Dissolve 5.0 g in 80 mL of dilute hydrochloric acid (1 + 3), and boil gently for 10 min. Cool, dilute with water to 100 mL, and use 10 mL of the solution.

SODIUM. Determine the sodium by the flame atomic absorption spectrophotometric method described on page 35.

> *Sample Stock Solution.* Dissolve 1.0 g of sample with water in a 100-mL volumetric flask and dilute to the mark with water (0.01 g/mL).

Analyze the solutions by means of a suitable atomic absorption spectrophotometer, using the conditions outlined in the following table. Calculate the metal content of the sample by the method of standard additions.

Element	Wavelength (nm)	Sample Wt (g)	Standard Added (mg)	Flame Type*	Background Correction
Na	589.0	0.1	0.01; 0.02	A/A	No

*A/A is air/acetylene.

Potassium Hydroxide

KOH   Formula Wt 56.11

CAS Number 1310–58–3

> *NOTE.* Reagent potassium hydroxide usually contains 10–15% water.

REQUIREMENTS

Assay and
   potassium carbonate (K$_2$CO$_3$) . . . . Not less than 85% KOH and not more than 2.0% K$_2$CO$_3$

MAXIMUM ALLOWABLE

Chloride (Cl) 0.01%
Nitrogen compounds (as N) 0.001%
Phosphate (PO_4) . . . . 5 ppm
Sulfate (SO_4) 0.003%
Ammonium hydroxide precipitate 0.02%
Heavy metals (as Ag) 0.001%
Iron (Fe) 0.001%
Nickel (Ni) 0.001%
Sodium (Na) 0.05%

TESTS

> *NOTE.* Special care must be taken in sampling to obtain a represen-
> tative sample and to avoid absorption of water and carbon dioxide
> by the sample taken.

ASSAY AND POTASSIUM CARBONATE. (By acid–base titrimetry).
Weigh rapidly 35–40 g, to within 0.1 g, dissolve, cool, and dilute to 1
L, using carbon dioxide-free water throughout. Dilute 50.0 mL of the
well-mixed solution to 200 mL with carbon dioxide-free water, add 5
mL of barium chloride reagent solution, shake, and allow to stand for
a few minutes. Titrate with 1 N hydrochloric acid to the phenolphthalein end point. One milliliter of 1 N hydrochloric acid corresponds
to 0.05611 g of KOH. Continue the titration with 1 N hydrochloric
acid to the methyl orange end point to determine carbonate. One
milliliter of 1 N hydrochloric acid corresponds to 0.06911 g of K_2CO_3.

> *Sample Solution A for the Determination of Chloride, Ni-
> trogen Compounds, Phosphate, Sulfate, Heavy Metals, and
> Iron.* Dissolve 50.0 g in carbon dioxide- and ammonia-free wa-
> ter, cool, and dilute with the water to 500 mL (1 mL = 0.1 g).

CHLORIDE. (Page 25). Use 1.0 mL of sample solution A (0.1-g
sample).

NITROGEN COMPOUNDS. (Page 30). Use 20 mL of sample solution
A (2-g sample). For the control use 10 mL of sample solution A and
0.01 mg of nitrogen (N).

PHOSPHATE. (Page 30, Method 1). To 40 mL of sample solution A
(4-g sample) add 10 mL of hydrochloric acid, and evaporate to dryness
on the steam bath. Dissolve the residue in 25 mL of approximately
0.5 N sulfuric acid, and continue as described. For the standard
include the residue from evaporation of 5 mL of hydrochloric acid.

SULFATE.   (Page 32, Procedure A, Method 1). Neutralize 16.7 mL of sample solution A, and evaporate to 10 mL. Allow 30 min for the turbidity to form.

AMMONIUM HYDROXIDE PRECIPITATE.   Weigh about 10 g and dissolve in about 100 mL of water. Cautiously add 12 mL of sulfuric acid to 12 mL of water, cool, add the mixture to the solution of the sample, and evaporate to dense fumes. Cool, dissolve the residue in 130 mL of hot water, and add ammonium hydroxide until the solution is just alkaline to methyl red. Heat to boiling, filter, wash with hot water, and ignite. The residue includes some, but not all, of the silica in the sample.

HEAVY METALS.   To 60 mL of sample solution A (6-g sample) cautiously add 15 mL of nitric acid. For the control add 0.05 mg of silver ion (Ag) to 10 mL of sample solution A and cautiously add 15 mL of nitric acid. Evaporate both solutions to dryness over a low flame or on an electric hot plate. Dissolve each residue in about 20 mL of water, filter if necessary through a chloride-free filter, and dilute with water to 25 mL. Adjust the pH of the control and sample solutions to between 3 and 4 (using a pH meter) with 1 N acetic acid or ammonium hydroxide (10% NH_3), dilute with water to 40 mL, and mix. Add 10 mL of freshly prepared hydrogen sulfide water to each and mix. Any yellow-brown color in the solution of the sample should not exceed that in the control.

IRON.   (Page 28, Method 1). Neutralize 10 mL of sample solution A (1-g sample) with hydrochloric acid, using phenolphthalein indicator, add 2 mL in excess, and dilute with water to 50 mL. Use this solution. For the standard use the residue obtained by evaporating the quantity of acid used to neutralize the sample solution.

NICKEL AND SODIUM.   Determine the nickel and sodium by the flame atomic absorption spectrophotometric method described on page 35.

> ***Sample Stock Solution.***   Dissolve 20.0 g of sample in 50 mL of water in a 200-mL volumetric flask. Neutralize with nitric acid, add 5 mL in excess, and dilute to the mark with water (0.1 g/mL).

Analyze the solutions by means of a suitable atomic absorption spectrophotometer, using the conditions outlined in the following

table. Calculate the metal content of the sample by the method of standard additions.

Element	Wavelength (nm)	Sample Wt (g)	Standard Added (mg)	Flame Type*	Background Correction
Ni	232.0	5.0	0.025; 0.05	A/A	Yes
Na	589.0	0.05	0.02; 0.04	A/A	No

*A/A is air/acetylene.

Potassium Iodate

KIO_3 Formula Wt 214.00

CAS Number 7758−05−6

REQUIREMENTS

Assay . . . . 99.4−100.4% KIO_3
pH of a 5% solution at 25°C . . . . 5.0−8.0

MAXIMUM ALLOWABLE

Insoluble matter 0.005%
Chloride and bromide (as Cl). 0.01%
Iodide (I) 0.001%
Nitrogen compounds (as N) 0.005%
Sulfate (SO_4). 0.005%
Heavy metals (as Pb) . . . . 5 ppm
Iron (Fe) 0.001%
Sodium (Na) 0.005%

TESTS

ASSAY. (By iodometric titration of oxidizing power). Weigh accurately 0.14 g and in a glass-stoppered conical flask dissolve in 50 mL of water. Add 3 g of potassium iodide and 1 mL of sulfuric acid. Stopper, swirl, and allow to stand for 3 min. Titrate the liberated iodine with 0.1 N sodium thiosulfate, adding 3 mL of starch indicator solution near the end of the titration. Correct for a blank. One milliliter of 0.1 N sodium thiosulfate corresponds to 0.003567 g of KIO_3.

pH OF A 5% SOLUTION.   (Page 43). The pH should be 5.0–8.0 at 25°C.

INSOLUBLE MATTER.   (Page 14). Use 20 g dissolved in 250 mL of water.

CHLORIDE AND BROMIDE.   Dissolve 1.0 g in 100 mL of water in a flask, filter if necessary through a chloride-free filter, and add 6 mL of hydrogen peroxide and 1 mL of phosphoric acid. Heat to boiling and boil gently until all the iodine is expelled and the solution is colorless. Cool, wash down the sides of the flask, and add 0.5 mL of hydrogen peroxide. If an iodine color develops, boil until the solution is colorless and for 10 min longer. If no color develops, boil for 10 min. Dilute with water to 100 mL, take 10 mL, and dilute with water to 23 mL. Prepare a standard containing 0.01 mg of chloride ion (Cl) and 0.6 mL of hydrogen peroxide in 23 mL of water. Add 1 mL of nitric acid and 1 mL of silver nitrate reagent solution to each. Any turbidity in the solution of the sample should not exceed that in the standard.

IODIDE.   Dissolve 11 g in 160 mL of water, and add 1 g of citric acid and 5 mL of chloroform. Prepare a control solution containing 1 g of the sample, 1 g of citric acid, and 0.1 mg of iodide ion (I) in 160 mL of water; add 5 mL of chloroform. Shake vigorously and allow the chloroform to separate. Any pink color in the chloroform layer from the sample should not exceed that in the chloroform layer from the control.

NITROGEN COMPOUNDS.   (Page 30). Dilute the distillate of a 1.0-g sample to 50 mL. Use 10 mL of the distillate. For the standard use 0.01 mg of nitrogen (N).

SULFATE.   (Page 32, Procedure A, Method 2). Use 3 mL of hydrochloric acid, and do three evaporations. Allow 30 min for the turbidity to form.

HEAVY METALS.   (Page 26, Method 1). Dissolve 7.0 g in 50 mL of water, heat to boiling, add 1.5 mL of dilute formic acid (1 + 9), and continue heating until the initial reaction ceases. Add dropwise 50 mL of dilute formic acid (1 + 9), and continue boiling until the solution appears to be free of iodine. Wash down the sides of the container, and boil again if the iodine color reappears. Repeat the washing down and boiling until the solution remains colorless. Evaporate to dryness on the steam bath. Dissolve the residue in about 20 mL of water, and dilute with water to 35 mL. Use 25 mL to prepare the sample solution, use 5.0 mL to prepare the control solution, and reserve the remaining 5.0 mL for the test for iron.

IRON. (Page 28, Method 2). Use 5.0 mL of the solution reserved from the test for heavy metals.

SODIUM. Determine the sodium by the flame atomic absorption spectrophotometric method described on page 35.

>   *Sample Stock Solution.* Dissolve 1.0 g of sample in water and dilute to 100 mL in a volumetric flask with water (0.01 g/mL).

Analyze the solutions by means of a suitable atomic absorption spectrophotometer, using the conditions outlined in the following table. Calculate the metal content of the sample by the method of standard additions.

Element	Wavelength (nm)	Sample Wt (g)	Standard Added (mg)	Flame Type*	Background Correction
Na	589.0	0.20	0.005; 0.01	A/A	No

*A/A is air/acetylene.

Potassium Iodide

KI   Formula Wt 166.00

CAS Number 7681−11−0

REQUIREMENTS

Assay . . . . ≥ 99.0% KI
pH of a 5% solution at 25°C . . . . 6.0−9.2

MAXIMUM ALLOWABLE

Insoluble matter 0.005%
Loss on drying at 150°C. 0.2%
Chloride and bromide (as Cl). 0.01%
Iodate (IO_3) . . . . 3 ppm
Nitrogen compounds (as N) 0.001%
Phosphate (PO_4) 0.001%
Sulfate (SO_4) 0.005%
Barium (Ba) 0.002%
Calcium, magnesium, and
    R_2O_3 precipitate 0.005%
Heavy metals (as Pb) . . . . 5 ppm
Iron (Fe) . . . . 3 ppm
Sodium (Na) 0.005%

TESTS

ASSAY. (By titration of reductive power). Weigh, to the nearest 0.1 mg, about 0.5 g of sample, and dissolve with 20 mL of water in a 200-mL glass-stoppered titration flask. Add 30 mL of hydrochloric acid and 5 mL of chloroform, cool if necessary, and titrate with 0.05 M potassium iodate solution until the iodine color disappears from the aqueous layer. Stopper, shake vigorously for 30 s, and continue the titration, shaking vigorously after each addition of the iodate, until the iodine color in the chloroform is discharged.

$$\% \text{ KI} = \frac{(\text{mL} \times \text{M KIO}_3) \times 33.20}{\text{Sample wt (g)}}$$

pH OF A 5% SOLUTION. (Page 43). The pH should be 6.0–9.2 at 25°C.

INSOLUBLE MATTER. (Page 14). Use 20 g dissolved in 150 mL of water. Retain the filtrate, without the washings, for the test for calcium, magnesium, and R_2O_3 precipitate.

LOSS ON DRYING. Weigh accurately about 2.0 g of a sample in which the large crystals have been crushed, but which has not been ground for very long in a mortar. Dry in a weighed dish for 6 h at 150°C.

CHLORIDE AND BROMIDE. Dissolve 1.0 g in 100 mL of water in a distilling flask. Add 1 mL of hydrogen peroxide and 1 mL of phosphoric acid, heat to boiling, and boil gently until all the iodine is expelled and the solution is colorless. Cool, wash down the sides of the flask, and add 0.5 mL of hydrogen peroxide. If an iodine color develops, boil until the solution is colorless and for 10 min longer. If no color develops, boil for 10 min, filter if necessary through a chloride-free filter, and dilute with water to 100 mL. Dilute 10 mL with water to 23 mL and add 1 mL of nitric acid and 1 mL of silver nitrate reagent solution. Any turbidity should not exceed that produced by 0.01 mg of chloride ion (Cl) in an equal volume of solution containing the quantities of nitric acid and silver nitrate used in the test.

IODATE. (Page 49). Determine iodate by differential pulse polarography. Use 10.0 g of sample in 25 mL of solution. For the standard add 0.03 mg of iodate (IO_3).

NITROGEN COMPOUNDS. (Page 30). Use 1.0 g. For the standard use 0.01 mg of nitrogen (N).

PHOSPHATE. (Page 30, Method 1). Dissolve 2.0 g in water, and add 10 mL of nitric acid and 5 mL of hydrochloric acid. Evaporate to dryness on the steam bath. Dissolve the residue in 25 mL of approximately 0.5 N sulfuric acid and continue as described. For the standard include the residue from evaporation of the quantities of acids used with the sample.

SULFATE. (Page 32, Procedure A, Method 1). Allow 30 min for the turbidity to form.

BARIUM. Dissolve 3.0 g of the sample in 10 mL of dilute hydrochloric acid (1 + 1) and add 5 mL of nitric acid. For the control dissolve 0.5 g of sample and 0.05 mg of barium ion (Ba) in 10 mL of dilute hydrochloric acid (1 + 1) and add 5 mL of nitric acid. Evaporate both solutions to dryness. Dissolve the residues in 10 mL of dilute hydrochloric acid (1 + 1), add 5 mL of nitric acid, and again evaporate to dryness. Dissolve each residue in water and dilute with water to 23 mL. To each solution add 2 mL of potassium dichromate reagent solution and 10% ammonium hydroxide reagent solution until the orange color is just dissipated and the yellow color persists. To each solution add with constant stirring 25 mL of methanol. Any turbidity in the solution of the sample should not exceed that in the control.

CALCIUM, MAGNESIUM, AND R_2O_3 PRECIPITATE. To the filtrate from the test for insoluble matter add 5 mL of ammonium oxalate reagent solution, 2 mL of ammonium phosphate reagent solution, and 20 mL of ammonium hydroxide. Allow to stand overnight, filter, wash with water containing 2.5% ammonia and about 0.1% ammonium oxalate, and ignite.

HEAVY METALS. (Page 26, Method 1). Dissolve 6.0 g in water, and dilute with water to 30 mL. Use 25 mL to prepare the sample solution, and use the remaining 5.0 mL to prepare the control solution.

IRON. (Page 28, Method 2). Dissolve 3.3 g in 50 mL of water, and add 0.1 mL of ammonium hydroxide.

SODIUM. Determine the sodium by the flame atomic absorption spectrophotometric method described on page 35.

> ***Sample Stock Solution.*** Dissolve 2.0 g of sample in water in a 100-mL volumetric flask, and dilute to the mark with water (0.02 g/mL).

Analyze the solutions by means of a suitable atomic absorption spectrophotometer, using the conditions outlined in the following table. Calculate the metal content of the sample by the method of standard additions.

Element	Wavelength (nm)	Sample Wt (g)	Standard Added (mg)	Flame Type*	Background Correction
Na	589.0	0.2	0.005; 0.01	A/A	No

*A/A is air/acetylene.

Potassium Nitrate

KNO$_3$ Formula Wt 101.10

CAS Number 7757–79–1

REQUIREMENTS

Assay . . . . $\geq$ 99.0% KNO$_3$
pH of a 5% solution at 25°C 4.5–8.5

MAXIMUM ALLOWABLE

Insoluble matter 0.005%
Chloride (Cl) 0.002%
Iodate (IO_3) . . . . 5 ppm
Iodate and nitrite . . . . Passes test (limit
                                                        about 5 ppm IO_3;
                                                        about 0.001%
                                                        NO_2)
Phosphate (PO_4) . . . . 5 ppm
Sulfate (SO_4) 0.003%
Calcium, magnesium, and
 R_2O_3 precipitate 0.01%
Heavy metals (as Pb) . . . . 5 ppm
Iron (Fe) . . . . 3 ppm
Sodium (Na) 0.005%

TESTS

ASSAY. (By acid–base titrimetry).

Ion-Exchange Column: Use Dowex HCR-W2, H$^+$ form, 16–40
mesh, or equivalent. Charge the resin to a column with a 2-cm
bore to a height of about 20 cm. If the resin is in the hydrogen ion
form, wash free of acid with carbon dioxide-free water. Otherwise
regenerate by passing 4 N hydrochloric acid through the column
and washing free of the hydrochloric acid.

*NOTE. All water used in this titration must be carbon dioxide- and
ammonia-free.*

Weigh, to the nearest 0.1 mg, 0.4 g of sample, and dissolve in 100 mL
of water. Pass the solution through a cation-exchange column with
water at a rate of about 5 mL/min, and collect the eluate in a 500-mL
titration flask. Then wash the resin in a column at a rate of about 10
mL/min, and collect in the same titration flask. Add 0.15 mL of
phenolphthalein to the flask and titrate with 0.1 N sodium hydroxide.
Continue the titration as the washing proceeds until 50 mL of the
eluate requires no further titration.

$$\% \ KNO_3 = \frac{(mL \times N \ NaOH) \times 10.11}{Sample \ wt \ (g)}$$

pH OF A 5% SOLUTION. (Page 43). The pH should be 4.5–8.5 at
25°C.

INSOLUBLE MATTER.   (Page 14). Use 20 g dissolved in 150 mL of water.

CHLORIDE.   (Page 25). Use 0.50 g.

IODATE.   (Page 49). Determine iodate by differential pulse polarography. Use 10.0 g of sample in 25 mL of solution. For the standard add 0.05 mg of iodate (IO_3).

IODATE AND NITRITE.   Dissolve 1.0 g in 10 mL of water and add 0.10 mL of 10% potassium iodide reagent solution, 1 mL of chloroform, and 1 mL of acetic acid. Shake gently for a few minutes. A pink or violet color should not be observed in the chloroform layer.

PHOSPHATE.   (Page 30, Method 1). Dissolve 4.0 g in 25 mL of approximately 0.5 N sulfuric acid and continue as directed.

SULFATE.   (Page 32, Procedure A, Method 2). Use 7 mL of dilute hydrochloric acid (1 + 1). Do two evaporations, and heat the last residue for 1 h at 120°C.

CALCIUM, MAGNESIUM, AND R_2O_3 PRECIPITATE.   Dissolve 10 g in 75 mL of water, filter, and add 5 mL of ammonium oxalate reagent solution, 2 mL of ammonium phosphate reagent solution, and 15 mL of ammonium hydroxide. Stir well and allow to stand overnight. If any precipitate forms, filter, wash with water containing 2.5% ammonia and about 0.1% ammonium oxalate, and ignite.

HEAVY METALS.   (Page 26, Method 1). Dissolve 6.0 g in about 20 mL of water, and dilute with water to 30 mL. Use 25 mL to prepare the sample solution, and use the remaining 5.0 mL to prepare the control solution.

IRON.   (Page 28, Method 1). Use 3.3 g.

SODIUM.   Determine the sodium by the flame atomic absorption spectrophotometric method described on page 35.

> ***Sample Stock Solution.***   Dissolve 2 g of sample in water in a 100-mL volumetric flask, and dilute to the mark with water (0.02 g/mL).

Analyze the solutions by means of a suitable atomic absorption spectrophotometer, using the conditions outlined in the following

table. Calculate the metal content of the sample by the method of standard additions.

Element	Wavelength (nm)	Sample Wt (g)	Standard Added (mg)	Flame Type*	Background Correction
Na	589.0	0.2	0.005; 0.01	A/A	No

*A/A is air/acetylene.

Potassium Nitrite

KNO_2 Formula Wt 85.10

CAS Number 7758–09–0

REQUIREMENTS

Assay . . . . $\geq$ 96.0% KNO_2

MAXIMUM ALLOWABLE

Insoluble matter 0.01%
Chloride (Cl) 0.03%
Sulfate (SO_4) 0.01%
Calcium, magnesium, and
    R_2O_3 precipitate 0.01%
Heavy metals (as Pb) 0.001%
Iron (Fe) 0.001%
Sodium (Na) 0.5%

TESTS

ASSAY. (By oxidation–reduction titrimetry). Weigh accurately 6–7 g, dissolve in water, and dilute with water to 1 L in a volumetric flask. Add 5 mL of sulfuric acid to 300 mL of water and, while the solution is still warm, add 0.1 N potassium permanganate until a faint pink color that persists for 2 min is produced. Then add 40.0 mL of 0.1 N potassium permanganate and mix gently. Add, slowly and with constant agitation, 25.0 mL of the sample solution, holding the tip of the pipet well below the surface of the liquid. Add 15.0 mL of

0.1 N ferrous ammonium sulfate, allow the solution to stand for 5 min, and titrate the excess of ferrous ammonium sulfate with 0.1 N potassium permanganate. Each milliliter of 0.1 N potassium permanganate consumed by the potassium nitrite corresponds to 0.004255 g of KNO_2.

INSOLUBLE MATTER. (Page 14). Use 20 g dissolved in 200 mL of water. Reserve the filtrate without the washings for calcium, magnesium, and R_2O_3 precipitate.

CHLORIDE. Dissolve 3.3 g in 20 mL of water, filter if necessary through a chloride-free filter, and dilute with water to 100 mL. Dilute 1.0 mL of this solution with 15 mL of water and slowly add 1 mL of acetic acid. Heat to boiling, boil gently for 5 min, cool, and dilute with water to 20 mL. Add 1 mL of nitric acid and 1 mL of silver nitrate reagent solution. Any turbidity should not exceed that produced by 0.01 mg of chloride ion (Cl) in an equal volume of solution containing the quantities of reagents used in the test.

> ***Sample Solution A for the Determination of Sulfate, Heavy Metals, and Iron.*** Dissolve 10 g in water and dilute with water to 100 mL (1 mL = 0.1 g).

SULFATE. (Page 32, Procedure A, Method 2). Use 5.0 mL of sample solution A (0.5-g sample) plus 1 mL of hydrochloric acid.

CALCIUM, MAGNESIUM, AND R_2O_3 PRECIPITATE. To the filtrate without the washings retained from the test for insoluble matter, add 10 mL of ammonium oxalate reagent solution, 5 mL of ammonium phosphate reagent solution, and 5 mL of ammonium hydroxide, and allow to stand overnight. If any precipitate is formed, filter, wash with water containing 2.5% ammonia and about 0.1% ammonium oxalate, and ignite.

HEAVY METALS. (Page 26, Method 1). To 30 mL of sample solution A (3-g sample) add 5 mL of hydrochloric acid, and evaporate to dryness on the steam bath. Add 5 mL of hydrochloric acid and again evaporate to dryness. Dissolve in about 20 mL of water, and dilute with water to 25 mL. For the control solution add 0.02 mg of lead ion (Pb) to 10 mL of sample solution A (1-g sample), add 5 mL of hydrochloric acid, and treat exactly as the 30 mL of sample solution A.

IRON. (Page 28, Method 1). To 10 mL of sample solution A (1-g sample) add 5 mL of hydrochloric acid, and evaporate to dryness on the steam bath. Dissolve the residue in 15–20 mL of water, add 2 mL of hydrochloric acid, filter if necessary, and dilute with water to 50 mL. Use the solution without further acidification.

SODIUM. Determine the sodium by the flame atomic absorption spectrophotometric method described on page 35.

> **Sample Stock Solution.** Dissolve 0.4 g of sample in water in a 100-mL volumetric flask, and dilute to the mark with water (0.004 g/mL).

Analyze the solutions by means of a suitable atomic absorption spectrophotometer, using the conditions outlined in the following table. Calculate the metal content of the sample by the method of standard additions.

Element	Wavelength (nm)	Sample Wt (g)	Standard Added (mg)	Flame Type*	Background Correction
Na	589.0	0.004	0.01; 0.02	A/A	No

*A/A is air/acetylene.

Potassium Oxalate Monohydrate
Oxalic Acid, Potassium Salt, Monohydrate

$(COOK)_2 \cdot H_2O$ Formula Wt 184.23

CAS Number 6487−48−5

REQUIREMENTS

Assay . . . . 98.5−101.0%
$K_2C_2O_4 \cdot H_2O$

Substances darkened by
  hot sulfuric acid . . . . Passes test
Neutrality . . . . Passes test

MAXIMUM ALLOWABLE

Insoluble matter 0.01%
Chloride (Cl) 0.002%
Sulfate (SO_4) 0.01%
Ammonium (NH_4) 0.002%
Heavy metals (as Pb) 0.002%
Iron (Fe) 0.001%
Sodium (Na) 0.02%

TESTS

ASSAY. (By oxidation–reduction titrimetry). Weigh accurately 1.8 g, and dissolve in water in a 250-mL volumetric flask. Dilute to volume and mix. To a 50-mL aliquot add 50 mL of water and 5 mL of sulfuric acid. Titrate slowly with 0.1 N potassium permanganate until about 25 mL has been added. Heat the solution to about 70°C, and complete the titration to a permanent faint pink color. One milliliter of 0.1 N potassium permanganate consumed corresponds to 0.009212 g of $K_2C_2O_4 \cdot H_2O$.

INSOLUBLE MATTER. (Page 14). Use 10 g dissolved in 100 mL of water.

NEUTRALITY. Dissolve 2.0 g in 200 mL of water, and add 10.0 mL of 0.01 N oxalic acid and 0.20 mL of a 1% solution of phenolphthalein. Boil the solution in a flask for 10 min, passing through it a stream of carbon dioxide-free air. Cool the solution rapidly to room temperature while keeping the flow of carbon dioxide-free air passing through it and titrate with 0.01 N sodium hydroxide. Not less than 9.2 nor more than 10.5 mL of 0.01 N sodium hydroxide should be required to match the pink color produced in a buffer solution containing 0.20 mL of a 1% solution of phenolphthalein. The buffer solution contains in 1 L 3.1 g of boric acid (H_3BO_3), 3.8 g of potassium chloride, and 5.90 mL of 1 N sodium hydroxide.

CHLORIDE. (Page 25). Dissolve 2.0 g in water plus 10 mL of nitric acid, filter if necessary through a chloride-free filter, and dilute with water to 100 mL. To 25 mL of the solution add 1 mL of silver nitrate reagent solution. Any turbidity should not exceed that produced by 0.01 mg of chloride ion (Cl) in an equal volume of solution containing the quantities of reagents used in the test.

SULFATE. (Page 32, Procedure A, Method 4).

AMMONIUM. Dissolve 1.0 g in 50 mL of ammonia-free water and add 2 mL of Nessler reagent. Any color should not exceed that produced by 0.02 mg of ammonium ion (NH_4) in an equal volume of solution containing 2 mL of Nessler reagent.

HEAVY METALS. (Page 26, Method 1). Mix 2.0 g with 2 mL of water, add 2 mL of nitric acid and 2 mL of 30% hydrogen peroxide, and digest in a covered beaker on the steam bath until reaction ceases. Remove the cover and evaporate to dryness. Dissolve the residue in 4 mL of dilute nitric acid $(1 + 1)$ and again evaporate to dryness. Dissolve the residue in about 20 mL of water, dilute with water to 32 mL, and use 24 mL to prepare the sample solution. Use the remaining 8.0 mL, plus the residue from evaporation of 1 mL of 30% hydrogen peroxide and 3 mL of nitric acid to dryness, to prepare the control solution.

IRON. (Page 28, Method 1). Mix 2.0 g with 2 mL of water, add 2 mL of nitric acid and 2 mL of 30% hydrogen peroxide, and digest in a covered beaker on the steam bath until reaction ceases. Remove the cover and evaporate to dryness. Add 5 mL of hydrochloric acid, and again evaporate to dryness. Dissolve the residue in a few milliliters of water, add 4 mL of hydrochloric acid, and dilute with water to 100 mL. Use 50 mL without further acidification. For the control use the residue from evaporation to dryness of 1 mL of nitric acid, 1 mL of 30% hydrogen peroxide, and 2.5 mL of hydrochloric acid.

SODIUM. Determine the sodium by the flame atomic absorption spectrophotometric method described on page 35.

> **Sample Stock Solution.** Dissolve 1.0 g of sample in water, and dilute to 100 mL in a volumetric flask with water (0.01 g/mL).

Analyze the solutions by means of a suitable atomic absorption spectrophotometer, using the conditions outlined in the following table. Calculate the metal content of the sample by the method of standard additions.

Element	Wavelength (nm)	Sample Wt (g)	Standard Added (mg)	Flame Type*	Background Correction
Na	589.0	0.05	0.01; 0.02	A/A	No

*A/A is air/acetylene.

SUBSTANCES DARKENED BY HOT SULFURIC ACID. Heat 1.0 g in a recently ignited test tube with 10 mL of sulfuric acid until the appearance of dense fumes. The acid when cooled should have no more color than a mixture of the following composition: 0.2 mL of cobalt chloride solution (5.95 g of $CoCl_2 \cdot 6H_2O$ and 2.5 mL of hydrochloric acid in 100 mL), 0.3 mL of ferric chloride solution (4.50 g of $FeCl_3 \cdot 6H_2O$ and 2.5 mL of hydrochloric acid in 100 mL), 0.3 mL of cupric sulfate solution (6.24 g of $CuSO_4 \cdot 5H_2O$ and 2.5 mL of hydrochloric acid in 100 mL), and 9.2 mL of water.

Potassium Perchlorate

KClO$_4$ Formula Wt 138.55

CAS Number 7778–74–7

REQUIREMENTS

Assay 99.0–100.5%
                                                                    KClO$_4$

MAXIMUM ALLOWABLE

Insoluble matter 0.005%
Chloride (Cl) 0.003%
Sulfate (SO$_4$) 0.001%
Calcium (Ca) 0.005%
Heavy metals (as Pb) . . . . 5 ppm
Iron (Fe) . . . . 5 ppm
Sodium (Na) 0.02%

TESTS

ASSAY.   (By argentimetric titration after decomposition to chloride). Weigh accurately about 0.25–0.30 g of sample into a platinum crucible or dish. Add 3 g of sodium carbonate, Na$_2$CO$_3$, and heat, gently at first, then to fusion until a clear melt is obtained. Leach with minimal dilute nitric acid, and boil gently to expel carbon dioxide. Cool to room temperature, add 10 mL of 30% ammonium acetate solution, and titrate with 0.1 N silver nitrate to the dichlorofluorescein end point. One milliliter of 0.1 N silver nitrate corresponds to 0.01385 g of KClO$_4$.

INSOLUBLE MATTER.   (Page 14). Use 50 g dissolved in 600 mL of hot water.

CHLORIDE.   (Page 25). Dissolve 0.33 g in 20 mL of hot water, filter, and cool.

SULFATE.   Dissolve 40 g in 400 mL of hot water, add 4 mL of hydrochloric acid, and heat the solution to boiling. Add 5 mL of barium chloride reagent solution, digest in a covered beaker on a hot plate at low setting for 2 h, and allow to stand overnight. Warm to dissolve any potassium perchlorate that may have crystallized. If a precipitate remains, filter, wash thoroughly, and ignite. The weight of the precipitate should not be more than 0.001 g greater than the weight obtained in a complete blank test.

CALCIUM AND SODIUM. Determine the calcium and sodium by the flame atomic absorption spectrophotometric method described on page 35.

> **Sample Stock Solution.** Dissolve 5.0 g of sample in water and dilute to 100 mL in a volumetric flask with water (0.05 g/mL).

Analyze the solutions by means of a suitable atomic absorption spectrophotometer, using the conditions outlined in the following table. Calculate the metal content of the sample by the method of standard additions.

Element	Wavelength (nm)	Sample Wt (g)	Standard Added (mg)	Flame Type*	Background Correction
Ca	422.7	0.5	0.025; 0.05	N/A	No
Na	589.0	0.05	0.01; 0.02	A/A	No

*A/A is air/acetylene; N/A is nitrous oxide/acetylene.

HEAVY METALS. (Page 26, Method 1). Dissolve 3.0 g in hot water, dilute with water to 60 mL, and use 50 mL to prepare the sample solution. For the control add 0.01 mg of lead ion (Pb) to the remaining 10 mL, and dilute with water to 50 mL.

IRON. (Page 28, Method 1). Use 2.0 g.

Potassium Periodate
Potassium Metaperiodate
(Suitable for determination of manganese)

KIO_4 Formula Wt 230.00

CAS Number 7790–21–8

REQUIREMENTS

Assay (dried basis) . . . . 99.8–100.3% KIO_4

MAXIMUM ALLOWABLE

Other halogens (as Cl) 0.01%
Manganese (Mn) . . . . 1 ppm

TESTS

ASSAY. (By titration of oxidative capacity). Weigh accurately 1 g, previously dried over magnesium perchlorate or phosphorus pentoxide for 6 h, dissolve in water, and dilute with water to 500 mL in a volumetric flask. To 50.0 mL in a glass-stoppered flask add 10 g of potassium iodide and 10 mL of a cooled solution of dilute sulfuric acid $(1 + 5)$. Allow to stand for 5 min, add 100 mL of cold water, and titrate the liberated iodine with 0.1 N sodium thiosulfate, adding starch indicator solution near the end point. Carry out a complete blank test and make any necessary correction. One milliliter of 0.1 N sodium thiosulfate corresponds to 0.002875 g of KIO_4.

OTHER HALOGENS. Dissolve 0.50 g in a mixture of 10 mL of water and 16 mL of sulfurous acid and boil for 3 min. Cool, add 5 mL of ammonium hydroxide and 20 mL of 2.5% silver nitrate solution, and filter. Dilute the filtrate with water to 100 mL, and to 20 mL of this solution add 1.5 mL of nitric acid. Any turbidity should not exceed that produced by 0.01 mg of chloride ion (Cl) in an equal volume of solution containing 0.5 mL of ammonium hydroxide, 1.5 mL of nitric acid, and 1 mL of silver nitrate reagent solution.

MANGANESE. Add 5.5 g to 85 mL of 10% sulfuric acid reagent solution. For the control add 0.5 g of the sample and 0.005 mg of manganese ion (Mn) to 85 mL of 10% sulfuric acid reagent solution. Add 10 mL of nitric acid and 5 mL of phosphoric acid to each, boil gently for 10 min, and cool. Any pink color in the solution of the sample should not exceed that in the control.

Potassium Permanganate

$KMnO_4$ Formula Wt 158.03

CAS Number 7722–64–7

REQUIREMENTS

Assay . . . . $\geq$ 99.0% $KMnO_4$

MAXIMUM ALLOWABLE

Insoluble matter 0.2%
Chloride and chlorate (as Cl) 0.005%
Nitrogen compounds (as N) 0.005%
Sulfate (SO$_4$) 0.02%

TESTS

ASSAY. (By titration of oxidizing power). Weigh, to the nearest 0.1 mg, about 1.5 g of sample. Transfer to a 500-mL volumetric flask, dissolve with water, then dilute to the mark with water, and mix thoroughly. Weigh accurately about 0.25 g of NIST or primary standard sodium oxalate (dried at 105°C), and dissolve in a 500-mL glass-stoppered flask. Add 15 mL of 25% sulfuric acid, then add from a buret 30 mL of the potassium permanganate solution, heat the solutions to about 70°C, and complete the titration with the permanganate solution.

$$\% \text{ KMnO}_4 = \frac{\text{g Na}_2\text{C}_2\text{O}_4 \times 23585}{\text{g sample} \times \text{mL KMnO}_4}$$

INSOLUBLE MATTER. (Page 14). Dissolve 2.0 g in 150 mL of warm water at steam-bath temperature. Filter at once through a tared filtering crucible, wash thoroughly, and dry at 105°C.

CHLORIDE AND CHLORATE. Dissolve 1.0 g in 75 mL of water and add 5 mL of nitric acid and 3 mL of 30% hydrogen peroxide. After reduction is complete, filter if necessary through a chloride-free filter, and dilute with water to 100 mL. To 20 mL of this solution add 1 mL of silver nitrate reagent solution. Any turbidity should not exceed that produced by 0.01 mg of chloride ion (Cl) in an equal volume of solution containing the quantities of reagents used in the test.

NITROGEN COMPOUNDS. (Page 30). Dilute the distillate of a 1.0-g sample, treated with 40 mL of 10% sodium hydroxide solution, to 100 mL and use 20 mL. For the standard use 0.01 mg of nitrogen (N).

SULFATE. Dissolve 5.0 g in 75 mL of water, and add 10 mL of alcohol and 10 mL of hydrochloric acid. Heat until colorless, adding more alcohol and a little hydrochloric acid if necessary. Evaporate to dryness. Dissolve the residue in 100 mL of dilute hydrochloric acid

(1 + 99), filter, and heat to boiling. Add 5 mL of barium chloride reagent solution, digest in a covered beaker on a hot plate at low setting for 2 h, and allow to stand overnight. Filter, wash thoroughly, and ignite. The weight of the precipitate should not be more than 0.0024 g greater than the weight obtained in a complete blank test.

Potassium Permanganate, Low in Mercury

KMnO$_4$ Formula Wt 158.03

CAS Number 7722-64-7

REQUIREMENTS

Assay . . . . ≥ 99.0% KMnO$_4$

MAXIMUM ALLOWABLE

Insoluble matter 0.2%
Chloride and chlorate (as Cl) 0.005%
Nitrogen compounds (as N) 0.005%
Sulfate (SO$_4$) 0.02%
Mercury (Hg) . . . . 0.05 ppm

TESTS

Assay and other tests except mercury are the same as for potassium permanganate, page 590.

MERCURY.   For the sample place 1.0 g in a 125-mL conical flask and add 10 mL of water. For the standard place 10 mL of water in a similar flask and add 0.05 μg of mercury (0.5 mL of a solution prepared freshly by diluting 1.0 mL of a standard mercury solution, page 84, with water to 500 mL). Add 2 mL of 4% potassium permanganate solution and 2 mL of 25% sulfuric acid to each flask. Heat on the steam bath for 30 min, then cool to room temperature. Transfer the contents of the sample flask to the aeration vessel (page 42), making the total volume about 50 mL. Rinse the flask with 4 mL of 10% hydroxylamine hydrochloride solution and add this to the aeration vessel also. Complete the reduction of manganese dioxide and mercury with 2 mL of 10% stannous chloride solution. Treat the standard solution in the same way. Obtain the peak height for the sample and standard solutions by flameless atomic absorption (page 41). The peak height for the sample should not exceed that for the standard.

Potassium Peroxydisulfate
Potassium Persulfate

$K_2S_2O_8$ Formula Wt 270.32

CAS Number 7727–21–1

REQUIREMENTS

Assay . . . . $\geq$ 99.0% $K_2S_2O_8$

MAXIMUM ALLOWABLE

Insoluble matter 0.005%
Chlorine compounds (as Cl) 0.001%
Nitrogen compounds (as N) 0.001%
Heavy metals (as Pb) 0.001%
Iron (Fe) . . . . 5 ppm
Manganese (Mn) . . . . 2 ppm

TESTS

ASSAY. (By titration of oxidizing power). Weigh accurately 0.5 g and add it to 25.0 mL of freshly prepared standard ferrous sulfate solution (about 0.2 N) in a glass-stoppered flask. Stopper the flask, allow it to stand for 1 h with frequent shaking, add 5 mL of phosphoric acid, and titrate the excess ferrous sulfate with 0.1 N potassium permanganate. Add 5 mL of phosphoric acid to 25.0 mL of the standard ferrous sulfate solution and titrate with 0.1 N potassium permanganate. The difference in the volume of permanganate consumed in the two titrations is equivalent to the potassium peroxydisulfate. One milliliter of 0.1 N potassium permanganate corresponds to 0.01352 g of $K_2S_2O_8$.

Standard Ferrous Sulfate Solution. Dissolve 7 g of clear crystals of ferrous sulfate heptahydrate, $FeSO_4 \cdot 7H_2O$, in 90 mL of freshly boiled and cooled distilled water and add sulfuric acid to make 100 mL. Standardize the solution with 0.1 N potassium permanganate. The solution must be freshly prepared.

INSOLUBLE MATTER. (Page 14). Use 20 g dissolved in 200 mL of water.

CHLORINE COMPOUNDS. Mix 1.0 g with 1 g of anhydrous sodium carbonate and fuse in a porcelain dish in a muffle furnace. Cool, dissolve the residue in 20 mL of water, neutralize with nitric acid, and add an excess of 1 mL of the acid. Filter if necessary through a chloride-free filter and add 1 mL of silver nitrate reagent solution. Any turbidity should not exceed that in a standard containing 0.01 mg of chloride ion (Cl) and 1 g of sodium carbonate fused and acidified as in the test.

NITROGEN COMPOUNDS. (Page 30). Use 1.0 g. For the standard use 0.01 mg of nitrogen (N).

HEAVY METALS. (Page 26, Method 1). Dissolve 2.0 g in 20 mL of hot water, add 5 mL of hydrochloric acid, and evaporate to about 5 mL. Cool, and dilute with water to 25 mL. For the standard use 0.02 mg of lead ion (Pb) in 5 mL of hydrochloric acid, add 10 mL of water and 0.25 mL of sulfuric acid, evaporate to fumes of sulfur trioxide, cool, and dilute with water to 25 mL.

IRON. (Page 28, Method 1). To 2.0 g add 20 mL of hot water and 5 mL of hydrochloric acid, and evaporate to dryness. Moisten the residue with 2 mL of hydrochloric acid, dilute with water to 50 mL, and use the solution without further acidification.

MANGANESE. To 5.0 g in a 150-mL beaker add 5 mL of sulfuric acid and 5 mL of phosphoric acid. Heat to dissolve the sample and continue heating to fumes of sulfur trioxide. Cool, dilute with water to 45 mL, add 10 mL of nitric acid, and boil for 5 min. Cool, add about 0.25 g of potassium periodate, and boil for 5 min. Cool and compare with a standard prepared by treating 0.01 mg of manganese ion (Mn) in the same manner, using the same quantities of reagents (with no sample), except that the standard solution need not be fumed before adding the nitric acid. Any color in the solution of the sample should not exceed that in the standard.

Potassium Phosphate, Dibasic
Dipotassium Hydrogen Phosphate

K_2HPO_4 Formula Wt 174.18

CAS Number 7758−11−4

REQUIREMENTS

Assay . . . . $\geq$ 98.0% K_2HPO_4
pH of a 5% solution at 25°C . . . . 8.5−9.6

MAXIMUM ALLOWABLE

Insoluble matter 0.01%
Loss on drying at 105°C. 1.0%
Chloride (Cl) 0.003%
Nitrogen compounds (as N) 0.001%
Sulfate (SO_4) 0.005%
Heavy metals (as Pb) . . . . 5 ppm
Iron (Fe) 0.001%
Sodium (Na) 0.05%

TESTS

ASSAY.   (By acid−base titrimetry). Weigh accurately 8.0 g of sample, add 50 mL of water and exactly 50.0 mL of 1 N hydrochloric acid, and stir until the sample is completely dissolved. Titrate the excess acid, constantly stirring, with 1 N sodium hydroxide to the inflection point occurring at pH 4, as measured with a pH meter and glass electrode system. Calculate (A), the volume of 1 N sodium hydroxide consumed by the sample. Continue the titration with 1 N sodium hydroxide to the inflection point occurring at about pH 8.8. Calculate (B), the volume of 1 N sodium hydroxide required in the titration between the two inflection points. One milliliter of 1 N sodium hydroxide corresponds to 0.1742 g of K_2HPO_4. If (A) is equal to or less than (B), calculate the K_2HPO_4 percentage from titration (A). If (A) is greater than (B), calculate the percentage from 2(B) − (A).

pH OF A 5% SOLUTION.   (Page 43). The pH should be 8.5−9.6 at 25°C.

INSOLUBLE MATTER. (Page 14). Use 10 g dissolved in 100 mL of water.

LOSS ON DRYING AT 105°C. Weigh accurately about 2 g. Dry to constant weight at 105°C.

CHLORIDE. (Page 25). Use 0.33 g.

NITROGEN COMPOUNDS. (Page 30). Use 1.0 g. For the standard use 0.01 mg of nitrogen (N).

SULFATE. (Page 32, Procedure B). Dissolve 5.0 g in 18 mL of water, heating on a hot plate at low setting if necessary. Cool, dilute with water to 20 mL, and transfer 2.0 mL to a reaction tube. Continue with the procedure.

HEAVY METALS. (Page 26, Method 1). Dissolve 6.0 g in about 20 mL of water, add 8 mL of 4 N hydrochloric acid, and dilute with water to 30 mL. Use 25 mL to prepare the sample solution, and use the remaining 5.0 mL to prepare the control solution.

IRON. (Page 28, Method 2). Use 1.0 g.

SODIUM. Determine the sodium by the flame atomic absorption spectrophotometric method described on page 35.

> *Sample Stock Solution.* Dissolve 2.0 g of sample in a 100-mL volumetric flask amd dilute to the mark with water (0.02 g/mL).

Analyze the solutions by means of a suitable atomic absorption spectrophotometer, using the conditions outlined in the following table. Calculate the metal content of the sample by the method of standard additions.

Element	Wavelength (nm)	Sample Wt (g)	Standard Added (mg)	Flame Type*	Background Correction
Na	589.0	0.02	0.01; 0.02	A/A	No

*A/A is air/acetylene.

Potassium Phosphate, Monobasic
Potassium Dihydrogen Phosphate

KH_2PO_4 Formula Wt 136.09

CAS Number 7778–77–0

REQUIREMENTS

Assay . . . . $\geq$ 99.0% KH_2PO_4
pH of a 5% solution at 25°C 4.1–4.5

MAXIMUM ALLOWABLE

Insoluble matter, calcium and ammonium
    hydroxide precipitate 0.01%
Loss on drying over sulfuric acid 0.2%
Chloride (Cl) 0.001%
Nitrogen compounds (as N) 0.001%
Sulfate (SO_4) 0.003%
Heavy metals (as Pb) 0.001%
Iron (Fe) 0.002%
Sodium (Na) 0.005%

TESTS

ASSAY. (By acid–base titrimetry). Weigh accurately 5.0 g of sample and dissolve in 50 mL of water. Titrate with 1 N sodium hydroxide to the inflection point at about pH 8.8 as measured with a pH meter and glass electrode system. One milliliter of 1 N sodium hydroxide corresponds to 0.1361 g of KH_2PO_4.

pH OF A 5% SOLUTION. (Page 43). The pH should be 4.1–4.5 at 25°C.

INSOLUBLE MATTER, CALCIUM AND AMMONIUM HYDROXIDE PRECIPITATE. Dissolve 10 g in 100 mL of water, add 5 mL of ammonium oxalate reagent solution, and add ammonium hydroxide until the solution is distinctly alkaline to litmus. Add an excess of 15 mL of ammonium hydroxide and allow to stand overnight. If a precipitate is formed, filter, wash with water containing 2.5% ammonia and about 0.1% ammonium oxalate, and ignite at 800 $\pm$ 25 °C for 15 min.

LOSS ON DRYING OVER SULFURIC ACID. Weigh accurately about 2 g, and dry for 24 h over sulfuric acid.

CHLORIDE. (Page 25). Use 1.0 g of sample and 2 mL of nitric acid.

NITROGEN COMPOUNDS. (Page 30). Use 1.0 g treated with 20 mL of freshly boiled 10% sodium hydroxide. For the standard use 0.01 mg of nitrogen (N).

SULFATE. Dissolve 13.7 g in 150 mL of water plus 2 mL of hydrochloric acid, filter if necessary, and heat to boiling. Add 5 mL of barium chloride reagent solution, digest in a covered beaker on a hot plate at low setting for 2 h, and allow to stand overnight. If a precipitate is formed, filter, wash thoroughly, and ignite. The weight of the precipitate should not be more than 0.001 g greater than the weight obtained in a complete blank test.

HEAVY METALS. (Page 26, Method 1). Dissolve 4.0 g in about 20 mL of water, and dilute with water to 32 mL. Use 24 mL to prepare the sample solution, and use the remaining 8.0 mL to prepare the control solution.

IRON. (Page 28, Method 2). Dissolve 1.0 g in water, dilute with water to 20 mL, and use 10 mL.

SODIUM. Determine the sodium by the flame atomic absorption spectrophotometric method described on page 35.

> *Sample Stock Solution.* Dissolve 1.0 g of sample in water and dilute to 100 mL in a volumetric flask with water (0.01 g/mL).

Analyze the solutions by means of a suitable atomic absorption spectrophotometer, using the conditions outlined in the following table. Calculate the metal content of the sample by the method of standard additions.

Element	Wavelength (nm)	Sample Wt (g)	Standard Added (mg)	Flame Type*	Background Correction
Na	589.0	0.2	0.01; 0.02	A/A	No

*A/A is air/acetylene.

Potassium Sodium Tartrate Tetrahydrate

2,3-Dihydroxybutanedioic Acid, Potassium Sodium Salt, Tetrahydrate

$$NaO-\overset{\overset{\displaystyle O}{\|}}{C}-\overset{\overset{\displaystyle OH}{|}}{CH}-\overset{\overset{\displaystyle OH}{|}}{CH}-\overset{\overset{\displaystyle O}{\|}}{C}-OK \cdot 4H_2O$$

$KNaC_4H_4O_6 \cdot 4H_2O$ Formula Wt 282.22

CAS Number 6381−59−5

REQUIREMENTS

Assay . . . . 99.0−102.0%
$KNaC_4H_4O_6 \cdot 4H_2O$
pH of a 5% solution at 25°C . . . . 6.0−8.5

MAXIMUM ALLOWABLE

Insoluble matter 0.005%
Chloride (Cl) 0.001%
Phosphate (PO_4) 0.002%
Sulfate (SO_4) 0.005%
Ammonium (NH_4) 0.002%
Calcium (Ca) 0.005%
Heavy metals (as Pb) . . . . 5 ppm
Iron (Fe) 0.001%

TESTS

ASSAY.   (Total alkalinity by nonaqueous titration). Weigh accurately 0.5 g and mix with 50 mL of glacial acetic acid, 30 mL of 96% formic acid, and 45 mL of acetic anhydride. Heat and stir until dissolution is complete, and titrate with 0.1 N perchloric acid in glacial acetic acid to a green end point with crystal violet indicator. Correct for a reagent blank. One milliliter of 0.1 N perchloric acid corresponds to 0.01411 g of $KNaC_4H_4O_6 \cdot 4H_2O$.

pH OF A 5% SOLUTION.   (Page 43). The pH should be 6.0−8.5 at 25°C.

INSOLUBLE MATTER.   (Page 14). Use 20 g in 200 mL of water.

CHLORIDE.   (Page 25). Use 1.0 g.

PHOSPHATE.   (Page 30, Method 1). Ignite 1.0 g in a platinum dish. Dissolve the residue in 5 mL of water, add 5 mL of nitric acid, and evaporate to dryness. Dissolve the residue in 25 mL of approximately 0.5 N sulfuric acid and continue as described.

SULFATE.   (Page 32, Procedure B). Dissolve 5 g in 18 mL of water, heating on a hot plate at low setting if necessary. Cool, dilute to 20 mL, transfer 2.0 mL to a reaction tube, and continue as described.

AMMONIUM.   Dissolve 1.0 g in 50 mL of water. To 25 mL of the solution add 2 mL of 10% sodium hydroxide reagent solution, dilute with water to 50 mL, and add 2 mL of Nessler reagent. Any color should not exceed that produced by 0.01 mg of ammonium ion (NH_4) in an equal volume of solution containing the quantities of reagents used in the test.

CALCIUM.   Determine the calcium by the flame atomic absorption spectrophotometric method described on page 35.

> *Sample Stock Solution.*   Dissolve and dilute 4.0 g to 100 mL with water in a volumetric flask (0.04 g/mL).

Analyze the solutions by means of a suitable atomic absorption spectrophotometer, using the conditions outlined in the following table. Calculate the metal content of the sample by the method of standard additions.

Element	Wavelength (nm)	Sample Wt (g)	Standard Added (mg)	Flame Type*	Background Correction
Ca	422.7	0.4 g	0.02; 0.04	N/A	No

*N/A is nitrous oxide/acetylene.

HEAVY METALS.   (Page 26, Method 1). Dissolve 8.0 g in about 20 mL of water, and dilute with water to 32 mL. Use 24 mL to prepare the sample solution, and use the remaining 8.0 mL to prepare the control solution.

IRON.   (Page 28, Method 1). Use 1.0 g.

Potassium Sulfate

K_2SO_4 Formula Wt 174.26

CAS Number 7778–80–5

REQUIREMENTS

Assay . . . . ≥ 99.0% K_2SO_4
pH of a 5% solution at 25°C 5.5–8.5

MAXIMUM ALLOWABLE

Insoluble matter 0.01%
Chloride (Cl) 0.001%
Nitrogen compounds (as N) . . . . 5 ppm
Arsenic (As) . . . . 2 ppm
Calcium, magnesium, and R_2O_3 precipitate 0.02%
Heavy metals (as Pb) . . . . 5 ppm
Iron (Fe) . . . . 5 ppm
Sodium (Na) 0.02%

TESTS

ASSAY.   (By acid–base titrimetry).

Ion-Exchange Column: Use Dowex HCR-W2, H^+ form, 16–40 mesh, or equivalent. Charge the resin to a column with a 2-cm bore to a height of about 20 cm. If the resin is in the hydrogen ion form, wash free of acid with carbon dioxide-free water. Otherwise regenerate by passing 4 N hydrochloric acid through the column and washing free of the hydrochloric acid.

NOTE. All water used in this titration must be carbon dioxide- and ammonia-free.

Weigh, to the nearest 0.1 mg, 0.4 g of sample and dissolve in 50 mL of water. Pass the solution through the cation-exchange column at a rate of about 5 mL/min, and collect the eluate in a 500-mL titration flask. Then wash the resin in the column with water at a rate of about 10 mL/min, and collect in the same titration flask. Add 0.15 mL of phenolpthalein indicator solution to the flask and titrate with 0.1 N sodium hydroxide. Continue the titration as the washing proceeds until 50 mL requires no further titration.

$$\% \ K_2SO_4 = \frac{(mL \times N \ NaOH) \times 8.715}{Sample \ wt \ (g)}$$

pH OF A 5% SOLUTION. (Page 43). The pH should be 5.5–8.5 at 25°C.

INSOLUBLE MATTER. (Page 14). Use 10 g dissolved in 150 mL of water.

CHLORIDE. (Page 25). Use 1.0 g.

NITROGEN COMPOUNDS. (Page 30). Use 2.0 g. For the standard use 0.01 mg of nitrogen (N).

ARSENIC. (Page 23). Use 1.5 g. For the standard use 0.003 mg of arsenic (As).

CALCIUM, MAGNESIUM, AND R_2O_3 PRECIPITATE. Dissolve 5.0 g in 75 mL of water, filter, and add 5 mL of ammonium oxalate reagent solution, 2 mL of ammonium phosphate reagent solution, and 10 mL of ammonium hydroxide. Allow to stand overnight. If any precipitate is formed, filter, wash with water containing 2.5% ammonia and about 0.1% ammonium oxalate, and ignite.

HEAVY METALS. (Page 26, Method 1). Dissolve 6.0 g in water, and dilute with water to 54 mL. Use 45 mL to prepare the sample solution, and use the remaining 9.0 mL to prepare the control solution.

IRON. (Page 28, Method 1). Use 2.0 g.

SODIUM. Determine the sodium by the flame atomic absorption spectrophotometric method described on page 35.

> *Sample Stock Solution.* Dissolve 1.0 g of sample with water in a 100-mL volumetric flask, and dilute to the mark with water (0.01 g/mL).

Analyze the solutions by means of a suitable atomic absorption spectrophotometer, using the conditions outlined in the following table. Calculate the metal content of the sample by the method of standard additions.

Element	Wavelength (nm)	Sample Wt (g)	Standard Added (mg)	Flame Type*	Background Correction
Na	589.0	0.1	0.01; 0.02	A/A	No

*A/A is air/acetylene.

Potassium Thiocyanate

KSCN   Formula Wt 97.18

CAS Number 333–20–0

REQUIREMENTS

Assay . . . . ≥ 99.0% KSCN
Appearance. . . . . Colorless or white
                                                             crystals
pH of a 5% solution at 25°C . . . . 5.3 – 8.7

MAXIMUM ALLOWABLE

Insoluble in water 0.005%
Insoluble in alcohol 0.01%
Chloride (Cl) 0.005%
Sulfate (SO_4) 0.005%
Ammonium (NH_4) 0.003%
Heavy metals (as Pb) . . . . 5 ppm
Iron (Fe) . . . . 2 ppm
Sodium (Na) 0.005%
Iodine-consuming substances . . . . Passes test (not
                                                             more than 0.2 mL
                                                             of 0.1 N iodine
                                                             solution per gram)

TESTS

CAUTION. Cyanide gases are given off during ignition and with addition of acids. Use a well-ventilated hood.

ASSAY. (By argentimetric titration of thiocyanate content). Weigh, to the nearest 0.1 mg, 7 g of sample, dissolve with 100 mL of water in a 1000-mL volumetric flask, dilute to the mark with water, and mix thoroughly. Transfer a 50.0-mL aliquot to a 250-mL glass-stoppered iodine-type flask, add 5 mL of 1 N nitric acid, then add with agitation exactly 50.0 mL of 0.1 N silver nitrate solution, and shake vigorously. Add 2 mL of ferric ammonium sulfate indicator solution, and chill in an ice bath to approximately 10°C or lower. Titrate the excess silver nitrate with 0.1 N ammonium thiocyanate solution. Near the end point, shake after the addition of each drop.

$$\% \text{ KCNS} = \frac{[(50.0 \times \text{N AgNO}_3) - (\text{mL} \times \text{N NH}_4\text{SCN})] \times 194.36}{\text{Sample wt (g)}}$$

pH OF A 5% SOLUTION.   (Page 43). The pH of the solution should be 5.3–8.7 at 25°C.

INSOLUBLE IN WATER.   Dissolve 20 g in 200 mL of water, heat to boiling, and digest in a covered beaker on the steam bath for 1 h. Filter through a tared filtering crucible, wash thoroughly, and dry at 105°C.

INSOLUBLE IN ALCOHOL.   Dissolve 10 g in 100 mL of alcohol in a conical flask provided with a reflux condenser and boil for 1 h. Filter while hot through a tared filtering crucible, wash with alcohol, and dry at 105°C.

CHLORIDE.   Dissolve 1.0 g in 20 mL of water, filter if necessary through a chloride-free filter, and add 10 mL of 25% sulfuric acid reagent solution and 7 mL of 30% hydrogen peroxide. Evaporate to 20 mL by boiling in a well-ventilated hood, add 15–20 mL of water, and evaporate again. Repeat until all the cyanide has been volatilized, cool, and dilute with water to 100 mL. To 20 mL of this solution add 1 mL of nitric acid and 1 mL of silver nitrate reagent solution. Any turbidity should not exceed that produced by 0.01 mg of chloride ion (Cl) in an equal volume of solution containing the quantities of reagents used in the test.

SULFATE.   (Page 32, Procedure A, Method 1). Allow 30 min for the turbidity to form.

AMMONIUM.   (Page 22). Dilute the distillate from a 1.0-g sample to 100 mL: 50 mL of the dilution shows not more than 0.015 mg of ammonium (NH$_4$).

HEAVY METALS.   (Page 26, Method 1). Dissolve 6.0 g in about 20 mL of water, and dilute with water to 30 mL. Use 25 mL to prepare the sample solution, and use the remaining 5.0 mL to prepare the control solution.

IRON.   (Page 28, Method 2). Dissolve 10 g in 60 mL of water. To 30 mL of the solution add 0.1 mL of hydrochloric acid, and continue as described.

SODIUM.   Determine the sodium by the flame atomic absorption spectrophotometric method described on page 35.

> ***Sample Stock Solution.***   Dissolve 1.0 g of sample with water in a 100-mL volumetric flask, and dilute to the mark with water (0.01 g/mL).

Analyze the solutions by means of a suitable atomic absorption spectrophotometer, using the conditions outlined in the following table. Calculate the metal content of the sample by the method of standard additions.

Element	Wavelength (nm)	Sample Wt (g)	Standard Added (mg)	Flame Type*	Background Correction
Na	589.0	0.1	0.005; 0.01	A/A	No

*A/A is air/acetylene.

IODINE-CONSUMING SUBSTANCES.   Dissolve 5.0 g in 50 mL of water and add 1.7 mL of 10% sulfuric acid reagent solution. Add 1 g of potassium iodide and 1 mL of starch indicator solution and titrate with 0.1 N iodine solution. Not more than 1.0 mL of the 0.1 N iodine solution should be required.

1,2-Propanediol
Propylene Glycol

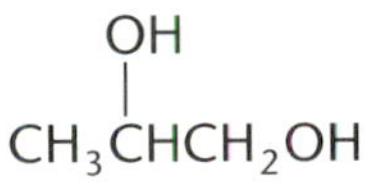

$$OH$$
$$|$$
$$CH_3CHCH_2OH$$

$CH_3CHOHCH_2OH$ Formula Wt 76.10

CAS Number 57−55−6

REQUIREMENTS

Assay . . . . ≥ 99.5%

$CH_3CHOHCH_2OH$

MAXIMUM ALLOWABLE

Color (APHA) 10
Residue after ignition 0.005%
Titrable acid . . . . 0.0005 meq / g
Chloride (Cl) . . . . 1 ppm
Water 0.2%

TESTS

ASSAY.   Analyze the sample by gas chromatography, using the general parameters cited on page 56. The following specific conditions are also required.

Column: Type III, polyethylene glycol

Measure the area under all peaks and calculate the 1,2-propanediol content in area percent. Correct for water content.

COLOR (APHA).   (Page 17).

RESIDUE AFTER IGNITION.   (Page 15). Use 50 g (48.5 mL).

TITRABLE ACID.   Place 60 g (58 mL) in a 250-mL conical flask. Add 0.15 mL of phenol red indicator, and titrate with 0.01 N sodium hydroxide to a pink color. Not more than 3.0 mL should be required.

CHLORIDE.   (Page 25). Use 10 g (9.7 mL).

WATER.   (Page 54, Method 1). Use 20 mL (19.4 g).

Propionic Acid
Propanoic Acid

$C_3H_6O_2$ Formula Wt 74.08

CAS Number 79–09–4

REQUIREMENTS

Assay . . . . $\geq$ 99.5%
$$CH_3CH_2COOH$$

MAXIMUM ALLOWABLE

Color (APHA) 20
Residue after evaporation 0.01%
Readily oxidizable substances (as HCOOH) 0.10%
Heavy metals (as Pb) 0.001%
Carbonyl compounds . . . . 0.002% of
formaldehyde, of
acetone, or of
acetaldehyde
plus propionalde-
hyde
(as the latter)
Water (H$_2$O) 0.15%

TESTS

ASSAY. (By acid–base titrimetry). Weigh accurately 3.0 g and dissolve in 100 mL of carbon dioxide-free water. Add 0.15 mL of phenolphthalein indicator solution and titrate with 1 N sodium hydroxide to the first appearance of a faint pink end point that persists for at least 30 s. One milliliter of 1 N sodium hydroxide corresponds to 0.07408 g of CH$_3$CH$_2$COOH.

COLOR (APHA). (Page 17).

RESIDUE AFTER EVAPORATION. (Page 14). Evaporate 20 g (20 mL) to dryness in a tared dish on the steam bath and dry the residue at 105°C for 30 min.

READILY OXIDIZABLE SUBSTANCES. Dissolve 7.5 g of sodium hydroxide in 50 mL of water, cool, add 3 mL of bromine, stir until dissolved, and dilute to 1000 mL with water. Transfer 25.0 mL of this solution to a glass-stoppered conical flask containing 100 mL of water, and add 10 mL of a 1 in 5 solution of sodium acetate trihydrate and 10.0 mL of the sample. Allow to stand for 15 min, then add 1.0 g of potassium iodide dissolved in 5 mL of water, mix, and add 10 mL of hydrochloric acid. Titrate with 0.1 N sodium thiosulfate just to the disappearance of the brown color. Perform a blank determination. One milliliter of 0.1 N sodium thiosulfate corresponds to 0.004603 g of formic acid (HCOOH). The difference between the blank and sample titrations should not exceed 2.2 mL.

HEAVY METALS. (Page 26, Method 1). Add about 25 mg of sodium carbonate to 3.0 g of the sample in a platinum dish and evaporate to dryness on the steam bath. Dissolve the residue in 1 mL of dilute hydrochloric acid (1 + 19) and dilute with water to 30 mL. Use 25 mL to prepare the sample solution, and use the remaining 5 mL to prepare the control solution.

CARBONYL COMPOUNDS. (Page 49). Use 1.0-g (1.0-mL) samples diluted with 4 mL of water and neutralized to a pH between 6.5 and 6.8 with 10% sodium hydroxide. For the standards use 0.02 mg each of formaldehyde (HCHO), propionaldehyde (CH_3CH_2CHO), and acetone (CH_3COCH_3). The peak potentials for acetaldehyde (CH_3CHO) and propionaldehyde overlap at about -1.15 to $-1.25V$.

WATER. (Page 54, Method 1). Use 10.0 mL (9.9 g) of the sample.

n-Propyl Alcohol
1-Propanol

$CH_3(CH_2)_2OH$ Formula Wt 60.10

CAS Number 71 –23 –8

REQUIREMENTS

Assay (GC) . . . . ≥ 99.5%
$CH_3(CH_2)_2OH$
Solubility in water . . . . Passes test

MAXIMUM ALLOWABLE

Color (APHA). 10
Residue after evaporation 0.001%
Water (H_2O) 0.2%
Titrable acid . . . . 0.0004 meq / g
Ethyl alcohol (CH_3CH_2OH) 0.01%
Methanol (CH_3OH) 0.01%
Isopropyl alcohol ($CH_3CHOHCH_3$) 0.05%
Carbonyl compounds (as C_2H_5CHO). 0.03%

TESTS

ASSAY, ETHYL ALCOHOL, METHANOL, AND ISOPROPYL ALCOHOL. Analyze the sample by gas chromatography, using the general parameters cited on page 56. The following specific conditions are also required.

Column: Type I, methyl silicone

Detector: Flame ionization

Measure the area under all peaks and calculate the area percent for methanol, ethyl alcohol, isopropyl alcohol, and *n*-propyl alcohol. Correct for water content.

COLOR. (APHA). (Page 17).

RESIDUE AFTER EVAPORATION. (Page 14). Evaporate 100 g (124 mL) to dryness in a tared dish on the steam bath and dry the residue at 105°C for 30 min.

SOLUBILITY IN WATER. Mix 10 mL with 40 mL of water and allow to stand for 1 h. The solution should be as clear as an equal volume of water.

WATER. (Page 54, Method 1). Use 10.0 mL (8.0 g) of the sample.

TITRABLE ACID. Add 10 mL to 30 mL of carbon dioxide-free water in a glass-stoppered flask and add 0.5 mL of phenolphthalein indicator solution. Add 0.01 N sodium hydroxide until a slight pink color persists after shaking for 30 s. Add 25 g (31 mL) of the sample, mix well, and titrate with 0.01 N sodium hydroxide until the pink color is restored. Not more than 1.0 mL of the sodium hydroxide solution should be required.

CARBONYL COMPOUNDS. (Page 49). Use 1.0 g (1.25 mL). For the standard, use 0.30 mg of propionaldehyde (C_2H_5CHO).

Pyridine

C_5H_5N Formula Wt 79.10

CAS Number 110–86–1

REQUIREMENTS

Assay . . . . ≥ 99.0% C_5H_5N
Solubility in water . . . . Passes test

MAXIMUM ALLOWABLE

Residue after evaporation 0.002%
Water (H_2O) 0.1%
Chloride (Cl) 0.001%
Sulfate (SO_4) 0.001%
Ammonia (NH_3) 0.002%
Copper (Cu) . . . . Passes test
                                                            (limit about
                                                            5 ppm)
Reducing substances . . . . Passes test

TESTS

ASSAY. Analyze the sample by gas chromatography, using the general parameters cited on page 56. The following specific conditions are also required.

Column: Type I, methyl silicone

Measure the area under all peaks and calculate the pyridine content in area percent. Correct for water content.

SOLUBILITY IN WATER. Dilute 10 mL with 90 mL of water. The solution should show no turbidity in 30 min.

RESIDUE AFTER EVAPORATION. (Page 14). Evaporate 50 g (51 mL) to dryness on the steam bath and dry the residue at 105°C for 30 min.

WATER.   (Page 54, Method 1). Use 20 mL (19.6 g) of the sample.

CHLORIDE.   (Page 25). Dilute 1.0 g (1.0 mL) with water to 25 mL.

SULFATE.   (Page 32, Procedure A, Method 3).

AMMONIA.   Add 2 g (2.0 mL) to 10 mL of carbon dioxide-free water. Add 0.10 mL of phenolphthalein indicator solution. If a pink color develops it should be discharged by not more than 0.24 mL of 0.01 N hydrochloric acid.

COPPER.   Add 5 g (5.0 mL) to 15 mL of water; add 2 mL of glacial acetic acid, 2 mL of ammonium thiocyanate reagent solution, and 5 mL of chloroform. Shake vigorously in a 60-mL separatory funnel and allow to separate. The chloroform layer should not be colored green and, at most, only faintly yellow.

REDUCING SUBSTANCES.   To 5 g (5.0 mL) add 0.50 mL of 0.1 N potassium permanganate solution. The pink color should not be entirely discharged in 30 min. If a brown color interferes, the solution should be centrifuged or filtered through a sintered glass filter. If a pink color persists, the sample passes.

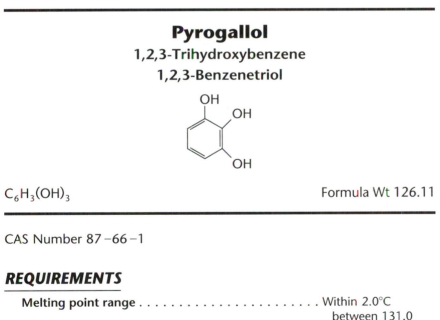

Pyrogallol
1,2,3-Trihydroxybenzene
1,2,3-Benzenetriol

$C_6H_3(OH)_3$ Formula Wt 126.11

CAS Number 87–66–1

REQUIREMENTS

Melting point range . . . . Within 2.0°C
                                                          between 131.0
                                                          and 135.0 °C

MAXIMUM ALLOWABLE

Residue after ignition 0.005%
Chloride (Cl) 0.001%
Sulfate (SO₄) 0.005%
Heavy metals (as Pb) . . . . 5 ppm
Iron (Fe) 0.001%

TESTS

RESIDUE AFTER IGNITION. (Page 15). Ignite 20 g, and use 1 mL of sulfuric acid. Retain the residue.

CHLORIDE. (Page 25). Dissolve 1.0 g in 20 mL of water, and add 3 mL of 10% nitric acid.

SULFATE. (Page 32, Procedure A, Method 1).

> *Sample Solution A.* To the residue retained from the test for residue after ignition, add 3 mL of hydrochloric acid and 1 mL of nitric acid, and evaporate to dryness on the steam bath. Take up the residue in 10 mL of water and 1 mL of hydrochloric acid (1 + 19), filter if necessary, and wash to a volume of 40 mL (1 mL = 0.5 g).

HEAVY METALS. (Page 27, Method 2). Use 4.0 g.

IRON. (Page 28, Method 1). Dilute 20 mL of sample solution A (10-g sample) with water to 100 mL, and use 10 mL.

8-Quinolinol
8-Hydroxyquinoline
Oxine

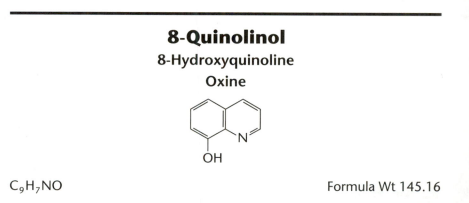

C₉H₇NO   Formula Wt 145.16

CAS Number 148−24−3

REQUIREMENTS

Melting point . . . . 72.5–74.0°C
Suitability for magnesium determination . . . . Passes test

MAXIMUM ALLOWABLE

Insoluble in alcohol 0.05%
Residue after ignition 0.05%
Sulfate (SO_4) . . . . Passes test
(limit about
0.02%)

TESTS

INSOLUBLE IN ALCOHOL.   Dissolve 3.0 g in 40 mL of alcohol, filter through a tared filtering crucible, wash with 95% alcohol, and dry at 105°C.

RESIDUE AFTER IGNITION.   (Page 15). Ignite 2.0 g. Use 1 mL of sulfuric acid.

SULFATE.   Dissolve 1.0 g in 1 mL of hydrochloric acid and dilute with 20 mL of water. Heat to boiling and add 1 mL of barium chloride reagent solution. No turbidity should be observed after standing 30 min.

SUITABILITY FOR MAGNESIUM DETERMINATION.   Dissolve 2.5 g in 5 mL of glacial acetic acid, warming if necessary, and dilute with water to 100 mL. Add 3.5 mL of this solution to 50 mL of a solution containing 6 mg of magnesium ion (Mg). Heat to 80°C and add with stirring 2 mL of ammonium hydroxide and allow to stand for 10 min. Filter, and to the filtrate, which should be alkaline and yellow in color, add a solution containing 3 mg of magnesium ion (Mg) and heat to 80°C. The characteristic yellow magnesium quinolate precipitate should form.

Reagent Alcohol
Alcohol, Reagent

NOTE. This material is a denatured form of ethyl alcohol, approved for sale under U.S. Regulations governing the stated description or equivalent, consisting of about 5 volumes of isopropyl alcohol and about 95 volumes of formula 3-A specially denatured alcohol (which consists of about 5 volumes of methanol and about 100 volumes of ethyl alcohol).

REQUIREMENTS

Assay . . . . Within 94.0–96.0% (v / v) methanol and ethyl alcohol and 4.0–6.0% (v / v) isopropyl alcohol

MAXIMUM ALLOWABLE

Water (H$_2$O) 0.5%
Color (APHA) 10
Residue after evaporation 0.001%

TESTS

ASSAY.   Analyze the sample by gas chromatography, using the general parameters cited on page 56. The following specific conditions are also required.

Column: Type I, methyl silicone

Measure the area under all peaks and calculate the ethyl alcohol plus methanol and isopropyl alcohol content, correcting for response factors. Correct for water content.

WATER.   (Page 54, Method 1). Use 10 mL (7.4 g) of the sample.

COLOR.   (APHA).   (Page 17).

RESIDUE AFTER EVAPORATION.   (Page 14). Evaporate 100 g (124 mL) to dryness in a tared dish on the steam bath and dry the residue at 105°C for 30 min.

Reinecke Salt

Ammonium Tetra(thiocyanato)diamminechromate(III) Monohydrate

Ammonium Diamminetetrakis(thiocyanato-N)chromate(1 −)

NH$_4$[Cr(SCN)$_4$(NH$_3$)$_2$] · H$_2$O   Formula Wt 354.45

CAS Number 13573−16−5

CAUTION. Aqueous solutions of Reinecke salt decompose slowly with the evolution of hydrogen cyanide gas, HCN. Decomposition is rapid above 65°C.

REQUIREMENTS

ASSAY . . . . ≥ 93.0%
                                NH$_4$[Cr(SCN)$_4$(NH$_3$)$_2$] · H$_2$O
Insoluble in dilute hydrochloric acid . . . . ≤ 0.05%
Sensitivity . . . . Passes test

TESTS

ASSAY. (By gravimetric determination as mercury salt). Weigh accurately 0.5 g and dissolve in 100 mL of water containing 0.5 mL of hydrochloric acid. Add a hot solution containing 0.5 g of mercuric acetate dissolved in a mixture of 140 mL of water and 10 mL of hydrochloric acid. Digest on a steam bath for 5 min, filter through a tared porcelain filtering crucible, and wash several times with hot water. The precipitate can be dried at 105°C and weighed as Hg[Cr(SCN)$_4$(NH$_3$)$_2$]$_2$:

$$\% \text{ Reinecke salt} = \frac{\text{Weight of precipitate} \times 85.67}{\text{Sample weight}}$$

Alternatively, the precipitate can be ignited in a hood and heated to constant weight at 800°C as Cr$_2$O$_3$:

$$\% \text{ Reinecke salt} = \frac{\text{Weight of Cr}_2\text{O}_3 \times 466.3}{\text{Sample weight}}$$

INSOLUBLE IN DILUTE HYDROCHLORIC ACID. Dissolve 2.0 g in 200 mL of dilute hydrochloric acid (1 + 99), heat to boiling, and digest in a covered beaker on the steam bath for 1 h. Filter through a tared filtering crucible, wash thoroughly with water, and dry at 105°C.

SENSITIVITY. Dissolve 50 mg in 10 mL of water. Add 0.2 mL of this solution to 1 mL of a solution of 10 mg of choline chloride in 20 mL of water, and shake gently. A distinct red precipitate should form within 10 s.

Salicylic Acid
2-Hydroxybenzoic Acid

$C_7H_6O_3$ Formula Wt 138.12

CAS Number 69–72–7

REQUIREMENTS

Assay . . . . ≥ 99.0% $C_7H_6O_3$
Melting point . . . . 158.0–161.0°C

MAXIMUM ALLOWABLE

Residue after ignition 0.01%
Chloride (Cl) 0.001%
Sulfate (SO_4) 0.003%
Heavy metals (as Pb) . . . . 5 ppm
Iron (Fe) . . . . 2 ppm
Substances darkened by sulfuric acid . . . . Passes test

TESTS

ASSAY. Analyze the sample by liquid chromatography, using the general method described on page 62. The parameters cited have given satisfactory results.

Weigh 0.5 g of sample in a vial and dissolve in 5 mL of acetonitrile plus 5 mL of mobile phase. Inject duplicate 10-μL aliquots into the chromatograph, and compare the peak areas with those obtained under the same conditions.

Mobile Phase: Acetonitrile–0.01 M phosphoric acid (25/75) at 2 mL/min. Prepare 1 M phosphoric acid by diluting 7 mL of the concentrated (85%) acid to 100 mL with water. Mix 10 mL of this solution with 740 mL of water and 250 mL of acetonitrile. De-gas under vacuum or with a sonic bath.

Column: Decyl (C-8 monomeric phase), 250 × 4.6 mm i.d., 5 μm, 9% C loading, end-capped.

Column Temperature: Ambient

Flow Rate: 2 mL/min

Sample Size: 10 μL

Detector: Ultraviolet at 280 nm

Approximate Retention Times (min): *p*-hydroxybenzoic acid, 1.3; *p*-hydroxyisophthalic acid, 1.8; phenol, 3.0; salicylic acid, 4.1. Actual times may vary, depending upon the age or condition of the column.

RESIDUE AFTER IGNITION. (Page 15). Ignite 20 g. Use 1 mL of sulfuric acid. Retain the residue to prepare sample solution A for the determination of heavy metals and iron.

CHLORIDE. (Page 25). Dissolve 1.0 g in 20 mL of 95% alcohol.

SULFATE. Mix 4.0 g with 1 g of sodium carbonate, add 30 mL of hot water in small portions, and evaporate to dryness. Ignite gently, taking care to protect the mixture from the sulfur in the flame. To the residue add 15 mL of water and 1 mL of 30% hydrogen peroxide, boil for 5 min, and add 2 mL of hydrochloric acid. Evaporate to dryness on a hot plate at low setting, cool, and add 10 mL of water. Filter, wash the filter with two 5-mL portions of water, and dilute the combined filtrate and washings with water to 25 mL. For the standard evaporate 1 mL of 30% hydrogen peroxide, 2 mL of hydrochloric acid, and

1 g of sodium carbonate to dryness on the hot plate. Take up the residue in 10 mL of water, add 0.12 mg of sulfate ion (SO_4), and dilute with water to 25 mL. To both sample and standard solutions add 0.5 mL of 1 N hydrochloric acid and 2 mL of barium chloride reagent solution. Any turbidity in the solution of the sample should not exceed that in the standard. Compare 10 min after adding the barium chloride to the sample and standard solutions.

> ***Sample Solution A for the Determination of Heavy Metals and Iron.*** To the residue retained from the test for residue after ignition, add 2 mL of hydrochloric acid and 0.5 mL of nitric acid, and evaporate to dryness on the steam bath. Warm the residue with 1 mL of 1 N hydrochloric acid, add 30 mL of hot water, cool, and dilute with water to 40 mL (1 mL = 0.5 g).

HEAVY METALS. (Page 27, Method 2). Use 8.0 mL of sample solution A (4-g sample).

IRON. (Page 28, Method 1). Use 10 mL of sample solution A (5-g sample).

SUBSTANCES DARKENED BY SULFURIC ACID. Dissolve 0.5 g in 10 mL of sulfuric acid. The solution should have not more than a pale yellow color.

Silica Gel Desiccant

CAS Number 7631-86-9

REQUIREMENTS

Suitability for moisture absorption . . . . Passes test

TESTS

SUITABILITY FOR MOISTURE ABSORPTION*. Weigh accurately about 5 g in a tared weighing dish. Place the dish for 24 h in a

---

*From *Reagent Chemicals and Standards* by J. Rosin. Copyright © 1967 by Litton Educational Publishing Inc. Reprinted by permission of Van Nostrand Reinhold Company.

desiccator in which the atmosphere possesses a relative humidity of 80%, maintained by equilibrium with sulfuric acid having a specific gravity of 1.19 (27% H_2SO_4). The increase in weight should not be less than 27%.

Silver Diethyldithiocarbamate

$(C_2H_5)_2NCS_2Ag$ Formula Wt 256.14

CAS Number 1470–61–7

> *NOTE.* To enhance stability, storage below 8°C is recommended.

REQUIREMENTS

Solubility in pyridine . . . . Passes test
Suitability for determination of arsenic . . . . Passes test

TESTS

SOLUBILITY IN PYRIDINE. Transfer 1 g of sample to a 200-mL volumetric flask and dilute to the mark with freshly distilled pyridine. The solution should be clear, bright yellow, and dissolution should be complete. (Save this solution for the suitability for determination of arsenic test.)

SUITABILITY FOR DETERMINATION OF ARSENIC. Transfer 0.002 mg of arsenic (As) to an arsine generator flask* and dilute with water to 35 mL. Add 20 mL of dilute sulfuric acid (1 + 4), 2 mL of 16.5% potassium iodide reagent solution, and 0.5 mL of 40% stannous chloride dihydrate reagent solution in concentrated hydrochloric acid. Mix and allow to stand for 30 min at room temperature. Pack the scrubber tube loosely with two pledgets of lead acetate cotton. Place 3.0 mL of sample solution, retained from the solubility in pyridine test, in the absorber tube. Quickly add 3.0 g of granulated (No. 20

---

*For a description of the apparatus used in this test, see the section on arsenic, page 23.

mesh) zinc to the arsine generator flask and immediately connect the flask to the scrubber–absorber assembly. Allow the evolution of hydrogen to proceed at room temperature for 45 min, swirling the solution every 10 min. Disconnect the absorber tube and determine the absorbance of the solution in a 1.00-cm cell at 525 nm, against a fresh portion of the sample solution in a similar matched cell set at zero absorbance as the reference liquid. The absorbance should not be less than 0.10.

Silver Nitrate

AgNO$_3$ Formula Wt 169.87

CAS Number 7761–88–8

REQUIREMENTS

Assay . . . . ≥ 99.0% AgNO$_3$
Clarity of solution . . . . Passes test

MAXIMUM ALLOWABLE

Chloride (Cl) . . . . 5 ppm
Free acid . . . . Passes test
Substances not precipitated by
  hydrochloric acid 0.01%
Sulfate (SO$_4$) 0.002%
Copper (Cu) . . . . 2 ppm
Iron (Fe) . . . . 2 ppm
Lead (Pb) 0.001%

TESTS

ASSAY.   (By argentometric titrimetry). Weigh, to the nearest 0.1 mg, about 0.65 g of sample in a 250-mL Erlenmeyer flask. Dissolve in 100 mL of water and add 5 mL of nitric acid. Add 2 mL of ferric ammonium sulfate indicator solution and titrate with 0.1 N ammonium thiocyanate to the first appearance of a reddish-brown end point. One milliliter of 1 N ammonium thiocyanate corresponds to 0.001699 g of AgNO$_3$.

CLARITY OF SOLUTION.   Dissolve 20 g in 100 mL of water in a 250-mL conical flask. The solution should be clear, and no significant amount of insoluble matter, such as minute fibers and/or other particles of any shape or color, should be observed. Reserve the solution for the test for substances not precipitated by hydrochloric acid.

CHLORIDE.   Dissolve 2.0 g in 40 mL of water, add ammonium hydroxide dropwise until the precipitate first formed is redissolved, and dilute with water to 50 mL. Transfer the solution to a 100-mL platinum dish that is made the cathode. Insert a rotating anode and electrolyze for 1 h, starting with a current of 1 ampere. Decant the solution and evaporate to approximately 25 mL. Neutralize the solution to the phenolphthalein end point with nitric acid, and add 1 mL of nitric acid in excess and 1 mL of silver nitrate reagent solution. Any turbidity should not exceed that produced by 0.01 mg of chloride ion (Cl) in an equal volume of solution containing 1 mL of nitric acid and 1 mL of silver nitrate reagent solution.

FREE ACID.   Dissolve 5.0 g in 50 mL of water, add 0.25 mL of bromcresol green indicator solution (0.04%), and mix well. The solution should be colored blue, not green or yellow.

SUBSTANCES NOT PRECIPITATED BY HYDROCHLORIC ACID.   Dilute the solution obtained in the test for clarity of solution to about 600 mL. Heat to boiling and add hydrochloric acid to precipitate the silver completely (about 11 mL). Allow to stand overnight and filter. Evaporate the filtrate to dryness, and add 0.15 mL of hydrochloric acid and 10 mL of water. Heat, filter, and wash with about 10 mL of water. Evaporate the filtrate to dryness in a tared dish or crucible and dry at 105°C. Correct for the weight obtained in a complete blank test. Reserve the residues for the preparation of sample solution A and blank solution B.

> *Sample Solution A and Blank Solution B.*   To each of the residues remaining from the test for substances not precipitated by hydrochloric acid add 3 mL of dilute hydrochloric acid (1 + 1), cover with a watch glass, and digest on the steam bath for 15–20 min. Cool and dilute with water to 100 mL. One is sample solution A (1 mL = 0.2 g); the other is blank solution B.

SULFATE.   (Page 32, Procedure A, Method 1). Use 12.5 mL of sample solution A. Use 12.5 mL of blank solution B to prepare the standard solution.

COPPER.   Add a slight excess of ammonium hydroxide to 25 mL of sample solution A (5-g sample). For the standard add 0.01 mg of copper ion (Cu) to 25 mL of blank solution B and make slightly alkaline with ammonium hydroxide. Add 10 mL of 0.1% solution of sodium diethyldithiocarbamate to each. Any yellow color in the solution of the sample should not exceed that in the standard. Estimate the copper content to aid in the test for lead.

IRON. (Page 28, Method 1). Use 25 mL of sample solution A (5-g sample). Add the standard to 25 mL of blank solution B.

LEAD. Dilute 25 mL of sample solution A (5-g sample) to 35 mL. For the standard add 0.05 mg of lead ion (Pb) and the amount of copper estimated to be present in the test for copper to 25 mL of blank solution B, and dilute with water to 35 mL. Adjust the pH of the standard and sample solutions to between 3 and 4 (using a pH meter) with 1 N acetic acid or ammonium hydroxide (10% NH_3), dilute with water to 40 mL, and mix. Add 10 mL of freshly prepared hydrogen sulfide water to each and mix. Any color in the solution of the sample should not exceed that in the standard.

Silver Sulfate

Ag_2SO_4 Formula Wt 311.80

CAS Number 10294−26−5

REQUIREMENTS

Assay . . . . ≥ 98.0% Ag_2SO_4

MAXIMUM ALLOWABLE

Insoluble matter and
   silver chloride. 0.02%
Nitrate (NO_3) 0.001%
Substances not precipitated by
   hydrochloric acid 0.03%
Iron (Fe) 0.001%

TESTS

ASSAY. (By argentometric titrimetry). Weigh, to the nearest 0.1 mg, about 0.5 g of sample in a 250-mL Erlenmeyer flask. Dissolve in 50 mL of water and 5 mL of nitric acid. Add 2 mL of ferric ammonium sulfate indicator solution and titrate with 0.1 N ammonium thiocyanate to the first appearance of a reddish-brown end point. One milliliter of 0.1 N ammonium thiocyanate corresponds to 0.001559 g of Ag_2SO_4.

INSOLUBLE MATTER AND SILVER CHLORIDE. Add 5.0 g of the powdered salt to 500 mL of boiling water and boil gently until the silver sulfate is dissolved. If any insoluble matter remains, filter while hot through a tared filtering crucible (retain the filtrate for the test for substances not precipitated by hydrochloric acid), wash thoroughly with hot water (discard the washings), and dry at 105°C.

NITRATE. To 0.50 g of the powdered salt add 2 mL of phenoldisulfonic acid reagent solution and heat on the steam bath for 15 min. Cool, dilute with 20 mL of water, filter, and make alkaline with ammonium hydroxide. Any yellow color should not exceed that produced when a solution containing 0.005 mg of nitrate ion (NO_3) is evaporated to dryness and the residue is treated in the same manner as the sample.

SUBSTANCES NOT PRECIPITATED BY HYDROCHLORIC ACID. Heat to boiling the filtrate obtained in the test for insoluble matter and silver chloride, add 5 mL of hydrochloric acid, and allow to stand overnight. Dilute with water to 500 mL and filter. Evaporate 250 mL of the filtrate to dryness, and add 0.15 mL of hydrochloric acid and 10 mL of water. Heat and filter. Add 0.10 mL of sulfuric acid to the resulting filtrate, evaporate to dryness in a tared dish, and ignite at 800 ± 25°C for 15 min. Correct for the weight obtained in a complete blank test. Retain the residue for the test for iron.

IRON. (Page 28, Method 1). To the residue obtained in the preceding test add 3 mL of dilute hydrochloric acid (1 + 1), cover with a watch glass, and digest on the steam bath for 15–20 min. Remove the watch glass and evaporate to dryness. Dissolve the residue in 35 mL of dilute hydrochloric acid (1 + 6), filter if necessary, and dilute with water to 50 mL. Use 20 mL of the solution without further acidification.

Soda Lime

CAS Number 8006-28-8

> *NOTE.* Soda lime is a mixture of variable proportions of sodium hydroxide with calcium oxide or hydroxide. This reagent may or may not include an indicator.

REQUIREMENTS

Carbon dioxide absorption capacity . . . . $\geq 19.0\%$

MAXIMUM ALLOWABLE

Loss on drying at 200°C. 7%
Fines. 1%

TESTS

CARBON DIOXIDE ABSORPTION CAPACITY.   Fill the lower transverse section of a U-shaped drying tube of about 15-mm internal diameter and 15-cm height with loosely packed glass wool. Place in one arm of the tube about 5 g of anhydrous calcium chloride, and accurately weigh the tube and its contents. Into the other arm of the tube place 9.5–10.5 g of soda lime, and again weigh accurately. Insert stoppers in the open arms of the U-tube, and connect the side tube of the arm filled with soda lime to a calcium chloride drying tube, which in turn is connected to a suitable source of carbon dioxide. Pass the carbon dioxide though the U-tube at a rate of 75 mL per minute for 30 min, accurately timed. Disconnect the U-tube, cool to room temperature, remove the stoppers, and weigh.

LOSS ON DRYING.   Weigh accurately about 10 g and dry at 200 °C for 18 h.

FINES.   Place 100 g on a clean U.S. No. 100 standard sieve nested in a receiving pan. Cover the sieve and shake on a mechanical shaker for 5 min. The weight of the fine material in the receiving pan should not exceed 1.0 g.

Sodium

Na   Atomic Wt 22.99

CAS Number 7440–23–5

REQUIREMENTS

MAXIMUM ALLOWABLE

Chloride (Cl) 0.002%
Nitrogen (N) . . . . Passes test (limit
                                                         about 0.003%)
Phosphate (PO_4) . . . . 5 ppm
Sulfate (SO_4) . . . . Passes test (limit
                                                         about 0.002%)
Heavy metals (as Pb) . . . . 5 ppm
Iron (Fe) 0.001%

TESTS

Sample Solution A. If the metal contains any adhering oil or other foreign material, shave off a thin layer and use only bright clean metal for the sample. Weigh 20 g and cut it into small pieces. Cool about 100 mL of water in a beaker in an ice bath. Add the small pieces of sodium one at a time to the ice-cold water. Keep the solution cool and do not add another piece until the preceding one has completely reacted and dissolved. If desired, a magnetic stirrer may be used to aid dissolution. After all the sample has been dissolved, cool, and dilute with water to 500 mL (1 mL = 0.04 g).

CHLORIDE. (Page 25). Neutralize 12.5 mL of sample solution A (0.5-g sample) with nitric acid.

NITROGEN. (Page 30). Use 12.5 mL of sample solution A (0.5-g sample). For the standard use 0.015 mg of nitrogen (N).

PHOSPHATE. (Page 30, Method 1). To 100 mL of sample solution A (4-g sample) add 15 mL of hydrochloric acid and evaporate to about 30 mL. Add 30 mL of hydrochloric acid, filter through a filtering crucible, and wash the precipitated sodium chloride twice with 5-mL portions of hydrochloric acid. Evaporate the filtrate and washings to dryness on the steam bath. Dissolve the residue in 25 mL of approximately 0.5 N sulfuric acid, and continue as described.

SULFATE.   Neutralize 125 mL of sample solution A with hydrochloric acid and evaporate to about 90 mL. Add 1 mL of hydrochloric acid and 5 mL of barium chloride reagent solution and allow to stand overnight. No turbidity or precipitate should be formed.

HEAVY METALS.   (Page 26, Method 1). To 125 mL of sample solution A add 25 mL of hydrochloric acid, and evaporate to dryness on the steam bath. Dissolve the residue in about 20 mL of water, and dilute with water to 25 mL. For the control add 0.02 mg of lead ion (Pb) to 25 mL of sample solution A, and treat exactly as the 125 mL of sample solution A.

IRON.   (Page 28, Method 1). Neutralize 25 mL of sample solution A (1-g sample) with hydrochloric acid, add 2 mL of hydrochloric acid in excess, and dilute with water to 50 mL. Use the solution without further acidification.

Sodium Acetate
Sodium Acetate, Anhydrous

CH_3COONa Formula Wt 82.03

CAS Number 127–09–3

REQUIREMENTS

Assay . . . . $\geq$ 99.0%
$C_2H_3O_2Na$
pH of a 5% solution at 25°C 7.0–9.2

MAXIMUM ALLOWABLE

Insoluble matter 0.01%
Loss on drying at 120°C. 1.0%
Chloride (Cl) 0.002%
Phosphate (PO_4) 0.001%
Sulfate (SO_4). 0.003%
Calcium, magnesium, and R_2O_3 precipitate 0.01%
Heavy metals (as Pb) 0.001%
Iron (Fe) 0.001%

TESTS

ASSAY. (Total alkalinity by nonaqueous titration). Weigh accurately 0.3 g of sample into a 125-mL flask, and dissolve in 50 mL of acetic acid and 5 mL of acetic anhydride as a blank. Allow to stand for 15 min. Add 0.10 mL of crystal violet indicator solution to both flasks, and titrate with 0.1 N perchloric acid in acetic acid until the solution color changes from violet to emerald green.

$$\% \ CH_3COONa = \frac{(V_s - V_b) \times N \ HClO_4 \times 8.203}{Sample \ wt \ (g)}$$

where

V_s, V_b = mL of perchloric acid to titrate sample and blank, respectively.

pH OF A 5% SOLUTION. (Page 43). The pH should be 7.0–9.2 at 25°C.

INSOLUBLE MATTER. (Page 14). Use 20 g dissolved in 150 mL of water.

LOSS ON DRYING AT 120°C. Weigh accurately about 2 g in a tared dish or crucible and dry at 120°C to constant weight.

CHLORIDE. (Page 25). Use 0.50 g.

PHOSPHATE. (Page 30, Method 2). Dissolve 2.0 g in 10 mL of nitric acid and evaporate to dryness on the steam bath. Add 10 mL of nitric acid and repeat the evaporation. Dissolve in 80 mL of water, add 0.5 g of ammonium molybdate, and adjust the pH to 1.8 (using a pH meter) with dilute hydrochloric acid (1 + 9). Heat to boiling, cool, add 10 mL of hydrochloric acid, and continue as described. Carry along a standard containing 0.02 mg of phosphate ion (PO_4) treated in the same manner as the sample after the addition of 80 mL of water.

SULFATE. (Page 32, Procedure A, Method 1). Allow 30 min for the turbidity to form.

CALCIUM, MAGNESIUM, AND R_2O_3 PRECIPITATE. Dissolve 10 g in 100 mL of water, filter, and add 5 mL of ammonium oxalate reagent solution, 2 mL of ammonium phosphate reagent solution, and 15 mL of ammonium hydroxide. Allow to stand overnight. Filter, wash with water containing 2.5% ammonia and about 0.1% ammonium oxalate, and ignite.

HEAVY METALS. (Page 26, Method 1). Dissolve 6.0 g in about 10 mL of water, add 15 mL of dilute hydrochloric acid (10%), and dilute with water to 60 mL. Use 25 mL to prepare the sample solution, and use 5.0 mL of the remaining solution to prepare the control solution.

IRON. (Page 28, Method 1). Dissolve 1.0 g in 50 mL of dilute hydrochloric acid (1 + 24), and use the solution without further acidification.

Sodium Acetate Trihydrate

$C_2H_3O_2 \cdot 3H_2O$ Formula Wt 136.08

CAS Number 6131–90–4

REQUIREMENTS

Assay 99.0–100.5%
$NaC_2H_3O_2 \cdot 3H_2O$
pH of a 5% solution at 25°C 7.5–9.2
Substances reducing permanganate . . . . Passes test

MAXIMUM ALLOWABLE

Insoluble matter 0.005%
Chloride (Cl) 0.001%
Phosphate (PO_4) . . . . 5 ppm
Sulfate (SO_4) 0.002%
Calcium, magnesium, and R_2O_3 precipitate 0.01%
Heavy metals (as Pb) . . . . 5 ppm
Iron (Fe) . . . . 5 ppm
Potassium (K) 0.005%

TESTS

ASSAY. (Total alkalinity by nonaqueous titration). Weigh accurately 0.5 g of sample into a 125-mL flask, and dissolve in 50 mL of acetic acid and 5 mL of acetic anhydride. Use a second flask containing 50 mL of acetic acid and 5 mL of acetic anhydride as a blank. Add 0.10 mL of crystal violet indicator solution to both flasks, and titrate with

0.1 N perchloric acid in acetic acid until the solution color changes from violet to emerald green.

$$\% \ CH_3COONa \cdot 3H_2O = \frac{(V_s - V_b) \times N \ HClO_4 \times 13.61}{Sample \ wt \ (g)}$$

where

V_s, V_b = mL of perchloric acid to titrate sample and blank, respectively

pH OF A 5% SOLUTION.   (Page 43). The pH should be 7.5–9.2 at 25°C.

INSOLUBLE MATTER.   (Page 14). Use 20 g dissolved in 150 mL of water.

CHLORIDE.   (Page 25). Use 1.0 g.

PHOSPHATE.   (Page 30, Method 2). Dissolve 2.0 g in 10 mL of nitric acid and evaporate to dryness on the steam bath. Add 10 mL of nitric acid and repeat the evaporation. Dissolve in 80 mL of water, add 0.5 g of ammonium molybdate tetrahydrate, $(NH_4)_6Mo_7O_{24} \cdot 4H_2O$, and adjust the pH to 1.8 (using a pH meter) with dilute hydrochloric acid $(1 + 9)$. Heat to boiling, cool, and add 10 mL of hydrochloric acid, and continue as described. Carry along a standard containing 0.01 mg of phosphate ion (PO_4) treated in the same manner as the sample after the addition of 80 mL of water.

SULFATE.   (Page 32, Procedure A, Method 1). Allow 30 min for the turbidity to form.

CALCIUM, MAGNESIUM, AND R_2O_3 PRECIPITATE.   Dissolve 10 g in 100 mL of water, filter, and add 5 mL of ammonium oxalate reagent solution, 2 mL of ammonium phosphate reagent solution, and 15 mL of ammonium hydroxide. Allow to stand overnight. Filter, wash with water containing 2.5% ammonia and about 0.1% ammonium oxalate, and ignite.

HEAVY METALS.   (Page 26, Method 1). Dissolve 6.0 g in about 10 mL of water, add 15 mL of dilute hydrochloric acid (10%), and dilute with water to 30 mL. Use 25 mL to prepare the sample solution, and use 5.0 mL of the remaining solution to prepare the control solution.

IRON. (Page 28, Method 1). Dissolve 2.0 g in 50 mL of dilute hydrochloric acid (1 + 24), and use the solution without further acidification.

SUBSTANCES REDUCING PERMANGANATE. Dissolve 5.0 g in 50 mL of water, and add 5 mL of 10% sulfuric acid reagent solution and 0.10 mL of 0.1 N potassium permanganate. The pink color should persist for at least 1 h.

POTASSIUM.   Determine the potassium by the flame atomic absorption spectrophotometric method described on page 35.

Sample Stock Solution.   Dissolve 10.0 g of sample in water in a 100-mL volumetric flask, and dilute to the mark with water (0.1 g/mL).

Analyze the solutions by means of a suitable atomic absorption spectrophotometer, using the conditions outlined in the following table. Calculate the metal content of the sample by the method of standard additions.

Element	Wavelength (nm)	Sample Wt (g)	Standard Added (mg)	Flame Type*	Background Correction
K	766.5	1.0	0.025; 0.05	A/A	No

*A/A is air/acetylene.

Sodium Arsenate Heptahydrate
Disodium Hydrogen Arsenate Heptahydrate

$Na_2HAsO_4 \cdot 7H_2O$ Formula Wt 312.01

CAS Number 10048–95–0

REQUIREMENTS

Assay 98.0–102.0%
$Na_2HAsO_4 \cdot 7H_2O$

MAXIMUM ALLOWABLE

Insoluble matter 0.005%
Arsenite (As_2O_3) 0.01%
Chloride (Cl) 0.001%
Nitrate (NO_3) 0.005%
Sulfate (SO_4) 0.01%
Heavy metals (as Pb) 0.002%
Iron (Fe) 0.001%

TESTS

ASSAY. (By titration of oxidizing power of arsenate). Weigh accurately 0.55 g and dissolve in 50 mL of water in a glass-stoppered conical flask. Heat to 80°C and add 10 mL of hydrochloric acid and 3 g of potassium iodide. Stopper the flask, swirl, and maintain at 80°C for 15 min. Cool to room temperature and titrate the liberated iodine with 0.1 N sodium thiosulfate, adding 3 mL of starch indicator solution near the end of the titration. One milliliter of 0.1 N sodium thiosulfate corresponds to 0.01560 g of $Na_2HAsO_4 \cdot 7H_2O$.

INSOLUBLE MATTER. (Page 14). Use 20 g dissolved in 200 mL of water.

ARSENITE. Dissolve 10 g in 75 mL of water, make the solution just acid to phenolphthalein with 10% sulfuric acid, and add 2 g of sodium bicarbonate. When dissolution is complete, add starch indicator solution, and titrate with 0.02 N iodine to a blue end point. Not more than 1.0 mL of 0.02 N iodine should be required.

CHLORIDE. (Page 25). Use 1.0 g of sample and 2 mL of nitric acid.

NITRATE

Sample Solution A. Dissolve 0.20 g in 3 mL of water by heating in a boiling-water bath. Dilute to 50 mL with brucine sulfate reagent solution.

Control Solution B. Dissolve 0.20 g in 2 mL of water and 1 mL of the standard nitrate solution containing 0.01 mg of nitrate ion (NO_3) per milliliter by heating in a boiling-water bath. Dilute to 50 mL with brucine sulfate reagent solution.

Continue with the procedure described on page 29, starting with the preparation of blank solution C.

SULFATE. Dissolve 10 g in 10 mL of water, add 5 mL of hydrochloric acid, and heat the solution to boiling. Add 5 mL of barium chloride reagent solution, digest in a covered beaker on a hot plate at low setting for 2 h, and allow to stand overnight. If any precipitate is formed, filter, wash thoroughly, and ignite. Correct for the weight obtained in a complete blank test.

HEAVY METALS. (Page 26, Method 1). Dissolve 2.5 g in 20 mL of water in a small dish. Add 2 g of potassium iodide, 10 mL of hydrobromic acid, and 0.10 mL of sulfuric acid. Evaporate to dryness on the steam bath, wash down the sides of the dish with a few milliliters of water, add 5 mL of hydrochloric acid, and again evaporate to dryness on the steam bath. Dissolve the residue in a few milliliters of water, neutralize to litmus with ammonium hydroxide (10% NH_3), and dilute with water to 50 mL. Use 30 mL to prepare the sample solution, and use 10 mL of the remaining solution to prepare the control solution.

IRON. (Page 28, Method 1). Use 1.0 g of sample and 5 mL of hydrochloric acid.

Sodium Bicarbonate
Sodium Hydrogen Carbonate

$NaHCO_3$ Formula Wt 84.01

CAS Number 144−55−8

REQUIREMENTS

Assay (dried basis) . . . . 99.7−100.3%
$NaHCO_3$

MAXIMUM ALLOWABLE

Insoluble matter 0.015%
Chloride (Cl) 0.003%
Phosphate (PO_4) 0.001%
Sulfur compounds (as SO_4) 0.003%
Ammonium (NH_4) . . . . 5 ppm
Calcium, magnesium, and
    R_2O_3 precipitate 0.02%
Heavy metals (as Pb) . . . . 5 ppm
Iron (Fe) 0.001%
Potassium (K) 0.005%

TESTS

ASSAY. (By acid–base titrimetry). Weigh accurately 3 g, previously dried over magnesium perchlorate or phosphorus pentoxide for 24 h, dissolve it in 50 mL of water, add methyl orange indicator solution, and titrate with 1 N hydrochloric acid. The sodium bicarbonate content calculated from the total alkalinity, as determined by the titration, should not be less than 99.7 nor more than 100.3% of the weight taken. One milliliter of 1 N hydrochloric acid corresponds to 0.08401 g of $NaHCO_3$.

INSOLUBLE MATTER. (Page 14). Use 10 g dissolved in 100 mL of hot water.

CHLORIDE. (Page 25). Use 0.33 g, and neutralize with nitric acid.

PHOSPHATE. (Page 30, Method 1). Dissolve 2.0 g in 15 mL of dilute hydrochloric acid (1 + 2), and evaporate to dryness on the steam bath. Dissolve the residue in 25 mL of approximately 0.5 N sulfuric acid, and continue as described.

SULFUR COMPOUNDS. Dissolve 10 g in 100 mL of hot water, add 0.25 mL of bromine water, and boil. Cool, neutralize with hydrochloric acid, add an excess of 1 mL of the acid, and filter. Heat the filtrate to boiling, add 5 mL of barium chloride reagent solution, digest in a covered beaker on the steam bath for 2 h, and allow to stand overnight. If a precipitate is formed, filter, wash thoroughly, and ignite. Correct for the weight obtained in a complete blank test.

AMMONIUM. Dissolve 2.0 g in 40 mL of ammonia-free water, and add 10 mL of 10% sodium hydroxide reagent solution and 2 mL of Nessler reagent. Any color should not exceed that produced by 0.01

mg of ammonium ion (NH_4) in an equal volume of solution containing the quantities of reagents used in the test.

CALCIUM, MAGNESIUM, AND R_2O_3 PRECIPITATE.
Dissolve 10 g in 100 mL of water and add 12 mL of hydrochloric acid. Boil to remove carbon dioxide and filter if necessary. Add 5 mL of ammonium oxalate reagent solution, 2 mL of ammonium phosphate reagent solution, and 10 mL of ammonium hydroxide. Stir well and allow to stand overnight. If any precipitate is formed, filter, wash with water containing 2.5% of ammonia and about 0.1% of ammonium oxalate, and ignite.

HEAVY METALS.
(Page 26, Method 1). To 5.0 g in a 150-mL beaker add 10 mL of water, mix, and cautiously add 10 mL of hydrochloric acid. Evaporate to dryness on the steam bath, dissolve residue in about 20 mL of water, and dilute with water to 25 mL. For the control solution add 0.02 mg of lead ion (Pb) to 1.0 g of sample, and treat exactly as the 5.0 g of sample.

IRON.
(Page 28, Method 1). Dissolve 1.0 g in 30 mL of dilute hydrochloric acid (1 + 9), dilute with water to 50 mL, and use the solution without further acidification.

POTASSIUM.
Determine the potassium by the flame atomic absorption spectrophotometric method described on page 35.

> ***Sample Stock Solution.*** Dissolve 10.0 g of sample with water in a 100-mL volumetric flask, and dilute to the mark with water (0.1 g/mL).

Analyze the solutions by means of a suitable atomic absorption spectrophotometer, using the conditions outlined in the following table. Calculate the metal content of the sample by the method of standard additions.

Element	Wavelength (nm)	Sample Wt (g)	Standard Added (mg)	Flame Type*	Background Correction
K	766.5	1.0	0.05; 0.10	A/A	No

*A/A is air/acetylene.

Sodium Bismuthate

NaBiO$_3$ Formula Wt 279.97

CAS Number 12232−99−4

REQUIREMENTS

Assay . . . . ≥ 80.0% NaBiO$_3$
Oxidizing efficiency . . . . ≥ 99.6%

MAXIMUM ALLOWABLE

Chloride (Cl) 0.002%
Manganese (Mn) . . . . 5 ppm

TESTS

ASSAY. (By titration of oxidizing power). Weigh accurately 0.7 g, place in a flask, add 25.0 mL of ferrous sulfate solution, and stopper the flask. Transfer 25.0 mL of the ferrous sulfate solution to another flask and stopper the flask. Allow each flask to stand for 30 min, shaking frequently, and titrate the ferrous sulfate in each with 0.1 N potassium permanganate. The difference in the volume of permanganate consumed in the two titrations is equivalent to the sodium bismuthate. One milliliter of 0.1 N potassium permanganate corresponds to 0.01400 g of NaBiO$_3$.

> ***Ferrous Sulfate Solution.*** Dissolve 7 g of clear crystals of ferrous sulfate, FeSO$_4$ · 7H$_2$O, in 90 mL of freshly boiled and cooled water, and add sulfuric acid to make 100 mL. The solution must be freshly prepared.

CHLORIDE. Add 1.0 g to 25 mL of water, heat to boiling, and keep at the boiling temperature for 10 min. Dilute with water to 50 mL and filter through a chloride-free filter. To 25 mL of the filtrate add 0.15 mL of 30% hydrogen peroxide to clear the solution, and then add 1 mL of nitric acid and 1 mL of silver nitrate reagent solution. Any turbidity should not exceed that produced by 0.01 mg of chloride ion (Cl) in an equal volume of solution containing the quantities of reagents used in the test.

MANGANESE. Dissolve 2.0 g in 35 mL of dilute nitric acid (5 + 2), heat to boiling, and boil gently for 5 min. Prepare a standard containing 0.01 mg of manganese ion (Mn) in 35 mL of dilute nitric acid (5 + 2). To each add 5 mL of sulfuric acid, 5 mL of phosphoric acid, and 0.5 mL of sulfurous acid. Boil gently to expel oxides of nitrogen, cool the solutions to 15°C, and add 0.5 g of sodium bismuthate to each. Allow to stand for 5 min with occasional stirring, dilute each with 25 mL of water, and filter through a filter other than paper. Any pink color in the solution of the sample should not exceed that in the standard.

> *Manganese Metal for Use as Oxidimetric Standard.* Assay a selected lot of commercial high-purity electrolytic manganese, previously screened through a number 10 and retained on a number 20 screen, by determining the concentration of impurities. Metals at levels below 0.03% can be evaluated with sufficient accuracy by the spectrographic semiquantitative method. Metals at higher concentrations are determined by suitable quantitative methods; carbon and sulfur can be determined by classical combustion methods. Oxygen, hydrogen, and nitrogen are determined by vacuum fusion analysis employing a 25-g iron bath containing 2–3 g of tin. The sample is placed in a tin capsule and dropped into the bath which is held at 1500–1550°C. Up to three samples can be analyzed before discarding the bath. The assay of the manganese metal for use as an oxidimetric standard should not be less than 99.8% Mn. (Because of the tendency of the metal to react with oxygen, it must be stored in a tightly sealed container after the assay has been performed.)

OXIDIZING EFFICIENCY. To 0.200 g of Oxidimetric Standard manganese metal in a 1-L conical flask, add 15 mL of dilute nitric acid (1 + 3), and heat cautiously until the manganese is dissolved. Add 8 mL of 70% perchloric acid and boil gently until the acid fumes strongly and manganese dioxide begins to separate. Cool, add 5 mL of water and 25 mL of dilute nitric acid (1 + 3), and boil for several minutes to expel free chlorine. Add sufficient sulfurous acid or sodium nitrite solution to just dissolve the manganese dioxide. Boil the solution to expel completely the oxides of nitrogen. Cool to room temperature, add 225 mL of colorless, dilute nitric acid (2 + 5) and sufficient water to bring the total volume to 250 mL, and cool to 10–15°C.

Add 7 g of sodium bismuthate (weighed to the nearest 10 mg) to the flask, agitate briskly for 1 min, dilute with 250 mL of cold water

(10–15°C), and filter immediately through a fine-porosity fritted glass filter (pretreat the frit in hot nitric acid and then wash it free of acid with hot water). The filter can be washed free of manganese more readily if not allowed to run dry during the filtering and washing. Wash the filter with cold, freshly boiled dilute nitric acid (3 + 97) until the washings are entirely colorless, and immediately treat the filtrate and washings as directed in the next paragraph.

Add 8.5 g of ferrous ammonium sulfate heptahydrate (weighed to the nearest mg) to the filtered solution of permanganic acid. Stir briskly. As soon as reduction is complete and all the salt is dissolved, add 0.01 M 1,10-phenanthroline indicator solution, and titrate the excess of ferrous ion with 0.1 N potassium permanganate to a clear green color that persists for at least 30 s. Standardize the 0.1 N potassium permanganate against Oxidimetric Standard sodium oxalate supplied by the NIST.

Determine the manganese equivalent of the ferrous ammonium sulfate heptahydrate by titrating 1.75 g of the salt with the 0.1 N potassium permanganate in 500 mL of cold, dilute nitric acid that has been pretreated with 2 g of sodium bismuthate under the conditions described.

Calculate the oxidizing efficiency of the sodium bismuthate as follows:

$$\text{Oxidizing efficiency} = \frac{(A - B) \times 0.00110 \times 100}{C}$$

where

A = milliliters of exactly 0.1 N potassium permanganate equivalent to the ferrous sulfate added
B = milliliters of exactly 0.1 N potassium permanganate required to titrate the excess ferrous ions
C = grams of manganese used, taking into account the assay of the metal

Sodium Bisulfite

CAS Number 7631 −90 −5 (NaHSO$_3$); 7681 −57 −4 (Na$_2$S$_2$O$_5$)

> *NOTE.* This product is usually a mixture of sodium bisulfite, NaHSO$_3$, and sodium metabisulfite, Na$_2$S$_2$O$_5$.

REQUIREMENTS

Assay . . . . ≥ 58.5% SO$_2$

MAXIMUM ALLOWABLE

Insoluble matter 0.005%
Chloride (Cl) 0.02%
Heavy metals (as Pb) 0.001%
Iron (Fe) 0.002%

TESTS

ASSAY. (Titration of reducing power). Weigh accurately 0.47 g and add to a mixture of 100.0 mL of 0.1 N iodine and 5 mL of 10% hydrochloric acid solution. Swirl gently until the sample is dissolved completely. Titrate the excess of iodine with 0.1 N sodium thiosulfate, adding 3 mL of starch indicator solution near the end of the titration. One milliliter of 0.1 N iodine consumed corresponds to 0.003203 g of SO$_2$.

INSOLUBLE MATTER. (Page 14). Use 20 g dissolved in 200 mL of water.

CHLORIDE. Dissolve 0.50 g in 100 mL of water and transfer 10 mL of the well-mixed solution to a platinum dish. Add 10% sodium hydroxide solution until the solution is slightly alkaline to litmus, making note of the volume of sodium hydroxide added. Prepare a standard containing 0.01 mg of chloride ion (Cl) in 10 mL of water, and add the same volume of 10% sodium hydroxide solution as was added to the sample solution. To each solution add, dropwise, 2 mL of 30% hydrogen peroxide and allow to stand at room temperature for 10 min. Evaporate the solutions to dryness on the steam bath, dissolve the residues in 10 mL of water, and add 1 mL of nitric acid and 1 mL of silver nitrate reagent solution to each. Any turbidity in the solution of the sample should not exceed that of the standard.

HEAVY METALS.   (Page 26, Method 1). Dissolve 3.0 g in a solution of 15 mL of water and 8 mL of hydrochloric acid. Evaporate to dryness on the steam bath, dissolve the residue in about 20 mL of water, and dilute with water to 25 mL. For the control solution add 0.02 mg of lead ion (Pb) to 1.0 g of sample, and treat exactly as the 5.0 g of sample.

IRON.   (Page 28, Method 1). Dissolve 1.0 g in 10 mL of water, add 2 mL of hydrochloric acid, and evaporate to dryness on the steam bath. Dissolve the residue in a mixture of 5 mL of water and 2 mL of hydrochloric acid, and again evaporate to dryness. Dissolve the residue in 4 mL of hydrochloric acid, dilute with water to 100 mL, and use 50 mL of the solution without further acidification.

Sodium Borate Decahydrate
Borax, Sodium Tetraborate Decahydrate

$Na_2B_4O_7 \cdot 10H_2O$ Formula Wt 381.37

CAS Number 1303–96–4

REQUIREMENTS

Assay	99.5–105.0%
	$Na_2B_4O_7 \cdot 10H_2O$
pH of a 0.01 *m* solution	9.15–9.20 at 25°C

MAXIMUM ALLOWABLE

Insoluble matter	0.005%
Chloride (Cl)	0.001%
Phosphate (PO_4)	0.001%
Sulfate (SO_4)	0.005%
Calcium (Ca)	0.005%
Heavy metals (as Pb)	0.001%
Iron (Fe)	5 ppm

TESTS

ASSAY. (By acid–base titrimetry). Weigh accurately 1 g, and dissolve in 50 mL of water. Make slightly acid to methyl red indicator with 1 N hydrochloric acid, cover with a watch glass, and boil gently for 2 min to expel any carbon dioxide. Cool, and adjust to the methyl red end point (pinkish yellow) with carbonate-free 0.5 N sodium hydroxide. Add phenolphthalein indicator and 8 g of mannitol, then titrate with 0.5 N sodium hydroxide through the development of a yellow color to a permanent pink end point. One milliliter of 0.5 N sodium hydroxide corresponds to 0.04767 g of $Na_2B_4O_7 \cdot 10H_2O$. The pH adjustment may be done potentiometrically. The end points are pH 5.4 and 8.5.

pH OF A 0.01 *m* SOLUTION. Dissolve 0.381 g in 100 g of carbon dioxide- and ammonia-free water (or 0.380 g in a volume of 100 mL). Standardize the pH meter and electrode at pH 9.18 at 25°C with 0.01 *m* NIST SRM Sodium Tetraborate Decahydrate. Determine the pH by the method described on page 43. The pH should be 9.15–9.20 at 25°C.

INSOLUBLE MATTER. (Page 14). Use 20 g dissolved in 300 mL of water.

CHLORIDE. (Page 25). Use 1.0 g.

PHOSPHATE. (Page 30, Method 1). Dissolve 2.0 g in 10 mL of warm water, add 2 mL of hydrochloric acid, and evaporate to dryness on the steam bath. Dissolve in 25 mL of approximately 0.5 N sulfuric acid, filter to remove any boric acid, and use the clear filtrate. Continue as described.

SULFATE. Dissolve 8.0 g in 120 mL of warm water plus 6 mL of hydrochloric acid. Filter and wash with 30 mL of water. Heat to boiling, add 5 mL of barium chloride reagent solution, digest in a covered beaker on a hot plate at low setting for 2 h, and allow to stand overnight. Heat to dissolve any boric acid that may be crystallized. If a precipitate is formed, filter, wash thoroughly, and ignite. Correct for the weight obtained in a complete blank test.

CALCIUM. Determine the calcium by the flame atomic absorption spectrophotometric method described on page 35.

Sample Stock Solution. Dissolve 10.0 g of sample with water in a 200-mL volumetric flask, add 10 mL of hydrochloric acid, and dilute to the mark with water (0.05 g/mL).

Analyze the solutions by means of a suitable atomic absorption spectrophotometer, using the conditions outlined in the following table. Calculate the metal content of the sample by the method of standard additions.

Element	Wavelength (nm)	Sample Wt (g)	Standard Added (mg)	Flame Type*	Background Correction
Ca	422.7	1.0	0.05; 0.10	N/A	No

*N/A is nitrous oxide/acetylene.

HEAVY METALS. (Page 26, Method 1). Dissolve 4.0 g in 40 mL of hot water, add 5 mL of glacial acetic acid, and dilute with water to 48 mL. Use 36 mL to prepare the sample solution, and use the remaining 12 mL to prepare the control solution.

IRON. (Page 28, Method 1). Use 2.0 g of sample and 3 mL of hydrochloric acid.

Sodium Bromide

NaBr   Formula Wt 102.89

CAS Number 7647–15–6

REQUIREMENTS

Assay (corrected) . . . . $\geq$ 99.0% NaBr
pH of a 5% solution at 25°C 5.0–8.8

MAXIMUM ALLOWABLE

Insoluble matter 0.005%
Bromate (BrO_3) 0.001%
Chloride (Cl) 0.2%
Nitrogen compounds (as N) . . . . 5 ppm
Sulfate (SO_4) 0.002%
Barium (Ba) 0.002%
Calcium, magnesium, and
  R_2O_3 precipitate 0.005%
Heavy metals (as Pb) . . . . 5 ppm
Iron (Fe) . . . . 5 ppm
Potassium (K) 0.1%

TESTS

ASSAY. (By argentimetric titration of bromide content). Weigh, to the nearest 0.1 mg, 0.4 g of sample. Transfer to a 250-mL titration flask and dissolve in 25 mL of water. Add slowly, while agitating, 50.0 mL of 0.1 N silver nitrate, then add 3 mL of nitric acid and 10 mL of benzyl alcohol, and shake vigorously. Add 2 mL of ferric ammonium nitrate and titrate the excess silver nitrate with 0.1 N ammonium thiocyanate.

% NaBr (uncorrected)

$$= \frac{[(mL \times N\, AgNO_3) - (mL \times N\, NH_4SCN)]10.29}{\text{Sample wt (g)}}$$

% NaBr (corrected) = [% NaBr (uncorrected)] − (2.90 × % Cl)

pH OF A 5% SOLUTION. (Page 43). The pH should be 5.0–8.8 at 25 °C.

INSOLUBLE MATTER. (Page 14). Use 20 g dissolved in 150 mL of water. Save the filtrate separate from the washings for the test for calcium, magnesium, and R_2O_3 precipitate.

BROMATE. (Page 49). Determine bromate by differential pulse polarography. Use 10.0 g of sample in 25 mL of solution. For the standard add 0.10 mg of bromate (BrO_3).

CHLORIDE. Dissolve 0.50 g in 15 mL of dilute nitric acid (1 + 2) in a small flask. Add 3 mL of 30% hydrogen peroxide and digest on the steam bath until the solution is colorless. Wash down the sides of the flask with a little water, digest for an additional 15 min, cool, and dilute with water to 200 mL. Dilute 2.0 mL with water to 20 mL, and add 1 mL of nitric acid and 1 mL of silver nitrate reagent solution. Any turbidity should not exceed that produced by 0.01 mg of chloride ion (Cl) in an equal volume of solution containing the quantities of reagents used in the test.

NITROGEN COMPOUNDS. (Page 30). Use 1.0 g. For the standard use 0.005 mg of nitrogen (N).

SULFATE. (Page 32, Procedure A, Method 1).

BARIUM. For the sample dissolve 6.0 g in 15 mL of water. For the control dissolve 1.0 g in 15 mL of water and add 0.1 mg of barium ion

(Ba). To each solution add 5 mL of acetic acid, 5 mL of 30% hydrogen peroxide, and 1 mL of hydrochloric acid. Digest in a covered beaker on the steam bath until reaction ceases, uncover, and evaporate to dryness. Dissolve the residues in 15 mL of water, filter if necessary, and dilute with water to 23 mL. Add 2 mL of 10% potassium dichromate reagent solution and add ammonium hydroxide until the orange color is just dissipated and the yellow color persists. Add 25 mL of methanol, stir vigorously, and allow to stand for 10 min. Any turbidity in the solution of the sample should not exceed that in the control.

CALCIUM, MAGNESIUM, AND R$_2$O$_3$ PRECIPITATE. To the filtrate from the test for insoluble matter, add 5 mL of ammonium oxalate reagent solution, 2 mL of ammonium phosphate reagent solution, and 10 mL of ammonium hydroxide. Allow to stand overnight, filter, wash with water containing 2.5% ammonia and about 0.1% ammonium oxalate, and ignite. The weight of the residue should not exceed 0.001 g.

HEAVY METALS. Dissolve 6.0 g in about 20 mL of water and dilute with water to 30 mL. For the control add 0.02 mg of lead ion (Pb) to 5.0 mL of the solution and dilute with water to 25 mL. For the sample use the remaining 25-mL portion. Adjust the pH of the control and sample solutions to between 3 and 4 (using a pH meter) with 1 N acetic acid or ammonium hydroxide (10% NH$_3$), dilute with water to 40 mL, and mix. Add 10 mL of freshly prepared hydrogen sulfide water to each and mix. Any color in the solution of the sample should not exceed that in the control.

IRON. Dissolve 2.0 g in 40 mL of water plus 2 mL of hydrochloric acid, and dilute with water to 50 mL. Add 30–50 mg of ammonium peroxydisulfate crystals and 3 mL of ammonium thiocyanate reagent solution. Any red color should not exceed that produced by 0.01 mg of iron (Fe) in an equal volume of solution containing the quantities of reagents used in the test.

POTASSIUM. Determine the potassium by the flame atomic absorption spectrophotometric method described on page 35.

> *Sample Stock Solution.* Dissolve 1.0 g of sample with water in a 100-mL volumetric flask, and dilute to the mark with water (0.01 g/mL).

Analyze the solutions by means of a suitable atomic absorption spectrophotometer, using the conditions outlined in the following

table. Calculate the metal content of the sample by the method of standard additions.

Element	Wavelength (nm)	Sample Wt (g)	Standard Added (mg)	Flame Type*	Background Correction
K	766.5	0.05	0.025; 0.05	A/A	No

*A/A is air/acetylene.

Sodium Carbonate
Sodium Carbonate, Anhydrous

Na_2CO_3 Formula Wt 105.99

CAS Number 497–19–8

REQUIREMENTS

Assay (dried basis) . . . . $\geq$ 99.5% Na_2CO_3

MAXIMUM ALLOWABLE

Insoluble matter 0.01%
Loss on heating at 285°C. 1.0%
Chloride (Cl) 0.001%
Nitrogen compounds (as N) 0.001%
Phosphate (PO_4) 0.001%
Silica (SiO_4). 0.005%
Sulfur compounds (as SO_4) 0.003%
Ammonium hydroxide precipitate 0.01%
Arsenic (As). . . . . 1 ppm
Calcium and magnesium precipitate 0.01%
Heavy metals (as Pb) . . . . 5 ppm
Iron (Fe) . . . . 5 ppm
Potassium (K) 0.005%

TESTS

ASSAY. (By acid–base titrimetry of carbonate). Weigh, to the nearest 0.1 mg, 2 g of the dried sample from the test for loss on heating at 285°C. Transfer to a 125-mL glass-stoppered flask, dissolve with 50 mL of water, add 0.10 mL of methyl orange indicator solution, and titrate with 1 N hydrochloric acid.

$$\% \text{ Na}_2\text{CO}_3 = \frac{(\text{mL} \times \text{N HCl}) \times 5.300}{\text{Sample wt (g)}}$$

INSOLUBLE MATTER. (Page 14). Use 10 g dissolved in 100 mL of water.

LOSS ON HEATING AT 285°C. Accurately weigh 1 g in a tared dish or crucible. Heat to constant weight at 270–300°C. Retain this sample for assay determination.

CHLORIDE. (Page 25). Use 1.0 g of sample and 2 mL of nitric acid.

NITROGEN COMPOUNDS. (Page 30). Use 1.0 g. For the standard use 0.01 mg of nitrogen (N).

PHOSPHATE. (Page 30, Method 2). Dissolve 1.0 g in 50 mL of water in a platinum dish and digest on the steam bath for 30 min. Cool, neutralize with dilute sulfuric acid (1 + 19) to a pH of about 4, and dilute with water to about 75 mL. Add 0.5 g of ammonium molybdate and adjust the pH to 1.8 (using a pH meter) with dilute hydrochloric acid (1 + 9). Heat to boiling, cool, add 10 mL of hydrochloric acid, and dilute with water to 100 mL. Continue as described, and concurrently prepare a standard containing 0.01 mg of phosphate ion (PO_4) and 0.05 mg of silica (SiO_2) in about 75 mL of water treated as the 75 mL of sample solution. Reserve the aqueous phase for the determination of silica.

SILICA. Add 10 mL of hydrochloric acid to the solutions reserved from the determination of phosphate and transfer to separatory funnels. Add 40 mL of butyl alcohol, shake vigorously, and allow to separate. Draw off and discard the aqueous phase. Wash the butyl alcohol three times with 20-mL portions of dilute hydrochloric acid (1 + 99), discarding the washings each time. Dilute each butyl alcohol solution with butyl alcohol to 50 mL, take 10 mL from each, and dilute each to 50 mL with butyl alcohol. Add 0.5 mL of a freshly prepared 2% solution of stannous chloride dihydrate in hydrochloric acid. The blue color in the extract from the sample should not exceed that in the standard. If the butyl alcohol extracts are turbid, wash with 10 mL of dilute hydrochloric acid (1 + 99).

SULFUR COMPOUNDS. (Page 32, Procedure B). Cautiously dissolve 1.0 g in 1 mL of hydrochloric acid plus 0.05 mL of bromine water in a reaction tube, and continue as described.

AMMONIUM HYDROXIDE PRECIPITATE.   Dissolve 10 g in 100 mL of water and filter if necessary. Cautiously add 30 mL of cooled dilute sulfuric acid (1 + 1) and evaporate until dense fumes of sulfur trioxide appear. Cool, dissolve the residue in 130 mL of hot water, and add ammonium hydroxide until the solution is just alkaline to methyl red. Heat to boiling, boil gently for 5 min, and filter, reserving the filtrate for the test for calcium and magnesium precipitate. Wash the precipitate with hot water, rejecting the washings, and ignite. The residue includes some, but not all, of the silica in the sample.

ARSENIC.   (Page 23). Dissolve 3.0 g in a small volume of water, neutralize with sulfuric acid, dilute with water to 35 mL, and proceed as directed. For the standard use 0.003 mg of arsenic (As).

CALCIUM AND MAGNESIUM PRECIPITATE.   Neutralize the filtrate from the test for ammonium hydroxide precipitate with hydrochloric acid and add an excess of 0.5 mL of the acid. Add 5 mL of ammonium oxalate reagent solution, 2 mL of ammonium phosphate reagent solution, and 10 mL of ammonium hydroxide. Stir well and allow to stand overnight. Filter, wash the precipitate with water containing 2.5% ammonia and about 0.1% ammonium oxalate, and ignite.

HEAVY METALS.   (Page 26, Method 1). To 5.0 g in a 150-mL beaker add 10 mL of water, mix, and cautiously add 10 mL of hydrochloric acid. Evaporate to dryness on the steam bath, dissolve the residue in about 20 mL of water, and dilute with water to 25 mL. For the control solution add 0.02 mg of lead ion (Pb) to 1.0 g of sample, and treat exactly as the 5.0 g of sample.

IRON.   (Page 28, Method 1). Dissolve 2.0 g in 50 mL of dilute hydrochloric acid (1 + 9). Use the solution without further acidification.

POTASSIUM.   Determine the potassium by the flame atomic absorption spectrophotometric method described on page 35.

>   ***Sample Stock Solution.***   Dissolve 10.0 g of sample with water in a 100-mL volumetric flask, and dilute to the mark with water (0.1 g/mL).

Analyze the solutions by means of a suitable atomic absorption spectrophotometer, using the conditions outlined in the following table. Calculate the metal content of the sample by the method of standard additions.

Element	Wavelength (nm)	Sample Wt (g)	Standard Added (mg)	Flame Type*	Background Correction
K	766.5	1.0	0.05; 0.10	A/A	No

*A/A is air/acetylene.

Sodium Carbonate, Alkalimetric Standard

Na_2CO_3 Formula Wt 105.99

CAS Number 497 –19 –8

NOTE. For use as an alkalimetric standard this material should be heated at 285°C for 2 h.

REQUIREMENTS

Assay (dried basis) . . . . 99.95 –100.05%
Na_2CO_3

MAXIMUM ALLOWABLE

Insoluble matter 0.01%
Loss on heating at 285°C. 1.0%
Chloride (Cl) 0.001%
Nitrogen compounds (as N) 0.001%
Phosphate (PO_4) 0.001%
Silica (SiO_4). 0.005%
Sulfur compounds (as SO_4) 0.003%
Ammonium hydroxide precipitate 0.01%
Arsenic (As). . . . . 1 ppm
Calcium and magnesium precipitate 0.01%
Heavy metals (as Pb) . . . . 5 ppm
Iron (Fe) . . . . 5 ppm
Potassium (K) 0.005%

TESTS

ASSAY. (By acid–base titrimetry of carbonate). Transfer a quantity of NIST SRM Potassium Hydrogen Phthalate to an agate or mullite mortar and grind to approximately 100-mesh fineness. Place 5.100 ± 0.003 g of the 100-mesh material in a weighing bottle, dry at 120°C for 2 h, and cool in a desiccator for at least 2 h. Weigh the bottle and contents accurately, transfer the contents to the titration flask*, and weigh the empty bottle to obtain the exact weight of the potassium hydrogen phthalate. Place 1.300 g of the alkalimetric standard sodium carbonate in a clean, dry weighing bottle. Heat at 285°C for 2 h, cool in a desiccator for at least 2 h, and weigh the bottle and contents accurately. Transfer the contents to a clean, dry 50-mL beaker and weigh the empty bottle to obtain the exact weight of the sodium carbonate.

Transfer the sodium carbonate to the titration flask by pouring it through a powder funnel. Rinse the beaker and funnel thoroughly with carbon dioxide-free water, using several small portions of the water to rinse the funnel and sides of the flask. Dilute the solution with carbon dioxide-free water to 90 mL, swirl to dissolve the sample, stopper the flask, and bubble carbon dioxide-free air through the solution during all subsequent operations. Boil the solution for 15 min, cool to room temperature in an ice bath, and dilute with carbon dioxide-free water to 75–80 mL. Titrate with 0.02 N sodium hydroxide, using a combination electrode (glass–calomel or glass–silver, silver chloride) for the measurement of E in millivolts or pH. The end point of the titration is determined by the second derivative method (page 51). Calculate the % Na_2CO_3 from the following formula:

$$A = \frac{25.949F}{B}$$

where

    A = % Na_2CO_3 in the sample
    B = weight, in grams, of the Na_2CO_3 sample
    F = calculation factor

The calculation factor, F, is obtained as follows:

$$F = (C \times D) - 0.20422EG$$

where

    C = weight, in grams, of the potassium hydrogen phthalate
    D = assay value of the potassium hydrogen phthalate
    E = normality of the sodium hydroxide solution, as determined against NIST standard potassium hydrogen phthalate, using the method in the NIST assay certificate
    G = volume, in milliliters, of sodium hydroxide consumed

Insoluble matter and other tests except assay are the same as for sodium carbonate, page 644.

---

    *A titration flask is a 500-mL conical flask fitted with a tube sealed into the side. The tube inside the flask is constricted and as close to the bottom as possible. The tube outside the flask is bent at a right angle to permit attachment of the carbon dioxide-free air supply.

Sodium Carbonate Monohydrate

$Na_2CO_3 \cdot H_2O$ Formula Wt 124.00

CAS Number 5968–11–6

REQUIREMENTS

Assay . . . . $\geq$ 99.5%
$Na_2CO_3 \cdot H_2O$
Loss on drying at 150°C. 13.0–15.0%

MAXIMUM ALLOWABLE

Insoluble matter 0.01%
Chloride (Cl) 0.001%
Nitrogen compounds (as N) 0.001%
Phosphate (PO_4) . . . . 5 ppm
Silica (SiO_4). 0.005%
Sulfur compounds (as SO_4) 0.004%
Ammonium hydroxide precipitate 0.01%
Arsenic (As). . . . . 1 ppm
Calcium and magnesium precipitate 0.01%
Heavy metals (as Pb) . . . . 5 ppm
Iron (Fe) . . . . 5 ppm
Potassium (K) 0.005%

TESTS

ASSAY. (By acid–base titrimetry of carbonate). Weigh, to the nearest 0.1 mg, 2.5 g of sample. Transfer to a 125-mL glass-stoppered flask, dissolve with 50 mL of water, add 0.10 mL of methyl orange indicator solution, and titrate with 1 N hydrochloric acid.

$$\% \, Na_2CO_3 \cdot H_2O = \frac{(mL \times N \; HCl) \times 6.201}{Sample \; wt \; (g)}$$

LOSS ON DRYING. Weigh accurately about 1 g and dry at 150°C to constant weight.

INSOLUBLE MATTER. (Page 14). Use 10 g dissolved in 100 mL of water.

CHLORIDE. (Page 25). Use 1.0 g of sample and 2 mL of nitric acid.

NITROGEN COMPOUNDS. (Page 30). Use 1.0 g. For the standard use 0.01 mg of nitrogen (N).

PHOSPHATE. (Page 30, Method 2). Dissolve 2.0 g in 50 mL of water in a platinum dish and digest on the steam bath for 30 min. Cool, neutralize with dilute sulfuric acid (1 + 19) to a pH of about 4, and dilute with water to about 75 mL. Add 0.5 g of ammonium molybdate and adjust the pH to 1.8 (using a pH meter) with dilute hydrochloric acid (1 + 9). Heat to boiling, cool, add 10 mL of hydrochloric acid, and dilute with water to 100 mL. Continue as described, and concurrently prepare a standard containing 0.01 mg of phosphate ion (PO_4) and 0.10 mg of silica (SiO_2) in about 75 mL of water treated as the 75 mL of sample solution. Reserve the aqueous phase for the determination of silica.

SILICA. Add 10 mL of hydrochloric acid to the solutions reserved from the determination of phosphate and transfer to separatory funnels. Add 40 mL of butyl alcohol, shake vigorously, and allow to separate. Draw off and discard the aqueous phase. Wash the butyl alcohol three times with 20-mL portions of dilute hydrochloric acid (1 + 99), discarding the washings each time. Dilute each butyl alcohol solution with butyl alcohol to 50 mL, take 10 mL from each, and dilute each to 50 mL with butyl alcohol. Add 0.5 mL of a freshly prepared 2% solution of stannous chloride dihydrate in hydrochloric acid. The blue color in the extract from the sample should not exceed that in the standard. If the butyl alcohol extracts are turbid, wash with 10 mL of dilute hydrochloric acid (1 + 99).

SULFUR COMPOUNDS. (Page 32, Procedure B). Cautiously dissolve 1.0 g in 1 mL of hydrochloric acid plus 0.05 mL of bromine water in a reaction tube, and continue as described.

AMMONIUM HYDROXIDE PRECIPITATE. Dissolve 10 g in 100 mL of water and filter if necessary. Cautiously add 30 mL of cooled dilute sulfuric acid (1 + 1). Evaporate until dense fumes of sulfur trioxide appear. Cool, dissolve the residue in 130 mL of hot water, and add ammonium hydroxide until the solution is just alkaline to methyl red. Heat to boiling, boil gently for 5 min, and filter, reserving the filtrate for the test for calcium and magnesium precipitate. Wash the precipitate with hot water, rejecting the washings, and ignite. The residue includes some, but not all, of the silica in the sample.

ARSENIC. (Page 23). Dissolve 3.0 g in a small volume of water, neutralize with sulfuric acid, dilute with water to 35 mL, and proceed. For the standard use 0.003 mg of arsenic (As).

CALCIUM AND MAGNESIUM PRECIPITATE. Neutralize the filtrate from the test for ammonium hydroxide precipitate with hydrochloric acid and add an excess of 0.5 mL of the acid. Add 5 mL of ammonium oxalate reagent solution, 2 mL of ammonium phosphate reagent solution, and 10 mL of ammonium hydroxide. Stir well and allow to stand overnight. Filter, wash the precipitate with water containing 2.5% ammonia and about 0.1% ammonium oxalate, and ignite.

HEAVY METALS. (Page 26, Method 1). To 5.0 g in a 150-mL beaker add 10 mL of water, mix, and cautiously add 10 mL of hydrochloric acid. Evaporate to dryness on the steam bath, dissolve the residue in about 20 mL of water, and dilute with water to 25 mL. For the control solution add 0.02 mg of lead ion (Pb) to 1.0 g of sample, and treat exactly as the 5.0 g of sample.

IRON. (Page 28, Method 1). Dissolve 2.0 g in 50 mL of dilute hydrochloric acid (1 + 9). Use the solution without further acidification.

POTASSIUM. Determine the potassium by the flame atomic absorption spectrophotometric method described on page 35.

Sample Stock Solution. Dissolve 10.0 g of sample with water in a 100-mL volumetric flask, and dilute to the mark with water (0.1 g/mL).

Analyze the solutions by means of a suitable atomic absorption spectrophotometer, using the conditions outlined in the following table. Calculate the metal content of the sample by the method of standard additions.

Element	Wavelength (nm)	Sample Wt (g)	Standard Added (mg)	Flame Type*	Background Correction
K	766.5	1.0	0.025; 0.05	A/A	No

*A/A is air/acetylene.

Sodium Chlorate

$NaClO_3$ Formula Wt 106.44

CAS Number 7775 –09 –9

REQUIREMENTS

Assay . . . . $\geq$ 99.0% $NaClO_3$

MAXIMUM ALLOWABLE

Insoluble matter 0.005%
Bromate (BrO_3) 0.015%
Chloride (Cl) 0.005%
Nitrogen compounds (as N) 0.001%
Sulfate (SO_4) 0.001%
Arsenic (As) . . . . 0.5 ppm
Calcium, magnesium, and
   R_2O_3 precipitate 0.005%
Heavy metals (as Pb) 0.001%
Iron (Fe) . . . . 5 ppm
Potassium (K) 0.01%

TESTS

ASSAY. (By argentimetric titration of chloride after reduction). Weigh accurately 0.5 g of sample, transfer to a glass-stoppered conical flask, and dissolve in 100 mL of water. Add 25 mL of fresh sulfurous acid (5% SO_2) and 5 mL of 25% nitric acid and heat for 30 min to remove the excess SO_2. Cool, add 50.0 mL of 0.1 N silver nitrate solution and 10 mL of toluene, and shake vigorously. Titrate the excess silver nitrate with 0.1 N ammonium thiocyanate, using ferric ammonium sulfate indicator solution. One milliliter of 0.1 N silver nitrate corresponds to 0.01064 g of $NaClO_3$.

INSOLUBLE MATTER. (Page 14). Use 20 g dissolved in 250 mL of water.

BROMATE. (Page 49). Determine bromate by differential pulse polarography. Use 1.0 g of sample and 0.5 g of calcium chloride dihydrate, $CaCl_2 \cdot 2H_2O$, in 25 mL of solution. For the standard add 0.15 mg of bromate ion (BrO_3). Also prepare a reagent blank.

CHLORIDE. (Page 25). Use 0.2 g. Use only nitric acid that is free from lower oxides of nitrogen in the test.

NITROGEN COMPOUNDS. (Page 30). Use 1.0 g. For the standard use 0.01 mg of nitrogen (N).

SULFATE. (Page 32, Procedure A, Method 1). Use 5.0 g.

ARSENIC. (Page 23). Dissolve 6.0 g in 30 mL of dilute hydrochloric acid (1 + 1) in a beaker, and evaporate to dryness on the steam bath. Transfer the residue to a generator flask with the aid of 35 mL of water and add 0.5 g of hydrazine sulfate. For the standard use 0.003 mg of arsenic (As).

CALCIUM, MAGNESIUM, AND R$_2$O$_3$ PRECIPITATE. Dissolve 10 g in 75 mL of dilute hydrochloric acid (1 + 2) and boil gently until no more chlorine is evolved. Dilute with water to about 80 mL, filter if necessary, heat to boiling, and add 5 mL of ammonium oxalate reagent solution, 2 mL of ammonium phosphate reagent solution, and 15 mL of ammonium hydroxide. Allow to stand overnight, filter, wash with water containing 2.5% of ammonia and about 0.1% of ammonium oxalate, and ignite.

HEAVY METALS. (Page 26, Method 1). Dissolve 2.5 g in 20 mL of dilute hydrochloric acid (1 + 1). Evaporate the solution to dryness on the steam bath, add 5 mL more of dilute hydrochloric acid (1 + 1), and again evaporate to dryness. Dissolve the residue in about 20 mL of water and dilute with water to 25 mL. For the standard–control solution add 0.02 mg of lead ion (Pb) to 0.5 g of sample and treat exactly as the 2.5-g sample.

IRON. (Page 28, Method 1). Dissolve 2.0 g in 20 mL of dilute hydrochloric acid (1 + 1) and evaporate to dryness on the steam bath. Add 5 mL of dilute hydrochloric acid (1 + 1) and again evaporate to dryness. Dissolve the residue in 50 mL of dilute hydrochloric acid (1 + 25) and use the solution without further acidification. In preparing the standard, use the residue from evaporation of 15 mL of hydrochloric acid.

POTASSIUM. Determine the potassium by the flame atomic absorption spectrophotometric method described on page 35.

> *Sample Stock Solution.* Dissolve 2.0 g in 25 mL of dilute hydrochloric acid (1 + 3) and digest in a covered beaker on the steam bath until the reaction ceases. Uncover the beaker and evaporate to dryness. Add 10 mL of dilute hydrochloric acid (1 + 3) and again evaporate to dryness. Dissolve in dilute hydro-

chloric acid (1 + 99) and dilute to 100 mL with dilute hydrochloric acid (1 + 99) (1 mL = 0.02 g).

Analyze the solutions by means of a suitable atomic absorption spectrophotometer, using the conditions outlined in the following table. Calculate the metal content of the sample by the method of standard additions.

Element	Wavelength (nm)	Sample Wt (g)	Standard Added (mg)	Flame Type*	Background Correction
K	766.5	0.4	0.02, 0.04	A/A	No

*A/A is air/acetylene.

Sodium Chloride

NaCl   Formula Wt 58.44

CAS Number 7647–14–5

REQUIREMENTS

Assay . . . . ≥ 99.0% NaCl
pH of a 5% solution at 25°C 5.0–9.0

MAXIMUM ALLOWABLE

Insoluble matter 0.005%
Iodide (I) 0.002%
Bromide (Br) 0.01%
Chlorate and nitrate (as NO_3) 0.003%
Nitrogen compounds (as N) 0.001%
Phosphate (PO_4) . . . . 5 ppm
Sulfate (SO_4) 0.004%
Barium (Ba) 0.001%
Calcium, magnesium, and
  R_2O_3 precipitate 0.005%
Heavy metals (as Pb) . . . . 5 ppm
Iron (Fe) . . . . 2 ppm
Potassium (K) 0.005%

TESTS

ASSAY. (By argentimetric titration of chloride content). Weigh accurately about 0.25 g and dissolve with 50 mL of water in a 250-mL glass-stoppered flask. Add, while agitating, exactly 50.0 mL of 0.1 N silver nitrate solution, then add 3 mL of nitric acid and 10 mL of benzyl alcohol. Shake vigorously, add 2 mL of ferric ammonium sulfate reagent solution (as indicator), and titrate the excess silver nitrate with 0.1 N ammonium thiocyanate solution from a 10-mL buret.

pH OF A 5% SOLUTION. (Page 43). The pH should be 5.0–9.0 at 25°C.

INSOLUBLE MATTER. (Page 14). Use 20 g dissolved in 200 mL of water. Save the filtrate separate from the washings for the test for calcium, magnesium, and R_2O_3 precipitate.

IODIDE. Dissolve 11 g in 50 mL of water. Prepare a control by dissolving 1 g of the sample, 0.2 mg of iodide ion (I), and 1.0 mg of bromide ion (Br) in 50 mL of water. To each solution, in a separatory funnel, add 2 mL of hydrochloric acid and 5 mL of ferric chloride reagent solution. Allow to stand for 5 min. Add 10 mL of carbon tetrachloride, shake for 1 min, allow the carbon tetrachloride to settle, and draw it off. Reserve the water solution for the test for bromide. Any violet color in the carbon tetrachloride extract from the solution of the sample should not exceed that in the extract from the control.

BROMIDE. Treat both the solution of the sample and the control, obtained in the test for iodide, as follows: wash twice by shaking with 10-mL portions of carbon tetrachloride. Each time allow to settle, then draw off and discard the carbon tetrachloride. To each of the water solutions add 10 mL of water, 65 mL of cool, dilute sulfuric acid (1 + 1), and 15 mL of a solution of chromic acid prepared by dissolving 10 g of chromium trioxide (CrO_3) in 100 mL of dilute sulfuric acid (1 + 3). Allow to stand for 5 min. Add 10 mL of carbon tetrachloride, shake for 1 min, allow to settle, and draw off. (Half of a 7-cm piece of filter paper rolled and placed in the stem of the separatory funnel will absorb any of the aqueous solution that may pass the stopcock and thus assure a clear extract.) Any yellow-brown color in the carbon tetrachloride extract from the solution of the sample should not exceed that in the extract from the control.

CHLORATE AND NITRATE

> *Sample Solution A.* Dissolve 0.50 g in 3 mL of water by heating in a boiling-water bath. Dilute to 50 mL with brucine sulfate reagent solution.

Control Solution B. Dissolve 0.50 g in 1.5 mL of water and 1.5 mL of the standard nitrate solution containing 0.01 mg of nitrate ion (NO_3) per mL by heating in a boiling-water bath. Dilute to 50 mL with brucine sulfate reagent solution.

Continue with the procedure described on page 29, starting with the preparation of blank solution C.

NITROGEN COMPOUNDS. (Page 30). Use 1.0 g. For the standard use 0.01 mg of nitrogen (N).

PHOSPHATE. (Page 30, Method 1). Dissolve 4.0 g in 25 mL of approximately 0.5 N sulfuric acid and continue as described.

SULFATE. (Page 32, Procedure B). Dissolve 1.0 g in 4 mL of water and transfer to a 40-mL reaction tube. Add 15 mL of reducing mixture and continue as described.

BARIUM. Dissolve 4.0 g in 20 mL of water, filter if necessary, and divide into two equal portions. To one portion add 2 mL of 10% sulfuric acid reagent solution and to the other 2 mL of water. The solutions should be equally clear at the end of 2 h.

CALCIUM, MAGNESIUM, AND R_2O_3 PRECIPITATE. To the filtrate (without washings) from the test for insoluble matter add 5 mL of ammonium oxalate reagent solution, 2 mL of ammonium phosphate reagent solution, and 30 mL of ammonium hydroxide, and allow to stand overnight. If a precipitate is formed, filter, wash with water containing 2.5% of ammonia and about 0.1% of ammonium oxalate, and ignite.

HEAVY METALS. (Page 26, Method 1). Dissolve 6.0 g in about 20 mL of water, and dilute with water to 30 mL. Use 25 mL to prepare the sample solution and use the remaining 5.0 mL to prepare the control solution.

IRON. (Page 28, Method 1). Use 5.0 g.

POTASSIUM. Determine the potassium by the flame atomic absorption spectrophotometric method described on page 35.

Sample Stock Solution. Dissolve 10.0 g of sample in 75 mL of water, transfer to a 100-mL volumetric flask, and dilute to the mark with water (0.1 g/mL).

Analyze the solutions by means of a suitable atomic absorption spectrophotometer, using the conditions outlined in the following

table. Calculate the metal content of the sample by the method of standard additions.

Element	Wavelength (nm)	Sample Wt (g)	Standard Added (mg)	Flame Type*	Background Correction
K	766,5	1.0	0.025; 0.05	A/A	No

*A/A is air/acetylene.

Sodium Citrate Dihydrate
2-Hydroxy-1,2,3-propanetricarboxylic Acid, Trisodium Salt, Dihydrate

$Na_3C_6H_5O_7 \cdot 2H_2O$ Formula Wt 294.10

CAS Number 6132–04–3

REQUIREMENTS

Assay . . . . ≥ 99.0%
$Na_3C_6H_5O_7 \cdot 2H_2O$
pH of a 5% solution at 25°C 7.0–9.0

MAXIMUM ALLOWABLE

Insoluble matter 0.005%
Chloride (Cl) 0.003%
Sulfate (SO_4) 0.005%
Ammonia (NH_3) 0.003%
Calcium (Ca) 0.005%
Heavy metals (as Pb) . . . . 5 ppm
Iron (Fe) . . . . 5 ppm

TESTS

ASSAY.   (Total alkalinity by nonaqueous titration). Weigh accurately 0.35 g and transfer to a 250-mL beaker. Dissolve in 100 mL of glacial acetic acid, stir until dissolution is complete, and titrate with 0.1 N perchloric acid in glacial acetic acid, determining the end point potentiometrically. Correct for a reagent blank. One milliliter of 0.1 N perchloric acid corresponds to 0.009803 g of $Na_3C_6H_5O_7 \cdot 2H_2O$.

pH OF A 5% SOLUTION.   (Page 43). The pH should be 7.0–9.0 at 25°C.

INSOLUBLE MATTER.   (Page 14). Use 20 g dissolved in 200 mL of water.

CHLORIDE.   (Page 25). Use 0.33 g.

SULFATE.   Drive off any moisture by heating a sample in a dish on a hot plate. Then carefully ignite 1 g in an electric muffle furnace until nearly free of carbon. Boil the residue with 10 mL of water and 0.5 mL of 30% hydrogen peroxide for 5 min. Add 5 mg of sodium carbonate and 1 mL of hydrochloric acid and evaporate on a hot plate at low setting to dryness. Dissolve the residue in 4 mL of hot water, to which has been added 1 mL of dilute hydrochloric acid (1 + 19), filter through a small filter, wash with 2-mL portions of water, and dilute the filtrate to 10 mL. Add 1 mL of barium chloride reagent solution and mix well. Any turbidity produced should not be greater than that in a standard prepared as described and to which the barium chloride solution is added at the same time as it is added to the sample solution. To 10 mL of water, add 5 mg of sodium carbonate, 1 mL of hydrochloric acid, 0.5 mL of 30% hydrogen peroxide, and 0.05 mg of sulfate (SO_4) and evaporate on a hot plate at low setting to dryness. Dissolve the residue in 9 mL of water and add 1 mL of dilute hydrochloric acid (1 + 19). If necessary, adjust with water to the same volume as the sample solution, add 1 mL of barium chloride reagent solution, and mix well.

AMMONIA.   Dissolve 1 g in 50 mL of ammonia-free water and add 2 mL of Nessler reagent. Any color should not exceed that produced by 0.03 mg of ammonia (NH_3) in an equal volume of solution containing 2 mL of Nessler reagent.

CALCIUM.   Determine the calcium by the flame atomic absorption spectrophotometric method described on page 35.

>   ***Sample Stock Solution.***   Dissolve 10.0 g of sample with water in a 100-mL volumetric flask, add 5 mL of hydrochloric acid, and dilute to the mark with water (0.1 g/mL).

Analyze the solutions by means of a suitable atomic absorption spectrophotometer, using the conditions outlined in the following table. Calculate the metal content of the sample by the method of standard additions.

Element	Wavelength (nm)	Sample Wt (g)	Standard Added (mg)	Flame Type*	Background Correction
Ca	422.7	1.0	0.05; 0.10	N/A	No

*N/A is nitrous oxide/acetylene.

HEAVY METALS. (Page 26, Method 1). Dissolve 6.0 g in water, add 7.5 mL of dilute hydrochloric acid (1 + 1), and dilute with water to 42 mL. Use 35 mL to prepare the sample solution and use the remaining 7.0 mL to prepare the control solution.

IRON.   (Page 28, Method 1). Use 2.0 g.

Sodium Cobaltinitrite
Trisodium Hexakis(nitrito-N)cobaltate(3 −)
Sodium Hexanitritocobaltate(III)
(For determination of potassium)

$Na_3Co(NO_2)_6$ Formula Wt 403.94

CAS Number 14649 −73 −1

REQUIREMENTS

Insoluble matter . . . . ≤ 0.02%
Suitability for determination of potassium . . . . Passes test

TESTS

INSOLUBLE MATTER. (Page 14). Use 5.0 g dissolved in 25 mL of dilute acetic acid (1 + 25), and allow to stand in a covered beaker overnight.

SUITABILITY FOR DETERMINATION OF POTASSIUM. Dissolve 1.583 g of potassium chloride in water and dilute with water to 500 mL. To 10.0 mL of this solution (20 mg of K_2O) in a 50-mL beaker, add 2 mL of 1 N nitric acid and 8 mL of water. Dissolve 5 g of the sodium cobaltinitrite in water, dilute with water to 25 mL, and filter. Cool both solutions to approximately 20°C and add 10 mL of the cobaltinitrite solution to the potassium chloride solution. Allow to stand for 2 h and filter through a sintered-glass crucible that has been washed with alcohol, dried at 105°C, and weighed. Wash the precipitate with cobaltinitrite wash solution, finally wash with 5–10 mL of alcohol, and dry at 105°C. The weight of the potassium sodium cobaltinitrite precipitate $[K_2NaCo(NO_2)_6 \cdot H_2O]$ should be between 0.0945 and 0.0985 g.

> ***Cobaltinitrite Wash Solution.*** (0.01 N nitric acid saturated with potassium sodium cobaltinitrite.) Transfer 75–100 mg of potassium sodium cobaltinitrite to a 125-mL glass-stoppered flask. Add 50 mL of 0.01 N nitric acid, shake on a mechanical shaker for 1 h, and filter through a fine-porosity sintered-glass funnel. The filtrate is cobaltinitrite wash solution.

> *NOTE.* If no potassium sodium cobaltinitrite is available, a sufficient amount may be prepared by following the suitability for determination of potassium test, but substituting 0.01 N nitric acid in place of the cobaltinitrite wash solution. When it is possible to follow the suitability procedure as written, the analytical precipitate of potassium sodium cobaltinitrite so obtained may be used to prepare the wash solution.

Sodium Cyanide

NaCN   Formula Wt 49.01

CAS Number 143–33–9

REQUIREMENTS

Assay . . . . ≥ 95.0% NaCN

MAXIMUM ALLOWABLE

Chloride (Cl) 0.15%
Phosphate (PO₄) 0.02%
Sulfate (SO₄) 0.05%
Sulfide (S) 0.005%
Thiocyanate (SCN) 0.02%
Iron, total (as Fe) 0.005%
Lead (Pb) . . . . 5 ppm

TESTS

CAUTION. Because of the extremely poisonous nature of sodium cyanide and of the hydrogen cyanide gas, HCN, evolved on treatment of the salt or solution of the salt with an acid, all tests must be made in a fume hood with a strong draft. Special care must be taken to avoid inhaling any of the fumes or allowing the salt or solution of the salt to come in contact with open cuts of the skin. If safety pipets are not available for measuring aliquots, use a graduated cylinder. Under no conditions should suction by mouth be used to fill an ordinary pipet.

ASSAY. (By argentimetric titration of cyanide content). Weigh accurately about 0.4 g and dissolve in 30 mL of water. Add 0.20 mL of 10% potassium iodide reagent solution and 1 mL of ammonium hydroxide and titrate with 0.1 N silver nitrate until a slight permanent yellowish turbidity forms. One milliliter of 0.1 N silver nitrate corresponds to 0.009802 g of NaCN.

> *Sample Solution A.* Dissolve 10 g in water, filter if necessary through a chloride-free filter, and dilute with water to 200 mL (1 mL = 0.05 g).

CHLORIDE. Dilute 1.0 mL of sample solution A (0.05-g sample) with water to 50 mL. To 10 mL of this solution add 5 mL of 30% hydrogen peroxide and cover the beaker until the reaction ceases. Digest in the covered beaker on the steam bath for about one-half hour, cool, and neutralize with nitric acid. Dilute with water to 25 mL and add 1 mL of nitric acid and 1 mL of silver nitrate reagent solution. The turbidity should not exceed that produced by 0.015 mg of chloride ion (Cl) in an equal volume of solution containing the quantities of reagents used in the test.

PHOSPHATE. (Page 30, Method 1). Dissolve 1.0 g in 5 mL of water, add 2 mL of hydrochloric acid, and evaporate to dryness in a well-ventilated hood. Add 5 mL of dilute hydrochloric acid (1 + 1) and evaporate to dryness again. Dissolve in 50 mL of approximately 0.5 N sulfuric acid. To 5.0 mL of the solution add 20 mL of approximately 0.5 N sulfuric acid and continue as described.

SULFATE. (Page 32, Procedure A, Method 2). Use 2 mL of sample solution A plus 1 mL of hydrochloric acid in a well-ventilated hood.

SULFIDE. To 10 mL of sample solution A (measured in a graduated cylinder) (0.5-g sample) add 10 mL of water and 0.15 mL of alkaline lead solution (made by adding 10% sodium hydroxide reagent solution to a 10% lead acetate reagent solution until the precipitate first formed is redissolved). The color should not be darker than that produced by 0.025 mg of sulfide ion (S) in an equal volume of solution when treated with 0.15 mL of the alkaline lead solution.

THIOCYANATE.   To 20 mL of sample solution A (measured in a graduated cylinder) (1-g sample) add 4 mL of hydrochloric acid and 0.20 mL of ferric chloride reagent solution. At the end of 5 min the solution should show no reddish tint when compared with 20 mL of water to which have been added the quantities of hydrochloric acid and ferric chloride used in the test.

IRON, TOTAL.   (Page 28, Method 1). Transfer 4.0 mL of sample solution A (0.2-g sample) to a platinum dish, add 3 mL of hydrochloric acid, and evaporate to dryness in a well-ventilated hood. Heat the residue at 650°C for 30 min, cool, and add 2 mL of hydrochloric acid and 10 mL of water. Cover with a watch glass, digest on the steam bath until dissolution is complete, and use the solution without further acidification.

LEAD.   Dissolve 0.50 g in 10 mL of water in a separatory funnel. Add 5 mL of ammonium citrate reagent solution (lead-free), 2 mL of hydroxylamine hydrochloride reagent solution for the dithizone test, and 0.10 mL of phenol red indicator solution, and make the solution alkaline if necessary by adding ammonium hydroxide. Add 5 mL of dithizone standard solution in chloroform, shake gently but well for 1 min, and allow the layers to separate. The intensity of the red color of the chloroform layer should be no greater than that of a control made with 0.002 mg of lead ion (Pb) and 0.1 g of the sample dissolved in 10 mL of water and treated exactly like the solution of 0.50 g of sample in 10 mL of water.

Sodium Dichromate Dihydrate

$Na_2Cr_2O_7 \cdot 2H_2O$ Formula Wt 298.00

CAS Number 7789−12−0

REQUIREMENTS

Assay . . . . 99.5−100.5%
$Na_2Cr_2O_7 \cdot 2H_2O$

MAXIMUM ALLOWABLE

Insoluble matter and ammonium
  hydroxide precipitate 0.005%
Chloride (Cl) 0.005%
Sulfate (SO_4) 0.01%
Calcium (Ca) 0.003%
Aluminum (Al) 0.002%

TESTS

ASSAY. (By titration of oxidizing power). Weigh accurately about 0.2 g and dissolve in 200 mL of water. Add 7 mL of hydrochloric acid and 3 g of potassium iodide, mix, and allow to stand in the dark for 10 min. Titrate the liberated iodine with 0.1 N sodium thiosulfate, using starch indicator near the end point, to a greenish blue color. One milliliter of 0.1 N sodium thiosulfate corresponds to 0.004967 g of $Na_2Cr_2O_7 \cdot 2H_2O$.

INSOLUBLE MATTER AND AMMONIUM HYDROXIDE PRECIPITATE.
Dissolve 20 g in 200 mL of water, add 2 mL of ammonium hydroxide, heat to boiling, and digest in a covered beaker on a steam bath for 1 h. Filter, wash thoroughly, and ignite. Retain the filtrate for the calcium test.

CHLORIDE. Dissolve 0.2 g in 10 mL of water, filter if necessary through a chloride-free filter, and add 1 mL of ammonium hydroxide and 1 mL of silver nitrate reagent solution. Prepare a standard containing 0.01 mg of chloride ion (Cl) in 10 mL of water and add 1 mL of ammonium hydroxide and 1 mL of silver nitrate reagent solution. Add 2 mL of nitric acid to each. The comparison is best made by the general method for chloride in colored solutions, page 25.

SULFATE.   Dissolve 10 g in 250 mL of water, filter if necessary, and heat to boiling. Add 25 mL of a solution containing 1 g of barium chloride and 2 mL of hydrochloric acid in 100 mL of solution. Digest in a covered beaker on a hot plate at low setting for 2 h and allow to stand overnight. If a precipitate forms, filter, wash thoroughly, and ignite. Fuse the residue with 1 g of sodium carbonate. Extract the fused mass with water and filter off the insoluble residue. Add 5 mL of hydrochloric acid to the filtrate, dilute with water to about 200 mL, heat to boiling, and add 10 mL of ethyl alcohol. Digest in a covered beaker on the hot plate until the reduction of chromate is complete, as indicated by the change to a clear green or colorless solution. Neutralize the solution with ammonium hydroxide and add 2 mL of hydrochloric acid. Heat to boiling, add 10 mL of barium chloride reagent solution, digest in a covered beaker on a hot plate at low setting for 2 h, and allow to stand overnight. Filter, wash thoroughly, and ignite. Correct for the weight obtained in a complete blank test. If the original precipitate of barium sulfate weighs less than the requirement permits, the fusion with sodium carbonate is not necessary.

CALCIUM.   To half of the filtrate from the test for insoluble matter and ammonium hydroxide precipitate, add 10 mL of ammonium oxalate reagent solution and 5 mL of ammonium hydroxide and allow to stand overnight. If a precipitate forms, filter, wash with a solution containing about 1% of ammonium oxalate, and ignite. Add 0.10 mL of sulfuric acid to the cooled residue and ignite again.

ALUMINUM.   Dissolve 20 g in 140 mL of water, filter, and add 5 mL of glacial acetic acid to the filtrate. Make alkaline with ammonium hydroxide. Digest for 2 h on a steam bath, filter, wash, ignite, and weigh.

Sodium Diethyldithiocarbamate Trihydrate

$(CH_3CH_2)_2NCSSNa \cdot 3H_2O$ Formula Wt 225.31

CAS Number 20624–25–3

REQUIREMENTS

Solubility in water . . . . Passes test
Sodium (as Na_2SO_4) 30.5%–32.5%
Sensitivity to copper . . . . Passes test

TESTS

SOLUBILITY IN WATER. Dissolve 1 g in 50 mL of water. There should be no undissolved residue and the solution should be substantially clear. Retain the solution for the test for sensitivity.

> *NOTE.* Because of inherent instability, this reagent may be expected to react with oxygen of the atmosphere to form insoluble oxidation products. After storage for some time the reagent may fail to meet the requirement for solubility in water.

SODIUM. Weigh accurately 2 g in a tared dish or crucible. Moisten the sample with concentrated sulfuric acid and ignite cautiously to char the material. Finally, ignite to constant weight at $800 \pm 25°C$.

SENSITIVITY TO COPPER

> *Solution A.* Dilute 5 mL of the solution retained from the test for solubility in water to 100 mL with water. For the sample prepare a solution containing 0.002 mg of copper ion (Cu) in 100 mL of dilute ammonium hydroxide (1 + 99). For the control use 100 mL of dilute ammonium hydroxide (1 + 99). Transfer the solutions to separatory funnels and add to each 10 mL of solution A and 10 mL of isopentyl alcohol. Shake for about 1 min and allow to separate (about 30 min). The isopentyl alcohol separated from the solution containing the 0.002 mg of copper should show a distinct yellow color when compared with the isopentyl alcohol from the control.

Sodium Fluoride

NaF   Formula Wt 41.99

CAS Number 7681 –49 –4

> *DANGER!* MAY BE FATAL IF SWALLOWED. AVOID BREATHING DUST
> AND WASH THOROUGHLY AFTER HANDLING.

REQUIREMENTS

Assay . . . . $\geq$ 99% NaF

MAXIMUM ALLOWABLE

Insoluble matter 0.02%
Loss on drying at 150°C. 0.3%
Chloride (Cl) 0.005%
Titrable acid . . . . 0.03 meq / g
Titrable base . . . . 0.01 meq / g
Sodium fluosilicate (Na_2SiF_6) 0.1%
Sulfate (SO_4) 0.03%
Sulfite (SO_2) 0.005%
Heavy metals (as Pb) 0.003%
Iron (Fe) 0.003%
Potassium (K) 0.02%

TESTS

ASSAY.   (By indirect acid–base titrimetry).

> *NOTE.* All water used in the assay must be carbon dioxide- and
> ammonia-free.

Weigh, to the nearest 0.1 mg, 1.6 g of sample into a 100-mL volumetric flask. Dissolve and dilute to the mark with water. Dilute 10.0 mL of the diluted sample with water to 50 mL. Pass the solution through a cation-exchange column at a rate of about 5 mL/min, and collect the eluate in a 500-mL titration flask. Then wash the resin in the column with water at a rate of about 10 mL/min and collect in the titration flask. Add 0.15 mL of phenolphthalein to the flask, and titrate with 0.1 N sodium hydroxide. Continue the titration, as the washing proceeds, until 50 mL of eluate requires no further titration.

Ion-Exchange Column: Use Dowex HCR-W2, H$^+$ form, 16–40 mesh, or equivalent. Charge the resin to a column with a 2-cm bore to a height of about 20 cm. If the resin is in the hydrogen ion form, wash free of acid with carbon dioxide-free water. Otherwise regenerate by passing 4 N hydrochloric acid through the column and washing free of the acid.

INSOLUBLE MATTER.   (Page 14). Use 5.0 g dissolved in 100 mL of warm water in a platinum dish.

LOSS ON DRYING AT 150°C.   Weigh accurately about 1 g in a tared platinum dish or crucible and dry to constant weight at 150°C.

CHLORIDE.   (Page 25). Dissolve 1.0 g plus 1 g of boric acid in 80 mL of water, filter if necessary through a chloride-free filter, and dilute with water to 100 mL. Use 20 mL of this solution.

TITRABLE ACID.   Dissolve 2.0 g in 50 mL of water (free from carbon dioxide) in a platinum dish, add 10 mL of a saturated solution of potassium nitrate, cool the solution to 0°C, and add 0.15 mL of phenolphthalein indicator solution. If a pink color is produced, omit the titration for free acid and reserve the solution for the test for titrable base. If no pink color is produced, titrate with 0.01 N sodium hydroxide until the pink color persists for 15 s while the temperature of the solution is near 0°C. Not more than 5.0 mL of the 0.01 N sodium hydroxide should be required. Reserve the solution for the test for sodium fluosilicate.

TITRABLE BASE.   If a pink color was produced in the solution prepared in the test for titrable acid, add 0.01 N hydrochloric acid, stirring the liquid only gently, until the pink color is discharged. Not more than 2.0 mL of the acid should be required. Reserve the solution for the test for sodium fluosilicate.

SODIUM FLUOSILICATE.   Heat the solution reserved from one of the preceding tests to boiling and titrate while hot with 0.01 N sodium hydroxide until a permanent pink color is obtained. Not more than 4.3 mL of the sodium hydroxide should be required.

SULFATE. (Page 32, Procedure A, Method 2). Use 2 mL of hydrochloric acid in a platinum dish. Repeat four times.

SULFITE. Dissolve 8.0 g in 150 mL of water, and add 2 mL of hydrochloric acid and 0.25 mL of starch indicator solution. Titrate immediately with 0.01 N iodine. Not more than 1.0 mL of the 0.01 N iodine should be required to produce a blue color.

HEAVY METALS. (Page 26, Method 1). Dissolve 1.0 g in about 25 mL of water, and dilute with water to 30 mL. Use 25 mL to prepare the sample solution, and use the remaining 5.0 mL to prepare the control solution. Use pH paper to adjust the pH.

IRON. (Page 28, Method 1). To 1.0 g in a platinum dish add 10 mL of hydrochloric acid, and evaporate to dryness. Dissolve the residue in 10 mL of hydrochloric acid, and repeat the evaporation to dryness. Warm the residue with a few drops of hydrochloric acid, dilute with water to 30 mL, and use 10 mL. In preparing the standard use the residue from evaporation of 7 mL of hydrochloric acid.

POTASSIUM. Determine the potassium by the flame atomic absorption spectrophotometric method described on page 35.

Sample Stock Solution. Dissolve 1.0 g of sample with water in a 100-mL volumetric flask, and dilute to the mark with water (0.01 g/mL).

Analyze the solutions by means of a suitable atomic absorption spectrophotometer, using the conditions outlined in the following table. Calculate the metal content of the sample by the method of standard additions.

Element	Wavelength (nm)	Sample Wt (g)	Standard Added (mg)	Flame Type*	Background Correction
K	766.5	0.1	0.02; 0.04	A/A	No

*A/A is air/acetylene.

Sodium Formate

HCOONa   Formula Wt 68.01

CAS Number 141−53−7

REQUIREMENTS

Assay . . . . ≥ 99.0% HCOONa

MAXIMUM ALLOWABLE

Insoluble matter 0.005%
Chloride (Cl) 0.001%
Sulfate (SO_4) 0.001%
Calcium (Ca) 0.005%
Heavy metals (as Pb) . . . . 5 ppm
Iron (Fe) . . . . 5 ppm

TESTS

ASSAY. (By titration of reducing power). Weigh accurately about 0.88 g, transfer to a 250-mL volumetric flask (or 0.70 g to a 200-mL flask), dissolve with 50 mL of water, dilute with water to volume, and mix thoroughly. To 25.0 mL of this solution, in a 250-mL glass-stoppered flask, add 3 mL of 10% sodium hydroxide reagent solution and 50.0 mL of 0.1 N potassium permanganate. Heat on the steam bath for 20 min, then cool completely to room temperature. Add 4 g of potassium iodide crystals, mix, and add 5 mL of 25% sulfuric acid. Titrate the liberated iodine, representing the excess permanganate, with 0.1 N sodium thiosulfate, using 5 mL of starch indicator solution near the end point. Run a blank on 25 mL of water with the same quantities of permanganate and other reagents, in the same manner as the sample. One milliliter of 0.1 N potassium permanganate consumed corresponds to 0.003401 g of HCOONa.

INSOLUBLE MATTER. (Page 14). Use 20 g dissolved in 200 mL of water.

CHLORIDE. (Page 25). Use 1.0 g.

SULFATE. (Page 32, Procedure A, Method 1). Allow 30 min for the turbidity to form.

CALCIUM.   Determine the calcium by the flame atomic absorption spectrophotometric method described on page 35.

> *Sample Stock Solution.*   Ignite 2.0 g of sample in a platinum crucible at 600°C. Dissolve the residue in 10 mL of 20% hydrochloric acid, dilute with water to 100 mL, and filter through a dry filter paper.

Analyze the solutions by means of a suitable atomic absorption spectrophotometer, using the conditions outlined in the following table. Calculate the metal content of the sample by the method of standard additions.

Element	Wavelength (nm)	Sample Wt (g)	Standard Added (mg)	Flame Type*	Background Correction
Ca	422.7	4.0	0.02; 0.04	N/A	No
Fe	248.3	4.0	0.02; 0.04	A/A	Yes

*A/A is air/acetylene; N/A is nitrous oxide/acetylene.

HEAVY METALS.   (Page 26, Method 1). Dissolve 6.0 g in water, and dilute with water to 30 mL. Use 25 mL to prepare the sample solution, and use the remaining 5.0 mL to prepare the control solution.

IRON.   (Page 28, Method 1). Ignite 2.0 g, cool, and dissolve the residue in 10 mL of dilute hydrochloric acid. Evaporate on the steam bath to dryness, dissolve the residue with 2 mL of hydrochloric acid and 10 mL of water, and use the solution without further acidification.

Sodium Hydrogen Sulfate, Fused

CAS Number 7681-38-1

NOTE. This product is usually a mixture of sodium pyrosulfate, $Na_2S_2O_7$, and sodium hydrogen sulfate, $NaHSO_4$.

REQUIREMENTS

Acidity (as H_2SO_4) 39.0–42.0%

MAXIMUM ALLOWABLE

Insoluble matter and ammonium
   hydroxide precipitate 0.01%
Chloride (Cl) 0.001%
Phosphate (PO_4) 0.001%
Calcium and magnesium precipitate 0.005%
Heavy metals (as Pb) . . . . 5 ppm
Iron (Fe) 0.002%

TESTS

ACIDITY.   Weigh accurately 4 g, dissolve in 50 mL of water, and titrate with 1 N sodium hydroxide, using methyl orange as indicator. One milliliter of 1 N sodium hydroxide corresponds to 0.04904 g of H_2SO_4.

INSOLUBLE MATTER AND AMMONIUM HYDROXIDE PRECIPITATE. Dissolve 20 g in 200 mL of water, add ammonium hydroxide until the solution is alkaline to methyl red, boil for 1 min, and digest in a covered beaker on the steam bath for 1 h. Filter through a tared filtering crucible, saving the filtrate separate from the washings for the test for calcium and magnesium precipitate. Wash thoroughly and dry at 105°C.

CHLORIDE.   (Page 25). Use 1.0 g.

PHOSPHATE.   Dissolve 2.0 g in 15 mL of dilute ammonium hydroxide (1 + 2) and evaporate to dryness. Dissolve the residue in 25 mL of approximately 0.5 N hydrochloric acid, add 1 mL of ammonium molybdate–nitric acid reagent solution and 1 mL of 4-(methylamino)phenol sulfate reagent solution, and allow to stand for 2 h at room temperature. Any blue color should not exceed that produced by 0.02 mg of phosphate ion (PO_4) in an equal volume of solution containing the quantities of reagents used in the test, including the residue from evaporation of 5 mL of ammonium hydroxide.

CALCIUM AND MAGNESIUM PRECIPITATE. To the filtrate from the test for insoluble matter and ammonium hydroxide precipitate (without the washings) add 5 mL of ammonium oxalate reagent solution, 3 mL of ammonium phosphate reagent solution, and 10 mL of ammonium hydroxide. If any precipitate forms on standing overnight, filter, wash thoroughly with water containing 2.5% ammonia and about 0.1% ammonium oxalate, and ignite.

HEAVY METALS. (Page 26, Method 1). Dissolve 6.0 g in about 25 mL of water, and dilute with water to 30 mL. Use 25 mL to prepare the sample solution, and use the remaining 5.0 mL to prepare the control solution.

IRON. (Page 28, Method 1). Dissolve 5.0 g in 80 mL of dilute hydrochloric acid (1 + 3), and boil gently for 10 min. Cool, dilute with water to 100 mL, and use 10 mL.

Sodium Hydroxide

NaOH   Formula Wt 40.00

CAS Number 1310–73–2

REQUIREMENTS

Assay . . . . ≥ 97.0% NaOH

MAXIMUM ALLOWABLE

Na_2CO_3	1.0%
Chloride (Cl)	0.005%
Nitrogen compounds (as N)	0.001%
Phosphate (PO_4)	0.001%
Sulfate (SO_4)	0.003%
Ammonium hydroxide precipitate	0.02%
Heavy metals (as Ag)	0.002%
Iron (Fe)	0.001%
Mercury (Hg)	0.1 ppm
Nickel (Ni)	0.001%
Potassium (K)	0.02%

TESTS

GENERAL. Special care must be taken in sampling to obtain a representative sample and to avoid absorption of water and carbon dioxide by the sample taken.

ASSAY AND SODIUM CARBONATE. (By acid–base titrimetry). Weigh accurately 35–40 g, dissolve and dilute to 1 L, using carbon dioxide-free water. Dilute 50.0 mL of this solution with carbon dioxide-free water to 200 mL. Add 5 mL of barium chloride reagent solution, stopper, and allow to stand for 5 min. Add 0.15 mL of phenolphthalein and titrate with 1 N hydrochloric acid to the end point to determine the hydroxide. One milliliter of 1 N hydrochloric acid corresponds to 0.04000 g of NaOH. Add 0.15 mL of methyl orange and continue the titration with 1 N hydrochloric acid to the end point to determine the carbonate. One milliliter of 1 N hydrochloric acid corresponds to 0.05300 g of Na_2CO_3.

> ***Sample Solution A for the Determination of Chloride, Nitrogen Compounds, Phosphate, Heavy Metals, Iron, and Nickel.*** Dissolve 50.0 g in carbon dioxide- and ammonia-free water, cool, and dilute with the water to 500 mL (1 mL = 0.1 g).

CHLORIDE. (Page 25). Use 2.0 mL of sample solution A (0.2-g sample) diluted with water to 20 mL.

NITROGEN COMPOUNDS. (Page 30). Use 20 mL of sample solution A. For the control use 10 mL of sample solution A and 0.01 mg of nitrogen (N) diluted to 20 mL.

PHOSPHATE. (Page 30, Method 1). To 20 mL of sample solution A (2-g sample) add 5 mL of hydrochloric acid, and evaporate to dryness on the steam bath. Dissolve the residue in 25 mL of approximately 0.5 N sulfuric acid and continue as described.

SULFATE.   (Page 32, Procedure B). Transfer 5.0 mL of a 20% (w/v) aqueous solution to a reaction tube. Add 2 mL of hydrochloric acid, swirl to mix, and cool to room temperature. Add 0.05 mL of methyl red indicator and neutralize with hydrochloric acid. Add 2 mL of 10% sodium pyrophosphate decahydrate and swirl. Add 3 mL of barium chloride reagent solution and swirl. Add sodium hydroxide dropwise to the yellow end point. Swirl, centrifuge for 2 min, and discard the supernatant liquid. Apply the procedure to the residue.

AMMONIUM HYDROXIDE PRECIPITATE.   Weigh about 10 g and dissolve in about 100 mL of water. Cautiously add 15 mL of sulfuric acid to 15 mL of water, cool, add the mixture to the solution of the sample, and evaporate to strong fuming. Cool, dissolve the residue in 130 mL of hot water, and add ammonium hydroxide until the solution is just alkaline to methyl red. Heat to boiling, filter, wash with hot water, and ignite. The residue includes some, but not all, of the silica in the sample.

HEAVY METALS.   To 30 mL of sample solution A (3-g sample) cautiously add 10 mL of nitric acid. For the control add 0.05 mg of silver ion (Ag) to 5.0 mL of sample solution A and cautiously add 10 mL of nitric acid. Evaporate both solutions to dryness over a low flame or on an electric hot plate. Dissolve each residue in about 20 mL of water, filter if necessary through a chloride-free filter, and dilute with water to 25 mL. Adjust the pH of the control and sample solutions to between 3 and 4 (using a pH meter) with 1 N acetic acid or ammonium hydroxide (10% NH_3), dilute with water to 40 mL, and mix. Add 10 mL of freshly prepared hydrogen sulfide water to each and mix. Any yellowish brown color in the solution of the sample should not exceed that in the control.

IRON.   (Page 28, Method 1). Neutralize 10 mL of sample solution A (1-g sample) with hydrochloric acid, using 0.15 mL of phenolphthalein indicator, add 2 mL in excess, dilute with water to 50 mL, and use this solution. In preparing the standard use the residue from evaporation of the quantity of acid used to neutralize the sample solution.

MERCURY.   To each of two 250-mL conical flasks add 20 mL of water and 1 mL of 4% potassium permanganate solution. For the sample, add to one flask 5.5 g of the sodium hydroxide. For the control, add to the other flask 0.5 g of sample and 0.5 μg of mercury (1.0 mL of a solution prepared freshly by diluting 1.0 mL of standard mercury solution, page 84, with water to 100 mL). To each flask add slowly, with constant swirling, 18 mL of hydrochloric acid. Heat to boiling, allow to cool, and dilute with water to 100 mL. Determine mercury in 10-mL aliquots by the flameless atomic absorption method (page 41), using 1 mL of 10% hydroxylamine hydrochloride solution and 2 mL of 10% stannous chloride solution for the reduction. The peak obtained for the sample should not exceed that obtained for the control.

NICKEL. Dilute 20 mL of sample solution A (2-g sample) to 50 mL with water and neutralize with hydrochloric acid. Dilute with water to 85 mL and adjust the pH to 8 with ammonium hydroxide. Add 5 mL of bromine water, 5 mL of 1% dimethylglyoxime solution in alcohol, and 5 mL of 10% sodium hydroxide reagent solution. Any red color should not exceed that produced by 0.02 mg of nickel ion (Ni) in an equal volume of solution containing the quantities of reagents used in the test.

POTASSIUM. Determine the potassium by the flame atomic absorption method described on page 35.

> ***Sample Stock Solution.*** Dissolve 10.0 g of sample in 50 mL of water in a 100-mL volumetric flask. Neutralize with nitric acid, add 5 mL in excess, and dilute to the mark with water (0.1 g/mL).

Analyze the solutions by means of a suitable atomic absorption spectrophotometer, using the conditions outlined in the following table. Calculate the metal content of the sample by the method of standard additions.

Element	Wavelength (nm)	Sample Wt (g)	Standard Added (mg)	Flame Type*	Background Correction
K	766.5	0.25	0.025; 0.05	A/A	No

*A/A is air/acetylene.

Sodium Iodide

NaI   Formula Wt 149.89

CAS Number 7681–82–5

REQUIREMENTS

Assay . . . . ≥ 99.5% NaI
pH of a 5% solution at 25°C 6.0–9.0

MAXIMUM ALLOWABLE

Insoluble matter 0.01%
Chloride and bromide (as Cl). 0.01%
Iodate (IO_3) . . . . 3 ppm
Nitrogen compounds (as N) 0.002%
Phosphate (PO_4) 0.001%
Sulfate (SO_4). 0.005%
Barium (Ba). 0.002%
Calcium (Ca) 0.001%
Copper (Cu) . . . . 1 ppm
Iron (Fe) . . . . 5 ppm
Lead (Pb) . . . . 1 ppm
Nickel (Ni). . . . . 1 ppm
Potassium (K) 0.01%

TESTS

ASSAY. (By titration of iodide content). Weigh accurately 0.5 g and dissolve in 20 mL of water in a glass-stoppered conical flask. Add 30 mL of hydrochloric acid and 5 mL of chloroform, cool if necessary, and titrate with standard 0.05 M potassium iodate until the iodine color disappears from the aqueous layer. Shake vigorously for 30 s and continue the titration, shaking vigorously after each addition, until the iodine color in the chloroform is discharged. One milliliter of 0.05 M potassium iodate corresponds to 0.01499 g of NaI.

INSOLUBLE MATTER. (Page 14). Use 20 g dissolved in 200 mL of water.

pH OF A 5% SOLUTION. (Page 43). The pH should be 6.0–9.0 at 25°C.

CHLORIDE AND BROMIDE. Dissolve 1.0 g in 100 mL of water in a distillation flask. Add 1 mL of hydrogen peroxide (30%) and 1 mL of phosphoric acid, heat to boiling, and boil gently until all the iodine is expelled and the solution is colorless. Cool, wash down the sides of the flask with water, and add 0.5 mL of hydrogen peroxide. If an iodine color develops, boil until the solution is colorless and for 10 min longer. If no color develops, boil for 10 min, filter if necessary through a chloride-free filter, and dilute with water to 100 mL. Dilute 10 mL with water to 23 mL and add 1 mL of nitric acid and 1 mL of silver nitrate reagent solution. Any turbidity should not exceed that produced by 0.01 mg of chloride ion (Cl) in an equal volume of solution containing the quantities of nitric acid and silver nitrate used in the test.

IODATE. (Page 49). Determine iodate by differential pulse polarography. Use 10.0 g of sample in 25 mL of solution. For the standard add 0.03 mg of iodate (IO_3).

NITROGEN COMPOUNDS. (Page 30). Use 1.0 g. For the standard use 0.02 mg of nitrogen (N).

PHOSPHATE. (Page 30, Method 1). Dissolve 2.0 g in water, and add 10 mL of nitric acid and 5 mL of hydrochloric acid. Evaporate to dryness on the steam bath. Dissolve the residue in 25 mL of approximately 0.5 N sulfuric acid and continue as described. For the standard include the residue from evaporation of the quantities of acids used with the sample.

SULFATE. (Page 32, Procedure A, Method 1). Use 1.0 g. Any yellow iodine color that develops can be discharged by the addition of a few milligrams of ascorbic acid.

BARIUM. Dissolve 3.0 g in 10 mL of dilute hydrochloric acid (1 + 1) and add 5 mL of nitric acid. For the control dissolve 0.5 g of sample and 0.05 mg of barium ion (Ba) in 10 mL of dilute hydrochloric acid (1 + 1) and add 5 mL of nitric acid. Evaporate both solutions to dryness. Dissolve the residues in 10 mL of dilute hydrochloric acid (1 + 1), add 5 mL of nitric acid, and again evaporate to dryness. Dissolve each residue in water and dilute with water to 23 mL. To each solution add 2 mL of potassium dichromate reagent solution and 10% ammonium hydroxide reagent solution until the orange color is just dissipated and the yellow color persists. To each solution add with constant stirring 25 mL of methanol. Any turbidity in the solution of the sample should not exceed that in the control.

IRON. (Page 28, Method 2). Use 2.0 g.

CALCIUM, COPPER, LEAD, NICKEL, AND POTASSIUM. Determine the calcium, copper, lead, nickel, and potassium by the flame atomic absorption spectrophotometric method descrtibed on page 35.

Sample Stock Solution. Dissolve 100 g in about 70 mL of water and dilute to 100 mL with water (1 mL = 1.0 g).

Analyze the solutions by means of a suitable atomic absorption spectrophotometer, using the conditions outlined in the following table. Calculate the metal content of the sample by the method of standard additions.

Element	Wavelength (nm)	Sample Wt (g)	Standard Added (mg)	Flame Type*	Background Correction
Ca	422.7	10.0	0.05; 0.10	N/A	No
Cu	324.7	10.0	0.01; 0.02	A/A	Yes
Pb	217.0	10.0	0.01; 0.02	A/A	Yes
Ni	232.0	10.0	0.01; 0.02	A/A	Yes
K	766.5	0.5	0.025; 0.05	A/A	No

*A/A is air/acetylene; N/A is nitrous oxide/acetylene.

Sodium Metabisulfite
Disodium Disulfite

$Na_2S_2O_5$ Formula Wt 190.11

CAS Number 7681–57–4

REQUIREMENTS

Assay . . . . ≥ 97.0% $Na_2S_2O_5$

MAXIMUM ALLOWABLE

Insoluble matter 0.005%
Chloride (Cl) 0.05%
Thiosulfute (S_2O_3) 0.05%
Arsenic (As) . . . . 1 ppm
Heavy metals (as Pb) 0.001%
Iron (Fe) 0.002%

TESTS

ASSAY. (By titration of reductive capacity). Weigh accurately 0.45 g and add to a mixture of 100.0 mL of 0.1 N iodine and 5 mL of 10% hydrochloric acid solution. Swirl gently until the sample is dissolved completely. Titrate the excess of iodine with 0.1 N sodium thiosulfate, adding 3 mL of starch indicator solution near the end of the titration. One milliliter of 0.1 N iodine corresponds to 0.004753 g of $Na_2S_2O_5$.

INSOLUBLE MATTER. (Page 14). Use 20 g dissolved in 200 mL of water.

CHLORIDE. Dissolve 1.0 g in 10 mL of water, filter if necessary through a small chloride-free filter, and add 6 mL of 30% hydrogen peroxide. Add 1 N sodium hydroxide until the solution is slightly alkaline to phenolphthalein and dilute with water to 100 mL. Dilute 2.0 mL of this solution with water to 20 mL, and add 1 mL of nitric acid and 1 mL of silver nitrate reagent solution. Any turbidity should not exceed that produced by 0.01 mg of chloride ion (Cl) in an equal volume of solution containing the quantities of reagents used in the test.

THIOSULFATE. Dissolve 2.2 g in 10 mL of 10% hydrochloric acid solution in a 50-mL beaker. Boil gently for 5 min, cool, and transfer to a small test tube. Any turbidity should not be greater than that produced by a standard of 1.0 mL of 0.01 N sodium thiosulfate solution carried through the entire procedure.

ARSENIC. (Page 23). Dissolve 3.0 g in 10 mL of water in a 150-mL beaker. Cautiously add 10 mL of nitric acid and 5 mL of sulfuric acid, and evaporate on the steam bath to a volume of about 5 mL. Transfer the beaker to a hot plate and heat just to dense fumes of sulfur trioxide. Cool, cautiously wash down the sides of the beaker with about 10 mL of water, and again heat to dense fumes. Cool, and repeat the washing and fuming. Cool and cautiously wash the solution into a generator flask with 50 mL of water. Omit the addition of 20 mL of dilute sulfuric acid (1 + 4). For the standard use 0.003 mg of arsenic (As).

HEAVY METALS. (Page 26, Method 1). Dissolve 3.0 g in 25 mL of dilute hydrochloric acid (2 + 3), and evaporate to dryness on the steam bath. Add 5 mL of hydrochloric acid, and again evaporate to dryness. Dissolve the residue in 20 mL of water, and dilute with water to 25 mL. For the control solution add 0.02 mg of lead ion (Pb) to 1.0 g of sample, and treat exactly as the 3.0 g of sample.

IRON. (Page 28, Method 1). Dissolve 0.50 g in 14 mL of dilute hydrochloric acid (2 + 5), and evaporate on the steam bath to dryness. Dissolve the residue in 7 mL of dilute hydrochloric acid (2 + 5), and again evaporate to dryness. Dissolve the residue in 2 mL of hydrochloric acid, dilute with water to 50 mL, and use the solution without further acidification.

Sodium Methoxide, 0.5 M Methanolic Solution
Sodium Methylate

CH_3ONa Formula Wt 54.02

CAS Number 124−41−4

REQUIREMENTS

Assay . . . . 0.48−0.52 mole
of CH_3ONa per
liter of solution
(25.9−28.1 g / L)
Clarity of solution . . . . Passes test

TESTS

ASSAY. (By acid−base titrimetry). Pipet 5.0 mL of the sample into a conical flask, and slowly add 100 mL of water, with stirring. Add 0.10 mL of phenolphthalein indicator solution and titrate with 0.1 N hydrochloric acid to a colorless end point. One milliliter of 0.1 N hydrochloric acid corresponds to 0.005402 g of CH_3ONa.

CLARITY OF SOLUTION. Place 50 mL in a 50-mL color comparison tube. The sample has not more than a trace of turbidity or insoluble matter.

Sodium Molybdate Dihydrate

$Na_2MoO_4 \cdot 2H_2O$ Formula Wt 241.95

CAS Number 10102−40−6

REQUIREMENTS

Assay 99.5–103.0%
$$Na_2MoO_4 \cdot 2H_2O$$
pH of 5% solution at 25°C 7.0–10.5

MAXIMUM ALLOWABLE

Insoluble matter 0.005%
Chloride (Cl) 0.005%
Phosphate (PO_4) . . . . 5 ppm
Sulfate (SO_4) 0.015%
Ammonium (NH_4) 0.001%
Heavy metals (as Pb) . . . . 5 ppm
Iron (Fe) 0.001%

TESTS

ASSAY. (By oxidative titration after reduction of Mo^{VI}). Weigh accurately 0.3 g and dissolve in 10 mL of water in a 150-mL beaker. Activate the zinc amalgam of a Jones reductor by passing 100 mL of 1 N sulfuric acid through the column. Discard this acid, and place 25 mL of ferric ammonium sulfate solution in the receiver under the column. To the sample solution add 100 mL of 1 N sulfuric acid, and pass this mixture through the reductor, followed by 100 mL of 1 N sulfuric acid and then 100 mL of water (using these to rinse the beaker). Add 5 mL of phosphoric acid to the solution in the receiver, and titrate with 0.1 N potassium permanganate. Run a blank and make any necessary correction. One milliliter of 0.1 N potassium permanganate corresponds to 0.008066 g of $Na_2MoO_4 \cdot 2H_2O$.

> ***Ferric Ammonium Sulfate Solution.*** Dissolve 50 g of ferric ammonium sulfate dodecahydrate, $FeNH_4(SO_4)_2 \cdot 12H_2O$, in 500 mL of solution containing 25 mL of sulfuric acid.

pH OF A 5% SOLUTION. (Page 43). The pH should be 7.0–10.5 at 25°C.

INSOLUBLE MATTER. (Page 14). Use 20 g dissolved in 200 mL of water.

CHLORIDE. Dissolve 1.0 g in 50 mL of water and mix. Pipet 10 mL of the solution into a color-comparison tube, add 2 mL of nitric acid, dilute with water to 25 mL, and add 1 mL of silver nitrate reagent solution. Any turbidity should not exceed that produced by 0.01 mg of chloride ion (Cl) in an equal volume of solution containing the quantities of reagents used in the test.

PHOSPHATE. (Page 30, Method 3).

SULFATE. (Page 32, Procedure A, Method 2). Use 5 mL of nitric acid.

AMMONIUM. Dissolve 1.0 g in 20 mL of water, add 10 mL of 10% sodium hydroxide solution, and dilute with water to 50 mL. For the standard add 0.01 mg of ammonium ion (NH_4) to 20 mL of water, add 10 mL of 10% sodium hydroxide solution, and dilute with water to 50 mL. To each solution add 2 mL of Nessler reagent. Any color produced in the sample solution should not exceed that in the standard.

HEAVY METALS. Dissolve 6.0 g in 35 mL of 10% sodium hydroxide solution, add 5 mL of ammonium hydroxide, and dilute with water to 42 mL. For the control add 0.02 mg of lead ion (Pb) to 7.0 mL of the solution and dilute with water to 40 mL. For the sample dilute the remaining 35-mL portion with water to 40 mL. Add 10 mL of freshly prepared hydrogen sulfide water to each and mix. Any color in the solution of the sample should not exceed that in the control.

IRON. Determine iron by the atomic absorption spectrophotometric method described on pages 35–41.

> ***Sample Stock Solution.*** Dissolve 10.0 g in water and dilute with water to 100 mL (0.10 g/mL).

Analyze the solution by means of a suitable atomic absorption spectrophotometer, using the conditions outlined in the following table. Calculate the metal content of the sample by the method of standard additions.

Element	Wavelength (nm)	Sample Wt (g)	Standard Added (mg)	Flame Type*	Background Correction
Fe	248.3	2.5	0.05; 0.10	A/A	Yes

*A/A is air/acetylene.

Sodium Nitrate

NaNO$_3$ Formula Wt 84.99

CAS Number 7631–99–4

REQUIREMENTS

Assay . . . . ≥ 99.0% KNO_3
pH of a 5% solution . . . . 5.5–8.3 at 25°C

MAXIMUM ALLOWABLE

Insoluble matter 0.005%
Chloride 0.001%
Iodate (IO_3) . . . . 5 ppm
Iodate and nitrite . . . . Passes test (limit
about 5 ppm
IO_3; about
0.001% NO_2)
Phosphate (PO_4) . . . . 5 ppm
Sulfate (SO_4) 0.003%
Calcium, magnesium, and R_2O_3 precipitate 0.005%
Heavy metals (as Pb) . . . . 5 ppm
Iron (Fe) . . . . 3 ppm

TESTS

ASSAY. (By indirect acid–base titrimetry).

> *NOTE.* All water used in the assay must be carbon dioxide- and ammonia-free.

Weigh, to the nearest 0.1 mg, 0.3 g of sample and dissolve in 100 mL of water. Pass the solution through a cation-exchange column at a rate of about 5 mL/min, and collect the eluate in a 500-mL titration flask. Then wash the resin in the column with water at a rate of about 10 mL/min, and collect in the same titration flask. Add 0.15 mL of phenolphthalein to the flask and titrate with 0.1 N sodium hydroxide. Continue the titration, as the washing proceeds, until 50 mL of eluate requires no further titration.

$$\% \ NaNO_3 = \frac{(mL \times N \ NaOH) \times 8.499}{Sample \ wt \ (g)}$$

Ion-Exchange Column. Use Dowex 50W-X8, 20–50 mesh, or ANGC-242, 16–50 mesh. Charge the resin to a column with a 2-cm bore to a height of about 20 cm. If the resin is in the hydrogen ion form, wash free of acid with carbon dioxide-free water. Otherwise regenerate by passing 4 N hydrochloric acid through the column and washing free of the acid.

pH OF A 5% SOLUTION.   (Page 43). The pH should be 5.5–8.3 at 25°C.

INSOLUBLE MATTER.   (Page 14). Use 20 g dissolved in 100 mL of water. Retain the filtrate without the washings for the test for calcium, magnesium, and R_2O_3 precipitate.

CHLORIDE.   (Page 25). Use 1.0 g.

IODATE.   (Page 49). Determine iodate by differential pulse polarography. Use 10.0 g of sample and dissolve in 25 mL. For the standard add 0.05 mg of iodate (IO_3).

IODATE AND NITRITE.   Dissolve 1.0 g in 10 mL of water, and add 0.10 mL of 10% potassium iodide reagent solution, 1 mL of chloroform, and 1 mL of acetic acid. Shake gently for 5 min. The chloroform should not acquire a pink or violet color.

PHOSPHATE.   (Page 30, Method 1). Dissolve 4.0 g in 25 mL of approximately 0.5 N sulfuric acid and continue as described.

SULFATE.   (Page 32, Procedure A, Method 2). Use 7 mL of dilute hydrochloric acid (1 + 1), and do two evaporations. Allow 30 min for the turbidity to form.

CALCIUM, MAGNESIUM, AND R_2O_3 PRECIPITATE.   To the filtrate, without washings, from the test for insoluble matter add 5 mL of ammonium oxalate reagent solution, 3 mL of ammonium phosphate reagent solution, and 15 mL of ammonium hydroxide. Allow to stand overnight. If any precipitate is formed, filter, wash with a solution containing 2.5% ammonia and about 0.1% ammonium oxalate, and ignite.

HEAVY METALS.   (Page 26, Method 1). Dissolve 6.0 g in about 20 mL of water, and dilute with water to 30 mL. Use 25 mL to prepare the sample solution, and use the remaining 5.0 mL to prepare the control solution.

IRON.   (Page 28, Method 1). Use 3.3 g.

Sodium Nitrite

NaNO$_2$ Formula Wt 69.00

CAS Number 7632 – 00 – 0

REQUIREMENTS

Assay . . . . ≥ 97.0% NaNO$_2$

MAXIMUM ALLOWABLE

Insoluble matter 0.01%
Chloride (Cl) 0.005%
Sulfate (SO$_4$) 0.01%
Calcium (Ca) 0.01%
Heavy metals (as Pb) 0.001%
Iron (Fe) 0.001%
Potassium (K) 0.005%

TESTS

Sample Solution A.   Weigh accurately 10 g, dissolve in water, and dilute with water to 100 mL in a volumetric flask (1 mL = 0.1 g).

ASSAY.   (By indirect titration of reductive capacity). Dilute a 10.0-mL aliquot of sample solution A to 100 mL in a volumetric flask. Add 5 mL of sulfuric acid to 300 mL of water, and while the solution is still warm, add 0.1 N potassium permanganate until a faint pink color that persists for 2 min is produced. Then add 40.0 mL of 0.1 N potassium permanganate and mix gently. Add, slowly and with constant agitation, 10.0 mL of the diluted sample solution, holding the tip of the pipet well below the surface of the liquid. Add 15.0 mL of 0.1 N ferrous ammonium sulfate, allow the solution to stand for 5 min, and titrate the excess of ferrous ammonium sulfate with 0.1 N potassium permanganate. Each milliliter of 0.1 N potassium permanganate consumed by the sodium nitrite corresponds to 0.003450 g of NaNO$_2$.

INSOLUBLE MATTER.   (Page 14). Use 10 g dissolved in 100 mL of water.

CHLORIDE.   (Page 25). To 2.0 mL of sample solution A (0.2-g sample) add 10 mL of water, slowly add 1 mL of glacial acetic acid, and boil gently for 5 min.

SULFATE.   (Page 32, Procedure A, Method 2). Use 5.0 mL of sample solution A plus 1 mL of hydrochloric acid.

CALCIUM AND POTASSIUM.   Determine the calcium and potassium by the flame atomic absorption spectrophotometric method described on page 35.

>*Sample Stock Solution.*   Dissolve 20.0 g of sample with water in a 200-mL volumetric flask, add 5 mL of nitric acid, and dilute to the mark with water (0.1 g/mL).

Analyze the solutions by means of a suitable atomic absorption spectrophotometer, using the conditions outlined in the following table. Calculate the metal content of the sample by the method of standard additions.

Element	Wavelength (nm)	Sample Wt (g)	Standard Added (mg)	Flame Type*	Background Correction
Ca	422.7	0.5	0.05; 0.10	N/A	No
K	766.5	1.0	0.05; 0.10	A/A	No

*A/A is air/acetylene; N/A is nitrous oxide/acetylene.

HEAVY METALS.   (Page 26, Method 1). To 30 mL of sample solution A (3-g sample) add 5 mL of hydrochloric acid and evaporate to dryness on a steam bath. Add 5 mL of hydrochloric acid and again evaporate to dryness. Dissolve in about 20 mL of water and dilute with water to 25 mL. For the control solution add 0.02 mg of lead ion (Pb) to 10 mL of sample solution A (1-g sample), add 5 mL of hydrochloric acid, and treat exactly as the 30 mL of sample solution A.

IRON.   (Page 28, Method 1). To 10 mL of sample solution A (1-g sample) add 5 mL of hydrochloric acid, and evaporate on the steam bath to dryness. Dissolve the residue in 2 mL of hydrochloric acid plus about 20 mL of water, filter if necessary, and dilute with water to 50 mL. Use the solution without further acidification.

Sodium Nitroferricyanide Dihydrate
Disodium Pentakis(cyano-C)nitrosylferrate(2 −) Dihydrate
Sodium Pentacyanonitrosoferrate(III) Dihydrate

$Na_2Fe(CN)_5NO \cdot 2H_2O$ Formula Wt 297.95

CAS Number 13755 − 38 − 9

REQUIREMENTS

Assay . . . . 99.0 − 102.0%
$Na_2Fe(CN)_5NO \cdot 2H_2O$

MAXIMUM ALLOWABLE

Insoluble matter 0.01%
Chloride (Cl) 0.02%
Sulfate (SO_4) . . . . Passes test (limit
about 0.01%)

TESTS

ASSAY.   (By argentimetric titration of ferricyanide content). Weigh accurately 1.5 g, and transfer to a 100-mL volumetric flask with about 80 mL of water. Dilute to the mark and mix after complete dissolution. To a 25-mL aliquot of this solution in a 250-mL conical flask, add 50 mL of water and 1 mL of 10% potassium chromate solution. Titrate with 0.1 N silver nitrate solution to a reddish brown color. One milliliter of 0.1 N silver nitrate corresponds to 0.01490 g of $Na_2Fe(CN)_5NO \cdot 2H_2O$.

INSOLUBLE MATTER.   Dissolve 10 g in 50 mL of water at room temperature, filter promptly through a tared filtering crucible, wash thoroughly, and dry at 105°C.

CHLORIDE.   Dissolve 1.0 g in 175 mL of water, add 1.25 g of cupric sulfate crystals dissolved in 25 mL of water, mix thoroughly, and allow to stand until the precipitate has settled. Filter through a chloride-free filter, rejecting the first 50 mL of the filtrate. Dilute 10 mL of the filtrate with water to 20 mL, and add 1 mL of nitric acid and 1 mL of silver nitrate reagent solution. Any turbidity should not exceed that produced by 0.01 mg of chloride ion (Cl) in an equal volume of solution containing the quantities of reagents used in the test, with enough cupric sulfate added to match the color of the test.

SULFATE. Dissolve 5.0 g in 100 mL of water without heating, filter, and to the filtrate add 0.25 mL of glacial acetic acid and 5 mL of barium chloride reagent solution. Stir and pour into a Nessler tube for observation. No turbidity should be produced in 10 min.

Sodium Oxalate
Ethanedioic Acid, Disodium Salt

$(COONa)_2$ Formula Wt 134.00

CAS Number 62–76–0

REQUIREMENTS

Assay . . . . $\geq$ 99.5%
$(COONa)_2$

MAXIMUM ALLOWABLE

Insoluble matter 0.005%
Loss on drying at 105°C. 0.01%
Neutrality . . . . Passes test
Chloride (Cl) 0.002%
Sulfate (SO_4) 0.002%
Ammonium (NH_4) 0.002%
Heavy metals (as Pb) 0.002%
Iron (Fe) 0.001%
Potassium (K) 0.005%
Substances darkened by
  hot sulfuric acid . . . . Passes test

TESTS

ASSAY. (By indirect titration of reducing power of oxalate against permanganate). Weigh, to the nearest 0.1 mg, 0.25 g of the dried sample, and dissolve with 100 mL of water in a 250-mL glass-stoppered flask. Add 3 mL of sulfuric acid and then slowly add 20 mL of 0.1 N potassium permanganate solution. Heat the solution to 70°C and complete the titration with the permanganate until a pale pink color persists for 30 s.

$$\% \, Na_2C_2O_4 = \frac{(mL \times N \, KMnO_4) \times 6.700}{Sample \, wt \, (g)}$$

INSOLUBLE MATTER.   (Page 14). Use 20 g dissolved in 500 mL of water.

LOSS ON DRYING AT 105°C.   Weigh accurately 10 g in a tared dish or crucible and dry to constant weight at 105°C.

NEUTRALITY.   Dissolve 2.0 g in 200 mL of water, and add 10.0 mL of 0.01 N oxalic acid and 0.15 mL of phenolphthalein solution. Boil the solution in a flask for 10 min, passing through it a stream of carbon dioxide-free air. Cool the solution rapidly to room temperature while keeping the flow of carbon dioxide-free air passing through it. Titrate with 0.01 N sodium hydroxide. Not less than 9.2 nor more than 10.5 mL of 0.01 N sodium hydroxide should be required to match the pink color of a buffer solution containing 0.15 mL of phenolphthalein solution. The buffer solution contains in 1 L 3.1 g of boric acid (H_3BO_3), 3.8 g of potassium chloride (KCl), and 5.90 mL of 1 N sodium hydroxide.

CHLORIDE.   Dissolve 2.0 g in water plus 10 mL of nitric acid, filter if necessary through a chloride-free filter, and dilute with water to 100 mL. To 25 mL of the solution add 1 mL of silver nitrate reagent solution. Any turbidity should not exceed that produced by 0.01 mg of chloride ion (Cl) in an equal volume of solution containing the quantities of reagents used in the test.

SULFATE.   Mix 20 g with 50 mL of water, add 25 mL of nitric acid and 25 mL of 30% hydrogen peroxide, and digest in a covered beaker on a hot plate at low setting until reaction ceases. Remove the cover and evaporate to dryness. Add 80 mL of dilute hydrochloric acid (1 + 1) and again evaporate to dryness. Repeat the evaporation with dilute hydrochloric acid. Dissolve the residue in 200 mL of water, add 2 mL of hydrochloric acid, and filter if necessary. Heat the filtrate to boiling, add 10 mL of barium chloride reagent solution, digest in a covered beaker on a hot plate at low setting for 2 h, and allow to stand overnight. If a precipitate is formed, filter, wash thoroughly, and ignite. Correct for the weight obtained in a complete blank test.

AMMONIUM. Dissolve 1.0 g in 50 mL of ammonia-free water and add 2 mL of Nessler reagent. Any color should not exceed that produced by 0.02 mg of ammonium ion (NH_4) in an equal volume of solution containing 2 mL of Nessler reagent.

HEAVY METALS. (Page 26, Method 1). Mix 2.0 g with 2 mL of water, add 2 mL of nitric acid and 2 mL of 30% hydrogen peroxide, and digest in a covered beaker on the steam bath until reaction ceases. Remove the cover and evaporate to dryness. Dissolve the residue in 2 mL of nitric acid plus 2 mL of water and again evaporate to dryness. Dissolve the residue in about 20 mL of water and dilute with water to 32 mL. Use 24 mL to prepare the sample solution. Use the remaining 8.0 mL, plus the residue from evaporating 1 mL of 30% hydrogen peroxide and 2 mL of nitric acid to dryness, to prepare the control solution.

IRON. (Page 28, Method 1). Mix 2.0 g with 2 mL of water, add 2 mL of nitric acid and 2 mL of 30% hydrogen peroxide, and digest in a covered beaker on the steam bath until reaction ceases. Remove the cover and evaporate to dryness. Add 5 mL of hydrochloric acid and again evaporate to dryness. Dissolve the residue in a few milliliters of water, add 4 mL of hydrochloric acid, and dilute with water to 100 mL. Use 50 mL of the solution without further acidification. Use the residue from evaporation of 1 mL of nitric acid, 1 mL of 30% hydrogen peroxide, and 2 mL of hydrochloric acid to prepare the standard solution.

POTASSIUM. Determine the potassium by the flame atomic absorption spectrophotometric method described on page 35.

> ***Sample Stock Solution.*** Dissolve 5.0 g of sample in 5 mL of nitric acid in a 200-mL volumetric flask, and dilute to the mark with water (0.025 g/mL).

Analyze the solutions by means of a suitable atomic absorption spectrophotometer, using the conditions outlined in the following table. Calculate the metal content of the sample by the method of standard additions.

Element	Wavelength (nm)	Sample Wt (g)	Standard Added (mg)	Flame Type*	Background Correction
K	766.5	0.5	0.025; 0.05	A/A	No

*A/A is air/acetylene.

SUBSTANCES DARKENED BY HOT SULFURIC ACID.   Heat 1.0 g in a recently ignited test tube with 10 mL of sulfuric acid until the appearance of dense fumes. The acid when cooled should have no more color than a mixture of the following composition: 0.2 mL of cobalt chloride solution (5.95 g of $CoCl_2 \cdot 6H_2O$ and 2.5 mL of hydrochloric acid in 100 mL), 0.3 mL of ferric chloride solution (4.50 g of $FeCl_3 \cdot 6H_2O$ and 2.5 mL of hydrochloric acid in 100 mL), 0.3 mL of cupric sulfate solution (6.24 g of $CuSO_4 \cdot 5H_2O$ and 2.5 mL of hydrochloric acid in 100 mL), and 9.2 mL of water.

Sodium Perchlorate

$NaClO_4$	Formula Wt 122.44
$NaClO_4 \cdot H_2O$	Formula Wt 140.46

CAS Number 7601–89–0 (anhydrous); 7791–07–3 (monohydrate)

NOTE. This specification includes the anhydrous form and the monohydrate.

REQUIREMENTS

Assay . . . . 98.0–102.0% $NaClO_4$ (anhydrous); 85.0–90.0% $NaClO_4$ (monohydrate)

pH of a 5% solution at 25°C 6.0–8.0

MAXIMUM ALLOWABLE

Insoluble matter 0.005%
Chloride (Cl) 0.003%
Sulfate (SO$_4$) 0.002%
Calcium (Ca) 0.02%
Heavy metals (as Pb) . . . . 5 ppm
Iron (Fe) . . . . 5 ppm
Potassium (K) 0.05%

TESTS

ASSAY.   (By argentimetric titration of chloride content after perchlorate reduction to chloride). Weigh accurately about 0.25–0.30 g of sample into a platinum crucible or dish. Add 3 g of sodium carbonate, Na_2CO_3, and heat, gently at first, then to fusion until a clear melt is obtained. Leach with minimal dilute nitric acid and boil gently to expel carbon dioxide. Cool to room temperature, add 10 mL of 30% ammonium acetate solution, and titrate with 0.1 N silver nitrate to the dichlorofluorescein end point. One milliliter of 0.1 N silver nitrate corresponds to 0.01225 g of $NaClO_4$ or 0.01405 g of $NaClO_4 \cdot H_2O$.

pH OF A 5% SOLUTION.   (Page 43). The pH should be 6.0–8.0 at 25°C.

INSOLUBLE MATTER.   (Page 14). Use 5 g dissolved in 50 mL of water.

CHLORIDE.   (Page 25). Dissolve 0.33 g in 20 mL of water.

SULFATE.   (Page 32, Method 1). Dissolve 2.5 g in 5 mL of water.

CALCIUM AND POTASSIUM.   Determine the calcium and potassium by the flame atomic absorption spectrophotometric method described on page 35.

> ***Sample Stock Solution.***   Dissolve 10.0 g of sample with water in a 100-mL volumetric flask, add 10 mL of hydrochloric acid, and dilute to the mark with water (0.1 g/mL).

Analyze the solutions by means of a suitable atomic absorption spectrophotometer, using the conditions outlined in the following table. Calculate the metal content of the sample by the method of standard additions.

Element	Wavelength (nm)	Sample Wt (g)	Standard Added (mg)	Flame Type*	Background Correction
Ca	422.7	0.25	0.05; 0.10	N/A	No
K	766.5	0.1	0.05; 0.10	A/A	No

*A/A is air/acetylene; N/A is nitrous oxide/acetylene.

HEAVY METALS. (Page 26, Method 1). Dissolve 3.0 g in water, dilute with water to 60 mL, and use 50 mL to prepare the sample solution. For the control add 0.01 mg of lead ion (Pb) to the remaining 10 mL and dilute with water to 50 mL.

IRON. (Page 28, Method 1). Use 2.0 g.

Sodium Periodate

Sodium Metaperiodate
Sodium Tetraoxoiodate(VII)
(Suitable for Determination of Manganese)

$NaIO_4$ Formula Wt 213.89

CAS Number 7790−28−5

REQUIREMENTS

Assay (dried basis) . . . . 99.8−100.3%
$NaIO_4$

MAXIMUM ALLOWABLE

Other halogens (as Cl) 0.02%
Manganese (Mn) . . . . 3 ppm

TESTS

ASSAY. (By indirect titration of reductive capacity). Dry a sample over phosphorus pentoxide or magnesium perchlorate for 6 h. Weigh accurately 1 g, dissolve in water, and transfer to a 500-mL volumetric flask. Dilute to volume with water and mix thoroughly. Place a 50.0-mL aliquot of this solution in a glass-stoppered flask and add 10 g of potassium iodide and 10 mL of a cooled solution of sulfuric acid (1 + 5). Stopper, swirl, allow to stand for 5 min, and add 100 mL of cold water. Titrate the liberated iodine with 0.1 N sodium thiosulfate, adding 3 mL of starch indicator solution near the end of the titration. Correct for a complete blank. One milliliter of 0.1 N sodium thiosulfate corresponds to 0.002674 g of $NaIO_4$.

OTHER HALOGENS.   Dissolve 0.50 g in 25 mL of a dilute solution of sulfurous acid (3 + 2). Boil for 3 min, cool, and add 5 mL of ammonium hydroxide and 20 mL of 2.5% silver nitrate solution. Allow the precipitate to coagulate, filter through a chloride-free filter, and dilute the filtrate with water to 200 mL. To 20 mL of this solution add 1.5 mL of nitric acid. Any turbidity should not exceed that produced by 0.01 mg of chloride ion (Cl) in an equal volume of solution containing 0.5 mL of ammonium hydroxide, 1.5 mL of nitric acid, and 1 mL of silver nitrate reagent solution.

MANGANESE.   Dissolve 2.5 g in 40 mL of 10% sulfuric acid reagent solution. For the control add 0.5 g of the sample and 0.006 mg of manganese ion (Mn) to 40 mL of 10% sulfuric acid reagent solution. Add 5 mL of nitric acid and 5 mL of phosphoric acid to each, boil gently for 10 min, and cool. Any pink color in the solution of the sample should not exceed that in the control.

Sodium Peroxide

Na_2O_2 Formula Wt 77.98

CAS Number 1313–60–6

CAUTION. This reagent should be stored in airtight containers in a cool place. Avoid contact with organic materials; fire or explosion may result.

NOTE. When dissolving sodium peroxide, the material should be added slowly and in small portions to well-cooled water; when neutralizing, the acid must be added cautiously in small portions, and the solution kept cool.

REQUIREMENTS

Assay . . . . $\geq$ 93.0% Na_2O_2

MAXIMUM ALLOWABLE

Chloride (Cl) 0.002%
Nitrogen compounds (as N) 0.003%
Phosphate (PO_4) . . . . 5 ppm
Sulfate (SO_4) 0.001%
Heavy metals (as Pb) 0.002%
Iron (Fe) 0.005%

TESTS

ASSAY. (By titration of reducing power against permanganate). Weigh accurately 0.7 g and add slowly to 400 mL of dilute sulfuric acid (1 + 99) that has been cooled to 10°C. Dilute to volume in a 500-mL volumetric flask with dilute sulfuric acid (1 + 99), and titrate 100.0 mL with 0.1 N potassium permanganate. One milliliter of 0.1 N potassium permanganate corresponds to 0.003899 g of Na_2O_2.

CHLORIDE. Add 1.0 g in small portions to 35 mL of water, cool, and slowly add 4 mL of nitric acid. Filter if necessary through a chloride-free filter and dilute the filtrate with water to 40 mL. To 20 mL of the filtrate add 1 mL of silver nitrate reagent solution. Any turbidity should not exceed that produced by 0.01 mg of chloride ion (Cl) in an equal volume of solution containing the quantities of reagents used in the test.

NITROGEN COMPOUNDS. (Page 30). Dissolve 1.0 g in 20 mL of ammonia-free water cooled to 10°C, add acetic acid until neutral plus an excess of 0.15 mL, and boil down to a volume of 10 mL. For the standard use 0.03 mg of nitrogen (N).

PHOSPHATE. (Page 30, Method 1). Dissolve 4.0 g in 50 mL of water, cautiously add 10 mL of nitric acid, and evaporate to dryness on the steam bath. Dissolve the residue in 25 mL of approximately 0.5 N sulfuric acid, and continue as described.

SULFATE. Completely dissolve 20 g in 300 mL of cold water (10 °C), neutralize with hydrochloric acid, and note the volume used. Add an excess of 2 mL of the acid and evaporate to a volume of about 200 mL. Filter if necessary, heat the filtrate to boiling, add 10 mL of barium chloride reagent solution, digest in a covered beaker on a hot plate at low setting for 2 h, and allow to stand overnight. If a precipitate is formed, filter, wash thoroughly, and ignite. Correct for the weight obtained in a complete blank test. For the blank add about 10 mg of sodium carbonate and 1 mL of bromine water to the volume of hydrochloric acid used to neutralize the sodium peroxide, and evaporate to dryness. Dissolve in 2 mL of hydrochloric acid, add 200 mL of water, boil, and filter if necessary. Add 10 mL of barium chloride reagent solution and carry along with the sample.

HEAVY METALS.   (Page 26, Method 1). Add 2.0 g cautiously to 15 mL of dilute hydrochloric acid (1 + 2) cooled to 10°C. Evaporate the solution to dryness on the steam bath, add 15 mL of dilute hydrochloric acid, and repeat the evaporation to dryness. Dissolve the residue in 20 mL of water, and dilute with water to 25 mL. For the control solution add 0.02 mg of lead ion (Pb) to 15 mL of dilute hydrochloric acid, cool the acid to 10°C, and cautiously add 1.0 g of the sample, continuing with the evaporations exactly as for the 2.0 g of sample.

IRON.   (Page 28, Method 1). Add 10 g, cautiously, to 75 mL of a cooled solution of dilute hydrochloric acid (1 + 2). Evaporate on the steam bath to dryness, add 10 mL of hydrochloric acid, and again evaporate to dryness. Dissolve the residue in 80 mL of dilute hydrochloric acid (1 + 40), dilute with water to 100 mL, and use 2.0 mL of the solution.

Sodium Phosphate, Dibasic
Disodium Hydrogen Phosphate
(Suitable for buffer solutions)

Na_2HPO_4 Formula Wt 141.96

CAS Number 7558–79–4

REQUIREMENTS

Assay . . . . ≥ 99.0% Na_2HPO_4
pH of a 5% solution at 25°C 8.7–9.3

MAXIMUM ALLOWABLE

Insoluble matter 0.01%
Loss on drying at 105°C. 0.2%
Chloride (Cl) 0.002%
Nitrogen compounds (as N) 0.002%
Sulfate (SO$_4$) 0.005%
Heavy metals (as Pb) 0.001%
Iron (Fe) 0.002%

TESTS

ASSAY. (By indirect acid–base titrimetry). Weigh accurately 6.0 g and dissolve in 50 mL of water. Add exactly 50.0 mL of 1 N hydrochloric acid, and stir until the sample is completely dissolved. Titrate the excess acid, constantly stirring, with 1 N sodium hydroxide to the inflection point occurring at about pH 4 as measured with a standardized pH meter and glass electrode system. Calculate A, the volume of 1 N sodium hydroxide consumed by the sample. Continue the titration with 1 N sodium hydroxide to the inflection point occurring at about pH 8.8. Calculate B, the volume of 1 N sodium hydroxide required in the titration between the two inflection points. One milliliter of 1 N sodium hydroxide corresponds to 0.1420 g of Na_2HPO_4. If A is equal to or less than B, calculate the Na_2HPO_4 percentage from the titration of A. If A is greater than B, calculate the percentage from $2B - A$.

pH OF A 5% SOLUTION. (Page 43). Use the sample retained from the test for loss on drying at 105°C. The pH should be 8.7–9.3.

INSOLUBLE MATTER. (Page 14). Use 10 g dissolved in 100 mL of water. Do not use a glass filtering crucible.

LOSS ON DRYING AT 105°C. Weigh accurately about 6 g and dry at 105°C to constant weight. Save the material for the determination of pH.

CHLORIDE. (Page 25). Use 0.50 g of sample and 1.5 mL of nitric acid.

NITROGEN COMPOUNDS. (Page 30). Use 0.50 g. For the standard use 0.01 mg of nitrogen (N).

SULFATE. Dissolve 10 g in 200 mL of water, add 14 mL of hydrochloric acid, and heat to boiling. Filter and reheat the filtrate to boiling. Add 10 mL of barium chloride reagent solution, digest in a covered beaker on a hot plate at low setting for 2 h, and allow to stand overnight. If a precipitate is formed, filter, wash thoroughly, and ignite. Correct for the weight obtained in a complete blank test.

HEAVY METALS. (Page 26, Method 1). Dissolve 4.0 g in about 20 mL of water, add 14.5 mL of 2 N hydrochloric acid, and dilute with water to 40 mL. Use 30 mL to prepare the sample solution, and use the remaining 10 mL to prepare the control solution.

IRON. (Page 28, Method 2). Dissolve 1.0 g in water, dilute with water to 20 mL, and use 10 mL of the solution.

Sodium Phosphate, Dibasic, Heptahydrate
Disodium Hydrogen Phosphate Heptahydrate

$Na_2HPO_4 \cdot 7H_2O$ Formula Wt 268.07

CAS Number 7782–85–6

REQUIREMENTS

Assay 98.0–102.0%
$Na_2HPO_4 \cdot 7H_2O$
pH of a 5% solution at 25°C 8.7–9.3

MAXIMUM ALLOWABLE

Insoluble matter 0.005%
Chloride (Cl) 0.001%
Nitrogen compounds (as N) 0.001%
Sulfate (SO_4) 0.005%
Arsenic (As) . . . . 5 ppm
Heavy metals (as Pb) 0.001%
Iron (Fe) 0.001%

TESTS

ASSAY. (By indirect acid–base titrimetry). Weigh accurately 10.0 g and dissolve in 50 mL of water. Add exactly 50.0 mL of 1 N hydrochloric acid and stir until the sample is completely dissolved. Titrate the excess acid, constantly stirring, with 1 N sodium hydroxide to the inflection point occurring at about pH 4 as measured with a standardized pH meter and glass electrode system. Calculate A, the volume of 1 N sodium hydroxide consumed by the sample. Continue the titration with 1 N sodium hydroxide to the inflection point occurring at about pH 8.8. Calculate B, the volume of 1 N sodium hydroxide required in the titration between the two inflection points. One milliliter of 1 N sodium hydroxide corresponds to 0.2681 g of $Na_2HPO_4 \cdot 7H_2O$. If A is equal to or less than B, calculate the $Na_2HPO_4 \cdot 7H_2O$ percentage from the titration of A. If A is greater than B, calculate the percentage from $2B - A$.

pH OF A 5% SOLUTION.   (Page 43). The pH should be 8.7–9.3 at 25°C.

INSOLUBLE MATTER.   (Page 14). Use 20 g dissolved in 200 mL of water.

CHLORIDE.   (Page 25). Use 1.0 g of sample and 3 mL of nitric acid.

NITROGEN COMPOUNDS.   (Page 30). Use 1.0 g. For the standard use 0.01 mg of nitrogen (N).

SULFATE.   Dissolve 10 g in 100 mL of water, add 7 mL of hydrochloric acid, and heat to boiling. Add 5 mL of barium chloride reagent solution, digest in a covered beaker on a hot plate at low setting for 2 h, and allow to stand overnight. If a precipitate is formed, filter, wash thoroughly and ignite. Correct for the weight obtained in a complete blank test.

ARSENIC.   (Page 23). Use 0.60 g. For the standard use 0.003 mg of arsenic (As).

HEAVY METALS.   (Page 26, Method 1). Dissolve 4.0 g in about 20 mL of water, add 8 mL of 2 N hydrochloric acid, and dilute with water to 32 mL. Use 24 mL to prepare the sample solution, and use the remaining 8.0 mL to prepare the control solution.

IRON.   (Page 28, Method 2). Use 1.0 g.

Sodium Phosphate, Monobasic, Monohydrate
Sodium Dihydrogen Phosphate Monohydrate

$NaH_2PO_4 \cdot H_2O$ Formula Wt 137.99

CAS Number 10049−21−5

REQUIREMENTS

Assay . . . . 98.0−102.0%
$NaH_2PO_4 \cdot H_2O$
pH of a 5% solution at 25°C . . . . 4.1−4.5

MAXIMUM ALLOWABLE

Insoluble matter, calcium and
   ammonium hydroxide precipitate 0.01%
Chloride (Cl) . . . . 5 ppm
Nitrogen compounds (as N) 0.001%
Sulfate (SO_4) 0.003%
Arsenic (As) . . . . 0.5 ppm
Heavy metals (as Pb) 0.001%
Iron (Fe) 0.001%

TESTS

ASSAY.   (By acid−base titrimetry). Weigh accurately 5 g and dissolve in 50 mL of water. Titrate with 1 N sodium hydroxide to the inflection point at about pH 8.8 as measured with a standardized glass electrode system. One milliliter of 1 N sodium hydroxide corresponds to 0.1380 g of $NaH_2PO_4 \cdot H_2O$.

pH OF A 5% SOLUTION.   (Page 43). The pH should be 4.1−4.5 at 25°C.

INSOLUBLE MATTER, CALCIUM AND AMMONIUM HYDROXIDE PRECIPITATE.   Dissolve 10 g in 100 mL of water. Add 5 mL of ammonium oxalate reagent solution and ammonium hydroxide until the solution is distinctly alkaline to litmus, add an excess of 15 mL of ammonium hydroxide, and allow to stand overnight. Filter, wash with water containing 2.5% ammonia and about 0.1% ammonium oxalate, and ignite at $800 \pm 25°C$ for 15 min.

CHLORIDE.   (Page 25). Use 2.0 g of sample and 2 mL of nitric acid.

NITROGEN COMPOUNDS.   (Page 30). Use 1.0 g. For the standard use 0.01 mg of nitrogen (N).

SULFATE.   Dissolve 15 g in 200 mL of water, add 2 mL of hydrochloric acid, heat to boiling, and filter. Heat the filtrate to boiling, add 10 mL of barium chloride reagent solution, digest in a covered beaker on a hot plate at low setting for 2 h, and allow to stand overnight. If a precipitate is formed, filter, wash thoroughly, and ignite. Correct for the weight obtained in a complete blank test.

ARSENIC.   (Page 23). Use 6.0 g. For the standard use 0.003 mg of arsenic (As).

HEAVY METALS.   (Page 26, Method 1). Dissolve 4.0 g in about 20 mL of water, and dilute with water to 32 mL. Use 24 mL to prepare the sample solution, and use the remaining 8.0 mL to prepare the control solution.

IRON.   (Page 28, Method 2). Use 1.0 g.

Sodium Phosphate, Tribasic, Dodecahydrate

$Na_3PO_4 \cdot 12H_2O$ Formula Wt 380.12

CAS Number 10101–89–0

> *NOTE.* This salt, when stored under ordinary conditions, may lose some water of crystallization. Any loss in water will result in an assay of more than 100% $Na_3PO_4 \cdot 12H_2O$ but does not affect the determination of the relative amount of free alkali present.

REQUIREMENTS

Assay 98.0–102.0%

$Na_3PO_4 \cdot 12H_2O$

MAXIMUM ALLOWABLE

Excess alkali (as NaOH) 2.5%
Insoluble matter 0.01%
Chloride (Cl) 0.001%
Nitrogen compounds (as N) 0.001%
Sulfate (SO_4) 0.01%
Arsenic (As) . . . . 5 ppm
Heavy metals (as PF) 0.001%
Iron (Fe) 0.001%

TESTS

ASSAY. (By acid–base titrimetry). Weigh accurately 13–14 g and dissolve in 50 mL of water. Add exactly 100.0 mL of 1 N hydrochloric acid and pass a stream of carbon dioxide-free air or nitrogen, in fine bubbles, through the solution for 30 min to expel carbon dioxide. The beaker must be covered with a perforated cover to prevent loss of any of the solution by spraying. Wash down the cover and sides of the beaker. Titrate the solution with 1 N sodium hydroxide to the inflection point occurring at about pH 4 as measured with a standardized pH meter and glass electrode system. Calculate A, the milliliters of 1 N hydrochloric acid consumed in this titration. Protect the solution from absorbing carbon dioxide from the air, and continue the titration with 1 N sodium hydroxide to the inflection point occurring at about pH 8.8. Calculate B, the volume of 1 N sodium hydroxide required in the titration between the two inflection points. One milliliter of 1 N sodium hydroxide corresponds to 0.3801 g of $Na_3PO_4 \cdot 12H_2O$.

If A is equal to or greater than $2B$, then

$$\frac{B \times 0.3801}{\text{Sample wt (g)}} \times 100 = \% \ Na_3PO_4 \cdot 12H_2O$$

If A is less than $2B$, then

$$\frac{(A - B) \times 0.3801}{\text{Sample wt (g)}} \times 100 = \% \ Na_3PO_4 \cdot 12H_2O$$

EXCESS ALKALI. Calculate the amount of excess alkali as NaOH from the titration values obtained in the test for assay. If A is equal to or less than $2B$, there is no excess alkali present when the salt is dissolved. If A is greater than $2B$, then

$$\frac{(A - 2B) \times 0.040}{\text{Sample wt (g)}} \times 100 = \% \text{ NaOH}$$

INSOLUBLE MATTER.   (Page 14). Dissolve 10 g in 100 mL of water, add 0.10 mL of methyl red indicator solution, add hydrochloric acid until the solution is slightly acid to the methyl red indicator, and continue as described.

CHLORIDE.   (Page 25). Use 1.0 g of sample and 3 mL of nitric acid.

NITROGEN COMPOUNDS.   (Page 30). Use 1.0 g. For the standard use 0.01 mg of nitrogen (N).

SULFATE.   Dissolve 5.0 g in 100 mL of water, filter if necessary, add 4 mL of hydrochloric acid, and heat to boiling. Add 5 mL of barium chloride reagent solution, digest in a covered beaker on a hot plate at low setting for 2 h, and allow to stand overnight. If a precipitate is formed, filter, wash thoroughly, and ignite. Correct for the weight obtained from a complete blank test.

ARSENIC.   (Page 23). Use 0.60 g. For the standard use 0.003 mg of arsenic (As).

HEAVY METALS.   (Page 26, Method 1). Dissolve 4.0 g in about 20 mL of water, add 6 mL of 2 N hydrochloric acid, and dilute with water to 32 mL. Use 24 mL to prepare the sample solution, and use the remaining 8.0 mL to prepare the control solution.

IRON.   (Page 28, Method 2). Dissolve 1.0 g in 10 mL of water, and add 3 mL of 1 N hydrochloric acid. Omit the addition of acid to the standard.

Sodium Pyrophosphate Decahydrate
Diphosphoric Acid, Tetrasodium Salt, Decahydrate

$Na_4P_2O_7 \cdot 10H_2O$ Formula Wt 446.06

CAS Number 13472–36–1

REQUIREMENTS

Assay	99.0–103.0%
	$Na_4P_2O_7 \cdot 10H_2O$
pH of a 5% solution at 25°C	9.5–10.5

MAXIMUM ALLOWABLE

Insoluble matter	0.01%
Chloride (Cl)	0.002%
Sulfate (SO$_4$)	0.005%
Nitrogen compounds (as N)	0.001%
Arsenic (As)	5 ppm
Heavy metals (as Pb)	0.001%
Iron (Fe)	0.001%

TESTS

ASSAY. (By acid–base titrimetry). Weigh accurately 8 g and dissolve in 100 mL of water. Titrate with 1 N hydrochloric acid to a green color with 0.15 mL of bromphenol blue indicator. One milliliter of 1 N hydrochloric acid corresponds to 0.2230 g of $Na_4P_2O_7 \cdot 10H_2O$.

pH OF A 5% SOLUTION.   (Page 43). The pH should be from 9.5–10.5 at 25°C.

INSOLUBLE MATTER.   (Page 14). Use 10 g dissolved in 150 mL of water.

CHLORIDE.   (Page 25). Use 0.50 g of sample and 3 mL of nitric acid.

SULFATE.   Dissolve 10 g in 100 mL of water, add 7 mL of hydrochloric acid, and heat to boiling. Add 5 mL of barium chloride reagent solution, digest in a covered beaker on a hot plate at low setting for 2 h, and allow to stand overnight. If a precipitate is formed, filter, wash thoroughly, and ignite. Correct for the weight obtained in a complete blank test.

NITROGEN COMPOUNDS.   (Page 30). Use 1.0 g. For the standard use 0.01 mg of nitrogen (N).

ARSENIC.   (Page 23). Use 0.60 g. For the standard use 0.003 mg of arsenic (As).

HEAVY METALS.   (Page 26, Method 1). Dissolve 3.0 g in about 20 mL of water, warming if necessary, add 8 mL of 2 N hydrochloric acid, and dilute with water to 42 mL. Use 35 mL to prepare the sample solution, and use the remaining 7.0 mL to prepare the control solution.

IRON.   (Page 28, Method 2). Use 1.0 g.

Sodium Sulfate
Sodium Sulfate, Anhydrous

Na_2SO_4 Formula Wt 142.04

CAS Number 7757–82–6

REQUIREMENTS

Assay . . . . ≥ 99.0% Na_2SO_4
pH of a 5% solution at 25°C 5.2–9.2

MAXIMUM ALLOWABLE

Insoluble matter 0.01%
Loss on ignition 0.5%
Chloride (Cl) 0.001%
Nitrogen compounds (as N) . . . . 5 ppm
Arsenic (As). . . . . 1 ppm
Calcium, magnesium, and
   R_2O_3 precipitate 0.02%
Heavy metals (as Pb) . . . . 5 ppm
Iron (Fe) 0.001%

TESTS

ASSAY. (By indirect acid–base titrimetry after ion exchange).

NOTE. All water used in the assay must be carbon dioxide- and ammonia-free.

Weigh, to the nearest 0.1 mg, 0.3 g of sample and dissolve in 50 mL of water. Pass the solution through a cation-exchange column at a rate of about 5 mL/min, and collect the eluate in a 500-mL titration flask. Then wash the resin in the column with water at a rate of about 10 mL/min, and collect in the same titration flask. Add 0.15 mL of phenolphthalein to the flask and titrate with 0.1 N sodium hydroxide. Continue the titration, as the washing proceeds, until 50 mL of eluate requires no further titration.

$$\% \ Na_2SO_4 = \frac{(mL \times N \ NaOH) \times 7.10}{Sample \ wt \ (g)}$$

Ion-Exchange Column: Use Dowex HCR-W2, H^+ form, 16–40 mesh, or equivalent. Charge the resin to a column with a 2-cm bore to a height of about 20 cm. If the resin is in the hydrogen ion form, wash free of acid and color with carbon dioxide-free water. Otherwise regenerate by passing 4 N hydrochloric acid through the column and washing free of the acid.

INSOLUBLE MATTER. (Page 14). Use 10 g dissolved in 100 mL of water.

LOSS ON IGNITION. Weigh accurately about 2 g in a tared dish or crucible and ignite at $800 \pm 25°C$ for 15 min.

pH OF A 5% SOLUTION. (Page 43). The pH should be 5.2–9.2 at 25°C.

CHLORIDE. (Page 25). Use 1.0 g.

NITROGEN COMPOUNDS. (Page 30). Use 2.0 g. For the standard use 0.01 mg of nitrogen (N).

ARSENIC. (Page 23). Use 3.0 g. For the standard use 0.003 mg of arsenic (As).

CALCIUM, MAGNESIUM, AND R_2O_3 PRECIPITATE. Dissolve 5.0 g in 75 mL of water, filter if necessary, and add 5 mL of ammonium oxalate reagent solution, 2 mL of ammonium phosphate reagent solution, and 10 mL of ammonium hydroxide. Allow to stand overnight. If a precipitate is formed, filter, wash with water containing 2.5% ammonia and about 0.1% ammonium oxalate, and ignite.

HEAVY METALS. (Page 26, Method 1). Dissolve 6.0 g in water, and dilute with water to 42 mL. Use 35 mL to prepare the sample solution, and use the remaining 7.0 mL to prepare the control solution.

IRON. (Page 28, Method 1). Use 1.0 g.

Sodium Sulfate Decahydrate

$Na_2SO_4 \cdot 10H_2O$ Formula Wt 322.19

CAS Number 7727–73–3

REQUIREMENTS

Assay . . . . $\geq$ 99%
$Na_2SO_4 \cdot 10H_2O$
pH of a 5% solution at 25°C 5.2–9.2

MAXIMUM ALLOWABLE

Insoluble matter 0.01%
Chloride (Cl) 0.001%
Nitrogen compounds (as N) . . . . 5 ppm
Phosphate (PO_4) 0.001%
Calcium (Ca) 0.002%
Heavy metals (as Pb) . . . . 5 ppm
Iron (Fe) 0.001%
Magnesium (Mg) 0.001%
Potassium (K) 0.002%

TESTS

ASSAY. (By indirect acid–base titrimetry after ion exchange).

NOTE. All water used in the assay must be free of carbon dioxide and ammonia.

Prepare an ion-exchange column by charging a glass column having a 2-cm bore to a height of about 20 cm with a suitable cation-exchange resin (Amberlite IR-120 analytical grade or equivalent). If the resin is in its hydrogen form, wash free of acid and color with water. If not, regenerate it by passing 50 mL of approximately 4 N hydrochloric acid through the column and then washing with water until neutral. Weigh 0.7 g of sample to the nearest 0.1 mg, and dissolve in about 50 mL of water. Pass the sample solution through the column at a rate of about 5 mL/min, collecting the eluate in a 500-mL titration flask. Wash the column with water at a rate of about 10 mL/min, collecting the washings in the same flask. Add 0.15 mL of phenolphthalein indicator solution, and titrate the eluate with 0.1 N sodium hydroxide. Continue to wash the column and to titrate the eluate until the passage of 50 mL of water requires no further titrant. One milliliter of 0.1 N sodium hydroxide corresponds to 0.01611 g of $Na_2SO_4 \cdot 10H_2O$.

INSOLUBLE MATTER.   (Page 14). Use 10 g dissolved in 100 mL of water.

pH OF A 5% SOLUTION.   (Page 43). The pH should be 5.2–9.2 at 25°C.

CHLORIDE.   (Page 25). Use 1.0 g.

NITROGEN COMPOUNDS.   (Page 30). Use 2.0 g. For the standard use 0.01 mg of nitrogen (N).

PHOSPHATE.   (Page 30, Method 1). Dissolve 2.0 g in 20 mL of 0.5 N sulfuric acid. For the standard, dilute 0.02 mg of phosphate ion (PO_4) to 20 mL with 0.5 N sulfuric acid.

HEAVY METALS.   (Page 26, Method 1). Dissolve 6.0 g in water and dilute with water to 42 mL. Use 35 mL to prepare the sample solution and use the remaining 7 mL to prepare the control solution.

CALCIUM, IRON, POTASSIUM, AND MAGNESIUM.   Determine the calcium, iron, potassium, and magnesium content by the flame atomic absorption spectrophotometric method described on page 35.

> ***Sample Stock Solution.***   Dissolve 20.0 g in 50 mL of water in a 100-mL volumetric flask. Dilute to the mark with water (1 mL = 0.2 g).

Analyze the solutions by means of a suitable atomic absorption spectrophotometer, using the conditions outlined in the following table. Calculate the metal content of the sample by the method of standard additions.

Element	Wavelength (nm)	Sample Wt (g)	Standard Added (mg)	Flame Type*	Background Correction
Ca	422.7	2.0	0.02; 0.04	N/A	No
Fe	248.3	5.0	0.025; 0.05	A/A	Yes
K	766.5	2.0	0.02; 0.04	A/A	No
Mg	285.2	1.0	0.005; 0.01	A/A	Yes

*A/A is air/acetylene; N/A is nitrous oxide/acetylene.

Sodium Sulfide Nonahydrate

$Na_2S \cdot 9H_2O$ Formula Wt 240.18

CAS Number 1313–84–4

REQUIREMENTS

Assay 98.0–103.0%
$Na_2S \cdot 9H_2O$
Appearance. . . . . Crystals, colorless
or with only a
slight yellow color

MAXIMUM ALLOWABLE

Ammonium (NH_4) 0.005%
Sulfite and thiosulfate (as SO_4) 0.1%
Iron . . . . Passes test

TESTS

ASSAY.   (By titration of reductive capacity). Weigh accurately 2.5 g and in a 250-mL volumetric flask dilute with water through which nitrogen gas has been freshly bubbled to expel oxygen. Dilute to volume with similar water and displace air from the headspace with nitrogen. Stopper and mix thoroughly. Place a 50.0-mL aliquot of this solution into a mixture of 50.0 mL of 0.1 N iodine and 25 mL of 0.1 N hydrochloric acid in 400 mL of water. Titrate the excess of iodine with 0.1 N sodium thiosulfate, adding 3 mL of starch indicator solution near the end of the titration. One milliliter of 0.1 N iodine consumed corresponds to 0.01201 g of $Na_2S \cdot 9H_2O$.

AMMONIUM.   (Page 22). Dissolve 1.0 g in 80 mL of water, add 20 mL of lead acetate reagent solution, and allow to stand until the precipitate has settled. Decant 50 mL of the clear supernatant liquid, and use the 50 mL after distillation. For the standard use 0.025 mg of ammonium (NH_4) treated exactly as the sample.

SULFITE AND THIOSULFATE.   (NOTE. The water used in this test must be free from dissolved oxygen.) Dissolve 3.0 g in 200 mL of water and add 100 mL of 5% zinc sulfate solution. Mix thoroughly;

allow to stand for 30 min. Filter and titrate 100 mL of the filtrate with 0.01 N iodine, using starch indicator. Not more than 3.0 mL of the iodine solution should be required.

IRON.   Dissolve 5 g in 100 mL of water. The solution should be clear and colorless.

Sodium Sulfite
Sodium Sulfite, Anhydrous

Na_2SO_3 Formula Wt 126.04

CAS Number 7757–83–7

REQUIREMENTS

Assay . . . . ≥ 98.0% Na_2SO_3

MAXIMUM ALLOWABLE

Insoluble matter 0.005%
Free acid . . . . Passes test
Titrable free base . . . . 0.03 meq / g
Chloride (Cl) 0.02%
Arsenic (As) . . . . 1 ppm
Heavy metals (as Pb) 0.001%
Iron (Fe) 0.001%

TESTS

ASSAY.   (By titration of reductive capacity). Place 0.5 g, accurately weighed, into 100 mL of 0.1 N iodine. Mix, allow to stand for 5 min, and add 1 mL of hydrochloric acid. Titrate the excess of iodine with 0.1 N sodium thiosulfate, adding 3 mL of starch indicator solution near the end of the titration. One milliliter of 0.1 N iodine consumed corresponds 0.006302 g of Na_2SO_3.

INSOLUBLE MATTER.   (Page 14). Use 20 g dissolved in 200 mL of water.

FREE ACID.   Dissolve 1.0 g in 10 mL of water and add 0.10 mL of phenolphthalein indicator solution. A pink color should be produced.

TITRABLE FREE BASE.   Dissolve 1.0 g in 10 mL of water and add 3 mL of 30% hydrogen peroxide that has been previously neutralized to 0.15 mL of methyl red. Shake well, allow to stand for 5 min, and titrate with 0.01 N hydrochloric acid. Not more than 3.0 mL of the acid should be required to neutralize the solution.

CHLORIDE.   Dissolve 1.0 g in 10 mL of water, filter if necessary through a small chloride-free filter, add 3 mL of 30% hydrogen peroxide, and dilute with water to 100 mL. Dilute 5 mL of this solution with water to 20 mL, and add 1 mL of nitric acid and 1 mL of silver nitrate reagent solution. Any turbidity should not exceed that produced by 0.01 mg of chloride ion (Cl) in an equal volume of solution containing the quantities of nitric acid and silver nitrate used in the test.

ARSENIC.   (Page 23). Dissolve 3.0 g in 10 mL of water in a 250-mL beaker, add 3 mL of sulfuric acid, evaporate in a hood to dense fumes of sulfur trioxide, and continue the fuming for 15 min. Cool, cautiously wash down the beaker with about 5 mL of water, add 3 mL of sulfuric acid, evaporate again to dense fumes of sulfur trioxide, and continue the fuming for 15 min. Cool, and cautiously wash the solution into a generator flask with sufficient water to make a volume of 35 mL, and use this solution. Omit the addition of 20 mL of dilute sulfuric acid (1 + 4). For the standard use 0.003 mg of arsenic (As).

HEAVY METALS.   (Page 26, Method 1). Dissolve 3.0 g in 20 mL of dilute hydrochloric acid (1 + 1), and evaporate the solution to dryness on the steam bath. Add 10 mL more of dilute hydrochloric acid (1 + 1), and again evaporate to dryness. Dissolve the residue in about 20 mL of water, and dilute with water to 25 mL. For the control solution add 0.02 mg of lead ion (Pb) to 1.0 g of sample, and treat exactly as the 3.0 g of sample.

IRON.   (Page 28, Method 1). Dissolve 1.0 g in 20 mL of dilute hydrochloric acid (1 + 9), and evaporate on the steam bath to dryness. Dissolve the residue in 5 mL of dilute hydrochloric acid (1 + 1), and again evaporate to dryness. Dissolve the residue in 4 mL of dilute hydrochloric acid (1 + 1), dilute with water to 50 mL, and use the solution without further acidification.

Sodium Tartrate Dihydrate
2,3-Dihydroxybutanedioic Acid, Disodium Salt, Dihydrate

$$NaO-\overset{O}{\overset{\|}{C}}-\overset{OH}{\overset{|}{CH}}-\overset{OH}{\overset{|}{CH}}-\overset{O}{\overset{\|}{C}}-ONa \cdot 2H_2O$$

$(CHOHCOONa)_2 \cdot 2H_2O$ Formula Wt 230.08

CAS Number 6106−24−7

> *NOTE.* This reagent is suitable for standardization of Karl Fischer reagent as used for the determination of trace amounts of water (less than 1%).

REQUIREMENTS

Assay	99.0–101.0% $Na_2C_4H_4O_6 \cdot 2H_2O$
Loss on drying at 150°C	15.61%–15.71%
pH of a 5% solution at 25°C	7.0–9.0

MAXIMUM ALLOWABLE

Insoluble matter	0.005%
Chloride (Cl)	5 ppm
Phosphate (PO_4)	5 ppm
Sulfate (SO_4)	0.005%
Ammonium (NH_4)	0.003%
Calcium (Ca)	0.01%
Heavy metals (as Pb)	5 ppm
Iron (Fe)	0.001%

TESTS

ASSAY. (Total alkalinity by nonaqueous titration). Weigh accurately 0.45 g, and dissolve in 100 mL of glacial acetic acid. Titrate with 0.1 N perchloric acid in glacial acetic acid to a green end point with crystal violet indicator. One milliliter of 0.1 N perchloric acid corresponds to 0.01151 g of $Na_2C_4H_4O_6 \cdot 2H_2O$.

LOSS ON DRYING AT 150°C. Weigh accurately about 3 g into a low-form weighing bottle and dry at 150°C to constant weight (minimum 4 hours).

pH OF A 5% SOLUTION. (Page 43). The pH should be 7.0–9.0 at 25°C.

INSOLUBLE MATTER. (Page 14). Use 20 g dissolved in 200 mL of water.

CHLORIDE. (Page 25). Use 2.0 g.

PHOSPHATE. (Page 30, Method 1). Ignite 4.0 g in a platinum dish. Dissolve the residue in 5 mL of water, add 5 mL of nitric acid, and evaporate to dryness. Dissolve the residue in 25 mL of approximately 0.5 N sulfuric acid and continue as described.

SULFATE. Drive off any moisture on a hot plate and then carefully ignite 1 g in an electric muffle furnace until it is nearly free of carbon. Boil the residue with 10 mL of water and 0.5 mL of 30% hydrogen peroxide for 5 min. Add 5 mg of sodium carbonate and 1 mL of hydrochloric acid and evaporate on a hot plate at low setting to dryness. Dissolve the residue in 4 mL of hot water to which has been added 1 mL of dilute hydrochloric acid (1 + 19), filter through a small filter, wash with two 2-mL portions of water, and dilute the filtrate to 10 mL. Add 1 mL of barium chloride reagent solution and mix well. Any turbidity produced should not be greater than that in a standard prepared as described and to which the barium chloride solution is added at the same time it is added to the sample solution. To 10 mL of water, add 5 mg of sodium carbonate, 1 mL of hydrochloric acid, 0.5 mL of 30% hydrogen peroxide, and 0.05 mg of sulfate (SO_4) and evaporate on a hot plate at low setting to dryness. Dissolve the residue in 9 mL of water and add 1 mL of dilute hydrochloric acid (1 + 19). If necessary, adjust with water to the same volume as the sample solution, add 1 mL of barium chloride reagent solution, and mix well.

AMMONIUM. Dissolve 1.0 g in 60 mL of water. To 20 mL add 1 mL of 10% sodium hydroxide reagent solution, dilute with water to 50 mL, and add 2 mL of Nessler reagent. Any color should not exceed that produced by 0.01 mg of ammonium ion (NH_4) in an equal volume of solution containing the quantities of reagents used in the test.

CALCIUM. Determine the calcium by the flame atomic absorption spectrophotometric method described on page 35.

Sample Stock Solution. Dissolve 10.0 g of sample with water in a 100-mL volumetric flask, add 5 mL of hydrochloric acid, and dilute to the mark with water (0.1 g/mL).

Analyze the solutions by means of a suitable atomic absorption spectrophotometer, using the conditions outlined in the following table. Calculate the metal content of the sample by the method of standard additions.

Element	Wavelength (nm)	Sample Wt (g)	Standard Added (mg)	Flame Type*	Background Correction
Ca	422.7	0.5	0.05; 0.10	N/A	No

*N/A is nitrous oxide/acetylene.

HEAVY METALS. (Page 26, Method 1). Dissolve 6.0 g in about 20 mL of water, and dilute with water to 30 mL. Use 25 mL to prepare the sample solution, and use the remaining 5.0 mL to prepare the control solution.

IRON. (Page 28, Method 1). Use 1.0 g.

Sodium Tetraphenylborate
Sodium Tetraphenylboron

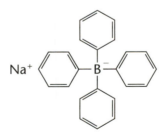

$NaB(C_6H_5)_4$

Formula Wt 342.22

CAS Number 143–66–8

REQUIREMENTS

Assay . . . . ≥ 99.5%
$$NaB(C_6H_5)_4$$

MAXIMUM ALLOWABLE

Loss on drying at 105°C. 0.5%
Clarity of solution . . . . Passes test

TESTS

ASSAY.   (By gravimetric determination of tetraphenylborate). Weigh accurately about 0.5 g, dissolve in 100 mL of water, and add 1 mL of acetic acid. Add, with constant stirring, 25 mL of 5% potassium hydrogen phthalate solution, and allow to stand for 2 h. Filter through a tared, fine-porosity filtering crucible, wash with three 5-mL portions of saturated potassium tetraphenylborate solution, and dry at 105°C for 1 h. The weight of the potassium tetraphenylborate multiplied by 0.9551 corresponds to the weight of the sodium tetraphenylborate.

LOSS ON DRYING AT 105°C.   Weigh accurately about 1.5 g and dry to constant weight at 105 ° C.

CLARITY OF SOLUTION.   Weigh 1.5 g, add 250 mL of water and 0.75 g of hydrated aluminum oxide, stir for 5 min, and filter through a filter paper. Filter the first 25-mL portion again. The filtrate should be clear.

Sodium Thiocyanate

NaSCN   Formula Wt 81.07

CAS Number 540–72–7

REQUIREMENTS

Assay . . . . ≥ 98.0% NaSCN

MAXIMUM ALLOWABLE

Insoluble matter 0.005%
Carbonate (as Na_2CO_3) 0.2%
Chloride (Cl) 0.01%
Sulfate (SO_4) 0.01%
Sulfide (S) 0.001%
Ammonium (NH_4) 0.002%
Heavy metals (as Pb) . . . . 5 ppm
Iron (Fe) . . . . 2 ppm

TESTS

ASSAY. (By indirect argentometric titration). Weigh, to the nearest 0.1 mg, 6 g of sample, dissolve with 100 mL of water in a 1000-mL volumetric flask, dilute to volume with water, and mix well. To 50.0 mL in a 250-mL glass-stoppered flask, add while agitating, exactly 50.0 mL of 0.1 N silver nitrate solution, then add 3 mL of nitric acid and 2 mL of ferric ammonium sulfate reagent solution (as indicator), and titrate the excess silver nitrate with 0.1 N ammonium thiocyanate solution.

$$\% \text{ NaSCN} = \frac{[(50.0 \times \text{N AgNO}_3) - (\text{mL} \times \text{N NH}_4\text{SCN})] \times 162.14}{\text{Sample wt (g)}}$$

INSOLUBLE MATTER. (Page 14). Use 20 g dissolved in 150 mL of water.

CARBONATE. Dissolve 10 g in 100 mL of carbon dioxide-free water, add 0.10 mL of methyl orange indicator, and titrate with 0.1 N sulfuric acid. Not more than 3.8 mL of the acid should be required.

CHLORIDE. Dissolve 0.50 g in 20 mL of water, filter if necessary through a chloride-free filter, and add 10 mL of 25% sulfuric acid solution and 7 mL of 30% hydrogen peroxide. Evaporate to 20 mL by boiling in a well-ventilated hood, add 15–20 mL of water, and evaporate again. Repeat until all of the cyanide has been volatilized, cool, and dilute with water to 100 mL. To 20 mL of this solution add 1 mL of nitric acid and 1 mL of silver nitrate reagent solution. Any turbidity should not exceed that produced by 0.01 mg of chloride ion (Cl) in an equal volume of solution containing the quantities of reagents used in the test.

SULFATE.   (Page 32, Procedure A, Method 1).

SULFIDE.   Add 5.0 g to 80 mL of solution prepared by mixing 10 mL of ammonium hydroxide and 10 mL of silver nitrate reagent solution in 60 mL of water. Heat on a hot plate at low setting with occasional shaking for 20 min and transfer to a color comparison tube. Any color should not exceed that produced by 0.05 mg of sulfide ion (S) in an equal volume of solution containing the quantities of reagents used in the test.

AMMONIUM.   (Page 22). Dilute the distillate from a 1.0-g sample to 100 mL, and use 50 mL of the dilution. For the standard use 0.01 mg of ammonium (NH_4).

HEAVY METALS.   (Page 26, Method 1). Dissolve 6.0 g in about 20 mL of water and dilute with water to 30 mL. Use 25 mL to prepare the sample solution, and use the remaining 5.0 mL to prepare the control solution.

IRON.   Dissolve 10 g in 60 mL of water. To 30 mL of the solution add 0.1 mL of hydrochloric acid, 6 mL of hydroxylamine hydrochloride reagent solution, and 4 mL of 1,10-phenanthroline reagent solution, and add ammonium hydroxide to bring the pH to approximately 5. Any red color should not exceed that produced by 0.01 mg of iron (Fe) in an equal volume of solution containing the quantities of reagents used in the test. Compare 1 h after adding the reagents to the sample and standard solutions.

Sodium Thiosulfate Pentahydrate

$Na_2S_2O_3 \cdot 5H_2O$ Formula Wt 248.19

CAS Number 10102−17−7

REQUIREMENTS

Assay . . . . 99.5−101.0%
$Na_2S_2O_3 \cdot 5H_2O$
pH of a 5% solution at 25°C . . . . 6.0−8.4

MAXIMUM ALLOWABLE

Insoluble matter 0.005%
Nitrogen compounds (as N) 0.002%
Sulfate and sulfite (as SO_4) 0.1%
Sulfide (S) . . . . Passes test (limit
about 1 ppm)

TESTS

ASSAY. (By titration of reductive capacity). Weigh accurately 1 g and dissolve in 30 mL of water. In the pH of a 5% solution test, if a value greater than 7.5 was obtained, add 0.05 mL of 1 N acetic acid. Titrate with 0.1 N iodine to a blue color with starch indicator. One milliliter of 0.1 N iodine corresponds to 0.02482 g of $Na_2S_2O_3 \cdot 5H_2O$.

INSOLUBLE MATTER. (Page 14). Use 10 g dissolved in 100 mL of water.

pH OF A 5% SOLUTION. (Page 43). The pH should be 6.0–8.4 at 25°C.

NITROGEN COMPOUNDS. (Page 30). Dissolve 1.5 g in 10 mL of water in a flask, and add 10 mL of 25% sulfuric acid reagent solution. Connect the flask to a water-cooled reflux condenser, heat to boiling, and reflux for about 5 min. Cool and rinse down the condenser with a small amount of water. Filter, using a filter that has been washed free of ammonia, into a flask, and wash the residue with 10–15 mL of water. Add to the flask 30 mL of freshly boiled 10% sodium hydroxide reagent solution and distil. To the distillate add 2 mL of 10% sodium hydroxide reagent solution, dilute with water to 100 mL, and use 50 mL of the dilution. For the control use 0.01 mg of nitrogen (N) and 0.25 g of sample.

SULFATE AND SULFITE. Dissolve 1.0 g in 50 mL of water and add approximately 0.1 N iodine until the liquid has a faint yellow color. Dilute with water to a volume of 100 mL and mix thoroughly. To 5.0 mL of the solution add 5 mL of water, 1 mL of 1 N hydrochloric acid, and 1 mL of barium chloride reagent solution. Any turbidity should not exceed that produced by 0.05 mg of sulfate ion (SO_4) in an equal volume of solution containing the quantities of reagents used in the test. Compare 10 min after adding the barium chloride to the sample and standard solutions.

SULFIDE. Dissolve 1.0 g in 10 mL of water and add 0.5 mL of alkaline lead solution (made by adding sufficient 10% sodium hydroxide reagent solution to 10% lead acetate reagent solution to redissolve the precipitate that is first formed). No dark color should be produced in 1 min.

Sodium Tungstate Dihydrate

$Na_2WO_4 \cdot 2H_2O$ Formula Wt 329.86

CAS Number 10213–10–2

REQUIREMENTS

Assay 99.0–101.0%
$Na_2WO_4 \cdot 2H_2O$

MAXIMUM ALLOWABLE

Insoluble matter 0.01%
Titrable free base . . . . 0.02 meq / g
Chloride (Cl) 0.005%
Molybdenum (Mo) 0.001%
Nitrogen compounds (as N) 0.001%
Sulfate (SO_4) 0.01%
Arsenic (As) . . . . 5 ppm
Heavy metals and iron (as Pb) 0.001%

TESTS

ASSAY.   (By gravimetry). Weigh accurately 1 g and dissolve in 50 mL of water in a 150-mL beaker. Add 2 mL of 30% hydrogen peroxide and 45 mL of nitric acid. Heat on a steam bath for 90 min, cool, and filter through a fine filter paper. Wash with 50 mL of hot 1% nitric acid, dry in a tared crucible, burn off the paper at a low temperature, and ignite at 825°C to constant weight. One gram of WO_3 corresponds to 1.4223 g of $Na_2WO_4 \cdot 2H_2O$.

INSOLUBLE MATTER.   (Page 14). Use 10 g dissolved in 100 mL of water.

TITRABLE FREE BASE.   Dissolve 2.0 g in 50 mL of cold water and add 0.10 mL of thymol blue indicator solution. A blue color should be produced, which is changed to yellow by the addition of not more than 4.0 mL of 0.01 N hydrochloric acid.

CHLORIDE. Dissolve 1.0 g in 20 mL of water, filter if necessary through a chloride-free filter, add 3 mL of phosphoric acid, mix well, and dilute with water to 50 mL. Dilute 10 mL of the solution with water to 25 mL and add 1 mL of nitric acid and 1 mL of silver nitrate reagent solution. Any turbidity should not exceed that produced by 0.01 mg of chloride ion (Cl) in an equal volume of solution containing the quantities of reagents used in the test.

MOLYBDENUM.

> *Sample Solution A.* Dissolve 0.50 g in 15 mL of ammonium citrate solution (300 g of dibasic ammonium citrate dissolved in 500 mL of water and filtered).

> *Standard Solution B.* Add a standard molybdenum solution containing 5 μg of molybdenum ion (Mo) to 15 mL of the ammonium citrate solution.

> *Blank Solution C.* Use 15 mL of the ammonium citrate solution.

Dilute the three solutions with water to 18 mL, then to each add the following reagents in order while stirring: 30 mL of dilute hydrochloric acid (1 + 1), 0.05 mL of 0.1 M cupric chloride in dilute hydrochloric acid (1 + 99), 3 mL of ammonium thiocyanate reagent solution (filtered if necessary), 3 mL of potassium iodide solution (50 g in 100 mL of water), and 2 mL of 1% sodium sulfite solution. Set a spectrophotometer at 460 nm and, using 1-cm cells, adjust the instrument to read 0 absorbance with blank solution C in the light path, then determine the absorbance of sample solution A and standard solution B. The absorbance of sample solution A should not exceed that of standard solution B. Compare within 30 min after development of the colors.

NITROGEN COMPOUNDS. (Page 30). Use 1.0 g. For the standard use 0.01 mg of nitrogen (N).

SULFATE. Dissolve 2.0 g in 100 mL of water and add slowly with stirring 5 mL of hydrochloric acid. Evaporate to dryness and heat for 20 min at 110°C. Add 30 mL of water, 2.5 mL of hydrochloric acid, and 2 mL of cinchonine solution [made by dissolving 5 g of cinchonine in 50 mL of dilute hydrochloric acid (1 + 3)] and heat just below the boiling point for 30 min. Dilute with water to 30 mL and allow to stand until cool. Filter, and to 15 mL of the filtrate add ammonium hydroxide drop by drop until a slight permanent precipitate forms. Add just enough hydrochloric acid to redissolve the precipitate and add 2 mL of barium chloride reagent solution. Any turbidity should not exceed that produced in a standard made as follows: to 1 mL of the cinchonine solution add 0.1 mg of sulfate ion (SO_4), dilute with water to 15 mL, and add 2 mL of barium chloride reagent solution. Compare 10 min after adding the barium chloride to the sample and standard solutions.

ARSENIC.   (Page 23). Transfer 0.6 g to a distilling flask fitted with a gas inlet tube, a dropping funnel, a thermometer inserted to within a few millimeters of the bottom, an outlet tube, and an efficient water condenser. Add 5 mL of phosphoric acid, 5 g of potassium bromide, 5 g of calcium chloride, and 0.1 g of cuprous chloride; then add 50 mL of hydrochloric acid through the dropping funnel. Pass a slow stream of carbon dioxide or nitrogen (2–3 bubbles per second) into the flask, boil the liquid gently, and absorb the distillate in 20 mL of cold water in a large test tube immersed in an ice bath. The reaction is complete when the temperature reaches 111°C. Neutralize the distillate with 50% sodium hydroxide solution. For the standard use 0.003 mg of arsenic (As). The standard should be prepared in a similar manner—including the preliminary preparation, distillation, and neutralization.

HEAVY METALS AND IRON.   Dissolve 2.0 g in 90 mL of water plus 2 mL of ammonium hydroxide, and dilute with water to 100 mL. To 10 mL of this solution in a separatory funnel add 10 mL of dithizone extraction solution in chloroform. Shake well for 20 s and discard the aqueous layer. Wash the chloroform layer with successive 10-mL portions of 0.5% ammonium hydroxide until the ammonium hydroxide solution remains nearly colorless. The pink color remaining in the chloroform should not exceed that produced by 0.002 mg of lead ion (Pb) in 10 mL of dilute ammonium hydroxide (1 + 99) treated exactly as the 10-mL portion of the solution of the sample.

Stannous Chloride Dihydrate
Tin(II) Chloride Dihydrate

$SnCl_2 \cdot 2H_2O$ Formula Wt 225.65

CAS Number 10025–69–1

REQUIREMENTS

Assay 98.0–103.0%
$SnCl_2 \cdot 2H_2O$

MAXIMUM ALLOWABLE

Solubility in hydrochloric acid . . . . Passes test
Sulfate (SO$_4$) . . . . Passes test (limit
about 0.003%)
Arsenic (As) . . . . 2 ppm
Substances not precipitated by
hydrogen sulfide (as sulfates) 0.05%
Iron (Fe) 0.003%
Other metals (as Pb) 0.01%

TESTS

ASSAY. (By titration of reductive capacity of Sn^{II}). In a weighing bottle, weigh accurately 2.0 g. Transfer to a 250-mL volumetric flask, first with three 5-mL portions of hydrochloric acid and then with water through which nitrogen gas has been freshly bubbled to expel oxygen. Dilute to volume with similar water. Displace air from the headspace with nitrogen. Stopper and mix thoroughly. Place a 50.0-mL aliquot of this solution in a 500-mL conical flask filled with nitrogen gas. Add 5 g of potassium sodium tartrate tetrahydrate and 60 mL of a cold saturated solution of sodium bicarbonate. Titrate promptly with 0.1 N iodine to a blue color with starch indicator. One milliliter of 0.1 N iodine corresponds to 0.01128 g of SnCl$_2$ · 2H$_2$O.

SOLUBILITY IN HYDROCHLORIC ACID. Dissolve 5 g in 5 mL of hydrochloric acid, heat to 40°C if necessary, and dilute with 5 mL of water. The salt should dissolve completely.

SULFATE. Dissolve 5.0 g in 5 mL of hydrochloric acid, dilute the solution with water to 50 mL, filter if necessary, and heat to boiling. Add 5 mL of barium chloride reagent solution, digest in a covered beaker on a hot plate at low setting for 2 h, and allow to stand overnight. No precipitate should be formed.

ARSENIC. (Page 23). Mix 1.5 g with 35 mL of water and proceed as directed, but use 4 g of granulated zinc instead of 3 g. For the standard use 0.003 mg of arsenic (As) and 3 g of granulated zinc.

SUBSTANCES NOT PRECIPITATED BY HYDROGEN SULFIDE. Dissolve 4.0 g in 5 mL of hydrochloric acid, dilute with water to 200 mL, and precipitate the tin with hydrogen sulfide. Filter without washing, and evaporate 100 mL of the filtrate in a tared dish to a few milliliters. Add 0.10–0.20 mL of sulfuric acid, evaporate to dryness, and ignite at 800 ± 25°C for 15 min. Retain the residue for the test for iron.

IRON. (Page 28, Method 1). To the residue from the preceding test, add 3 mL of dilute hydrochloric acid (1 + 1), cover with a watch glass, and digest on the steam bath for 15 min. Remove the cover, evaporate to dryness, and dissolve the residue in water plus 12 mL of hydrochloric acid. Dilute with water to 150 mL, and use 25 mL of the solution without further acidification.

OTHER METALS. Dissolve 1.0 g in 5 mL of aqua regia (2 mL of hydrochloric acid plus 3 mL of nitric acid). Boil until dissolution is complete and brown fumes are no longer given off in abundance. Cool and dilute with water to 50 mL. To 10 mL of the solution add 8 mL of 10% sodium hydroxide reagent solution, dilute with water to 40 mL, and add 10 mL of freshly prepared hydrogen sulfide water. Any brown color should not exceed that produced upon the addition of 10 mL of hydrogen sulfide water to a solution that contains in a volume of 40 mL, 8 mL of 10% sodium hydroxide reagent solution and 0.02 mg of lead ion (Pb).

Starch, Soluble
(For iodometry)

CAS Number 9005–84–9

REQUIREMENTS

Solubility . . . . Passes test
pH of a 2% solution . . . . 5.0–7.0 at 25°C
Residue after ignition . . . . ≤ 0.4%
Sensitivity . . . . Passes test

TESTS

SOLUBILITY. Prepare a paste of 2 g of the sample with a little cold water and add to it with stirring 100 mL of boiling water. The solution should be no more than opalescent, and on cooling it should remain liquid and not increase in opalescence.

pH OF A 2% SOLUTION.   (Page 43). Use the solution obtained in the test for solubility. The pH should be 5.0–7.0 at 25°C.

RESIDUE AFTER IGNITION.   (Page 15). Ignite 1.0 g and moisten the char with 1 mL of sulfuric acid.

SENSITIVITY.   Prepare a paste of 1.0 g of the sample with a little cold water and add it with stirring to 200 mL of boiling water. Cool and add 5 mL of this solution to 100 mL of water containing 50 mg of potassium iodide, and add 0.05 mL of 0.1 N iodine solution. A deep blue color should be produced, which will be discharged by 0.05 mL of 0.1 N sodium thiosulfate solution.

Strontium Chloride Hexahydrate

$SrCl_2 \cdot 6H_2O$ Formula Wt 266.62

CAS Number 10025–70–4

REQUIREMENTS

Assay	99.0–103.0% $SrCl_2 \cdot 6H_2O$
pH of a 5% solution at 25°C	5.0–7.0

MAXIMUM ALLOWABLE

Insoluble matter	0.005%
Sulfate (SO_4)	0.001%
Barium (Ba)	0.05%
Calcium (Ca)	0.05%
Heavy metals (as Pb)	5 ppm
Iron (Fe)	5 ppm
Magnesium (Mg)	2 ppm

TESTS

ASSAY.   (By argentimetric titration of chloride). Weigh accurately 2.7 g, dissolve in water in a 500-mL volumetric flask, and dilute to volume with water. Pipet 100 mL into a 500-mL conical flask, add 100 mL of water, and titrate with 0.1 N silver nitrate, using dichlorofluorescein as indicator. One milliliter of 0.1 N silver nitrate corresponds to 0.01333 g of $SrCl_2 \cdot 6H_2O$.

pH OF A 5% SOLUTION.   (Page 43). The pH should be 5.0–7.0 at 25°C.

INSOLUBLE MATTER.   (Page 14). Use 20 g dissolved in 150 mL of water.

SULFATE.   (Page 32, Procedure A, Method 1). Use a 5-g sample. Allow 30 min for the turbidity to form.

BARIUM, CALCIUM, AND MAGNESIUM.   Determine the barium, calcium, and magnesium by the flame atomic absorption spectrophotometric method described on page 35.

> *Sample Stock Solution.*   Dissolve 10 g of sample with water in a 100-mL volumetric flask, add 5 mL of nitric acid, and dilute to the mark with water (0.1 g/mL).

Analyze the solutions by means of a suitable atomic absorption spectrophotometer, using the conditions outlined in the following table. Calculate the metal content of the sample by the method of standard additions.

Element	Wavelength (nm)	Sample Wt (g)	Standard Added (mg)	Flame Type*	Background Correction
Ba	553.6	0.2	0.10; 0.20	N/A	No
Ca	422.7	0.1	0.05; 0.10	N/A	No
Mg	285.2	2.5	0.005; 0.01	A/A	No

*A/A is air/acetylene; N/A is nitrous oxide/acetylene.

HEAVY METALS.   (Page 26, Method 1). Dissolve 6.0 g in about 20 mL of water, and dilute with water to 30 mL. Use 25 mL to prepare the sample solution, and use the remaining 5.0 mL to prepare the control solution.

IRON.   Dissolve 2.0 g in 25 mL of water, add 2 mL of hydrochloric acid and 0.10 mL of 0.1 N potassium permanganate, and allow to stand for 5 min. Add 3 mL of ammonium thiocyanate reagent solution. Any red color should not exceed that produced by 0.01 mg of iron

(Fe) in an equal volume of solution containing the quantities of reagents used in the test.

Strontium Nitrate

$Sr(NO_3)_2$
Formula Wt 211.63

CAS Number 10042−76−9

REQUIREMENTS

Assay . . . . ≥ 99.0% $Sr(NO_3)_2$
pH of a 5% solution at 25°C . . . . 5.0−7.0

MAXIMUM ALLOWABLE

Insoluble matter 0.01%
Loss on drying at 105°C. 0.1%
Chloride (Cl) 0.002%
Sulfate (SO_4) 0.005%
Barium (Ba). 0.05%
Calcium (Ca) 0.05%
Magnesium and alkali salts (as sulfates) 0.15%
Heavy metals (as Pb) . . . . 5 ppm
Iron (Fe) . . . . 5 ppm

TESTS

ASSAY. (By complexometric titration). Weigh accurately 0.8 g, transfer to a 250-mL beaker, and dissolve in 50 mL of water. Add 5 mL of *N*-ethylethanamine (i.e., diethylamine) and 25 mg of methylthymol blue indicator mixture. Titrate immediately with standard 0.1 M EDTA until the blue color just turns to colorless or grey. One milliliter of 0.1 M EDTA corresponds to 0.02116 g of $Sr(NO_3)_2$.

pH OF A 5% SOLUTION. (Page 43). The pH should be 5.0–7.0 at 25°C.

INSOLUBLE MATTER. (Page 14). Use 10 g dissolved in 100 mL of water.

LOSS ON DRYING AT 105°C. Weigh accurately about 2 g and dry at 105°C for 4 h.

CHLORIDE. (Page 25). Use 0.50 g.

SULFATE. (Page 32, Procedure A, Method 2). Use 6.0 mL of dilute hydrochloric acid (1 + 1), and do two evaporations. Allow 30 min for the turbidity to form.

BARIUM AND CALCIUM. Determine the barium and calcium by the flame atomic absorption spectrophotometric method described on page 35.

> *Sample Stock Solution.* Dissolve 1.0 g of sample with water in a 100-mL volumetric flask, add 5 mL of nitric acid, and dilute to the mark with water (0.01 g/mL).

Analyze the solutions by means of a suitable atomic absorption spectrophotometer, using the conditions outlined in the following table. Calculate the metal content of the sample by the method of standard additions.

Element	Wavelength (nm)	Sample Wt (g)	Standard Added (mg)	Flame Type*	Background Correction
Ba	553.6	0.20	0.10; 0.20	N/A	No
Ca	422.7	0.05	0.02; 0.04	N/A	No

*N/A is nitrous oxide/acetylene.

MAGNESIUM AND ALKALI SALTS. Dissolve 2.0 g in 80 mL of water plus 0.15 mL of hydrochloric acid, heat to boiling, and add 20 mL of dilute sulfuric acid (1 + 9). Cool, dilute with water to 100 mL, add 100 mL of ethyl alcohol, mix, and allow to stand overnight. Decant through a dry filtering crucible or paper. Evaporate 100 mL of the filtrate to dryness in a tared dish and ignite at 800 ± 25°C for 15 min.

HEAVY METALS. (Page 26, Method 1). Dissolve 6.0 g in about 20 mL of water, and dilute with water to 30 mL. Use 25 mL to prepare the sample solution, and use the remaining 5.0 mL to prepare the control solution.

IRON. Dissolve 2.0 g in 10 mL of dilute hydrochloric acid (1 + 1) and evaporate to dryness. Repeat the evaporation. Dissolve the residue in 20 mL of water and add 2 mL of hydrochloric acid. Add 0.10 mL of 0.1 N potassium permanganate, dilute with water to 50 mL, and allow to stand for 5 min. Add 3 mL of ammonium thiocyanate reagent solution. Any red color should not exceed that produced by 0.01 mg of iron (Fe) treated exactly as the sample.

Succinic Acid

Butanedioic Acid

$$HO-\overset{\displaystyle O}{\overset{\displaystyle \|}{C}}-CH_2CH_2-\overset{\displaystyle O}{\overset{\displaystyle \|}{C}}-OH$$

$HOOCCH_2CH_2COOH$ Formula Wt 118.09

CAS Number 110–15–6

REQUIREMENTS

Assay . . . . ≥ 99.0%
$HOOCCH_2CH_2COOH$
Melting range . . . . Within 2°C
between 185.0
and 191.0°C

MAXIMUM ALLOWABLE

Insoluble matter 0.01%
Residue after ignition 0.02%
Chloride (Cl) 0.001%
Phosphate (PO_4) 0.001%
Sulfate (SO_4) 0.003%
Nitrogen compounds (as N) 0.001%
Heavy metals (as Pb) . . . . 5 ppm
Iron (Fe) . . . . 5 ppm

TESTS

ASSAY.   (By acid–base titrimetry). Weigh accurately 0.3 g, dissolve in 25 mL of water in a conical flask, add 0.15 mL of phenolphthalein indicator solution, and titrate with 0.1 N sodium hydroxide. One milliliter of 0.1 N sodium hydroxide corresponds to 0.005905 g of $HOOCCH_2CH_2COOH$.

MELTING RANGE. (Page 20).

INSOLUBLE MATTER. (Page 14). Use 10 g dissolved in 150 mL of water.

RESIDUE AFTER IGNITION. (Page 15). Ignite 10 g, omitting the use of sulfuric acid.

CHLORIDE. (Page 25). Use 1.0 g.

> *Sample Solution A for the Determination of Phosphate and Heavy Metals.* To 10 g add about 10 mg of sodium carbonate, 5 mL of nitric acid, and 5 mL of 30% hydrogen peroxide. Digest in a covered beaker on the steam bath until the reaction ceases. Wash down the cover glass and the sides of the beaker, and evaporate to dryness. Repeat the treatment with nitric acid and peroxide, and again evaporate to dryness. Dissolve the residue in about 30 mL of water, filter if necessary, and dilute with water to 50 mL (1 mL = 0.2 g).

PHOSPHATE. (Page 30, Method 1). Dilute 10 mL of sample solution A (2-g sample) with water to 20 mL, add 5 mL of 2.5 N sulfuric acid, and continue as directed.

SULFATE. Dissolve 15 g in 250 mL of hot water, add 1 mL of hydrochloric acid, filter if necessary, and heat to boiling. Add 10 mL of barium chloride reagent solution, and allow to stand overnight. Filter through a tared filtering crucible, wash thoroughly, dry at 105°C, cool, and weigh. Correct for the weight obtained in a complete blank test.

NITROGEN COMPOUNDS. (Page 30). Use 1.0 g. For the standard, use 0.01 mg of nitrogen (N).

HEAVY METALS. (Page 26, Method 1). Dilute 20 mL of sample solution A (4-g sample) with water to 25 mL.

IRON. (Page 28, Method 1). Use 2.0 g.

Sucrose

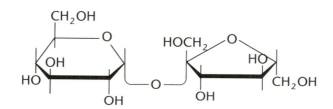

$C_{12}H_{22}O_{11}$ Formula Wt 342.30

CAS Number 57–50–1

REQUIREMENTS

Specific rotation $[\alpha]_D^{25°C}$ . . . .+66.3° to +66.8°

MAXIMUM ALLOWABLE

Insoluble matter 0.005%
Loss on drying at 105°C. 0.03%
Residue after ignition 0.01%
Titrable acid . . . . 0.0008 meq / g
Chloride (Cl) 0.005%
Sulfate and sulfite (as SO_4) 0.005%
Heavy metals (as Pb) . . . . 5 ppm
Iron (Fe) . . . . 5 ppm
Invert sugar. 0.05%

TESTS

SPECIFIC ROTATION. Weigh accurately 26 g and dissolve in 90 mL
of water in a 100-mL volumetric flask. Dilute with water to volume at
25°C. Observe the optical rotation in a polarimeter at 25°C using
sodium light, and calculate the specific rotation.

INSOLUBLE MATTER. (Page 14). Use 40 g dissolved in 150 mL of
water. Reserve the filtrate, without the washings, for preparation of
sample solution A.

LOSS ON DRYING AT 105°C. Weigh accurately 4.9–5.1 g in a tared
dish or crucible and dry at 105°C for 2 h.

RESIDUE AFTER IGNITION. (Page 15). Ignite 10 g and moisten the char with 1 mL of sulfuric acid.

TITRABLE ACID. To 100 mL of carbon dioxide-free water add 0.10 mL of phenolphthalein indicator solution and 0.01 N sodium hydroxide until a pink color is produced. Dissolve 10 g of the sample in this solution and titrate with 0.01 N sodium hydroxide to the same end point. Not more than 0.83 mL should be required.

Sample Solution A. Transfer the filtrate from the test for insoluble matter to a volumetric flask and dilute with water to 200 mL.

CHLORIDE. Dilute 10 mL of sample solution A with water to 50 mL. Dilute 5 mL of this solution with water to 25 mL and add 1 mL of nitric acid and 1 mL of silver nitrate reagent solution. Any turbidity should not exceed that produced by 0.01 mg of chloride ion (Cl) in an equal volume of solution containing the quantities of reagents used in the test.

SULFATE AND SULFITE. (Page 32, Procedure A, Method 2). Use 5.0 mL of sample solution A, add 1 mL of bromine water, and boil.

HEAVY METALS. (Page 26, Method 1). Use 25 mL of sample solution A to prepare the sample solution, and use 5.0 mL of sample solution A to prepare the control solution.

IRON. (Page 28, Method 1). Use 10 mL of sample solution A (2-g sample).

INVERT SUGAR. Prepare a reagent solution containing in 1 L 150 g of potassium bicarbonate, $KHCO_3$, 100 g of potassium carbonate, K_2CO_3, and 6.928 g of cupric sulfate pentahydrate, $CuSO_4 \cdot 5H_2O$. Transfer 50 mL of this reagent to a 400-mL beaker, cover, heat to boiling, and allow to boil for 1 min. Dissolve 10 g of the sample in water, dilute with water to 50 mL, and add this solution to the solution in the 400-mL beaker. Heat to boiling and boil for 5 min. At the end of this period stop the reaction by adding 100 mL of cold, recently boiled water. Filter through a tared filtering crucible, wash thoroughly, and dry at 105°C. The weight of the precipitate should not exceed 0.0277 g.

Sulfamic Acid

NH_2SO_3H Formula Wt 97.09

CAS Number 5329–14–6

REQUIREMENTS

Assay (dried basis) 99.3–100.3%
NH_2SO_3H

MAXIMUM ALLOWABLE

Insoluble matter 0.01%
Residue after ignition 0.01%
Chloride (Cl) 0.001%
Sulfate (SO$_4$) 0.05%
Heavy metals (as Pb) 0.001%
Iron (Fe) . . . . 5 ppm

TESTS

ASSAY. (By acid–base titrimetry). Weigh accurately 0.4 g, previously dried over sulfuric acid for 2 h, and dissolve in about 30 mL of
water. Add 0.15 mL of phenolphthalein indicator solution and titrate
with 0.1 N sodium hydroxide. One milliliter of 0.1 N sodium hydroxide corresponds to 0.009709 g of NH_2SO_3H.

INSOLUBLE MATTER. (Page 14). Use 10 g dissolved in 200 mL of
water.

RESIDUE AFTER IGNITION. (Page 15). Ignite 10.0 g without addition of sulfuric acid.

CHLORIDE. (Page 25). Use 1.0 g.

SULFATE. (Page 32, Procedure A, Method 1). Dissolve 1.0 g of
sample in 100 mL of water and use 10 mL (0.1 g) of this solution for
the test.

HEAVY METALS. (Page 26, Method 1). Dissolve 4.0 g in 30 mL of
water, neutralize to litmus with ammonium hydroxide, and dilute

with water to 40 mL. Use 30 mL to prepare the sample solution and use the remaining 10 mL to prepare the control solution.

IRON. (Page 28, Method 1). Use 2.0 g.

Sulfanilic Acid

4-Aminobenzenesulfonic Acid

$NH_2C_6H_4SO_3H$ Formula Wt 173.19

$NH_2C_6H_4SO_3H \cdot H_2O$ Formula Wt 191.21

CAS Number 121−57−3

NOTE. This reagent is available in both the anhydrous and monohy-drate forms. The identity should be indicated on the label.

REQUIREMENTS

Assay . . . . 98.0−102.0% of
                                                                                    the form offered

MAXIMUM ALLOWABLE

Residue after ignition 0.01%
Insoluble in sodium carbonate solution 0.02%
Chloride (Cl) 0.002%
Nitrite (NO_4) . . . . 0.5 ppm
Sulfate (SO_4) 0.01%

TESTS

ASSAY. (By acid−base titrimetry). Weigh accurately 0.7 g, and dissolve by warming gently in 50 mL of water. Cool and titrate with

0.1 N sodium hydroxide, using 0.15 mL of phenolphthalein indicator. One milliliter of 0.1 N sodium hydroxide corresponds to 0.01912 g of $NH_2C_6H_4SO_3H \cdot H_2O$ and to 0.01732 g of $NH_2C_6H_4SO_3H$.

RESIDUE AFTER IGNITION.   Gently ignite 10 g in a tared crucible or dish until charred. Slowly raise the temperature until all carbon is removed, and finally heat at $800 \pm 25°C$ for 15 min.

INSOLUBLE IN SODIUM CARBONATE SOLUTION.   Dissolve 5.0 g in 50 mL of a clear 5% solution of sodium carbonate and allow to stand in a covered beaker for 1 h. If an insoluble residue remains, filter through a tared porous porcelain or a platinum filtering crucible, wash with cold water, and dry at 105°C.

CHLORIDE.   Boil 5.0 g with 100 mL of water until dissolved. Cool, dilute with water to 100 mL, mix well, and filter through a chloride-free filter. Dilute 10 mL of the filtrate with water to 20 mL, and add 1 mL of nitric acid and 1 mL of silver nitrate reagent solution. Any turbidity should not exceed that produced by 0.01 mg of chloride ion (Cl) in an equal volume of solution containing the quantities of reagents used in the test. Save the remaining filtrate for the test for sulfate.

NITRITE.   Dissolve 0.70 g in 100 mL of water, warming if necessary but keeping the temperature below 30°C. For the control dissolve 0.20 g in about 75 mL of water, add 0.00025 mg of nitrite ion (NO_2), and dilute with water to 100 mL. To each solution add 5 mL of sulfanilic-1-naphthylamine reagent solution and allow to stand for 10 min. Any pink color produced in the solution of the sample should not exceed that in the control.

> *Sulfanilic-1-Naphthylamine Solution.*   Dissolve 0.5 g of sulfanilic acid in 150 mL of 36% acetic acid. Dissolve 0.1 g of 1-naphthylamine hydrochloride in 150 mL of 36% acetic acid. Mix the two solutions. If a pink color develops on standing, it may be discharged with a little zinc dust. (Substitute, if necessary, N-naphthylethylenediamine dihydrochloride for 1-naphthylamine hydrochloride, which may be unavailable.)

SULFATE.   Cool about 30 mL of the solution reserved from the test for chloride to about 0°C and filter. To 10 mL of the filtrate add 1 mL of dilute hydrochloric acid (1 + 19) and 1 mL of barium chloride reagent solution. Any turbidity should not exceed that produced by 0.05 mg of sulfate ion (SO_4) in an equal volume of solution containing the quantities of reagents used in the test. Compare 10 min after adding the barium chloride to the sample and standard solutions.

5-Sulfosalicylic Acid Dihydrate
2-Hydroxy-5-Sulfobenzoic Acid Dihydrate

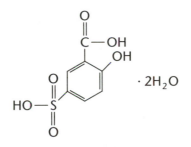

$HOC_6H_3(COOH)SO_3H \cdot 2H_2O$ Formula Wt 254.22

CAS Number 5965-83-3

REQUIREMENTS

Assay 99.0–101.0%
$HOC_6H_3(COOH)SO_3H \cdot 2H_2O$

MAXIMUM ALLOWABLE

Insoluble matter 0.02%
Residue after ignition 0.1%
Chloride (Cl) 0.001%
Salicylic acid (HOC_6H_4COOH) 0.04%
Sulfate (SO_4) 0.02%
Heavy metals (as Pb) 0.002%
Iron (Fe) 0.001%

TESTS

ASSAY. (By acid–base titrimetry). Weigh accurately 5 g, dissolve in 50 mL of water, add 0.15 mL of phenolphthalein indicator solution, and titrate with 1 N sodium hydroxide. One milliliter of 1 N sodium hydroxide corresponds to 0.1271 g of $HOC_6H_3(COOH)SO_3H \cdot 2H_2O$.

INSOLUBLE MATTER. (Page 14). Use 5.0 g dissolved in 50 mL of water.

RESIDUE AFTER IGNITION. (Page 15). Ignite 1.0 g in a dish, other than platinum, and moisten the char with 1 mL of sulfuric acid.

CHLORIDE. (Page 25). Use 1.0 g.

SALICYLIC ACID. Dissolve 5.0 g in 15 mL of cold water, add 10 mL of dilute hydrochloric acid (1 + 1), and extract with 50 mL of benzene in a 250-mL separatory funnel. Shake for 1.5 min, allow the layers to separate, and discard the aqueous layer. Add about 2 g of anhydrous sodium sulfate to the benzene layer, shake for 1 min, and allow the sodium sulfate to settle. Pour some of the clear benzene extract into a beaker, transfer 10 mL to a test tube, and add 10 mL of ferric ammonium sulfate solution. Shake the tube vigorously for 15 s and centrifuge at about 2500 rpm (about 1700 rcf) for 3 min. The color in the clear lower aqueous layer should not exceed that produced by 2.0 mg of salicyclic acid treated in exactly the same manner as the sample. In cases of borderline results or apparent nonconformity, aspirate off the benzene layer and determine the absorbance of the aqueous layer from the sample and the standard in 1-cm cells at 540 nm, using the ferric ammonium sulfate solution as the reference liquid in the spectrophotometer. Calculate the percent of salicylic acid as follows:

$$\% \text{ Salicylic acid} = \frac{C \times \dfrac{A_u}{A_s}}{W \times 10}$$

where

C = weight, in milligrams, of salicylic acid in standard
A_u = absorbance of sample solution
A_s = absorbance of standard solution
W = weight, in grams, of sample

Ferric Ammonium Sulfate Solution. To 200 mL of water in a beaker add 10% sulfuric acid until the pH is about 3, dissolve 200 mg of ferric ammonium sulfate dodecahydrate, $FeNH_4(SO_4)_2 \cdot 12H_2O$, in the solution, and adjust the pH to 2.45 (using a pH meter) with 10% sulfuric acid.

SULFATE. (Page 32, Procedure A, Method 1). Neutralize with dilute ammonium hydroxide (1 + 9) to phenolphthalein (0.15 mL of indicator).

Sample Solution A. Gently ignite 4.0 g in a crucible or dish, other than platinum, until charred. Cool, moisten the char with

1 mL of sulfuric acid, and ignite again slowly until all the carbon and excess sulfuric acid have been volatilized. Add 3 mL of hydrochloric acid and 0.5 mL of nitric acid, cover, and digest on a hot plate at low setting until dissolved. Remove the cover, evaporate to dryness on a hot plate at low setting, dissolve the residue in 4 mL of 1 N acetic acid, and dilute with water to 100 mL (1 mL = 0.04 g).

HEAVY METALS. (Page 26, Method 1). Use 25 mL of sample solution A (1.0 g) to prepare the sample solution.

IRON. (Page 28, Method 1). Use 25 mL of sample solution A (1.0 g).

Sulfuric Acid

H_2SO_4 Formula Wt 98.08

CAS Number 7664–93–9

REQUIREMENTS

Appearance . . . . Free from suspended or insoluble matter
Assay . . . . 95.0–98.0% H_2SO_4

MAXIMUM ALLOWABLE

Color (APHA). 10
Residue after ignition . . . . 5 ppm
Chloride (Cl) . . . . 0.2 ppm
Nitrate (NO_3) . . . . 0.5 ppm
Ammonium (NH_4) . . . . 2 ppm
Substances reducing permanganate . . . . 2 ppm as SO_2
Arsenic (As). . . . . 0.01 ppm
Heavy metals (as Pb) . . . . 1 ppm
Iron (Fe) . . . . 0.2 ppm
Mercury (Hg) . . . . 5 ppb

TESTS

APPEARANCE.   Mix the material in the original container, pour 10 mL into a test tube (20 × 150 mm), and compare with distilled water in a similar tube. The liquids should be equally clear and free from suspended matter.

ASSAY.   (By acid–base titrimetry). Tare a small glass-stoppered flask, add about 1 mL of the sample, and weigh accurately. Cautiously add 30 mL of water, cool, add 0.15 mL of methyl orange indicator solution, and titrate with 1 N sodium hydroxide. One milliliter of 1 N sodium hydroxide corresponds to 0.04904 g of H_2SO_4.

COLOR (APHA).   (Page 17).

RESIDUE AFTER IGNITION.   (Page 15). Evaporate 200 g (110 mL) to dryness in a tared platinum dish and ignite at $800 \pm 25°C$ for 15 min.

CHLORIDE.   Place 40 mL of water in each of three beakers. To one, carefully add 50 g (27.2 mL) of sample, and to the others, carefully add 50 g (27.2 mL) of chloride-free sulfuric acid. To the second beaker, add 0.01 mg of chloride (Cl). Cool to room temperature, mix, and to each beaker add 1 mL of nitric acid and 1 mL of silver nitrate reagent solution. Mix well, and if necessary, make the volume of each solution identical by adding water. Transfer equal portions of each solution to Nessler tubes. After 10 min, any turbidity of the sample solution should not exceed that of the standard. The blank should be free of turbidity. The comparison can be aided by use of a nephelometer.

NITRATE.

> ***Sample Solution A.***   Cautiously add 40 g (22 mL) to 2.0 mL of water, dilute to 50 mL with brucine sulfate reagent solution, and mix.

> ***Control Solution B.***   Cautiously add 40 g (22 mL) to 2.0 mL of the standard nitrate solution containing 0.01 mg of nitrate ion (NO_3) per milliliter, dilute to 50 mL with brucine sulfate reagent solution, and mix.

Continue with the procedure described on page 29, starting with the preparation of blank solution C.

AMMONIUM.   (Page 48). Determine ammonium by differential pulse polarography. Carefully add 2.0 g (1.1 mL) to 4 mL of water in a

25-mL volumetric flask in an ice bath. Use 9.5 mL of ammonia-free 6 N sodium hydroxide for neutralization and 0.004 mg of ammonium (NH_4) for the standard.

SUBSTANCES REDUCING PERMANGANATE.   Carefully dilute 80 g (43.5 mL) with 60 mL of carbon dioxide-free ice-cold water, keeping the solution cool during the addition. Add 0.50 mL of 0.01 N potassium permanganate. The solution should remain pink for not less than 5 min.

ARSENIC.   (Page 23). To 300 g (165 mL) add 3 mL of nitric acid, and evaporate on a hot plate to about 10 mL. Cool, cautiously dilute with about 20 mL of water, and evaporate just to dense fumes of sulfur trioxide. Cool, and cautiously wash the solution into a generator flask with the aid of 50 mL of water. Omit the addition of dilute sulfuric acid. For the standard use 0.003 mg of arsenic (As).

HEAVY METALS.   (Page 26, Method 1). Add 20 g (11 mL) to about 10 mg of sodium carbonate dissolved in a small quantity of water. Heat over a low flame until nearly dry, then add 1 mL of nitric acid. Evaporate to dryness, add about 20 mL of water, and dilute with water to 25 mL.

IRON.   (Page 28, Method 1). Add 50 g (27 mL) to about 10 mg of sodium carbonate dissolved in a small quantity of water. Evaporate to dryness by heating on an electric hot plate. Cool, add 5 mL of dilute hydrochloric acid (1 + 1), cover with a watch glass, and digest on the steam bath for 15 min. Cool, and dilute with water to 25 mL.

MERCURY.   To each of two 125-mL conical flasks add 20 mL of water and 5 mL of 4% potassium permanganate solution. For the sample add to one flask, slowly and with cooling, 11 g (6.0 mL) of the sulfuric acid. For the control add to the second flask 1 g (0.5 mL) of the sulfuric acid and 0.05 μg of mercury (Hg) (0.5 mL of a solution freshly prepared by diluting 1 mL of standard mercury solution, page 84, with water to 500 mL). Place both flasks on the steam bath for 15 min, then cool to room temperature. Determine the mercury in each by flameless atomic absorption, page 41, using 1 mL of 10% hydroxylamine hydrochloride solution and 2 mL of 10% stannous chloride solution for the reduction. Any mercury found in the sample should not exceed that in the control.

Sulfuric Acid, Fuming

CAS Number 8014–95–7

NOTE. This specification applies to fuming sulfuric acid with nominal contents of 15, 20, or 30% free SO_3.

REQUIREMENTS

Appearance . . . . Colorless to very light brown color
Assay (free SO_3) . . . . 12.0–17.0%, 18.0–24.0%, or 27.0–33.0%

MAXIMUM ALLOWABLE

Residue after ignition 0.002%
Nitrate (NO_3) . . . . 1 ppm
Ammonium (NH_4) . . . . 3 ppm
Arsenic (As) . . . . 0.03 ppm
Iron (Fe) . . . . 2 ppm

TESTS

ASSAY. (By acid–base titrimetry). Weigh accurately 4 g in a tared Dely weighing tube. Carefully transfer to a casserole containing 100 mL of carbon dioxide-free water by placing the tip of the tube beneath the surface and flushing the tube with carbon dioxide-free water. Cool, add 0.15 mL of phenolphthalein indicator solution, and titrate with 1 N sodium hydroxide.

$$\% \, H_2SO_4 = \frac{4.904V}{W}$$

where

$$\% \text{ free } SO_3 = 4.445 \, (\% \, H_2SO_4 - 100)$$
V = volume, in milliliters, of 1 N sodium hydroxide
W = weight, in grams, of sample

NOTE. For accurate results a weight buret should be used. The Dely weighing tube and its use are described in standard reference books, such as *Standard Methods of Analysis,* 6th ed., Vol. IIA, F. J. Welcher, Ed., Van Nostrand, 1963, pp. 537–538.

RESIDUE AFTER IGNITION.   (Page 15). Evaporate 50 g (27 mL) to dryness in a tared platinum dish and ignite at $800 \pm 25°C$ for 15 min.

NITRATE.

Sample Solution A.   Cautiously add 10 g (5.0 mL) to 1.0 mL of water, dilute to 50 mL with brucine sulfate reagent solution, and mix.

Control Solution B.   Cautiously add 10 g (5.0 mL) to 1.0 mL of the standard nitrate solution containing 0.01 mg of nitrate ion (NO_3) per mL, dilute to 50 mL with brucine sulfate reagent solution, and mix.

Continue with the procedure described on page 29, starting with the preparation of blank solution C.

AMMONIUM.   (Page 48). Determine ammonium by differential pulse polarography. Carefully add 2.0 g (1.1 mL) to 4 mL of water in a 25-mL volumetric flask in an ice bath. Use approximately 10 mL of ammonia-free 6 N sodium hydroxide for neutralization and 0.006 mg of ammonium (NH_4) for the standard. The volume of buffer solution may be lowered to 4 mL.

ARSENIC.   (Page 23). To 100 g (52 mL) add 3 mL of nitric acid, and evaporate on a hot plate to about 10 mL. Cool, cautiously dilute with about 20 mL of water, and evaporate to about 5 mL. Cool, cautiously dilute again with about 20 mL of water, and evaporate just to dense fumes of sulfur trioxide. Cool, and cautiously wash the solution into a generator flask with the aid of 50 mL of water. Omit the addition of 20 mL of dilute sulfuric acid (1 + 4). For the standard use 0.003 mg of arsenic (As).

IRON.   (Page 28, Method 1). Cautiously add 5 g (2.6 mL) to about 10 mg of sodium carbonate, and evaporate to dryness by heating with a small flame or on an electric hot plate. Cool, add 5 mL of dilute hydrochloric acid (1 + 1), cover with a watch glass, and digest on the steam bath for 15 min. Cool, and dilute with water to 25 mL.

Sulfurous Acid
(A solution of SO$_2$ in water)

CAS Number 7782–99–2

REQUIREMENTS

Assay . . . . $\geq$ 6.0% SO$_2$

MAXIMUM ALLOWABLE

Residue after ignition 0.005%
Chloride (Cl) . . . . 5 ppm
Arsenic (As) . . . . 0.5 ppm
Heavy metals (as Pb) . . . . 2 ppm
Iron (Fe) . . . . 5 ppm

TESTS

ASSAY. (By iodometric titration). Tare a glass-stoppered conical flask containing 50.0 mL of 0.1 N iodine. Quickly introduce about 2 mL of the sample, stopper, and weigh again. Titrate the excess iodine with 0.1 N sodium thiosulfate, adding 3 mL of starch indicator solution near the end of the titration. One milliliter of 0.1 N iodine consumed corresponds to 0.003203 g of SO$_2$.

RESIDUE AFTER IGNITION. Evaporate 20 g (20 mL) to dryness on the steam bath in a tared crucible or dish and ignite at 800 $\pm$ 25°C for 15 min.

CHLORIDE. Digest 10 g (10 mL) with 2 mL of nitric acid on the steam bath for 1 h. Cool and dilute with water to 100 mL. To 20 mL of the solution add 1 mL of nitric acid and 1 mL of silver nitrate reagent solution. Any turbidity should not exceed that produced by 0.01 mg of chloride ion (Cl) in an equal volume of solution containing the quantities of reagents used in the test.

ARSENIC. (Page 23). To 6 g (6 mL) in a 250-mL beaker add 3 mL of sulfuric acid, evaporate in a hood to dense fumes of sulfur trioxide, and continue the fuming for 15 min. Cool, cautiously rinse down the beaker walls with about 5 mL of water, add 3 mL of sulfuric acid, and repeat the evaporation and fuming. Cool, and cautiously wash the solution into a generator flask with sufficient water to make 35 mL. Omit the addition of dilute sulfuric acid. For the standard use 0.003 mg of arsenic (As).

HEAVY METALS. (Page 26, Method 1). To 10 g (10 mL) add 10 mL of water, boil to expel the sulfur dioxide, and dilute with water to 25 mL.

IRON. (Page 28, Method 1). To 2 g (2.0 mL) add about 10 mg of sodium carbonate, and evaporate to dryness. Dissolve the residue with 0.5 mL of hydrochloric acid, add 0.5 mL of nitric acid, and evaporate again to dryness. Dissolve the residue in 2 mL of hydrochloric acid, dilute with water to 50 mL, and use the solution without further acidification.

Tannic Acid

CAS Number 1401 –55 –4

> *NOTE.* For analytical purposes, tannic acid should be the hydrolyzable type, such as that isolated from nutgalls, sumac, or seed pods of Tara.

REQUIREMENTS

Identification. . . . . Passes test

MAXIMUM ALLOWABLE

Loss on drying 12.0%
Residue after ignition 0.5%
Heavy metals (as Pb) 0.003%
Zinc (Zn). 0.005%
Sugars, dextrin . . . . Passes test

TESTS

IDENTIFICATION. To 2 mL of a 10% aqueous solution of tannic acid add 0.1 mL of a 5% aqueous lead nitrate solution. A precipitate forms immediately, but completely dissolves on continued swirling to yield a clear solution.

LOSS ON DRYING.   Weigh accurately 1 g and dry to constant weight at 105°C.

RESIDUE AFTER IGNITION.   (Page 15). Ignite 5.0 g and moisten the char with 1 mL of sulfuric acid. Retain the residue.

HEAVY METALS.   (Page 26, Method 1). Transfer 0.67 g into a 150-mL beaker and cautiously add 15 mL of nitric acid and 5 mL of 70% perchloric acid. Evaporate the mixture to dryness on a hot plate in a suitable hood, cool, add 2 mL of hydrochloric acid, and wash down the sides of the beaker with water. Carefully evaporate the solution to dryness on the hot plate, rotating the beaker to avoid spattering. Repeat the addition of 2 mL of hydrochloric acid, wash down the sides of the beaker with water, and evaporate to dryness. Cool the residue, take up in 1 mL of hydrochloric acid and 10 mL of water, and dilute with water to 25 mL.

ZINC.   Dissolve the residue from the residue after ignition test in 2 mL of glacial acetic acid, dilute with 8 mL of water, and filter if necessary. Add 0.5 g of sodium acetate and 5 mL of hydrogen sulfide water and mix. Any white turbidity should not exceed that produced by 0.25 mg of zinc ion (Zn) treated exactly as the sample residue.

SUGARS, DEXTRIN.   Dissolve 2.0 g in 10 mL of water and add 20 mL of alcohol. The mixture should be clear and should remain clear after standing for 1 h. Add 0.5 mL of ether. No turbidity should be produced.

Tartaric Acid
2,3-Dihydroxybutanedioic Acid

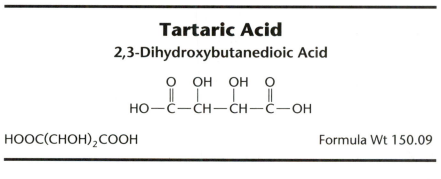

HOOC(CHOH)$_2$COOH   Formula Wt 150.09

CAS Number 87–69–4

REQUIREMENTS

MAXIMUM ALLOWABLE

Insoluble matter 0.005%
Residue after ignition 0.02%
Chloride (Cl) 0.001%
Oxalate (C_2O_4) . . . . Passes test (limit
  about 0.1%)
Phosphate (PO_4) 0.001%
Sulfur compounds (as S) 0.002%
Heavy metals (as Pb) . . . . 5 ppm
Iron (Fe) . . . . 5 ppm

TESTS

INSOLUBLE MATTER.   (Page 14). Use 20 g dissolved in 200 mL of water.

RESIDUE AFTER IGNITION.   (Page 15). Ignite 5.0 g, and moisten the char with 2 mL of sulfuric acid.

CHLORIDE.   (Page 25). Use 1.0 g.

OXALATE.   Dissolve 5.0 g in 30 mL of water and divide into two equal portions. Neutralize one portion with ammonium hydroxide, using litmus paper as indicator. Add the other portion and dilute with water to 40 mL. Shake well, cool, and allow to stand at 15°C for 15 min. Filter, and to 20 mL of the filtrate add an equal volume of a saturated, filtered solution of calcium sulfate. No turbidity or precipitate should appear in 2 h.

> ***Sample Solution A.***   To 10 g add about 10 mg of sodium carbonate, 5 mL of nitric acid, and 5 mL of 30% hydrogen peroxide. Digest in a covered beaker on a hot plate at low setting until the reaction ceases. Wash down the cover glass and the sides of the beaker and evaporate to dryness. Repeat the treatment with nitric acid and peroxide and again evaporate to dryness. Dissolve the residue in about 30 mL of water, filter if necessary, and dilute with water to 50 mL (1 mL = 0.2 g).

PHOSPHATE.   (Page 30, Method 1). Dilute 10 mL of sample solution A (2-g sample) with water to 20 mL, add 5 mL of 2.5 N sulfuric acid, and continue as described.

SULFUR COMPOUNDS. (Page 32, Procedure A, Method 1). Use 4.2 mL of sample solution A.

HEAVY METALS. (Page 26, Method 1). Dilute 20 mL of sample solution A (4-g sample) with water to 25 mL.

IRON. (Page 28, Method 1). Use 2.0 g.

Tetrabromophenolphthalein Ethyl Ester

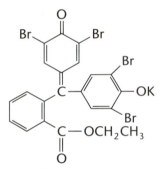

Formula Wt 661.97

CAS Number 1176–74–5

REQUIREMENTS

Appearance . . . . Yellow to red
                                                          powder
Melting point . . . . 208.0–211.0°C
Assay . . . . ≥ 95.0%
Solubility . . . . Passes test

TESTS

ASSAY. (By spectrophotometry). Weigh accurately 50.0 mg, transfer to a 100-mL volumetric flask, dissolve in about 75 mL of a 50/50 mixture of alcohol and buffer solution, and dilute to volume with the mixture. Pipet 2 mL of this solution into a 100-mL volumetric flask and dilute to volume with buffer solution. Determine the absorbance of this solution in 1.00-cm cells at the wavelength of maximum absorption, approximately 593 nm, using the buffer solution as the blank. Determine the molar absorptivity by dividing the absorbance by 1.5×10^{-5} (that is, the molar concentration of the tetrabromophenolphthalein ethyl ester) and calculate the assay value as follows:

$$\frac{\text{Molar absorptivity at the maximum}}{76{,}922} \times 100$$

$$= \% \text{ Tetrabromophenolphthalein ethyl ester}$$

Buffer Solution (pH = 7.00). Add 291 mL of 0.1 M sodium hydroxide to 500 mL of 0.1 M monobasic potassium phosphate, KH_2PO_4, and dilute with water to 1 L.

NOTE. Because the solutions are light-sensitive, low-actinic glassware should be used.

SOLUBILITY. Dissolve 0.1 g in 100 mL of toluene. The solution should be clear and dissolution should be complete.

Tetrahydrofuran

C_4H_8O Formula Wt 72.11

CAS Number 109–99–9

WARNING. Tetrahydrofuran tends to form an explosive peroxide and is also very flammable. No distillation or evaporation should be performed unless peroxide is shown to be absent. Generally, a stabilizer is present to retard peroxide formation.

REQUIREMENTS

Assay . . . . $\geq$ 99.0% C_4H_8O

MAXIMUM ALLOWABLE

Color (APHA). 20
Peroxide (as H_2O_2) 0.015%
Residue after evaporation 0.03%
Water 0.05%

TESTS

ASSAY.   Analyze the sample by gas chromatography, using the general parameters cited on page 56. The following specific conditions are also required.

Column: Type I, methyl silicone

Measure the area under all peaks and calculate the tetrahydrofuran content in area percent. Correct for water content.

COLOR (APHA).   (Page 17).

PEROXIDE (AS H_2O_2).   To 44 g (50 mL) in a conical flask, add 10 mL of 10% potassium iodide solution and 2 mL of dilute sulfuric acid. Titrate the liberated iodine with 0.1 N sodium thiosulfate. Not more than 3.9 mL should be required.

> *CAUTION.* If peroxide is present, do not make the test for residue after evaporation.

RESIDUE AFTER EVAPORATION.   Evaporate 20 g (17.8 mL) to dryness in a tared dish on a steam bath and dry the residue at 105°C for 30 min.

WATER.   (Page 55, Method 2). Use 100 μL (88 mg).

Tetramethylsilane

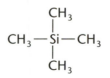

$(CH_3)_4Si$ Formula Wt 88.22

CAS Number 75–76–3

REQUIREMENTS

Residue after evaporation . . . . ≤ 0.05%
Suitability for proton NMR position reference . . . . . Passes test

TESTS

RESIDUE AFTER EVAPORATION. (Page 14). Evaporate 2 g (3.1 mL) to dryness in a tared dish and dry the residue at 105°C for 30 min.

SUITABILITY FOR PROTON NMR POSITION REFERENCE. An undiluted sample should display no impurity peak in its proton NMR reference spectrum more intense than 20% of the ^{13}C satellite line at ±59.1 Hz from tetramethylsilane.

Thioacetamide

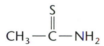

$$CH_3CSNH_2 \qquad\qquad\qquad \text{Formula Wt } 75.13$$

CAS Number 62–55–5

NOTE. This reagent, when stored under ordinary conditions, may decompose slightly and fail the test for clarity of a 2% solution.

REQUIREMENTS

Assay . . . . ≥ 99.0%
CH_3CSNH_2
Melting point . . . . 111–114°C
Clarity of a 2% solution . . . . Passes test

MAXIMUM ALLOWABLE

Residue after ignition 0.05%

TESTS

ASSAY. (By argentimetric titration). Weigh accurately 1.5 g, dissolve in water, and dilute with water to 500 mL in a volumetric flask.

To 50.0 mL of this solution add 1 mL of ammonium hydroxide and 50.0 mL of 0.1 N silver nitrate. Allow to stand for 20 min, carefully filter through a filtering crucible or a sintered glass funnel that has been cleaned with dilute nitric acid, and wash the funnel and flask well with water. To the clear solution add 5 mL of nitric acid and 2 mL of ferric ammonium sulfate indicator solution, and titrate with 0.1 N potassium thiocyanate. One milliliter of 0.1 N silver nitrate corresponds to 0.003757 g of CH_3CSNH_2.

RESIDUE AFTER IGNITION.   (Page 15). Ignite 2.0 g, and moisten the char with 1 mL of sulfuric acid.

CLARITY OF A 2% SOLUTION.   Dissolve 2 g in 100 mL of water. The solution should be clear and colorless.

Thiourea

$$H_2N-\overset{\overset{\displaystyle S}{\|}}{C}-NH_2$$

NH_2CSNH_2 Formula Wt 76.12

CAS Number 62–56–6

REQUIREMENTS

Assay (dried basis) . . . . $\geq$ 99.0%
NH_2CSNH_2
Melting point . . . . 174–177°C
Sensitivity to bismuth . . . . Passes test

MAXIMUM ALLOWABLE

Solubility in water . . . . Passes test
Residue after ignition 0.1%
Loss on drying 0.5%

TESTS

ASSAY. (By argentimetric titration). Weigh accurately 1 g, dissolve in water, and dilute to 250 mL in a volumetric flask. To 20.0 mL of this solution in a glass-stoppered flask, add 25.0 mL of 0.1 N silver nitrate and 10 mL of ammonium hydroxide (10% NH_3). Stopper the flask, shake vigorously for 2 min, heat to boiling, and cool. To the cooled solution add 60 mL of dilute nitric acid, shake vigorously, and filter through a filtering crucible or a sintered glass funnel that has been cleaned with dilute nitric acid, and wash the funnel and flask well with water. To the filtrate plus washings add 2 mL of ferric ammonium sulfate indicator solution, and titrate with 0.1 N potassium thiocyanate. One milliliter of 0.1 N silver nitrate corresponds to 0.003806 g of NH_2CSNH_2.

SOLUBILITY IN WATER. Dissolve 1 g in 20 mL of water. The solution should be clear and colorless.

RESIDUE AFTER IGNITION. (Page 15). Ignite 10 g.

LOSS ON DRYING. Weigh accurately about 1.5 g and dry to constant weight at 105°C.

SENSITIVITY TO BISMUTH. Dissolve 0.10 g in 10 mL of water. Add 1.0 mL of this solution to 10 mL of bismuth solution. A distinct yellow color is produced immediately.

> *Bismuth Solution.* Dissolve 0.23 g of bismuth subnitrate in a mixture of 12 mL of nitric acid and 80 mL of water. When dissolution is complete, dilute with water to 2000 mL.

Thorium Nitrate Tetrahydrate

$Th(NO_3)_4 \cdot 4H_2O$ Formula Wt 552.12

CAS Number 13470–07–0

REQUIREMENTS

Assay 98.0–102.0%
$Th(NO_3)_4 \cdot 4H_2O$

MAXIMUM ALLOWABLE

Insoluble matter 0.01%
Chloride (Cl) 0.002%
Sulfate (SO_4) 0.01%
Substances not precipitated by
   ammonium hydroxide 0.2%
Heavy metals (as Pb) 0.002%
Iron (Fe) 0.002%
Rare earth elements (as La) 0.2%
Titanium (Ti) 0.01%

TESTS

ASSAY. (By gravimetry). Weigh accurately 1 g, and dissolve in 100 mL of water in a beaker. Add 2 mL of 10% sulfuric acid, heat to boiling, and add 20 mL of a hot 10% solution of oxalic acid. Cool, filter, wash with a little water, and ignite the precipitate to constant weight. Multiply the oxide weight by 209.0, and divide by the weight of the sample to obtain % $Th(NO_3)_4 \cdot 4H_2O$.

INSOLUBLE MATTER. (Page 14). Use 10 g dissolved in 100 mL of water. Reserve the filtrate and washings for the preparation of sample solution A.

> ***Sample Solution A for the Determination of Chloride, Sulfate, Substances Not Precipitated by Ammonium Hydroxide, Heavy Metals, Iron, Rare Earth Elements, and Titanium.*** Transfer the filtrate and washings reserved from the test for insoluble matter to a 200-mL volumetric flask and dilute with water to 200 mL (1 mL = 0.05 g).

CHLORIDE. (Page 25). Dilute 10 mL (0.5-g sample) of sample solution A with water to 20 mL.

SULFATE. Dilute 10 mL (0.5-g sample) of sample solution A with water to 20 mL. Add 5 mL of 2 N ammonium acetate and transfer the solution to a separatory funnel. Extract with successive 10-mL portions of 0.2 M N-phenylbenzohydroxamic acid in chloroform until the color of the chloroform layer remains unchanged. Wash the aqueous layer twice with 5-mL portions of chloroform. Transfer the aqueous solution to a beaker and evaporate on a hot plate at low setting to 2–3 mL. Add 2 mL of nitric acid and 2 mL of hydrochloric acid, cover, and digest on the hot plate at low setting until any reaction ceases.

Uncover, wash down the sides of the beaker, and evaporate to dryness. Dissolve the residue in 4 mL of water plus 1 mL of dilute hydrochloric acid (1 + 19). Filter if necessary through a small filter, wash with two 2-mL portions of water, and dilute with water to 10 mL. Add 1 mL of barium chloride reagent solution. Any turbidity should not exceed that produced by 0.05 mg of sulfate ion (SO_4) in an equal volume of solution that has been treated exactly like the sample. Compare 10 min after adding the barium chloride to the sample and standard solutions.

SUBSTANCES NOT PRECIPITATED BY AMMONIUM HYDROXIDE. Dilute 40 mL (2-g sample) of sample solution A with water to 100 mL. Add 0.10 mL of methyl red indicator solution and add ammonium hydroxide until the solution is alkaline to methyl red. Heat the solution to boiling and digest on the steam bath for 15 min. Dilute the solution with water to 150 mL, mix thoroughly, and filter. Evaporate 75 mL (1-g sample) to dryness in a tared dish on the steam bath. Moisten the residue with 0.10 mL of sulfuric acid. Heat gently to volatilize the excess salts and acid, and finally ignite at $800 \pm 25°C$ for 15 min.

HEAVY METALS. (Page 26, Method 1). Use 30 mL of sample solution A (1.5-g sample) to prepare the sample solution, and use 10 mL of sample solution A to prepare the control solution.

IRON. (Page 28, Method 1). Use 10 mL of sample solution A (0.5-g sample).

RARE EARTH ELEMENTS. Dilute 5.0 mL (0.25-g sample) of sample solution A to 90 mL with water. Use a pH meter to adjust to pH 4.5 with sodium acetate–acetic acid buffer solution and dilute with water to 100 mL. To 10 mL in a separatory funnel add 10 mL of N-phenylbenzohydroxamic acid reagent solution, shake vigorously, allow the layers to separate, and draw off and discard the chloroform layer. Repeat the extraction to 5-mL portions of N-phenylbenzohydroxamic acid reagent solution two more times, drawing off and discarding the chloroform layer each time. Wash the aqueous layer with 5 mL of chloroform and draw off and discard the chloroform. Add 2 mL of alizarin red S solution and dilute with water to 50 mL. Adjust the pH of the solution to 4.68 with 0.5 N sodium acetate. Any red color should not exceed that produced by 0.05 mg of lanthanum ion (La) that has been treated exactly like the 10 mL of the sample.

> *Sodium Acetate–Acetic Acid Buffer.* Dissolve 54.4 g of sodium acetate in water, add 23 mL of acetic acid, and dilute with water to 200 mL.

N-Phenylbenzohydroxamic Acid Reagent Solution. Dissolve 4.26 g of *N*-phenylbenzohydroxamic acid in chloroform and dilute to 100 mL with chloroform.

Alizarin Red S Solution. Dissolve 0.10 g of 3,4-dihydroxy-2-anthraquinonesulfonic acid, sodium salt, in water and dilute with water to 100 mL.

TITANIUM. Dilute 10 mL (0.5-g sample) of sample solution A with water to 20 mL and add 0.5 mL of 30% hydrogen peroxide. Any yellow color should not exceed that produced by 0.05 mg of titanium ion (Ti) in an equal volume of solution containing the quantities of reagents used in the test.

Thymol Blue
Thymolsulfonphthalein
4,4'-(3 H-2,1-Benzoxathiol-3-ylidene)bis-5-[5-methyl-2-(1-methylethyl)phenol] S, S-Dioxide

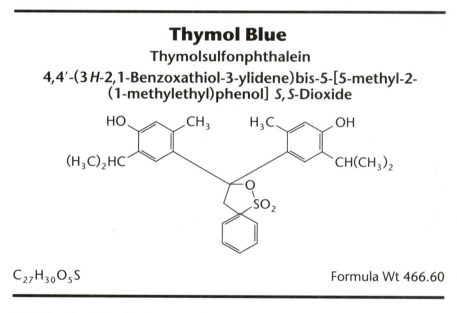

$C_{27}H_{30}O_5S$ Formula Wt 466.60

CAS Number 76–61–9

NOTE. This specification applies to both the free acid form and the salt form of this indicator.

REQUIREMENTS

Clarity of solution . . . . Passes test
Visual transition interval (acid range) . . . . From pH 1.2 (red)
                 to pH 2.8 (yellow)
Visual transition interval (alkaline range) . . . . From pH 8.0
                 (yellow) to
                 pH 9.2 (blue)

TESTS

CLARITY OF SOLUTION. If the indicator is the acid form, dissolve 0.1 g in 100 mL of alcohol. If the indicator is a salt form, dissolve 0.1 g in 100 mL of water. Not more than a faint trace of turbidity or insoluble matter should remain. Reserve the solution for the tests for visual transition interval.

VISUAL TRANSITION INTERVAL (ACID RANGE). Dissolve 1 g of potassium chloride in 100 mL of water. Adjust the pH of the solution to 1.20 (using a pH meter as described on page 43) with 1.0 N hydrochloric acid. Add 0.1–0.3 mL of the 0.1% solution reserved from the test for insoluble matter. The color of the solution should be pink. Titrate the solution with 1.0 N sodium hydroxide until the pH is 2.2 (using the pH meter). The color of the solution should be orange. Continue the titration until the pH is 2.8. The color of the solution should be yellow. Not more than 8.6 mL of the 1.0 N sodium hydroxide should be consumed in the titration.

VISUAL TRANSITION INTERVAL (ALKALINE RANGE). Dissolve 1 g of potassium chloride in 100 mL of water. Adjust the pH of the solution to 8.0 (using a pH meter as described on page 43) by adding 0.01 N hydrochloric acid or sodium hydroxide. Add 0.1–0.3 mL of the 0.1% solution reserved from the test for insoluble matter. The color of the solution should be yellow. Titrate the solution with 0.01 N sodium hydroxide until the pH is 8.4 (using the pH meter). The color of the solution should be green. Continue the titration until the pH is 9.2. The color of the solution should be blue. Not more than 0.50 mL of the 0.01 N sodium hydroxide should be consumed in the titration.

Thymolphthalein

5',5"-Diisopropyl-2',2"-dimethylphenolphthalein

3,3-Bis[4-hydroxy-2-methyl-5-(1-methylethyl)phenyl]-
1(3*H*)-isobenzofuranone

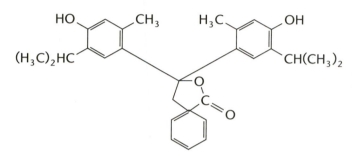

$C_{29}H_{29}O_4$ Formula Wt 430.54

CAS Number 125–20–2

REQUIREMENTS

Clarity of solution . . . . Passes test
Visual transition interval . . . . From pH 8.8
(colorless) to
pH 10.5 (blue)

TESTS

CLARITY OF SOLUTION.   Dissolve 0.1 g in 100 mL of alcohol. Not more than a faint trace of turbidity or insoluble matter should remain. Reserve the solution for the test for visual transition interval.

VISUAL TRANSITION INTERVAL.   Dissolve 1 g of potassium chloride in 100 mL of water. Adjust the pH of the solution to 8.80 (using a pH meter as described on page 43) with 0.01 N sodium hydroxide. Add 0.5–1.0 mL of the 0.1% solution reserved from the test for insoluble matter. The solution should be colorless. Titrate the solution with 0.01 N sodium hydroxide to pH 9.4 (using the pH meter). The solution should have a pale grayish-blue color. Continue the titration to pH 9.9. The solution should have a blue color. Not more than 2.0 mL of

0.01 N sodium hydroxide should be consumed in the titration. Each additional 0.05 mL of 0.01 N sodium hydroxide should increase the amount of blue color until at pH 10.5 the color should be an intense blue.

Tin

Sn   Atomic Wt 118.71

CAS Number 7440–31–5

REQUIREMENTS

Assay . . . . ≥ 99.5% Sn

MAXIMUM ALLOWABLE

Antimony (Sb) 0.02%
Arsenic (As). . . . . 1 ppm
Copper (Cu) 0.005%
Iron (Fe) 0.01%
Lead (Pb) 0.005%

TESTS

ASSAY.   (By complexometric titration). Weigh, to the nearest 0.1 mg, 0.25 g of sample, and digest on the steam bath in a covered beaker with 5 mL of hydrochloric acid until the sample is dissolved. Add 30.0 mL of standardized 0.1 M EDTA and heat the solution almost to boiling. Cool and adjust the pH to 5.5, using a pH meter, with a saturated hexamethylenetetramine solution. Cool, dilute with water to about 250 mL, and add a few milligrams of xylenol orange indicator. Titrate with 0.1 M lead solution to a change from yellow to reddish purple.

$$\% \text{ Sn} = \frac{[(30.0 \times \text{M EDTA}) - (\text{mL} \times \text{M Pb})] \times 11.87}{\text{Sample wt (g)}}$$

ANTIMONY, COPPER, IRON, AND LEAD.   Determine the antimony, copper, iron, and lead by the flame atomic absorption spectrophotometric method described on page 35.

Sample Stock Solution.   Dissolve 10.0 g of sample in 30 mL of hydrochloric acid and 3 mL of nitric acid by allowing to stand in a covered beaker until dissolution is complete. Transfer to a 100-mL volumetric flask and dilute to the mark with water (0.1 g/mL).

Analyze the solutions by means of a suitable atomic absorption spectrophotometer, using the conditions outlined in the following table. Calculate the metal content of the sample by the method of standard additions.

Element	Wavelength (nm)	Sample Wt (g)	Standard Added (mg)	Flame Type*	Background Correction
Sb	217.6	1.0	0.2; 0.4	A/A	Yes
Cu	324.8	0.5	0.025; 0.05	A/A	Yes
Fe	248.3	0.5	0.05; 0.10	A/A	Yes
Pb	217.0	1.0	0.05; 0.10	A/A	Yes

*A/A is air/acetylene.

ARSENIC.   Dissolve 1.0 g in a mixture of 5 mL of water and 10 mL of nitric acid in a generator flask in a hood. Add 10 mL of dilute sulfuric acid (1 + 1), evaporate on a steam bath to about 5 mL, transfer the flask to a hot plate, and heat just to fumes of sulfur trioxide. (At this point the metastannic acid should go into solution.) Cool, cautiously wash down the flask with 5–10 mL of water, and again heat just to fumes of sulfur trioxide. Cool, cautiously wash down the flask with 5–10 mL of water, and repeat the fuming. Cool, dilute with water to 55 mL, and proceed as described in the general method for arsenic on page 23, omitting the addition of the dilute sulfuric acid and using 2 mL of hydrobromic acid (48%) instead of the potassium iodide solution specified under procedure (page 23). Swirl the contents of the flask occasionally to break up the zinc mass. When evolution of gas stops, open the generator flask, and add 1-g portions of granulated zinc until the solution is clear and colorless. Any red color in the silver diethyldithiocarbamate solution of the sample should not exceed that in a standard containing 0.001 mg of arsenic (As).

o-Tolidine Dihydrochloride

3,3'-Dimethyl-[1,1'-biphenyl]-4,4'-diamine Dihydrochloride

[-C$_6$H$_3$(CH$_3$)-4-NH$_2$]$_2$ · 2HCl   Formula Wt 285.22

CAS Number 612–82–8

REQUIREMENTS

Sensitivity to chlorine . . . . Passes test (0.1 ppm free chlorine)

Solubility . . . . Passes test

MAXIMUM ALLOWABLE

Residue after ignition 0.1%

TESTS

RESIDUE AFTER IGNITION.   Place 1.0 g in a tared porcelain crucible, add 0.5 mL of sulfuric acid, and slowly ignite in a hood. Heat in a furnace at 800 ± 25°C to constant weight.

SENSITIVITY TO CHLORINE.   Dissolve 0.7 g in 50 mL of water. Add this solution, with constant stirring, to 50 mL of dilute hydrochloric acid (3 + 7). Add 1 mL of this solution to 20 mL of water containing 5 mL of chlorine standard. A distinct yellow color is produced that, when measured in a 1-cm cell against the chlorine standard at 430 nm, has an absorbance of not less than 0.42. (Absorbance readings should be made within 10 min.)

Chlorine Standard.   Pipet 1.0 mL of a 2.5% sodium hypochlorite solution into a 1-L volumetric flask, and dilute with carbon dioxide-free water to volume. (Standardize the 2.5% sodium hypochlorite solution, and in the dilution correct for any departure from 2.5%.)

SOLUBILITY. Dissolve 0.50 g in 50 mL of water. The resulting solution is complete and colorless.

Toluene

Methylbenzene

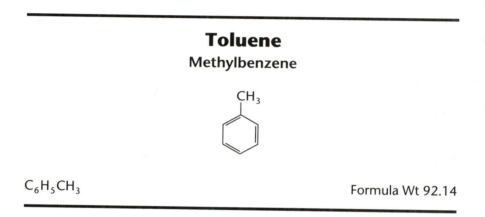

$C_6H_5CH_3$ Formula Wt 92.14

CAS Number 108–88–3

Suitable for use in ultraviolet spectrophotometry or general use. Product labeling shall designate the one or both of these uses for which suitability is represented on the basis of meeting the relevant requirements and tests. The ultraviolet spectrophotometry requirements include all of the requirements for general use.

REQUIREMENTS

General Use

Assay . . . . ≥ 99.5%
$C_6H_5CH_3$

MAXIMUM ALLOWABLE

Color (APHA). 10
Residue after evaporation 0.001%
Substances darkened by sulfuric acid . . . . Passes test
Sulfur compounds (as S) 0.003%
Water (H_2O) 0.03%

Specific Use

ULTRAVIOLET SPECTROPHOTOMETRY

Wavelength (nm)	Absorbance (AU)
350–400	0.01
335	0.02
310	0.05
300	0.10
293	0.20
288	0.50
286	1.00

TESTS

ASSAY.   Analyze the sample by gas chromatography, using the general parameters cited on page 56. The following specific conditions are also required.

   Column: Type I, methyl silicone

Measure the area under all peaks and calculate the toluene content in area percent. Correct for water content.

COLOR (APHA).   (Page 17).

RESIDUE AFTER EVAPORATION.   (Page 14). Evaporate 100 g (115 mL) to dryness in a tared dish on the steam bath and dry the residue at 105°C for 30 min.

SUBSTANCES DARKENED BY SULFURIC ACID.   Shake 15 mL with 5 mL of sulfuric acid for 15–20 s and allow to stand for 15 min. The toluene layer should be colorless and the color of the acid should not exceed that of a color standard composed of 2 volumes of water plus 1 volume of a color standard containing 5 g of $CoCl_2 \cdot 6H_2O$, 40 g of $FeCl_3 \cdot 6H_2O$, and 20 mL of hydrochloric acid in a liter.

SULFUR COMPOUNDS.   Place 30 mL of approximately 0.5 N potassium hydroxide in methanol in a conical flask, add 5.0 g (6.0 mL) of the sample, and boil the mixture gently for 30 min under a reflux condenser, avoiding the use of a rubber stopper or connection. Detach the condenser, dilute with 50 mL of water, and heat on the steam bath until the toluene and methanol are evaporated. Add 50 mL of bromine water and heat for 15 min longer. Transfer the solution to a beaker, neutralize with dilute hydrochloric acid (1 + 3), add an excess of 1 mL of the acid, and evaporate to about 50 mL. Filter if necessary, heat the filtrate to boiling, add 5 mL of barium chloride reagent solution, digest in a covered beaker on the steam bath for 2 h, and allow to stand overnight. If a precipitate is formed, filter, wash thoroughly, and ignite. Correct for the weight obtained in a complete blank test.

WATER. (Page 55, Method 2). Use 100 μL (87 g) of the sample.

ULTRAVIOLET SPECTROPHOTOMETRY. Use the procedure on page 39 to determine the absorbance.

---

para-Toluenesulfonic Acid Monohydrate
4-Methylbenzenesulfonic Acid Monohydrate

$CH_3C_6H_4SO_3H \cdot H_2O$ Formula Wt 190.22

CAS Number 6192–52–5

REQUIREMENTS

Assay . . . . $\geq$ 98.5%
$CH_3C_6H_4SO_3H \cdot H_2O$
Water (H_2O) 9.5–11.5%

MAXIMUM ALLOWABLE

Clarity of solution . . . . Passes test
Residue after ignition 0.1%
Sulfate (SO_4) 0.3%
Heavy metals (as Pb) 0.001%
Iron (Fe) 0.01%
Sodium (Na) 0.002%

TESTS

ASSAY. (By acid–base titrimetry). Weigh accurately about 0.8 g, dissolve in 100 mL of water, add 0.10 mL of phenolphthalein indicator solution, and titrate with 0.1 N sodium hydroxide to a pink end point. One milliliter of 0.1 N sodium hydroxide corresponds to 0.01902 g of $CH_3C_6H_4SO_3H \cdot H_2O$.

CLARITY OF SOLUTION. Dissolve 10 g in 50 mL of water. The solution should be clear and complete.

WATER. (Page 54, Method 1). Use 0.5 g.

RESIDUE AFTER IGNITION. (Page 15). Use 5.0 g. Retain the residue to prepare sample solution A.

SULFATE. Dissolve 0.7 g in 180 mL of water, add 10 mL of hydrochloric acid, and heat to boiling. Add 10 mL of barium chloride reagent solution, digest in a covered beaker on a hot plate at low setting for 2 h, and allow to stand overnight. If a precipitate is formed, filter, wash thoroughly, and ignite. The weight of the precipitate should not be more than 0.005 g greater than the weight obtained in a complete blank test.

> ***Sample Solution A.*** Dissolve the residue from the test for residue after ignition with a mixture of 2 mL of nitric acid and 4 mL of hydrochloric acid, and evaporate to dryness on the steam bath. Take up the residue with 1 mL of dilute hydrochloric acid (1 + 1) and 8 mL of hot water. Cool and dilute to 50 mL with water.

HEAVY METALS. (Page 26, Method 1). Use 20 mL of sample solution A (2.0 g).

IRON AND SODIUM. Determine the iron and sodium by the flame atomic absorption spectrophotometric method described on page 35.

> ***Sample Stock Solution.*** Dissolve 10.0 g in water and dilute with water to 100 mL (1 mL = 0.1 g).

Analyze the solutions by means of a suitable atomic absorption spectrophotometer, using the conditions outlined in the following table. Calculate the metal content of the sample by the method of standard additions.

Element	Wavelength (nm)	Sample Wt (g)	Standard Added (mg)	Flame Type*	Background Correction
Fe	248.3	1.0	0.05; 0.10	A/A	Yes
Na	589.0	0.5	0.005; 0.01	A/A	No

*A/A is air/acetylene.

Trichloroacetic Acid

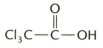

CCl$_3$COOH   Formula Wt 163.39

CAS Number 76–03–9

REQUIREMENTS

Assay . . . . $\geq$ 99.0%
CCl$_3$COOH

MAXIMUM ALLOWABLE

Residue after ignition 0.03%
Chloride (Cl) 0.001%
Nitrate (NO$_3$) 0.002%
Phosphate (PO$_4$) . . . . 5 ppm
Sulfate (SO$_4$) 0.02%
Heavy metals (as Pb) 0.002%
Iron (Fe) 0.001%
Substances darkened by sulfuric acid . . . . Passes test

TESTS

ASSAY. (By acid–base titrimetry). Weigh accurately 5 g, previously dried for 24 h over an efficient desiccant, and dissolve in 20 mL of water in a 500-mL reflux flask. Add 0.15 mL of methyl red indicator solution and titrate with 1 N sodium hydroxide to a distinct yellow color, then add 50.0 mL of 1 N sulfuric acid. Add sufficient dioxane to make the final solution at least 55% dioxane. Add silicon carbide boiling chips and boil gently under complete reflux for 1 h. Cool, add 0.15 mL of methyl red indicator solution, and titrate the excess of sulfuric acid with 1 N sodium hydroxide, the end point being the distinct transition from a light orange color to a definite yellow color. Correct for the acid present in the dioxane by running a complete blank, containing 2.5 mL of chloroform.

$$\% \ CCl_3COOH = \frac{V \times 0.1634 \times 100}{Wt \ of \ sample}$$

where V is the volume, in milliliters, of 1 N sulfuric acid consumed by the sample. This quantity is the volume of 1 N sodium hydroxide consumed by the blank minus the volume of 1 N sodium hydroxide consumed by the sample.

RESIDUE AFTER IGNITION.   (Page 15). Ignite 5.0 g.

CHLORIDE.   (Page 25). Use 1.0 g.

NITRATE.   (Page 29). For sample solution A use 0.25 g. For control solution B use 0.25 g and 0.5 mL of standard nitrate solution.

> *Sample Solution A for the Determination of Phosphate, Sulfate, Heavy Metals, and Iron.*   Evaporate 5.0 g to dryness in a beaker on a hot plate at low setting. Dissolve in about 10 mL of water, filter if necessary, and dilute with water to 100 mL (1 mL = 0.05 g).

PHOSPHATE.   (Page 30, Method 2). Dilute 40 mL of sample solution A (2-g sample) with water to about 80 mL. Add 0.5 g of ammonium molybdate and adjust the pH to 1.8 (using a pH meter) with dilute hydrochloric acid (1 + 9). Heat to boiling, cool to room temperature, dilute with water to 90 mL, and add 10 mL of hydrochloric acid. Continue as described, and carry along a standard containing 0.01 mg of phosphate ion (PO_4) treated in the same manner as the 40 mL of sample solution A.

SULFATE.   (Page 32, Procedure A, Method 1). Use 5.0 mL of sample solution A (0.25-g sample).

HEAVY METALS.   (Page 26, Method 1). Use 20 mL of sample solution A (1-g sample) with water to 25 mL.

IRON.   (Page 28, Method 1). Use 20 mL of sample solution A (1-g sample).

SUBSTANCES DARKENED BY SULFURIC ACID.   Dissolve 1 g in 10 mL of sulfuric acid and digest in a covered beaker on the steam bath for 15 min. Any color should not exceed that produced when 10 mL of sulfuric acid is added to 1 mL of water containing 0.1 mg of sucrose and heated in the same way as the sample.

1,1,1-Trichloroethane
Methyl Chloroform

CH_3CCl_3 Formula Wt 133.41

CAS Number 79 –01 –6

> *NOTE.* This solvent usually contains additives to retard decomposi-
> tion or to inhibit the corrosion of metals, particularly aluminum, with
> which it might come into contact. The nature and concentrations of
> such compounds should be stated on the label. Typical stabilizers
> include dioxane and alcohols such as 2-butanol.

REQUIREMENTS

Assay (CH_3CCl_3 + active additives) . . . . 98.5 –101.0%

MAXIMUM ALLOWABLE

Color (APHA). 10
Water 0.10%
Residue after evaporation 0.001%
Titrable acid . . . . 0.00030 meq / g

TESTS

ASSAY.   Analyze the sample by gas chromatography, using the general parameters cited on page 56. The following specific conditions are
also required.

Column: 6 m × 3.2 mm stainless steel, packed with 15% SP
1000 on Supelcoport

Column Temperature: 70°C for 8 min, then programmed at
8°C/min to 200°C

Injection Port Temperature: 200°C

Detector Temperature: 250°C

Carrier Gas: Nitrogen at 18 mL/min

Sample Size: 1 μL

Detector: Flame ionization

Range: 100

Approximate Retention Times (Min):   Vinylidene chloride, 4.8; butylene oxide, 7.9; 1,1-dichloroethane, 8.9; carbon tetrachloride, 9.0; 1,1,1-trichloroethane, 9.8; 2-propanol, 10.5; dichloromethane, 10.8; 1,1,2-trichloroethylene, 13.5; chloroform, 14.2; *n*-propanol, 14.7; perchloroethylene, 15.2; 1,2-dichloroethane, 16.0; dioxane, 16.2; 2-methyl-1-propanol, 17.0; *n*-butanol, 17.8; 1,1,2-trichloroethane, 21.3. Carbon tetrachloride and 1,1-dichloroethane are not resolved from 1,1,1-trichloroethane.

Measure the area under all peaks and calculate the 1,1,1-trichloroethane content in area percent plus intentional additives. Correct for water content.

COLOR (APHA).   (Page 17).

WATER.   (Page 54). Use 13.3 g (10 mL) of the sample.

RESIDUE AFTER EVAPORATION.   Evaporate 100 g (75 mL) to dryness in a tared dish on the steam bath, and dry the residue at 105°C for 30 min.

TITRABLE ACID.   Measure 100 g (75 mL) into a dry conical 250-mL flask. Add 0.2 mL of bromthymol blue indicator, and titrate with 0.02 N sodium hydroxide in methanol to a greenish-blue color. Not more than 1.50 mL should be required.

Trichloroethylene
Trichloroethene

$$Cl_2C\!=\!CHCl$$

CHCl:CCl$_2$ Formula Wt 131.39

CAS Number 79−01−6

NOTE. This material usually contains a stabilizer. If a stabilizer is present, the amount and type should be stated on the label.

REQUIREMENTS

Assay . . . . $\geq$ 99.5%
$$CHCl:CCl_2$$

MAXIMUM ALLOWABLE

Color (APHA). 10
Residue after evaporation 0.001%
Titrable acid . . . . 0.0001 meq / g
Titrable base . . . . 0.0003 meq / g
Water (H_2O) 0.02%
Heavy metals (as Pb) . . . . 1 ppm
Free halogens . . . . Passes test

TESTS

ASSAY. Analyze the sample by gas chromatography, using the general parameters cited on page 56. The following specific conditions are also required.

Column: Type I, methyl silicone

Measure the area under all peaks and calculate the trichloroethylene content in area percent. Correct for water content.

COLOR (APHA). (Page 17).

RESIDUE AFTER EVAPORATION. (Page 14). Evaporate 100 g (69 mL) to dryness in a tared dish on the steam bath in a well-ventilated hood and dry the residue at 105°C for 30 min.

TITRABLE ACID AND TITRABLE BASE. To 25 mL of water and 0.10 mL of phenolphthalein indicator solution in a 250-mL glass-stoppered flask, add 0.01 N sodium hydroxide until a slight pink color appears. Add 36 g (25 mL) of sample and shake for 30 s. If the pink color persists, titrate with 0.01 N hydrochloric acid, shaking repeatedly, until the pink color just disappears. Not more than 0.90 mL of 0.01 N hydrochloric acid should be required (titrable base). If the pink color is discharged when the sample is added, titrate with 0.01 N sodium hydroxide until the pink color is restored. Not more than 0.50 mL of 0.01 N sodium hydroxide should be required (titrable acid).

WATER. (Page 55, Method 2). Use 100 μL (140 mg) of the sample.

HEAVY METALS. Evaporate 10 g (7.0 mL) to dryness in a glass evaporating dish on the steam bath inside a well-ventilated hood. For the standard evaporate a solution containing 0.01 mg of lead ion (Pb) to dryness. Cool, add 2 mL of hydrochloric acid to each, and slowly evaporate to dryness on the steam bath. Moisten the residues with 0.05 mL of hydrochloric acid, add 10 mL of hot water, and digest for 2 min. Filter if necessary through a small filter, wash the evaporating dish and the filter with about 10 mL of water, and dilute each with water to 25 mL. Adjust the pH of the standard and sample solutions to between 3 and 4 (using a pH meter) with 1 N acetic acid or ammonium hydroxide (10% NH_3), dilute with water to 40 mL, and mix. Add 10 mL of freshly prepared hydrogen sulfide water to each and mix. Any color in the solution of the sample should not exceed that in the standard.

FREE HALOGENS.   Shake 10 mL for 2 min with 10 mL of water to which 0.10 mL of 10% potassium iodide reagent solution has been added and allow to separate. The lower layer should not show a violet tint.

2,2,4-Trimethylpentane

Isooctane

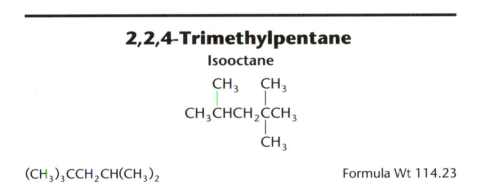

$(CH_3)_3CCH_2CH(CH_3)_2$ Formula Wt 114.23

CAS Number 540–84–1

Suitable for use in ultraviolet spectrophotometry or general use. Product labeling shall designate the one or both of these uses for which suitability is represented on the basis of meeting the relevant requirements and tests. The ultraviolet spectrophotometry requirements include all of the requirements for general use.

REQUIREMENTS

General Use

Assay . . . . ≥ 99.0%
$(CH_3)_3CCH_2CH(CH_3)_2$

MAXIMUM ALLOWABLE

Color (APHA). 10
Residue after evaporation 0.001%
Water-soluble titrable acid . . . . 0.0003 meq / g
Sulfur compounds (as S) 0.005%

Specific Use

ULTRAVIOLET SPECTROPHOTOMETRY

Wavelength (nm)	Absorbance (AU)
250 –400	0.01
240	0.04
230	0.10
220	0.20
210	1.0

TESTS

ASSAY.   Analyze the sample by gas chromatography, using the general parameters cited on page 56. The following specific conditions are also required.

Column: Type I, methyl silicone

Measure the area under all peaks and calculate the 2,2,4-trimethylpentane content in area percent. Correct for water content.

COLOR (APHA).   (Page 17).

RESIDUE AFTER EVAPORATION.   (Page 14). Evaporate 100 g (145 mL) to dryness in a tared dish on the steam bath and dry the residue at 105°C for 30 min.

WATER-SOLUBLE TITRABLE ACID.   To 30 g (44 mL) in a separatory funnel, add 50 mL of water, and shake vigorously for 2 min. Allow the layers to separate, draw off the aqueous layer, and add 0.15 mL of phenolphthalein indicator solution to the aqueous layer. Not more than 1.0 mL of 0.01 N sodium hydroxide should be required to produce a pink color.

SULFUR COMPOUNDS.   To 30 mL of 0.5 N potassium hydroxide in methanol in a conical flask, add 5.5 g (8 mL) of the sample, and boil the mixture gently for 30 min under a reflux condenser, avoiding the use of a rubber stopper or connection. Detach the condenser, dilute with 50 mL of water, and heat on a hot plate at low setting until the

2,2,4-trimethylpentane and methanol are evaporated. Add 50 mL of bromine water, heat for 15 min longer, and transfer to a beaker. Neutralize with dilute hydrochloric acid (1 + 3), add an excess of 1 mL of the acid, evaporate to about 50 mL, and filter if necessary. Heat the filtrate to boiling, add 5 mL of barium chloride reagent solution, digest in a covered beaker on a hot plate at low setting for 2 h, and allow to stand overnight. If a precipitate is formed, filter, wash thoroughly, and ignite. Correct for the weight obtained in a complete blank test.

ULTRAVIOLET SPECTROPHOTOMETRY. Use the procedure on page 67 to determine the absorbance.

Tris(hydroxymethyl)aminomethane
2-Amino-2-(hydroxymethyl)-1,3-propanediol
"tris"

$$HOCH_2\overset{\overset{\displaystyle CH_2OH}{|}}{\underset{\underset{\displaystyle NH_2}{|}}{C}}CH_2OH$$

$NH_2C(CH_2OH)_3$ Formula Wt 121.14

CAS Number 77–86–1

REQUIREMENTS

Assay (dry basis) 99.8–100.1%
$C_4H_{11}NO_3$

MAXIMUM ALLOWABLE

Absorbance . . . . Passes test
Water (H_2O) 2%
Insoluble matter 0.005%
Heavy metals (as Pb) . . . . 5 ppm
Iron (Fe) . . . . 5 ppm

TESTS

ABSORBANCE.   Determine the absorbance of a 40% solution of the sample in water in a 1.00-cm cell at 290 nm against water in a similar matched cell set at zero absorbance as the reference liquid. The absorbance should not exceed 0.2 absorbance units.

WATER.   (Page 54, Method 1). Use 1.0 g of the sample.

ASSAY.   (By acid–base titrimetry). Weigh accurately 0.5 g of sample previously dried at 105°C for 3 hours. Transfer to a 200-mL beaker, dissolve in 50 mL of carbon dioxide- and ammonia-free water, and titrate the solution with 0.1 N hydrochloric acid to a pH of 4.7, using a suitable pH meter. [A few drops of a 0.1% solution of 4-(4-dimethyl-amino-1-naphthylazo)-3-methoxybenzenesulfonic acid may be used as an indicator.] One milliliter of 0.1 N hydrochloric acid corresponds to 0.012114 g of $C_4H_{11}NO_3$.

INSOLUBLE MATTER.   Dissolve 20 g in 200 mL of water. Filter through a tared filtering crucible, wash thoroughly, and dry at 105°C.

HEAVY METALS.   (Page 26, Method 1). Dissolve 6.0 g in about 20 mL of water, and dilute with water to 30 mL. Use 25 mL to prepare the sample solution, and use the remaining 5.0 mL to prepare the control solution.

IRON.   (Page 28, Method 2). Use a 6.0-g sample.

Uranyl Acetate Dihydrate

$UO_2(CH_3COO)_2 \cdot 2H_2O$ Formula Wt 424.15

CAS Number 6159−44−0

> NOTE. The formula weight of this reagent is likely to deviate from the value cited since the natural distribution of uranium isotopes is often altered in current sources of uranium compounds.

REQUIREMENTS

Assay 98.0–102.0%
$UO_2(CH_3COO)_2 \cdot 2H_2O$

MAXIMUM ALLOWABLE

Insoluble matter 0.01%
Chloride (Cl) 0.003%
Sulfate (SO_4) 0.01%
Alkalies and alkaline earths (as sulfates) 0.05%
Heavy metals (as Pb) 0.002%
Iron (Fe) 0.001%
Substances reducing permanganate
  (as U^{IV}) 0.06%

TESTS

ASSAY. (Total uranium by oxidimetry). Weigh accurately 0.6 g, dissolve in 100 mL of 1 N sulfuric acid, and add enough 0.1 N potassium permanganate to give a distinct pink color. Pass 100 mL of 1 N sulfuric acid through a Jones reductor, followed by 100 mL of water, and discard. Pass the sample through the reductor, followed in succession by 100 mL each of 1 N sulfuric acid and water, and collect the effluent in a titration flask. Bubble air through the solution for 5–10 min, then titrate with 0.1 N potassium permanganate to a pink color. Run a reagent blank and subtract it from the sample titration. One milliliter of 0.1 N potassium permanganate corresponds to 0.02121 g of $UO_2(CH_3COO)_2 \cdot 2H_2O$.

INSOLUBLE MATTER.   Dissolve 10 g in 185 mL of water plus 5 mL of glacial acetic acid at room temperature. Filter through a tared filtering crucible, wash thoroughly with water, and dry at 105°C.

> ***Sample Solution A.***   Dissolve 15 g in 275 mL of water plus 5 mL of glacial acetic acid, add 10 mL of 30% hydrogen peroxide, and heat to coagulate the precipitate. Decant through a fritted-glass filter, without washing, and dilute with water to 300 mL (1 mL = 0.05 g).

CHLORIDE.   (Page 25). Dilute 6.7 mL of sample solution A (0.33-g sample) with water to 20 mL.

SULFATE.   (Page 32, Procedure A, Method 1). Use 10 mL of sample solution A (0.5-g sample).

ALKALIES AND ALKALINE EARTHS.   To 40 mL of sample solution A (2-g sample) add 0.10 mL of sulfuric acid, evaporate to dryness, and gently ignite at about 450°C. Digest the residue with 25 mL of hot water, filter, evaporate the filtrate to dryness in a tared dish, and ignite at 800 ± 25°C for 15 min.

HEAVY METALS. (Page 26, Method 1). Evaporate 20 mL of sample solution A (1-g sample) to dryness, dissolve the residue in about 20 mL of water, and dilute with water to 25 mL.

IRON. (Page 28, Method 1). Evaporate 20 mL of sample solution A (1-g sample) to dryness, dissolve the residue in 2 mL of hydrochloric acid, and dilute with water to 50 mL. Use the solution without further acidification.

SUBSTANCES REDUCING PERMANGANATE. Prepare two solutions, each containing 3.0 g in 200 mL of dilute sulfuric acid (1 + 99). Titrate one solution with 0.1 N potassium permanganate. Not more than 0.20 mL of permanganate should be required to cause a color change that can be observed by comparison with the solution not titrated. The 0.20 mL includes the 0.05 mL allowed to produce the color change in the absence of uranous compounds.

Uranyl Nitrate Hexahydrate

$UO_2(NO_3)_2 \cdot 6H_2O$ Formula Wt 502.13

CAS Number 13520−83−7

NOTE. The formula weight of this reagent is likely to deviate from the value cited, since the natural distribution of uranium isotopes is often altered in current sources for uranium compounds.

REQUIREMENTS

Assay . . . . 98.0−102.0%
$UO_2(NO_3)_2 \cdot 6H_2O$

MAXIMUM ALLOWABLE

Insoluble matter 0.005%
Chloride (Cl) 0.002%
Sulfate (SO₄) 0.005%
Alkalies and alkaline earths (as sulfates) 0.1%
Heavy metals (as Pb) 0.002%
Iron (Fe) 0.002%
Substances reducing permanganate
  (as U^{IV}) 0.06%

TESTS

ASSAY. (By gravimetry for uranium). Weigh accurately 0.5 g and dissolve in 100 mL of water in a 250-mL beaker. Heat the solution to boiling and add ammonium hydroxide, free from carbonate, until precipitation is complete. Filter through a tared Gooch crucible, wash with 1% ammonium nitrate solution, and ignite gently with free access of air, to constant weight. Multiply the oxide weight by 178.9 and divide by the sample weight to obtain %$UO_2(NO_3)_2 \cdot 6H_2O$.

INSOLUBLE MATTER. (Page 14). Use 20 g dissolved in 200 mL of water.

> *Sample Solution A.* Dissolve 15 g in 275 mL of water, add 7 g of ammonium acetate and 10 mL of 30% hydrogen peroxide, and heat to coagulate the precipitate. Decant through a fritted-glass filter, without washing, and dilute with water to 300 mL (1 mL = 0.05 g).

CHLORIDE. (Page 25). Dilute 10 mL of sample solution A (0.5-g sample) with water to 20 mL.

SULFATE. (Page 32, Procedure A, Method 1). Evaporate 20 mL of sample solution A (1-g sample) to 10 mL.

ALKALIES AND ALKALINE EARTHS. To 20 mL of sample solution A (1-g sample) add 0.10 mL of sulfuric acid, evaporate to dryness in a tared evaporating dish, and ignite at $800 \pm 25°C$ for 15 min.

HEAVY METALS. (Page 26, Method 1). Evaporate 20 mL of sample solution A (1-g sample) to dryness, dissolve the residue in about 20 mL of water, and dilute with water to 25 mL.

IRON. (Page 28, Method 1). Use 10 mL of sample solution A (0.5-g sample.

SUBSTANCES REDUCING PERMANGANATE. Prepare two solutions, each containing 3.0 g in 200 mL of dilute sulfuric acid (1 + 99). Titrate one solution with 0.1 N potassium permanganate. Not more than 0.20 mL of permanganate should be required to cause a color change that can be observed by comparison with the solution not titrated. The 0.20 mL includes 0.05 mL allowed to produce the color change in the absence of uranous compounds.

Urea

$$H_2N - \overset{\overset{\textstyle O}{\|}}{C} - NH_2$$

NH$_2$CONH$_2$ Formula Wt 60.06

CAS Number 57–13–6

REQUIREMENTS

Melting point . . . . Not below 132°C
                                                                    nor above 135°C

MAXIMUM ALLOWABLE

Insoluble matter 0.01%
Residue after ignition 0.01%
Chloride (Cl) . . . . 5 ppm
Sulfate (SO$_4$) 0.001%
Heavy metals (as Pb) 0.001%
Iron (Fe) 0.001%

TESTS

INSOLUBLE MATTER. (Page 14). Use 10 g dissolved in 100 mL of water.

RESIDUE AFTER IGNITION. (Page 15). Ignite 10 g.

CHLORIDE. (Page 25). Use 2.0 g.

SULFATE. Dissolve 4.0 g and about 10 mg of sodium carbonate in 10 mL of hydrochloric acid. Add 10 mL of nitric acid and digest in a covered beaker on a hot plate at low setting until reaction ceases. Add 5 mL of hydrochloric acid and 10 mL of nitric acid and again digest in a covered beaker on a hot plate at low setting until reaction ceases. Remove the cover and evaporate to dryness. Dissolve the residue in 4 mL of water plus 1 mL of dilute hydrochloric acid (1 + 19) and filter through a small filter. Wash with two 2-mL portions of water, dilute with water to 10 mL, and add 1 mL of barium chloride reagent solution. Any turbidity should not exceed that produced when a solution containing 0.04 mg of sulfate ion (SO_4) is treated exactly like the 4.0-g sample. Compare 10 min after adding the barium chloride to the sample and standard solutions.

HEAVY METALS. (Page 26, Method 1). Dissolve 2.0 g in about 20 mL of water, and dilute with water to 25 mL.

IRON.   (Page 28, Method 1). Use 1.0 g.

Water, Reagent
Distilled Water, Deionized Water

H_2O Formula Wt 18.02

CAS Number 7732–18–5

REQUIREMENTS

MAXIMUM ALLOWABLE

Specific conductance at 25°C . . . . 2.0×10^{-6}
$\text{ohm}^{-1} \text{ cm}^{-1}$
Silicate (as SiO_2) . . . . 0.01 ppm
Heavy metals (as Pb) . . . . 0.01 ppm
Substances reducing permanganate . . . . Passes test

TESTS

SPECIFIC CONDUCTANCE. Fill a conductance cell that has been properly standardized. Take precautions to prevent the water from absorbing carbon dioxide, ammonia, hydrogen chloride, or any other gases commonly present in a laboratory. Immerse in a constant-temperature bath at 25°C, equilibrate, and measure the resistance with a suitable conductance bridge. The calculated specific conductance should not exceed 2.0×10^{-6} ohm^{-1} cm^{-1}.

SILICATE.   Evaporate 1500 mL to 80 mL in a platinum dish on the steam bath (in portions, if necessary). Add 10 mL of silica-free ammonium hydroxide,* cool, and add 5 mL of a 10% solution of ammonium molybdate. Adjust the pH to between 1.7 and 1.9 (using a pH meter) with hydrochloric acid or silica-free ammonium hydroxide. Heat the solution just to boiling, cool to room temperature, and dilute with water to 90 mL. Transfer the solution to a separatory funnel, add 10 mL of hydrochloric acid and 35 mL of ethyl ether, and shake vigorously for a few minutes. Allow the two layers to separate, draw off the aqueous layer into another separatory funnel, and add 10 mL of hydrochloric acid and 50 mL of butyl alcohol. Shake vigorously, allow to separate, draw off the aqueous phase, and discard. Wash the butyl alcohol phase three times with 20 mL of dilute hydrochloric acid (1 + 99), discarding the aqueous solution each time. Add 0.5 mL of freshly prepared 2% stannous chloride dihydrate solution in hydrochloric acid to the butyl alcohol. Any blue color should not exceed that produced by 0.015 mg of silica (SiO_2) when treated exactly like the sample. Because the blue color fades on standing, compare immediately after the reduction with stannous chloride. Treatment again with stannous chloride will restore a fading blue color to its original intensity.

Silicate Solution in Ammonium Hydroxide.   (0.01 mg of SiO_2 in 1 mL). To prepare a stock solution of silica dissolve 9.46 g of assayed sodium silicate, $Na_2SiO_3 \cdot 9H_2O$, in 9 mL of silica-free ammonium hydroxide and dilute to 200 mL (199.4 g at 25°C) with water. Dilute 1.0 mL of this solution plus 5 mL of silica-free ammonium hydroxide to 1 L (997.1 g at 25°C) with water. If volumetric polyolefin ware is unavailable, dilute the silica solution by weight.

HEAVY METALS.   (Page 26, Method 1). Add 10 mg of sodium chloride and 1 mL of 1 N hydrochloric acid to 2 L, and evaporate to dryness on the steam bath. Dissolve the residue in about 20 mL of water and dilute with water to 25 mL. For the control solution evaporate 1 mL of 1 N hydrochloric acid to dryness on the steam bath, cool, add 0.02 mg of lead ion (Pb), and dilute with water to 25 mL.

---

*AMMONIUM HYDROXIDE (SILICA-FREE). Aqueous ammonia in contact with glass, even momentarily, will dissolve sufficient silica to make the reagent useless for this test. Because ammonium hydroxide is ordinarily supplied in glass containers by commercial sources, silica-free ammonium hydroxide must be prepared in the laboratory. The most convenient method is the saturation of water with ammonia gas from a cylinder of compressed anhydrous ammonia. Plastic tubing and bottles—for example, polyethylene—must be used throughout.

SUBSTANCES REDUCING PERMANGANATE. To 500 mL add 1 mL of sulfuric acid and 0.30 mL of 0.01 N potassium permanganate and allow to stand for 1 h at room temperature. The pink color should not be entirely discharged.

Xylenes
Dimethylbenzenes

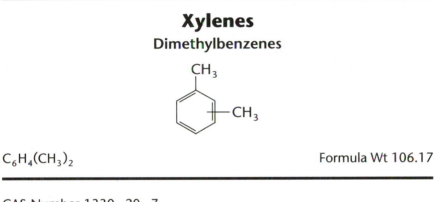

$C_6H_4(CH_3)_2$ Formula Wt 106.17

CAS Number 1330–20–7

> *NOTE.* This reagent is generally a mixture of the ortho, meta, and para isomers and may contain some ethylbenzene.

REQUIREMENTS

Assay . . . . ≥ 98.5% xylene isomers plus ethylbenzene (ethylbenzene not to exceed 25%)

MAXIMUM ALLOWABLE

Color (APHA) 10
Residue after evaporation 0.002%
Substances darkened by sulfuric acid . . . . Passes test
Sulfur compounds (as S) 0.003%
Water (H_2O) 0.05%

TESTS

ASSAY. Analyze the sample by gas chromatography, using the general parameters cited on page 56. The following specific conditions are also required.

Column: Type I, methyl silicone

Measure the area under all peaks and calculate the area percent of each peak. The total percent of xylenes is the sum of the percents of ortho, meta, and para isomers and that of ethylbenzene. Correct for water content and the area percent of ethylbenzene.

COLOR (APHA).   (Page 17).

RESIDUE AFTER EVAPORATION.   (Page 14). Evaporate 100 g (115 mL) to dryness in a tared dish on the steam bath and dry the residue at 105°C for 30 min.

SUBSTANCES DARKENED BY SULFURIC ACID.   Shake 15 mL with 5 mL of sulfuric acid for 15–20 s and allow to stand for 15 min. The xylene layer should be colorless, and the color of the acid should not exceed that of a color standard composed of 1 volume of water and 3 volumes of a color standard containing 5 g of $CoCl_2 \cdot 6H_2O$, 40 g of $FeCl_3 \cdot 6H_2O$, and 20 mL of hydrochloric acid in a liter.

SULFUR COMPOUNDS.   Place 30 mL of approximately 0.5 N potassium hydroxide in methanol in a conical flask, add 6.0 mL of the sample, and boil the mixture gently for 30 min under a reflux condenser, avoiding the use of a rubber stopper or connection. Detach the condenser, dilute with 50 mL of water, and heat on a hot plate at low setting until the xylenes and methanol have evaporated. Add 50 mL of bromine water and heat for 15 min longer. Transfer the solution to a beaker, neutralize with dilute hydrochloric acid (1 + 3), add an excess of 1 mL of the acid, and concentrate to about 50 mL. Filter if necessary, heat the filtrate to boiling, add 5 mL of barium chloride reagent solution, digest in a covered beaker on a hot plate at low setting for 2 h, and allow to stand overnight. If a precipitate is formed, filter, wash thoroughly, and ignite. Correct for the weight obtained in a complete blank test.

WATER.   (Page 55, Method 2). Use 100 μL (87 mg) of the sample.

Xylenol Orange

CAS Number 1611−35−4 (acid); 3618−43−7 (salt)

NOTE. This standard applies to both the free acid form and the salt form of this reagent.

REQUIREMENTS

Clarity of solution . . . . Passes test
Suitability for zinc titration . . . . Passes test

TESTS

CLARITY OF SOLUTION. If the indicator is the acid form, dissolve 0.1 g in 100 mL of alcohol. If the indicator is a salt form, dissolve 0.1 g in 100 mL of water. Not more than a faint trace of turbidity or insoluble matter should remain. Reserve the solution for the test for suitability for zinc titration.

SUITABILITY FOR ZINC TITRATION. Transfer 25.00 mL of 0.05 M zinc sulfate volumetric solution to a 250-mL conical flask. Add 1 mL of 0.1% sample solution reserved from the test for clarity of solution, 1 mL of 5 N nitric acid, 2.5 g of hexamethylenetetramine, and 25 mL of water. Upon addition of an equivalent amount of 0.05 M EDTA solution, a distinct change from purple to clear yellow should occur.

Zinc

Zn   Atomic Wt 65.39

CAS Number 7440 –66 –6

REQUIREMENTS

Assay . . . . $\geq$ 99.8% Zn
Suitability for determination of arsenic (As) . . . . . . Passes test

MAXIMUM ALLOWABLE

Iron (Fe) 0.01%
Lead (Pb) 0.01%

TESTS

ASSAY. (By complexometric titration for zinc). Weigh, to the nearest 0.1 mg, 0.2–0.25 g of sample, and dissolve with 10 mL of (1 + 1) nitric acid. When completely dissolved, dilute with water to about 30 mL, and with the aid of magnetic stirring, neutralize with 10% sodium hydroxide to methyl red. Add a few milligrams of ascorbic acid, 15 mL of pH 10 buffer solution, and about 50 mg of Eriochrome Black T indicator mixture. Titrate immediately with 0.1 M EDTA to a blue end point.

$$\% \text{ Zn} = \frac{(\text{mL} \times \text{M EDTA}) \times 6.539}{\text{Sample wt (g)}}$$

SUITABILITY FOR DETERMINATION OF ARSENIC. Note that the apparatus used for this test is described in the general method for arsenic on page 23. Add 13 g of the sample to a generator flask containing a mixture of 35 mL of water, 10 mL of hydrochloric acid, 2 mL of potassium iodide solution (16.5% KI in water), and 0.5 mL of stannous chloride solution (40% $SnCl_2 \cdot 2H_2O$ in concentrated HCl). Immediately connect the scrubber–absorber assembly to the generator flask. When the gas evolution stops, disconnect the flask, and add successive 5-mL portions of hydrochloric acid. Immediately connect the scrubber–absorber assembly to the flask after each addition. After four or five portions of acid have been added, decant the spent acid and add a mixture of 35 mL of water, 10 mL of hydrochloric acid, 2 mL of potassium iodide solution (16.5% KI in water), and 0.5 mL of stannous chloride solution (40% $SnCl_2 \cdot 2H_2O$ in concentrated HCl). Immediately reconnect the scrubber–absorber assembly to the flask. Continue the addition of 5-mL portions of hydrochloric acid and the decantations, if necessary, until all of the zinc sample is dissolved. Disconnect the tubing from the generator flask, and transfer the silver diethyldithiocarbamate solution to a suitable color-comparison tube. Any red color in the silver diethyldithiocarbamate solution of the sample should not exceed that in a control prepared with 3 g of the zinc sample and containing 0.001 mg of arsenic (As).

Sample Solution A. Dissolve 2.0 g in 15 mL of dilute hydrochloric acid (1 + 1). When dissolution is nearly complete, add 1 mL of nitric acid and heat to boiling, or until any residue from the zinc is dissolved. Cool and dilute with water to 100 mL (1 mL = 0.02 g).

IRON AND LEAD. Determine the iron and lead by the flame atomic absorption spectrophotometric method described on page 35.

> *Sample Stock Solution.* Dissolve 10.0 g of sample in 75 mL of (1 + 1) nitric acid. When dissolution is nearly complete, add 5 mL of nitric acid, and heat to boiling, or until any residue from the zinc is dissolved. Cool, transfer with water to a 100-mL volumetric flask, and dilute to the mark with water (0.1 g/mL).

Analyze the solutions by means of a suitable atomic absorption spectrophotometer, using the conditions outlined in the following table. Calculate the metal content of the sample by the method of standard additions.

Element	Wavelength (nm)	Sample Wt (g)	Standard Added (mg)	Flame Type*	Background Correction
Fe	248.3	1.0	0.05; 0.1	A/A	Yes
Pb	217.0	1.0	0.05; 0.1	A/A	Yes

*A/A is air/acetylene.

Zinc Acetate Dihydrate

$(CH_3COO)_2Zn \cdot 2H_2O$ Formula Wt 219.51

CAS Number 5970–45–6

REQUIREMENTS

Assay 98.0–101.0%
$(CH_3COO)_2Zn \cdot 2H_2O$
pH of a 5% solution at 25°C 6.0–7.0

MAXIMUM ALLOWABLE

Insoluble matter 0.005%
Chloride (Cl) . . . . 5 ppm
Sulfate (SO_4) 0.005%
Arsenic (As) . . . . 0.5 ppm
Calcium (Ca) 0.005%
Iron (Fe) . . . . 5 ppm
Lead (Pb) 0.002%
Magnesium (Mg) 0.005%
Potassium (K) 0.01%
Sodium (Na) 0.05%

TESTS

ASSAY. (By complexometric titration for zinc). Weigh accurately 0.8 g, transfer to a 500-mL beaker, and dilute to about 300 mL with water. Add 15 mL of saturated aqueous hexamethylenetetramine solution and 50 mg of xylenol orange indicator mixture. Titrate with standard 0.1 M EDTA to a color change of purple-red to lemon-yellow. One milliliter of 0.1 M EDTA corresponds to 0.02195 g of $(CH_3COO)_2Zn \cdot 2H_2O$.

pH OF A 5% SOLUTION. (Page 43). The pH should be 6.0–7.0 at 25°C.

INSOLUBLE MATTER. (Page 14). Use 20 g dissolved in a mixture of 2 mL of glacial acetic acid and 200 mL of water.

CHLORIDE. (Page 25). Use 2.0 g.

SULFATE. Dissolve 20 g in 200 mL of water, add 1 mL of hydrochloric acid, filter, heat to boiling, and add 10 mL of barium chloride reagent solution. Digest in a covered beaker on a hot plate at low setting for 2 h and allow to stand overnight. If a precipitate forms, filter, wash with water containing about 0.2% hydrochloric acid, wash finally with plain water, and ignite. Correct for a complete blank determination. One gram of $BaSO_4$ so obtained corresponds to 0.41 g of sulfate (SO_4).

ARSENIC. (Page 23). Use 6.0 g. For the standard use 0.003 mg of arsenic (As).

CALCIUM, MAGNESIUM, POTASSIUM, AND SODIUM. Determine the calcium, magnesium, potassium, and sodium by the flame atomic absorption spectrophotometric method described on page 35.

> *Sample Stock Solution.* Dissolve 10.0 g in water in a 100-mL volumetric flask, add 0.5 mL of hydrochloric acid, and dilute with water to the mark (1 mL = 0.1 g).

Analyze the solutions by means of a suitable atomic absorption spectrophotometer, using the conditions outlined in the following table. Calculate the metal content of the sample by the method of standard additions.

Element	Wavelength (nm)	Sample Wt (g)	Standard Added (mg)	Flame Type*	Background Correction
Ca	422.7	0.8	0.02; 0.04	N/A	No
Mg	285.2	0.2	0.005; 0.01	A/A	Yes
K	766.5	0.2	0.01; 0.02	A/A	No
Na	589.0	0.04	0.01; 0.02	A/A	No

*A/A = Air Acetylene; N/A is nitrous oxide/acetylene.

IRON. (Page 28, Method 1). Use 2.0 g.

LEAD. (Page 46). Determine lead by differential pulse polarography. Use 5.0 g of sample and 0.25 mL of hydrochloric acid. For the standard add 0.10 mg of lead ion (Pb).

Zinc Chloride

$ZnCl_2$ Formula Wt 136.30

CAS Number 7646–85–7

REQUIREMENTS

Assay . . . . $\geq$ 97.0% $ZnCl_2$

MAXIMUM ALLOWABLE

Oxychloride. . . . . Passes test
Insoluble matter 0.005%
Nitrate (NO_3) 0.003%
Sulfate (SO_4) 0.01%
Ammonium (NH_4) 0.005%
Calcium (Ca) 0.06%
Iron (Fe) 0.001%
Lead (Pb) 0.005%
Magnesium (Mg) 0.01%
Potassium (K) 0.02%
Sodium (Na) 0.05%

TESTS

ASSAY. (By argentometric titration). Using precautions to avoid absorption of moisture, weigh accurately 0.3 g, dissolve in about 100 mL of water in a 200-mL volumetric flask, and add 5 mL of nitric acid and 50.0 mL of 0.1 N silver nitrate. Shake vigorously, dilute to volume, mix well, and filter through a dry paper into a dry flask or beaker, rejecting the first 20 mL of the filtrate. To 100 mL of the filtrate subsequently collected, add 2 mL of ferric ammonium sulfate indicator solution, and titrate the excess of silver nitrate with 0.1 N ammonium thiocyanate. One milliliter of 0.1 N silver nitrate consumed corresponds to 0.006815 g of $ZnCl_2$.

OXYCHLORIDE. Dissolve 20 g in 200 mL of water. On the addition of 6 mL of 1 N hydrochloric acid, any flocculent precipitate should entirely dissolve. Retain the solution for the test for insoluble matter.

INSOLUBLE MATTER. To the solution obtained in the test for oxychloride add 6 mL of 1 N hydrochloric acid. If insoluble matter is present, filter through a tared filtering crucible, wash thoroughly with water containing about 0.2% hydrochloric acid, and dry at 105°C.

NITRATE. (Page 29). For sample solution A use 0.50 g. For control solution B use 0.50 g and 1.5 mL of standard nitrate solution.

SULFATE. (Page 32, Procedure A, Method 1). Allow 30 min for the turbidity to form.

AMMONIUM. Dissolve 1.0 g in water and dilute with water to 50 mL. Pour 10 mL of the solution into 10 mL of freshly boiled 10% sodium hydroxide reagent solution. Dilute with water to 50 mL and add 2 mL of Nessler reagent. Any color should not exceed that produced by 0.01 mg of ammonium ion (NH_4) in an equal volume of solution containing the quantities of reagents used in the test.

CALCIUM, IRON, LEAD, MAGNESIUM, POTASSIUM, AND SODIUM. Determine the calcium, iron, lead, magnesium, potassium, and sodium by the flame atomic absorption spectrophotometric method described on page 35.

>*Sample Stock Solution.* Dissolve 10.0 g in water in a 100-mL volumetric flask, add 0.5 mL of hydrochloric acid, and dilute with water to the mark (1 mL = 0.1 g).

Analyze the solutions by means of a suitable atomic absorption spectrophotometer, using the conditions outlined in the following table. Calculate the metal content of the sample by the method of standard additions.

Element	Wavelength (nm)	Sample Wt (g)	Standard Added (mg)	Flame Type*	Background Correction
Ca	422.7	0.1	0.03; 0.06	N/A	No
Fe	248.3	2.0	0.02; 0.04	A/A	Yes
Pb	217.0	2.0	0.05; 0.10	A/A	Yes
Mg	285.2	0.1	0.005; 0.01	A/A	Yes
K	766.5	0.1	0.01; 0.02	A/A	No
Na	589.0	0.02	0.005; 0.01	A/A	No

*A/A is air/acetylene; N/A is nitrous oxide/acetylene.

Zinc Oxide

ZnO

Formula Wt 81.39

CAS Number 1314−13−2

REQUIREMENTS

Assay . . . . ≥ 99.0% ZnO

MAXIMUM ALLOWABLE

Insoluble in dilute sulfuric acid. 0.01%
Alkalinity. . . . . Passes test
Chloride (Cl) 0.001%
Nitrate (NO_3) 0.003%
Sulfur compounds (as SO_4) 0.01%
Arsenic (As). . . . . 2 ppm
Calcium (Ca) 0.005%
Iron (Fe) 0.001%
Lead (Pb) 0.005%
Magnesium (Mg) 0.005%
Manganese (Mn) . . . . 5 ppm
Potassium (K) 0.01%
Sodium (Na) 0.05%

TESTS

ASSAY. (By complexometric titration of zinc). Weigh accurately 0.25 g and transfer to a 500-mL beaker. Dissolve in 15 mL of hydrochloric acid (1 + 1) and dilute to about 300 mL with water. Add 0.15 mL of methyl red indicator solution and neutralize (yellow color) by dropwise addition of 10% sodium hydroxide solution. Add 15 mL of saturated aqueous hexamethylenetetramine solution and 50 mg of xylenol orange indicator mixture. Titrate with standard 0.1 M EDTA to a color change of purple-red to lemon-yellow. One milliliter of 0.1 M EDTA corresponds to 0.008139 g of ZnO.

INSOLUBLE IN DILUTE SULFURIC ACID. Dissolve 10 g by heating in a covered beaker on a steam bath with 160 mL of dilute sulfuric acid (1 + 15) for 1 h. Filter through a tared filtering crucible, wash thoroughly, and dry at 105°C.

ALKALINITY. Suspend 2 g in 20 mL of water, boil for 1 min, and filter. Add 0.10 mL of phenolphthalein indicator solution to 10 mL of the filtrate. No red color should be produced.

CHLORIDE. (Page 25). Suspend 1.0 g in 20 mL of water, and dissolve by adding 3 mL of nitric acid.

NITRATE. (Page 29). For sample solution A use 0.50 g suspended in 3 mL of water. For control solution B use 0.50 g and 1.5 mL of standard nitrate solution.

SULFUR COMPOUNDS. Suspend 5.0 g in 50 mL of water and add about 50 mg of sodium carbonate and 1 mL of bromine water. Boil for 5 min, cautiously add hydrochloric acid in small portions until the zinc oxide is dissolved, then add 1 mL more of the acid. Filter, wash, and dilute the filtrate with water to about 150 mL. Heat the filtrate to boiling, add 5 mL of barium chloride reagent solution, digest in a covered beaker on a hot plate at low setting for 2 h, and allow to stand overnight. If a precipitate is formed, filter, wash two or three times with dilute hydrochloric acid (1 + 99), complete the washing with water alone, and ignite. Correct for the weight obtained in a complete blank test.

ARSENIC. (Page 23). Dissolve 1.5 g in 25 mL of dilute sulfuric acid (1 + 4) in a generator flask, and add 10 mL of water. For the standard use 0.003 mg of arsenic (As), and omit the addition of additional dilute sulfuric acid.

CALCIUM, IRON, LEAD, MAGNESIUM, MANGANESE, POTASSIUM, AND SODIUM.
Determine the calcium, iron, lead, magnesium, manganese, potassium, and sodium by the flame atomic absorption spectrophotometric method described on page 35.

> *Sample Stock Solution.* Dissolve 10.0 g in 10 mL of hydrochloric acid in a 100-mL volumetric flask, and dilute with water to volume (1 mL = 0.1 g).

Analyze the solutions by means of a suitable atomic absorption spectrophotometer, using the conditions outlined in the following table. Calculate the metal content of the sample by the method of standard additions.

Element	Wavelength (nm)	Sample Wt (g)	Standard Added (mg)	Flame Type*	Background Correction
Ca	422.7	0.8	0.02; 0.04	N/A	No
Fe	248.3	2.0	0.02; 0.04	A/A	Yes
Pb	217.0	2.0	0.05; 0.10	A/A	Yes
Mg	285.2	0.2	0.005; 0.01	A/A	Yes
Mn	279.5	2.0	0.01; 0.02	A/A	Yes
K	766.5	0.2	0.01; 0.02	A/A	No
Na	589.0	0.04	0.01; 0.02	A/A	No

*A/A is air/acetylene; N/A is nitrous oxide/acetylene.

Zinc Sulfate Heptahydrate

$ZnSO_4 \cdot 7H_2O$ Formula Wt 287.56

CAS Number 7446–20–0

REQUIREMENTS

Assay 99.0–103.0%
$ZnSO_4 \cdot 7H_2O$

pH of a 5% solution at 25°C 4.4–6.0

MAXIMUM ALLOWABLE

Insoluble matter 0.01%
Chloride (Cl) . . . . 5 ppm
Nitrate (NO_3) 0.002%
Ammonium (NH_4) 0.001%
Arsenic (As) . . . . 1 ppm
Calcium (Ca) 0.005%
Iron (Fe) 0.001%
Lead (Pb) 0.003%
Magnesium (Mg) 0.005%
Manganese (Mn) . . . . 3 ppm
Potassium (K) 0.01%
Sodium (Na) 0.05%

TESTS

ASSAY. (By complexometric determination of zinc). Weigh accurately 1 g, transfer to a 500-mL beaker, and dissolve in about 300 mL of water. Add 15 mL of saturated aqueous hexamethylenetetramine solution and 50 mg of xylenol orange indicator mixture. Titrate with standard 0.1 M EDTA to a color change of purple-red to lemon-yellow. One milliliter of 0.1 M EDTA corresponds to 0.02876 g of $ZnSO_4 \cdot 7H_2O$.

INSOLUBLE MATTER. (Page 14). Use 10 g dissolved in 100 mL of water.

pH OF A 5% SOLUTION. (Page 43). The pH should be 4.4–6.0 at 25°C.

CHLORIDE. (Page 25). Use 2.0 g.

NITRATE. (Page 29). For sample solution A use 1.0 g. For control solution B use 1.0 g and 2 mL of standard nitrate solution.

AMMONIUM. Dissolve 1.0 g in 25 mL of water, pour into 25 mL of freshly boiled and cooled 10% sodium hydroxide reagent solution, and add 2 mL of Nessler reagent. Any color should not exceed that produced by 0.01 mg of ammonium ion (NH_4) in an equal volume of solution containing the quantities of reagents used in the test.

ARSENIC. (Page 23). Use 3.0 g. For the standard use 0.003 mg of arsenic (As).

CALCIUM, IRON, LEAD, MAGNESIUM, MANGANESE, POTASSIUM, AND SODIUM.   Determine the calcium, iron, lead, magnesium, manganese, potassium, and sodium by the flame atomic absorption spectrophotometric method described on page 35.

> ***Sample Stock Solution.***   Dissolve 10.0 g of sample in water in a 100-mL volumetric flask, add 0.5 mL of hydrochloric acid, and dilute with water to the mark (1 mL = 0.1 g).

Analyze the solutions by means of a suitable atomic absorption spectrophotometer, using the conditions outlined in the following table. Calculate the metal content of the sample by the method of standard additions.

Element	Wavelength (nm)	Sample Wt (g)	Standard Added (mg)	Flame Type*	Background Correction
Ca	422.7	0.8	0.02; 0.04	N/A	No
Fe	248.3	2.0	0.02; 0.04	A/A	Yes
Pb	217.0	2.0	0.06; 0.12	A/A	Yes
Mg	285.2	0.2	0.005; 0.01	A/A	Yes
Mn	279.5	2.0	0.01; 0.02	A/A	Yes
K	766.5	0.2	0.01; 0.02	A/A	No
Na	589.0	0.04	0.01; 0.02	A/A	No

*A/A is air/acetylene; N/A is nitrous oxide/acetylene.

INDEX

Index

1*H*-Indene-1,2,3-trione monohydrate, 491
Indicators, unlisted impurity, 9
Indigo carmine, preparation, 75
Insoluble matter
  foreign substance, 14
  separation procedure, 14
Instrumentation, atomic absorption
  spectroscopy, 36–38
Interferences
  arsenic determination, 23
  atomic absorption spectroscopy, 38
International Atomic Weights, 1989, inside
  front cover
Interpretation of requirements, 4–9
Iodate
  polarographic analysis, 49
  preparation (0.10 mg of IO_3 in 1 mL), 83
Iodic acid, 381–383
Iodine
  preparation
    0.01 mg of I in 1 mL, 83
    0.1 N, 92
  specifications, 383–384
Iodine monochloride, 384–385
Ion chromatography, separation mechanism, 58
Iron
  low in magnesium and manganese, 385–386
  preparation (0.01 mg of Fe in 1 mL), 83
  procedures for determination, 28–29
Iron(III) chloride hexahydrate, 336–339
Iron(III) nitrate nonahydrate, 339–340
Iron(II) sulfate heptahydrate, 343–345
Isoamyl alcohol, 388–390
1,3-Isobenzofurandione, 528–529
Isobutyl alcohol, 387–388
Isooctane, 769–771
Isopentyl alcohol, 388–390
Isopropyl alcohol, 390–391
Isopropyl ether, 392–393

J

Jones reductor, preparation, 75

K

Karl Fischer method
  volumetric procedure
    apparatus, 52–53
    reagents, 53
  water content of organic reagents, 52–55
Karl Fischer reagent
  components and use, 52
  preparation, 53, 75

Karl Fischer reagent—*Continued*
  standardization, 54
  water determination procedure, 54–55

L

Lactic acid, 85%, 393–394
Lactose monohydrate, 395–396
Lanthanum chloride, hydrated, 396–397
LC, *see* Liquid chromatography
Lead, 398–400
Lead acetate, preparation, 75
Lead acetate trihydrate, 400–401
Lead carbonate, 402–403
Lead chromate, 404–405
Lead dioxide, 405–408
Lead in zinc compounds, polarographic
  analysis, 50
Lead monoxide, 408–410
Lead nitrate, 410–411
Lead(II) oxide, 408–410
Lead(IV) oxide, 405–408
Lead perchlorate trihydrate, 411–412
Lead standard
  dithizone test (0.001 mg of Pb in 1 mL), 83
  heavy metals test (0.01 mg of Pb in 1 mL), 83
Lead stock solution, preparation (0.1 mg of Pb
  in 1 mL), 83
Lead subacetate, 412–414
Light, purity of reagent chemicals, 1
Ligroin, 511–512
Liquid chromatography
  description and mechanisms, 57–59
  procedure, 64
  solvent extraction, 64–66
  solvent suitability, 62–64
  specialized detectors, 59
Litharge, 408–410
Lithium carbonate, 414–417
Lithium chloride, 417–419
Lithium hydroxide monohydrate, 419–420
Lithium metaborate, 420–425
Lithium perchlorate, 425–426
Lithium sulfate monohydrate, 426–428
Lithium tetraborate, 429–433
Litmus paper, 433–434

M

Magnesium, preparation (0.01 or 0.002 mg of
  Mg in 1 mL), 83
Magnesium acetate tetrahydrate, 434–436
Magnesium chloride hexahydrate, 436–438
Magnesium nitrate hexahydrate, 439–441
Magnesium oxide, 441–444

Committee liaison, editor, and indexer: Colleen P. Stamm
Production Manager: Robin Giroux

Text typeset by Technical Typesetters, Inc., Baltimore, MD
Printed & bound by Maple Press, York, PA